BIOLOGICAL MACROMOLECULES AND ASSEMBLIES

Volume 3: Active Sites of Enzymes

BIOLOGICAL MACROMOLECULES AND ASSEMBLIES

Volume 3: Active Sites of Enzymes

Editors

Frances A. Jurnak and Alexander McPherson

*University of California
Riverside*

A Wiley-Interscience Publication

JOHN WILEY & SONS
New York   Chichester   Brisbane   Toronto   Singapore

Copyright © 1987 by John Wiley & Sons, Inc.

Library of Congress Cataloging in Publication Data:

Biological macromolecules and assemblies.

   "A Wiley-Interscience publication."
   Includes bibliographies and indexes.
   Contents: v. 1. Virus structures—v. 2. Nucleic acids
and interactive proteins—v. 3. Active sites of enzymes.
   1. Macromolecules—Collected works. I. Jurnak,
Frances A. II. McPherson, Alexander, 1944-
   [DNLM: 1. Macromolecular systems.   QD 381 B615]

QP801.P64B554   1984   574.19'24   83-21732
ISBN 0-471-85142-6 (v.3)

Printed in the United States of America

10 9 8 7 6 5 4 3 2 1

Contributors

EDWARD N. BAKER, Department of Chemistry and Biochemistry, Massey University, Palmerston North, New Zealand

CARL-IVAR BRÄNDEN, Department of Chemistry and Molecular Biology, Swedish University of Agricultural Sciences Uppsala Biomedical Center, Uppsala, Sweden

B. W. DIJKSTRA, Laboratory of Chemical Physics, University of Groningen, Groningen, The Netherlands

JAN DRENTH, Laboratory of Chemical Physics, University of Groningen, Groningen, The Netherlands

HANS EKLUND, Department of Chemistry and Molecular Biology, Swedish University of Agricultural Sciences Uppsala Biomedical Center, Uppsala, Sweden

OTTO EPP, Max-Planck-Institut für Biochemie, Munich, West Germany

WIM G. J. HOL, Laboratory of Chemical Physics, Department of Chemistry, University of Groningen, Groningen, The Netherlands

MICHAEL N. G. JAMES, Medical Research Council of Canada, Group in Protein Structure and Function, Department of Biochemistry, University of Alberta, Edmonton, Alberta, Canada

J. N. JANSONIUS, Department of Structural Biology, Biozentrum University, Basel, Switzerland

ANTHONY A. KOSSIAKOFF, Genentech, Inc., South San Francisco, California

JOSEPH KRAUT, Department of Chemistry, University of California–San Diego, La Jolla, California

RUDOLF LADENSTEIN, Max-Planck-Institut für Biochemie, Munich, West Germany

DAVID A. MATTHEWS, Department of Chemistry, University of California–San Diego, La Jolla, California

HILARY MUIRHEAD, Department of Biochemistry, University of Bristol, Bristol, England

R. RENETSEDER, Laboratory of Chemical Physics, University of Groningen, Groningen, The Netherlands

ANITA R. SIELECKI, Medical Research Council of Canada, Group in Protein Structure and Function, Department of Biochemistry, University of Alberta, Edmonton, Alberta, Canada

M. G. VINCENT, Department of Structural Biology, Biozentrum University, Basel, Switzerland

Preface

The advent and development of genetic engineering techniques have transformed the once visionary concept of creating novel enzymes of unique specificities, properties, and power into a realistic enterprise. At the same time, a dramatic increase in the power of analytical methods, principally x-ray crystallography that allows visualization of the active sites of enzymes and the events that occur, have made possible the synthesis of ligands systematically designed to affect the behavior of target proteins. The creation and engineering of enzymes will lead ultimately to the utilization of biosynthetic proteins for new industrial processes, unique medical applications, and an array of imaginative products that will impact on countless aspects of our daily lives. The ability to rationally design and synthesize conventional molecules to interact with enzymes will significantly speed our ability to devise, optimize, and put into service pharmacological and biochemical agents for defense against disease and the alleviation of suffering. The potential offered by an increased understanding of enzyme structure, mechanism, dynamics, and regulation is enormous. It provides a striking opportunity for the biochemical sciences to significantly alter in a positive way our quality of life.

Given our newfound capacity to create, alter, and analyze biochemical structures and events, then the immediate question is what shall guide us in this work? How do we go about the task? The answer, it seems, is clear. We must turn to the wealth of information currently available that can assist us in delineating the precise structural properties, discovering the cunning distributions of chemical groups, and deducing the synergistic chemical relationships that generate and govern the astonishing power of enzyme catalysis.

It was our intention in producing this volume to gather together in a comprehensive manner the available structural information for a cross section of enzyme molecules representing a diverse set of catalytic purposes. The authors of the individual chapters, who were in all cases principals both in the origination and conduct of the research, have endeavored to emphasize the salient structural assemblies and the essential dynamic elements that characterize the mechanism of each active site. They have sought, with great diligence, to review the evidence from all sources that bear directly on their assertions. They have attempted to correlate observations and measurements from a broad spectrum of methods to establish for each case a chemical basis for the binding of specific substrates and the catalytic events that transform them during the reaction.

The primary sources of structural information are the three-dimensional, high-resolution structure determinations derived from x-ray diffraction analyses of crystalline enzymes. Difference Fourier analyses were used to extend these investigations to enzyme-substrate and enzyme-inhibitor complexes. These directly show the ligand as it is bound at the active site of the enzyme as well as those protein conformational changes that its presence induces. When correlated with dynamic studies based on spectroscopic methods and combined with the contributions of classical enzymology, the chapters provide convincing descriptions of the fundamental bases of enzyme catalysis.

Three of the chapters in this volume deal with the hydrolysis of polypeptides by different classes of enzymes characterized by their mechanisms as serine, thiol, and aspartic acid proteases. These represent some of the most accurate and thoroughly studied enzymes in terms of both structure and mechanism. Comparison of their active sites suggests those essential similarities that must be present to bring about peptide bond cleavage, while the differences point up the structural basis of their diverse specificities.

Dihydrofolate reductase is a focus of major attention in the area of drug design and a prime example of the extension of protein structure analysis to medical and pharmaceutical application. Alcohol dehydrogenase, studied by x-ray diffraction not only in its native state but as a series of substrate complexes, illustrates well the elucidation of reaction pathways and catalytic events by combining crystallographic with classical enzymological techniques. The chapter dealing with aspartate amino transferase represents one of the most comprehensive and detailed descriptions of an enzyme's properties and mechanisms that has yet been compiled, and pyruvate kinase offers an exceptional opportunity to review an enzyme of great regulatory complexity.

Finally, the chapters describing rhodanese and glutathione reductase present unique and fascinating mechanisms that illustrate the extensive variety of chemical interactions employed by nature to achieve enzyme catalysis. We would by no means claim that this volume exhausts the possibilities for enzyme catalysis; it is but a beginning, but we hope a useful and stimulating start.

FRANCES A. JURNAK
ALEXANDER McPHERSON

Riverside, California
January 1987

Contents

1 Dihydrofolate Reductase **1**
Joseph Kraut and David A. Matthews

2 Alcohol Dehydrogenase **73**
Hans Eklund and Carl-Ivar Brändén

3 Pyruvate Kinase **143**
Hilary Muirhead

4 Structural Basis for Catalysis by Aspartate Aminotransferase **187**
J. N. Jansonius and M. G. Vincent

5 Catalysis by Phospholipase A2 **287**
Jan Drenth, B. W. Dijkstra, and R. Renetseder

6 The Thiol Proteases: Structure and Mechanism **313**
Edward N. Baker and Jan Drenth

7 Catalytic Properties of Trypsin **369**
Anthony A. Kossiakoff

8 Aspartic Proteinases and Their Catalytic Pathway **413**
Michael N. G. James and Anita R. Sielecki

9   Structural Basis for Catalysis by Rhodanese   **483**
      Wim G. J. Hol

10   Catalysis by Seleno Glutathione Peroxidase   **523**
      Rudolf Ladenstein and Otto Epp

Index   **559**

Dihydrofolate Reductase

JOSEPH KRAUT
DAVID A. MATTHEWS
Department of Chemistry
University of California, San Diego
La Jolla, California

CONTENTS

1. INTRODUCTION AND PERSPECTIVES

2. CRYSTAL STRUCTURES
  2.1. Accuracy of Structure Determinations
  2.2. Description of Structures

3. GEOMETRY OF COFACTOR BINDING
  3.1. Conformational States of Pyridine Dinucleotides
  3.2. Cofactor Binding Domain
  3.3. DHFR-Cofactor Interactions

4. STEREOCHEMISTRY OF INHIBITOR BINDING
  4.1. Overview
  4.2. Methotrexate
  4.3. Trimethoprim
  4.4. Active Site Structural Differences and Inhibitor
        Selectivity

**5. STRUCTURAL CHANGES RESULTING FROM LIGAND
     BINDING**
  5.1. Definitions and General Considerations
  5.2. NMR Studies
  5.3. Kinetic Studies
  5.4. X-ray Structural Results

**6. ENZYME-KINETIC PROPERTIES AND
     SUBSTRATE/COFACTOR BINDING**
  6.1. Some Questions and Problems
  6.2. Hysteretic Phenomena
  6.3. Specificity for Substrate and Cofactor
  6.4. Activation
  6.5. Cofactor Binding

7. STEREOCHEMISTRY OF THE CATALYTIC MECHANISM
  7.1. The Kinetic Mechanism
  7.2. Stereochemistry of Dihydrofolate Binding
  7.3. Stereochemistry of Cofactor Binding

8. MUTANTS AND MUTAGENESIS
  8.1. Perspectives
  8.2. Spontaneously Occuring Mutants
  8.3. Preview of Findings From Mutagenesis Studies

REFERENCES

1. INTRODUCTION AND PERSPECTIVES

Dihydrofolate reductase (DHFR; tetrahydrofolate:NADP$^+$ oxidoreductase, EC 1.5.1.3) is a very widely occurring enzyme that catalyzes the NADPH-dependent reduction of 7,8-dihydrofolate (FAH$_2$) to 5,6,7,8-tetrahydrofolate (FAH$_4$)

$$FAH_2 + NADPH + H^+ \longrightarrow FAH_4 + NADP^+$$

Structural formulas for folate, dihydrofolate, and tetrahydrofolate are given in Figure 1. DHFR activity was first detected in crude preparations of pigeon liver by Greenberg (1954), and the enzyme was partially purified from chicken liver by Futterman (1957) and by Zakrzewski and Nichol (1958). Osborn and Huennekens (1958) further purified chicken DHFR 100-fold, defined the enzyme-catalyzed reaction, and reported the first definitive enzymological characterization.

DHFR plays a vital role in the metabolism of proliferating cells because it is required for the maintenance of adequate levels of fully reduced folate. Figure 2 depicts the metabolic cycle, in which a molecule of tetrahydrofolate is oxidized to dihydrofolate for each molecule of thymidylate produced in the thymidylate synthase reaction. Owing to the peculiarity of the biosynthesis of thymidylate, inhibition of DHFR, in a line of rapidly multiplying cells, blocks the re-reduction of dihydrofolate and thereby results in stasis or death due to depletion of the metabolic pool of tetrahydrofolate. The latter, in its various derivative forms, is needed to act as the coenzyme that carries one-carbon fragments in the biosynthesis not only of thymidylate, but of

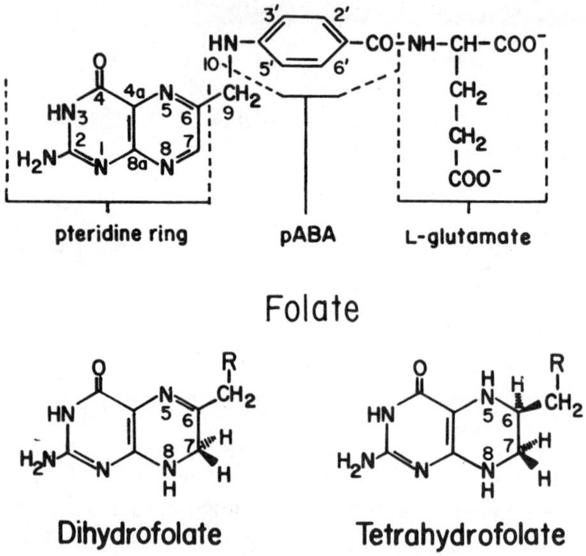

Figure 1. Structural formulas for folic acid and its reduced derivatives. The overall structure of folic acid is commonly discussed in terms of the three components indicated. *R* corresponds to the *p*-aminobenzoyl-L-glutamate portion of folic acid. Hydrogens with dashed bonds are transferred from C4 of NADPH.

purine nucleotides, methionine, and other essential metabolites as well (Blakley, 1969; Hitchings, 1983; Kisliuk, 1984).

DHFR occurs in almost every type of living cell throughout the biological world, with the sole known exceptions of the archaebacteria (J. J. Burchall, personal communication) and of a few parasitic protozoa (Wang et al., 1983). It is particularly intriguing to note that the archaebacterium *Methanobacterium thermoautotrophicum* utilizes a curious variation on the pterin cofactor theme, 5,10-methenyl-5,6,7,8-tetrahydromethanopterin, instead of the corresponding tetrahydrofolate as a one-carbon carrier in methanogenesis (Van Beelen et al., 1984).

There are several fundamental as well as practical reasons for focusing on DHFR as a model for investigating broad questions concerning the relationship between molecular structure and functional properties in enzymes, and DHFR has consequently been the subject of increasing study in recent years. One advantage of DHFR as a model enzyme is that the molecule is quite small, with a sequence length ranging from 159–189 amino acid residues. DHFR, therefore is, relatively convenient to study by X-ray

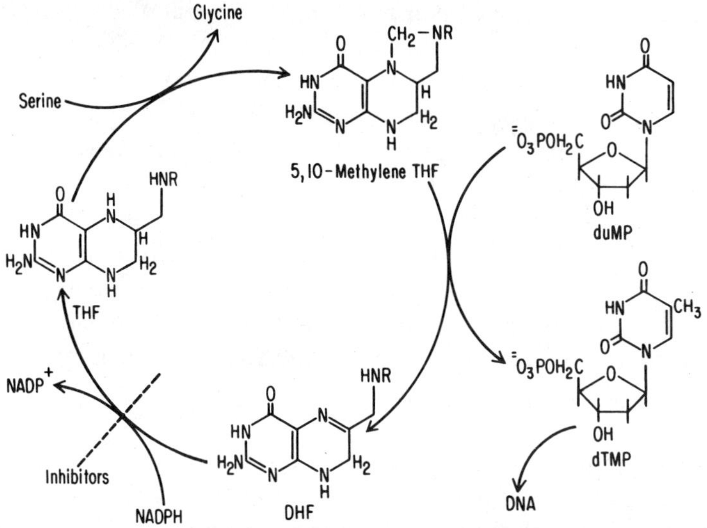

Figure 2. The relationship between dihydrofolate reduction and thymidylate synthesis.

crystallographic and nuclear magnetic resonance (NMR) techniques in comparison with, for example, other familiar nicotinamide nucleotide-dependent reductases and dehydrogenases, which are much larger multisubunit and multidomain molecules. At present, refined high-resolution X-ray structures are available for three diverse species of DHFR: those of the two bacteria, *Escherichia coli* and *Lactobacillus casei* (Bolin et al., 1982), and of chicken DHFR (Matthews et al., 1985a), which is representative of a closely related family of vertebrate DHFRs. A preliminary report has also appeared on the structure of a second vertebrate DHFR, isolated from mouse L1210 cells (Stammers et al., 1983); a model of human DHFR has been constructed using the chicken structure, but incorporating the amino acid sequence of the human enzyme. Very recently, a crystal structure for the R67 plasmid DHFR, which is unrelated to the chromosomal DHFRs, has also been determined (Matthews et al., 1986c). Description of the DHFR structures and discussion of their implications for enzyme mechanism will occupy a substantial portion of this review.

From the perspective of pharmacology and medicinal chemistry, DHFR is also an interesting model system. It has been identified quite early as the target for inhibition by a number of folate analogs and related compounds that interfere with one-carbon metabolism. Several of these agents, now known collectively as folate antagonists or antifols, have become important in the treatment of certain cancers, bacterial infections, malaria, and other diseases. Examples are methotrexate (MTX), trimethoprim (TMP), and pyrimethamine (Roth and Cheng, 1982; Jolivet et al., 1983; Hitchings and Baccanari, 1984). Structural formulas for MTX and TMP are shown in Figure 3. A significant early finding was that some of these DHFR inhibitors, notably TMP, showed a marked species selectivity, that is, they inhibit DHFR from some species much more strongly than others. This latter property renders TMP a useful antibiotic in the treatment of bacterial infections. X-ray methods have now furnished solid structural information on which it may be possible to base the design of still more sophisticated DHFR inhibitors for use in chemotherapy.

Another attraction of DHFR as an object of study is the availability of powerful inhibitors, which furnish an almost indispensable tool to the enzymologist and molecular biologist. Although levels of DHFR are usually very low in wild-type cell cultures and tissue preparations, the enzyme can often be extensively purified in one step by the application of affinity chromatography utilizing an MTX-bound Sepharose matrix or similar techniques. Moreover, the level of expression of DHFR can be amplified by a factor of several thousand by growing successive generations of cells in the presence

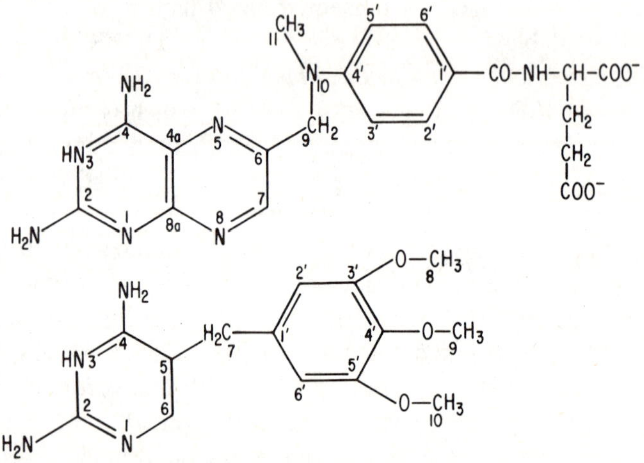

Figure 3. Covalent structure and atom numbering for methotrexate (top) and trimethoprim (bottom).

of increasing concentrations of inhibitors such as TMP or MTX (Freisheim and Matthews, 1984). It was the latter strategy that made it possible to prepare large quantities of various species of DHFR for intensive study prior to the introduction of contemporary cloning and amplification methods.

A recent development in the study of biological macromolecules that holds great promise for rapidly increasing the understanding not only of DHFR, but of enzymes in general, is the application of directed mutagenesis techniques to modify the molecule by altering the gene. Several mutant DHFRs have already been constructed, and a few have been subject to intensive study by both solution and X-ray crystallographic methods. These advances will be described in Section 8.

A number of earlier review articles concerned with various aspects of folate biochemistry and related subjects are listed in references 1–10 of the recent comprehensive review of DHFR by Blakley (1984). Most, however, were written from more of a pharmacological perspective than Blakley's, which is particularly valuable for its complete and critical tabulations of all DHFR isolations and characterizations to date, for its comprehensive survey of kinetic and thermodynamic parameters relating to substrates and inhibitors, and for its extensive list of references. Several other useful surveys of folate biochemistry, particularly as it relates to the properties of existing folate antagonists and to the design of new drugs, are collected in the volume edited by Sirotnak et al. (1984). Additional pertinent recent overviews of the enzymology and biochemistry of DHFR, including structural aspects, have been presented by Burchall (1983) and by Freisheim and Matthews (1984). The perspective in this chapter will differ somewhat from that taken by most previous reviewers in that it will focus principally on the interpretation of the enzymic properties of DHFR in terms of its molecular structure.

2. CRYSTAL STRUCTURES

2.1. Accuracy of Structure Determinations

Anticipating the central importance of X-ray diffraction results to the following discussion, it is worthwhile to summarize briefly the criteria by which the accuracy of any X-ray structure determination can be judged. In general, the accuracy of derived atomic coordinates depends on two principal factors: (1) the resolution of the diffraction data, and (2) the level of structure refinement. All DHFR structures discussed here, with the single exception of the chicken apoenzyme, have been determined at 2.2 Å resolution or higher and exhaustively refined to crystallographic R factors of 0.20 or less. These statistics correspond to a mean uncertainty in atomic position of about 0.20 Å (method of Luzzati, 1952), which is an average value. Larger errors may be present in the model for surface side chains or other "loose" portions of the structure characterized by high thermal motion or static disorder. For *E. coli* DHFR, an independent estimate of positional errors in well-ordered regions of the structure can be obtained by least-squares superposition of α-carbon coordinates for corresponding atoms in the two crystallographically independent molecules. When coordinates for 15 residues near the dimer interface are excluded from such a comparison, to allow for the obvious conformational differences in the two molecules, the average difference between corresponding α-carbons is 0.25 Å (Matthews et al., 1985a). This average deviation is an upper bound for the true error, because part of the discrepancy probably is real and occurs because corresponding atoms in the two DHFR molecules occupy slightly different crystallographic environments.

2.2. Description of Structures

2.2.1. Summary

X-ray structural studies of DHFR were first carried out on crystals of the *E. coli* enzyme containing bound MTX (Matthews et al., 1977). Shortly thereafter, crystallographic results were reported for a *L. casei* DHFR ternary complex with MTX and NADPH (Matthews et al., 1978, 1979); more recently, the structure of the chicken enzyme has been determined as well (Volz et al., 1982). Subsequently, all three enzyme structures were refined to a resolution of 1.6–1.7 Å (Bolin et al., 1982; Matthews et al., 1985a). A second vertebrate DHFR, the L1210 mouse lymphoma enzyme, has been crystallized at the Wellcome Laboratories in England, and a preliminary report on its structure has appeared (Stammers et al., 1983). Recent crystallographic studies of a novel plasmid-encoded DHFR from *E. coli* (R67) reveal a tertiary structure unrelated to that for the known chromosomally encoded DHFRs (Matthews et al., 1986). In this section, the folding of the polypeptide backbone of bacterial and vertebrate DHFRs will be described and compared. In the following discussion, the symbols ec, lc, and cl are appended to residue numbers to indicate the *E. coli*, *L. casei*, and chicken enzymes, respectively, wherever ambiguity might arise.

2.2.2. *Bacterial DHFRs*

Backbone folding for the *L. casei* DHFR-NADPH-MTX ternary complex is diagrammed in Figure 4*a*, which may be seen as representative of all the

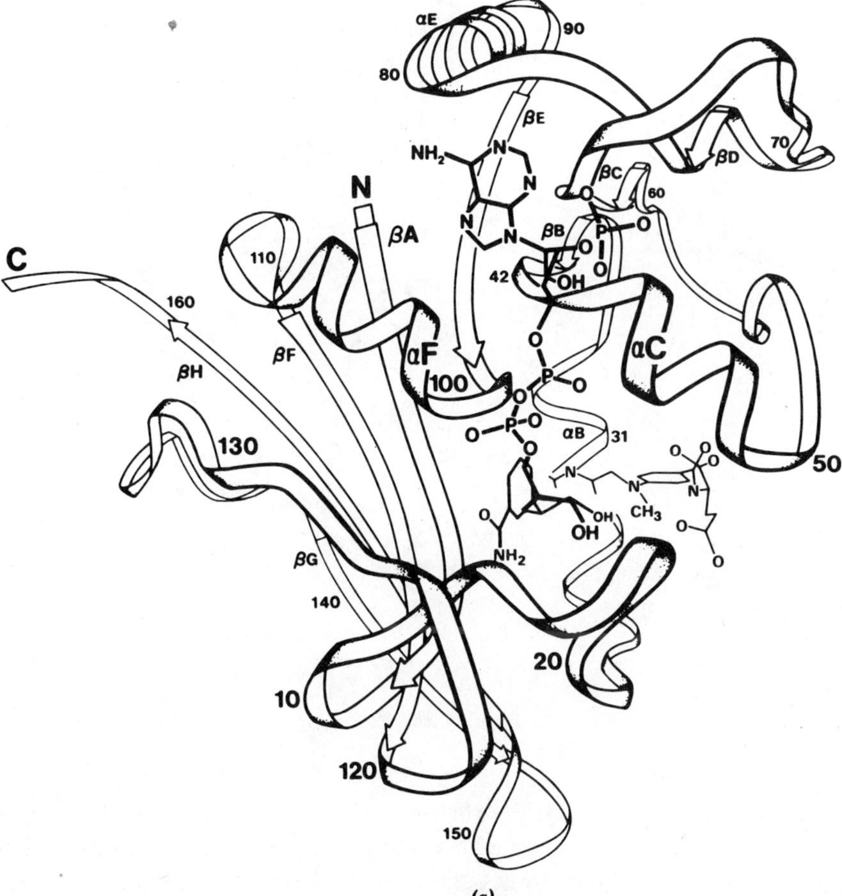

(a)

Figure 4. Backbone ribbon representations of the dihydrofolate reductase molecule. Figure *a* shows the *L. casei* enzyme ternary complex with NADPH and methotrexate. Figure *b* shows the chicken enzyme ternary complex with NADPH and trimethoprim. The α helices and β strands (arrows) are labeled. The approximate position of every tenth residue is indicated.

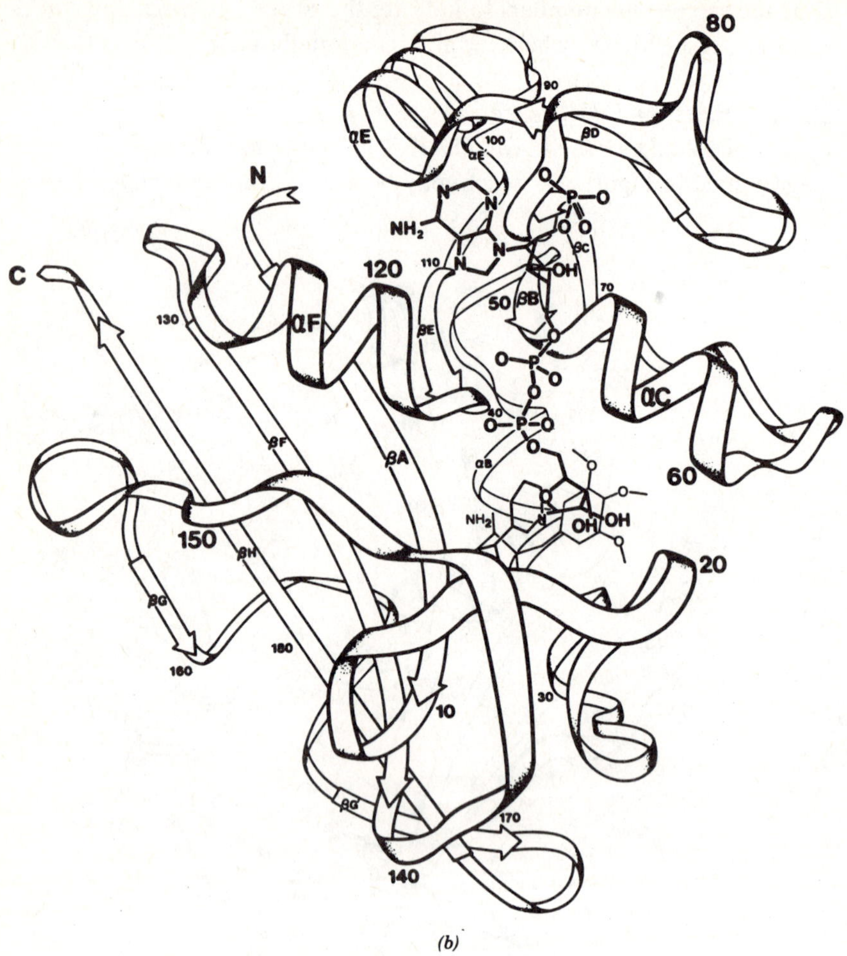

(b)

Figure 4. (b) *(continued)*

known chromosomal DHFR structures. The central architectural feature is
an 8-stranded mixed β sheet with a single antiparallel strand leading to the
C-terminus. This sheet shows the usual twist between adjacent strands
(Chothia, 1973) amounting to approximately 130° from one edge to the
other. Four α helices pack against the central β sheet, while the remaining
residues lie within loops, some quite lengthy, connecting the elements of
secondary structure. The active site containing bound MTX is seen as a 15 Å

deep cavity cutting across one whole face of the enzyme. In subsequent discussions, individual strands of β sheet are referred to as βA through βH, beginning at the N-terminal strand. Each α helical segment, αA, αC, αE, or αF, is lettered according to the β strand immediately following it in the linear amino acid sequence.

Although differing in total number of residues by only three, DHFRs from *E. coli* and *L. casei* have an amino acid sequence homology of less than 30%. In contrast to this variation in primary sequence, X-ray diffraction investigations reveal that overall backbone geometries for the two bacterial reductases are almost identical despite the presence of a bound cofactor molecule in the *L. casei* ternary complex that is not present in the *E. coli* DHFR-MTX binary complex (Matthews et al., 1978). When 142 of 159 α-carbon coordinates in *E. coli* DHFR are matched by least-squares to structurally equivalent α-carbon coordinates for the *L. casei* enzyme, the root-mean square deviation is only 1.07 Å (Bolin et al., 1982). In three instances, a single amino acid insertion or deletion occurs in *L. casei* DHFR relative to the *E. coli* enzyme. These differences appear in loops connecting elements of secondary structure; and it is in these turns that one finds the greatest structural variability between the two bacterial reductases. A two-residue insertion in the *L. casei* molecule provides an extra half-turn of helix at the carboxyl end of αE, as compared with the corresponding α helix in *E. coli* DHFR.

An interesting disruption of β-sheet hydrogen bonding is caused in both bacterial DHFRs by a parallel β bulge located at the carboxyl end of βE. The backbone carbonyl of Ile 94, ec, points out of the plane of the sheet into the substrate-binding site, positioning the carbonyl oxygen of Gly 95 to complete the interstrand hydrogen bonding at this edge of the β sheet. Immediately following the β bulge is an unusual *cis* peptide bond between Gly 95 and Gly 96, ec. An analogous β bulge followed by a *cis* gly-gly linkage is observed in chicken DHFR as well; from sequence alignments this structural feature appears to be conserved in all chromosomal DHFRs of known sequence. The possible mechanistic significance of this peculiar geometry is not yet known.

2.2.3. Chicken DHFR

A ribbon diagram of the backbone folding of chicken DHFR-NADPH-TMP ternary complex is shown in Figure 4*b*. The highly homologous vertebrate DHFRs of known sequence are 25−30 residues longer than their bacterial counterparts. However, the overall backbone folding for chicken DHFR is very similar to that observed in the *E. coli* and *L. casei* enzymes. All four α helices and seven of the eight β strands present in the bacterial enzymes are conserved in the chicken reductase (Volz et al., 1982). It is notable, however, that the relative positionings of secondary structural elements in the vertebrate DHFRs, particularly in the neighborhood of the substrate binding site, differ by as much as 1–3 Å from their relative positioning in the bacterial DHFRs. Except for six residues that lengthen the N- and C-termini of chicken DHFR relative to those of *E. coli* and *L. casei*, over 70% of the extra residues in the chicken enzyme occur in three loops that are far removed from the active site. For each insertion, the number of residues involved and its location in terms of secondary structure are summarized in Table 1. The conformation of each insertion (or deletion) present in the chicken enzyme can be seen in Figure 5, where these "extra" loops are added onto a schematic drawing of the polypeptide backbone folding of *L. casei* DHFR.

A thorough discussion of structural differences between the chicken and the bacterial enzymes is given by Volz et al. (1982), whose conclusions can be summarized as follows: (1) Insertions occur in loop regions that connect elements of secondary structure. In three of the loop regions, some or all of

Table 1. Summary of Major Backbone Conformational Differences Between Avian and Bacterial Dihydrofolate Reductases

Insertion	Number of Residues Inserted in Avian DHFR Relaitve to *L. casei* and *E. coli* DHFR[a]	Location of the Insertion[b]
I	+1	βA → αB
II	+7	αB(+1) → βB
III	+2, +3	αC → βC
IV	+1, +2	βC → βD
V	+4, +6	αE(+2) → βE
VI	+2	αF(+2) → βF
VII	+6	βG
VIII	−2, −3	βG → βH

[a]When two numbers are listed, they refer to *L. casei* and *E. coli* DHFRs, respectively.
[b]An arrow designates a loop region connecting the two elements of secondary structure. In parentheses are the number of *inserted* residues that extend a helix in avian dihydrofolate reductase compared with the corresponding helix in the *E. coli* enzyme. (From Freisheim and Matthews, 1984.)

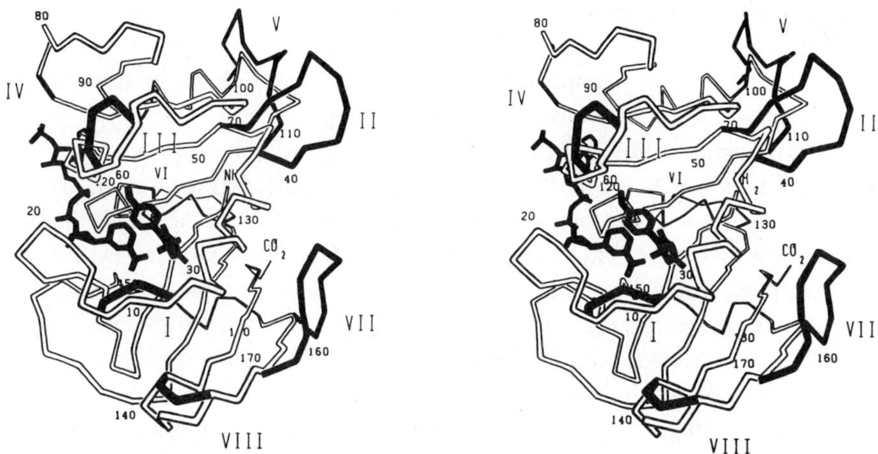

Figure 5. Schematic drawing comparing α-carbon backbone folding for ternary complexes of chicken dihydrofolate reductase (enzyme-NADPH-phenyltriazine) and *L. casei* dihydrofolate reductase (enzyme-NADPH-MTX). Backbone conformation is for the bacterial enzyme, with "insertions" present in the chicken enzyme shown in black. (Adapted from Volz et al. 1982.)

the inserted residues found in the chicken enzyme are in an α helical conformation. Thus, the lengths of α helices B, E, and F are extended in comparison with the corresponding helices in bacterial DHFRs, and an extra helix, αE′, is formed that is unique to the vertebrate enzymes. The single exception to this generalization occurs as a result of a six amino acid insertion in the long β strand, βG, at one edge of the central β sheet, causing a major disruption of the antiparallel interchain hydrogen bonding between βG and βH. Interestingly, the disruption of the βG strand in chicken DHFR does not significantly alter the overall tertiary structure of the enzyme, but may contribute to the 1−3 Å differences in the relative placement of secondary structural elements between chicken and bacterial DHFRs referred to above. (2) Except for a three amino acid insertion in the loop connecting αC to βC, the additional residues characteristic of vertebrate DHFRs are located more than 12 Å from the substrate binding site.

2.2.4. A Model for Human DHFR

For receptor-based drug design research, it would be useful to determine the clinically relevant human DHFR structure, but (as yet) no crystallographic results for this species of the enzyme have been reported. Evidence from amino acid homologies and inhibitor binding studies suggest that an accurate model of human DHFR can be constructed directly from the chicken structure by simply changing amino acid side chains in accordance with known differences in the primary sequences.

Two bacterial DHFRs with only 26% sequence homology nevertheless have a very similar polypeptide backbone folding (Matthews et al., 1978; Bolin et al., 1982). Vertebrate DHFRs of known sequence have highly homologous amino acid sequences (73–90% identities), suggesting that the overall molecular architecture may be even more highly conserved. Further evidence comes from inhibitor binding studies (e.g., see, Roth and Cheng, 1982; Hansch et al., 1984, and references therein). For a given DHFR, measurement of binding constants for a large number of suitably chosen inhibitor analogs, in principle, can provide a sensitive "fingerprinting" or mapping of the enzyme's active site, because the strength of the noncovalent interactions—and therefore the binding constants themselves—will depend on, among other things, the geometrical and chemical complementarity between inhibitor and receptor. The interpretation of these results in structural terms is not usually straightforward, but in many cases the data can be correlated by linear regression in terms of a quantitative structure activity relationship (QSAR). The latter provides a mathematical expression relating enzyme inhibition to certain descriptor variables, such as hydrophobicity, steric parameters, and electronic effects (Yoshimoto and Hansch, 1976). If binding data for the same inhibitor series are available for different DHFRs, the derived QSAR equations should be quite sensitive to differences in active site topography. This approach has been used by Hansch and his collaborators to study DHFRs from various organisms, including vertebrate reductases from chicken, human, cow, and mouse (Hansch et al., 1984, and references therein; Hathaway et al., 1984). For a series of 60 2,4-diamino-5-(3′-substituted phenyl)-s-triazines, the QSAR equations for all four vertebrate reductases agree quite well, except that the cow enzyme is, on the average, approximately 10 times more sensitive to the inhibitors than the other three. In addition, the human and mouse DHFR data are best correlated by including an extra "indicator" variable that affects the fit of a certain subset of inhibitors tested (Hathaway et al., 1984; Hansch et al., 1984). These considerations suggest that inhibition of chicken and human DHFRs by phenyltriazines with a wide variety of 3′ substituents is influenced by similar stereochemical factors, in conformity with the hypothesis that the respective enzyme active sites are structurally highly homologous.

A model for human DHFR has been constructed by changing appropriate side chains in the high-resolution refined structure of chicken DHFR to agree with the known primary sequence of the human enzyme (D. Matthews, unpublished results). The two species of DHFR have different amino acids at 51 of the 186 sites. Of these 51 sequence variations, 40 occur on fully exposed portions of the protein surface where the changed side chains are easily accommodated in low-energy conformations without altering other portions of the three-dimensional structure. The remaining changes involve residues in chicken DHFR having side chains that are partially buried, so that if everything else were equal, one might expect some main chain structural readjustments when these amino acids are replaced by sterically dissimilar ones at corresponding structural locations in human DHFR. However, what is found is that an amino acid substitution at one of these partially buried sites is invariably associated with chemically and sterically compensating changes at other neighboring residues that directly interact with the mutated residue. For example, the β carbon of Ala 49 in chicken DHFR is in van der Waals contact with the side chain of Leu 100. In human DHFR, residue 49 is changed to leucine and residue 100 to threonine, so that mutation to a larger side chain at one location is compensated by a second mutation to a smaller side chain. Thus, the net effect on tertiary structure of such mutations is predicted to be quite small. These observations support the idea that a hypothetical model for human DHFR, constructed in this way from the chicken enzyme structure, should closely approximate the true molecular structure.

2.2.5. *Plasmid-Encoded Type II R67 DHFR*

In 1972 Fleming et al. reported that certain clinical strains of *E. coli* and *Klebsiella aerogenes* carry *R* factors that confer high levels of resistance to TMP.

Subsequent studies by Pattishall et al. (1977) showed that *R* factor-specified DHFRs fall into at least two distinct groups designated type I and type II, while Joyner et al. (1984) have characterized a type III enzyme. Burchall (1984) recently classified the *R* factor-containing bacteria into two groups: those that are insensitive to TMP and those that are completely resistant. Type I and type III DHFRs confer insensitivity to TMP, but probably not for identical reasons. The type I enzymes are dimeric molecules of molecular weight 35,000 having K_i values for TMP of about 10 μM (Pattishall et al., 1977) compared to 0.4 nM for *E. coli* chromosomal DHFR.

Type III DHFRs are monomeric proteins with molecular weights of around 18,000, K_i's for TMP of 19 nM, and K_m's for dihydrofolate almost 10-fold lower than those observed for chromosomal DHFR (Joyner et al., 1984). Consequently, whereas type I enzymes are insensitive to TMP primarily because of a lowered affinity for the drug, type III DHFR insensitivity to TMP results from a reduced binding of TMP (but a higher affinity than that for the type I enzyme) augmented by tighter binding of substrate. Key amino acids important for substrate and cofactor binding in chromosomal DHFRs are conserved in both type I and type III plasmid-encoded DHFRs, suggesting that these TMP-insensitive R factor enzymes are genetically related to normal cellular DHFR.

The type II enzyme is 1000-fold less sensitive to TMP than even the type I DHFR and confers resistance to host bacteria. Furthermore, unlike all other known DHFRs, it is only very weakly inhibited by MTX ($K_i = 0.9 \times 10^{-4}M$) despite having a K_m for dihydrofolate close to that for chromosomal *E. coli* DHFR (Pattishall et al., 1977). The type II plasmid enzyme is reported to have a molecular weight of approximately 35,000 and to consist of four identical monomeric subunits of molecular weight 8500 (Smith et al., 1979). Each individual subunit is a single polypeptide chain of 78 residues with no apparent sequence homology to any other known DHFR (Stone and Smith, 1979).

The recent crystallographic analysis of the R67 apoenzyme DHFR, together with symmetry arguments, suggests that the apoenzyme molecule is probably a dimer rather than a tetramer (Matthews et al., 1986). Each 78-residue subunit is folded into a 6-stranded antiparallel β barrel as shown in Figure 6. Corresponding β strands from the two subunits form a third β barrel at the dimer interface. Model building suggests that NADPH binds along a lengthwise "slit" in the intersubunit barrel, while two molecules of dihydrofolate bind at opposite ends of the barrel. If subsequently confirmed by diffraction experiments with the holoenzyme, this geometry represents a novel mode of protein-cofactor interaction in which unsatisfied hydrogen bonding along one edge of an antiparallel β barrel structure provides the locus for cofactor binding. Although the overall tertiary structure of type II DHFR is completely different from that of the chromosomal enzyme, a conserved threonine (Thr 113, ec) and a conserved carboxylate (Asp 27, ec) each has its structural counterpart in the postulated active site of the type II R factor enzyme (Matthews et al., 1986).

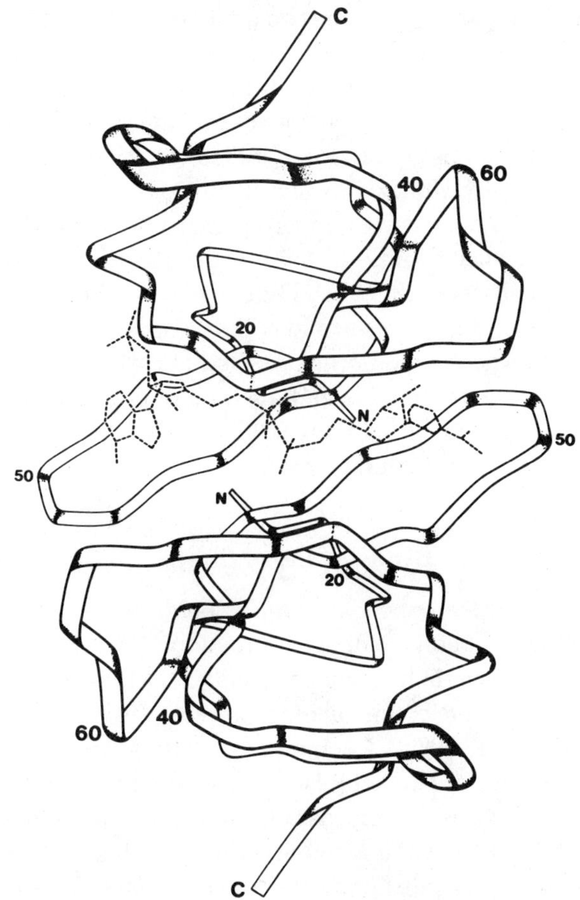

Figure 6. Ribbon representation of α-carbon backbone conformation of R67 dihydrofolate reductase showing hypothetical NADPH binding between symmetry-related subunits.

3. GEOMETRY OF COFACTOR BINDING

3.1. Conformational States of Pyridine Dinucleotides

It has been proposed that $NADP^+$ exists in solution as an equilibrium mixture of, on the one hand, folded conformations having the adenine and nicotinamide rings stacked on top of one another and, on the other hand, an extended or open form in which the two bases are unstacked (Weber, 1957); Velick, 1961). More recent NMR studies of NAD^+ have yielded results that appear to be inconsistent with these earlier assertions (Reddy et al., 1981). What is clear from crystallographic investigations is that both folded and extended conformations occur for pyridine dinucleotides bound to enzymes. In complexes with DHFR and five NAD^+-dependent dehydrogenases, these cofactors exist in extended conformations with their respective nicotinamide and adenine rings 14–17 Å apart. Recently, folded conformations have been observed for NADH bound to crystalline phosphorylase b, both at the allosteric effector site and at the nucleoside inhibitor site (Stura et al., 1983). Dinucleotide conformations at these two sites are different, but in both cases, the nicotinamide and adenine rings are in close proximity (3–6 Å) to one another. A partially folded conformation in which the two bases are 12 Å apart is found in crystals of lithium NAD^+ (Reddy et al., 1981).

3.2. The Cofactor Binding Domain

3.2.1. *The Dehydrogenase Dinucleotide Fold*

The polypeptide backbone folding pattern of DHFR resembles the pattern found in the coenzyme-binding domains of the NAD^+-dependent dehydrogenases and other nucleotide-binding proteins (Rossmann et al., 1975; Brändén and Eklund, 1980). A common structural feature among the dehydrogenases is a central parallel β sheet flanked on both sides by α helices whose axes are approximately parallel to nearby β strands. The basic mononucleotide-binding unit is a continuous stretch of amino acids folded into three strands of parallel β structure connected in a right-handed sense (Richardson, 1976) by loops having several turns of α helix. The dinucleotide, or Rossmann, fold is then obtained by assembling two of the β-α-β-α-β units adjacent to one another in such a way that an approximate twofold axis is generated between the innermost strands in the plane of the sheet. Differences in the relative positions of the various structural elements among the dehydrogenases led Rossmann et al. (1975) to conclude that the four central strands and two connecting helices have been more strongly conserved during the course of evolution than the extremities of the structure.

3.2.2. *Structural and Topological Comparison with the Dehydrogenases*

Figure 7 is a schematic drawing of the backbone helix and β sheet connecti-

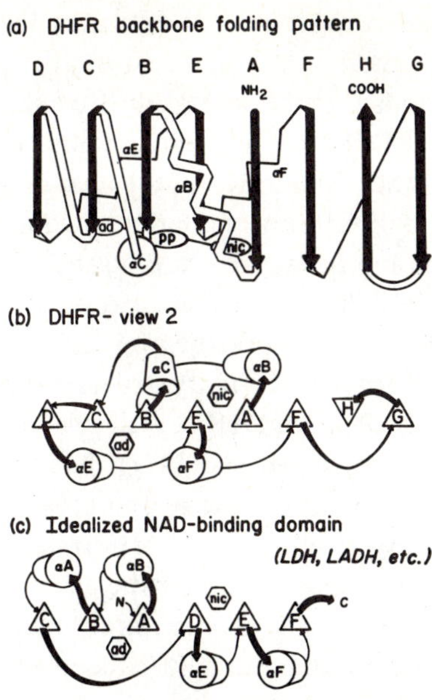

Figure 7. Schematic drawings of the DHFR structure and of an idealized NAD binding domain. In (*a*), the solid arrows are β strands, with the head of the arrow at the carboxy end of the strand. The zigzags represent helices that lie parallel to the sheet. Helix αC, which is approximately perpendicular to the sheet, is shown as a circle. The approximate positions of the nicotinamide (nic), pyrophosphate (pp), and adenine (ad) portions of the bound NADPH are indicated. In (*b*) and (*c*), β strands are shown by triangles and helices by cylinders. The orientation of the triangle for βH is inverted because βH is antiparallel to the other strands. In both (*b*) and (*c*), the view is from the carboxy edge of the β sheet. (From Bolin, 1982.)

vities for DHFR and for the NAD$^+$-binding domain found in five dehydrogenases. It is apparent that cofactor binding to DHFR and to the dehydrogenases of known structure is qualitatively similar. In both cases, dinucleotides bind at the carboxyl edges of central β sheets, with the nicotinamide and adenine rings situated on opposite sides and with the crossover occurring at the pyrophosphate-binding site. A key feature of this binding geometry is the close proximity of the cofactor's pyrophosphate group to the amino end of nearby α helices, where the negatively charged oxygens can be stabilized by α helix dipoles (Hol et al., 1978).

Despite these qualitative similarities in cofactor binding to DHFR, as compared with the dehydrogenases, a closer examination reveals a number of interesting differences. If cofactor position on the β sheet is used as a criterion for structurally aligning the dinucleotide binding domains of

DHFR and lactate dehydrogenase (LDH), then the geometrical equivalence between corresponding structural features of these folds is poor. For example, when β strands B and C of DHFR are aligned structurally with β strands A and B of LDH, the respective connecting helices are nearly orthogonal to one another. Similarly, for the nicotinamide-binding sites, αF in DHFR is positioned differently in relation to adjacent β strands E and A than is the corresponding helix, αE, to β strands D and E in LDH. It is noteworthy that the dehydrogenase dinucleotide fold can be geometrically aligned almost exactly with other structural features in DHFR by simply equivalencing the elements of secondary structure differently, that is, aligning βB-αB-βE and βA-αF-βF of DHFR with βA-αB-βB and βD-αE-βE of LDH, respectively (Matthews et al., 1977); but in this case, there is no correspondence in the location of the bound cofactors. Apparently something other than geometrical equivalence of β strands and adjacent α helices is of critical importance in accounting for the exact geometry of nucleotide binding to proteins containing such structural features.

Brändén has pointed out that the right-handed connectivity of the classical β-α-β unit is topologically favorable for forming binding crevices in specific regions at the carboxyl end of β sheets (Brändén, 1980). According to the argument, the pyrophosphate of a dinucleotide could be expected to bind in such a groove at the carboxyl end of the sheet formed by two β-α-β units. These considerations correctly predict that βA-αF-βF-βE-αB-βB in DHFR is not the dinucleotide-binding site despite close structural equivalence with the known dinucleotide fold in the dehydrogenases. Note that for this alignment, the corresponding interstrand connectivities of DHFR and LDH are different; and because of this, DHFR has no right-handed connection between the first β strand and helix in each β-α-β subunit. The importance of such connectivities is underscored by the fact that the alternative alignment (Fig. 7) in which right-handed connections are conserved, but geometrical equivalence with the dinucleotide fold of LDH is greatly reduced, correctly predicts the NADPH binding site of DHFR.

3.3. DHFR–Cofactor Interactions

Specific hydrophobic and hydrogen bonding interactions between NADPH and *L. casei* DHFR based on analysis of the highly refined 1.7 Å X-ray structure are discussed in Filman et al. (1982). Several generalizations can be made on the basis of these detailed studies. Adenine binds in a hydrophobic pocket in van der Waals contact with five residues, none of which are conserved among DHFRs of known sequence. No direct hydrogen bonding occurs between protein and *exo-* or *endo*cyclic adenine nitrogens; however, N1, N6, and N7 are hydrogen bonded to fixed water molecules, which, in two instances, also make hydrogen bonds with backbone carbonyl oxygen atoms (see Fig. 8). This is a specific example of a more general phenomenon that was fully appreciated only after high-resolution refinement of the structure was completed; namely, that fixed solvent molecules are involved

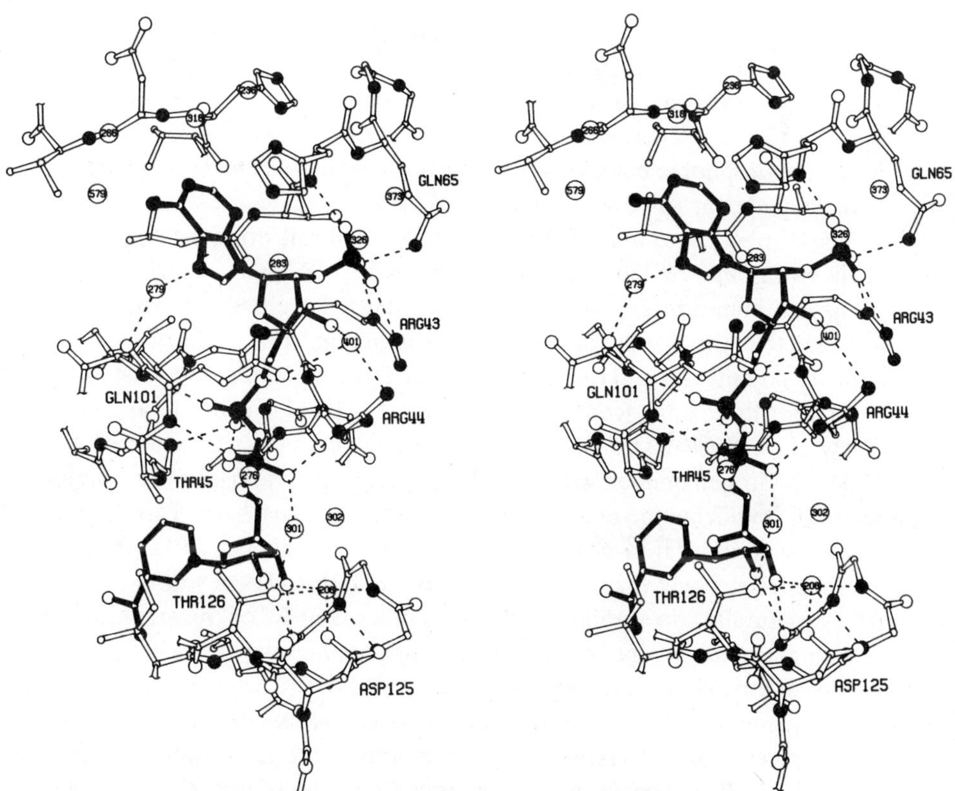

Figure 8. Stereo diagram of the hydrogen bonds between NADPH and *L. casei* dihydrofolate reductase. Hydrogen bonds involving the nicotinamide group are depicted in Figure 9. NADPH is indicated by solid bonds and protein by open bonds. Carbon atoms are represented by smaller open circles, oxygen atoms by larger open circles, and nitrogen atoms by solid circles. Large numbered circles represent fixed solvent molecules. (From Filman et al., 1982.)

in almost half (14 of 31) of the identifiable hydrogen bonding interactions with enzyme-bound NADPH.

The pyrophosphate bridge of NADPH resides at the C-terminal ends of β strands B and E, where in each case the chain turns sharply at a conserved glycine residue and immediately folds into α helices C and F, respectively. Backbone amido groups and side chain atoms from residues at the N-terminal ends of these helices make numerous hydrogen bonds with pyrophosphate oxygen atoms.

Both ribosyl groups of the dinucleotide occupy relatively solvent-exposed parts of the cofactor cleft. Binding of the NMN ribose appears to be fairly specific, in that it makes hydrophobic or hydrogen bonding interactions with six residues, five of which are either absolutely invariant among the known primary sequences or identically conserved in all but one species of DHFR. The only hydrogen bond between the AMN ribose and the protein is mediated by a fixed water molecule held by nonconserved residues; although hydrophobic contacts with four protein residues have been identified, they appear to be weak and nonspecific.

The 2′-phosphate makes hydrogen bonds and salt bridges to four side chain atoms and the backbone amide of His 64. ^{31}P-NMR experiments on coenzyme bound to *L. casei* and *E. coli* DHFR have been interpreted to indicate that the 2′-phosphate is in the dianionic state in both enzyme-cofactor complexes, and that the pK_a of the 2′-phosphate group must differ by at least three units from its value in free NADPH (Feeney et al., 1975; Caley et al., 1980). The positively charged guanidinium side chain of Arg 43,lc is hydrogen bonded with two phosphate oxygens and must be a major factor in stabilizing the unusual doubly charged 2′-phosphate. This interaction must account, in large part, for the observation that NADPH binds 100 times more strongly than NADH to *L. casei* DHFR (Feeney et al., 1975). The structurally corresponding residue in other DHFRs of known sequence is also arginine, except in the chicken and human enzymes where lysine serves an analogous function and in T4 phage DHFR, which, surprisingly, has an alanine at this geometrical location. The absence of a charged side chain corresponding to Arg 43,lc in the vicinity of the 2′-phosphate binding site of T4 DHFR suggests that this particular species of DHFR may be less discriminating in its preference for NADPH over NADH. In fact, Mosher et al. 1977) report that when assayed with NADH, the T4 enzyme catalyzes dihydrofolate turnover at fully 20% of the rate observed with NADPH.

Nicotinamide binds in a pocket formed by residues from two central strands of pleated sheet (βA and βE) and by the twisted loop connecting βA to αB (Fig. 9). Many residues in the region that directly contact nicotin-amide are conserved in all chromosomal DHFRs. On the B side of the nicotinamide ring, a cluster of hydrophobic side chains excludes solvent, whereas the A side is proximate to the pteridine-binding site. An interesting feature of the nicotinamide-binding pocket is that each of three highly conserved residues positions a main chain or side chain oxygen atom nearly coplanar with the nicotinamide ring of bound NADPH at distances from ring carbons C2, C4, and C6 that are slightly too short for standard non-bonded contacts. Filman et al. (1982) have argued that such CH . . . O polar interactions may be important for stabilizing a positively charged nicotinamide in the enzymic transition state.

Individual torsion angles for NADPH bound to *L. casei* DHFR are reported in Filman et al. (1982). In general, these angles fall near the minimum energy values (Arnott and Hukins, 1969), with one significant exception. The dinucleotide conformational angle O_n, which characterizes rotation about the C5′—O5′ bond of the NMN, assumes a value of 129°.

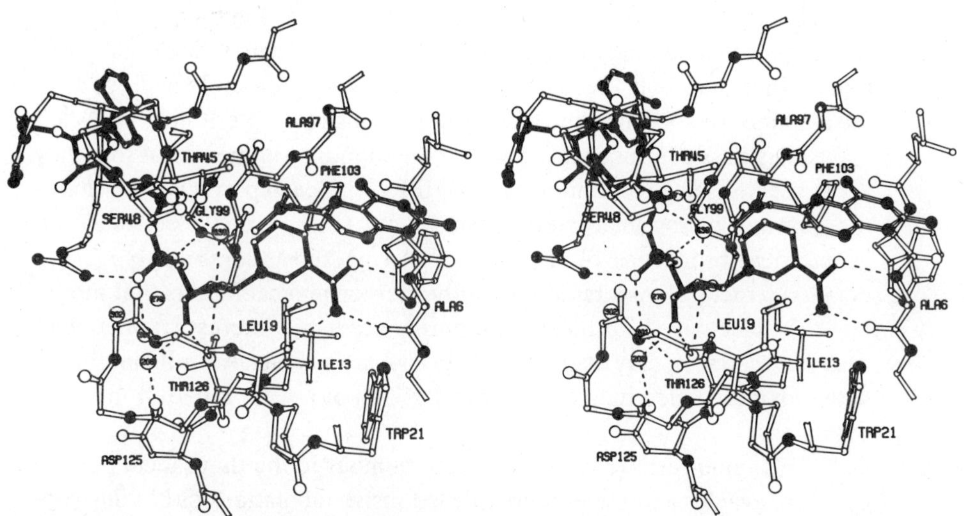

Figure 9. Stereo diagram of the nicotinamide binding site of *L. casei* dihydrofolate reductase. NADPH is indicated by solid bonds, methotrexate by striped bonds, and protein by open bonds. Carbon atoms are represented by smaller open circles, oxygen atoms by larger open circles, and nitrogen atoms by solid circles. Large numbered circles represent fixed solvent molecules. (From Filman et al., 1982.)

This deviates by 51° from the minimum energy value, but is in good agreement with results from solution NMR measurements on the *L. casei* holoenzyme complex, in which ^{31}P—O—C5'—1H5' spin coupling was detected for one pyrophosphate resonance and interpreted to indicate that either O_a or O_n changes by about 60° when NADPH binds to the enzyme (Feeney et al., 1975). Correlation of the X-ray structural result with the NMR spectra permits assignment of this ^{31}P resonance to P_n (Hyde et al., 1980a, 1980b).

4. STEREOCHEMISTRY OF INHIBITOR BINDING

4.1. Overview

From the perspective of the scientist interested in unraveling the intricacies of ligand binding to macromolecules, it would be difficult to envision a more appealing and potentially informative protein for study than the enzyme DHFR. Inhibitors of DHFR include compounds such as MTX that are close structural analogs of folate, but act as extremely potent competitive inhibitors of chromosomal DHFRs, binding as much as 10^6 times more tightly to the reductase than folate itself. A second group of inhibitors comprises compounds that bear no obvious structural relationship to folate or dihydrofolate other than that they contain a 2,4-diaminoheterocycle analogous to the 2-amino-4-oxopyrimidine portion of the substrate's pteridine ring. What makes this class of structurally diverse compounds so interesting is that they act as species-selective inhibitors of DHFR, with each particular species of the enzyme having characteristic sensitivity to the exact electronic and geometrical properties of the inhibitor's ring substituents (Table 2). A detailed structural understanding of the stereochemical factors that modulate this selectivity is of more than purely academic interest. Armed with such knowledge, it may be possible to design new, clinically useful inhibitors with enhanced selectivity against the DHFRs of a wide range of invading pathogens.

DHFR inhibitors synthesized to date number in the thousands. Several excellent articles and reviews provide extensive tabulations of binding constants, and in most instances a particular compound has been tested against various DHFRs of different species (e.g., Roth and Cheng, 1982; Hansch et al., 1984, and references therein). This section focuses primarily on a structural characterization of inhibitor binding to bacterial and vertebrate

 25

Table 2. Comparative IC$_{50}$ Values for Various Enzymes and Inhibitors Values $\times$ 10^8 M

Source of Enzyme	E. coli[a]	S. aureus[a]	Rat liver[a]	P. berghei[b]	T. equiperdum[c]
Compound[d]					
MTX	0.1	0.2	0.21	0.07	0.02
PYR	250	300	70	0.05	20
TMP	0.5	1.5	26,000	7.0	100
BuPP	50	4	46	1.7	50
MeBuPP	2	7	26	1	—
BuDHT	65,000	50,000	14	0.8	2000

[a]Burchall and Hitchings (1965).
[b]Ferone et al. (1969).
[c]Jaffe (1972).
[d]PYR, 2,4-diamino-5-p-chlorophenyl-6-ethylpyrimidine (pyrimethamine); BuPP, 2,4-diamino-6-butylpyrido[2,3-d]pyrimidine; MeBuPP, 2,4-diamino-5-methyl-6-butylpyrido[2,3-d]pyrimidine; BuDHT, 1-(p-butyphenyl)-1,2-dihydro-2,2-dimethyl-4,6-diamino-s-triazine.
Source: Adapted from Hitchings and Smith (1980).

DHFRs in order to understand, at the molecular level, the geometrical and electronic factors that account for the unusually potent binding of antifolates such as MTX. Also considered are the stereochemical factors that influence species-selective inhibition of DHFR by compounds such as TMP.

4.2. Methotrexate

Methotrexate (MTX) binds to DHFR in an open conformation, as shown in Figure 10, with its pteridine ring nearly perpendicular to the aromatic ring of its *p*-aminobenzoyl group (Matthews et al., 1977, 1978). The inhibitor-binding cleft is a 15 Å deep crevice formed primarily by residues from helix αB, the central β strands A and F, and the loops connecting βA to αB and αC to βC. The pyrimidine portion of MTX is deeply buried in a portion of the pocket that is formed by highly conserved residues, including a key acidic amino acid (aspartate or glutamate depending on the enzyme species) that is positioned at the N-terminal end of αB. The *p*-aminobenzoylglutamate moiety leads away from the interior of the protein to the surface, where its α carboxylate interacts with the guanidinium side chain of an invariant arginine residue (Arg 57,lc). MTX binding to DHFR has been studied crystallographically with enzymes from *E. coli* and *L. casei*. The

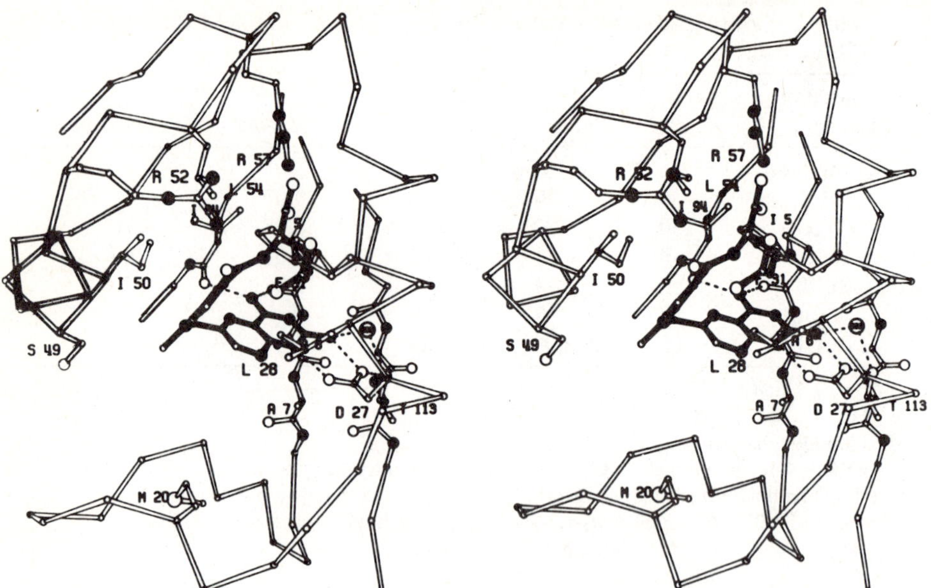

Figure 10. Stereo diagram of the binding of methotrexate to *E. coli* dihydrofolate reductase. Methotrexate is indicated by solid bonds and protein by open bonds. Carbon atoms are represented by smaller open circles, oxygen atoms by larger open circles, and nitrogen atoms by solid circles. Fixed solvent molecules have been omitted. Hydrogen bonds between the pyrimidine portion of methotrexate and the protein are indicated by dashed lines. (From Matthews et al., 1985a.)

conformation of the bound γ carboxylate side chain of MTX in the two DHFRs is different owing to species-specific variations in what appears to be a nonconserved weak interaction between this group and the particular species of DHFR. For the *L. casei* enzyme, inhibitor binding is reduced ninefold when the γ carboxylate of MTX is changed to an amide (Antonjuk et al., 1984). A complete description of MTX binding based on refined crystallographic structures is given in Bolin et al. (1982).

As shown in Figure 11, the pteridine ring of MTX makes extensive nonpolar contacts with protein and side chain atoms of residues 4, 5, 6, 19, 27, 30, and 97, all in the *L. casei* numbering system. Ala 6,lc and Phe 30,lc are invariant in all DHFRs of known sequence and, although the other five residues are not strictly conserved, the corresponding side chains are always hydrophobic. In the *L. casei* ternary complex containing NADPH, addi-

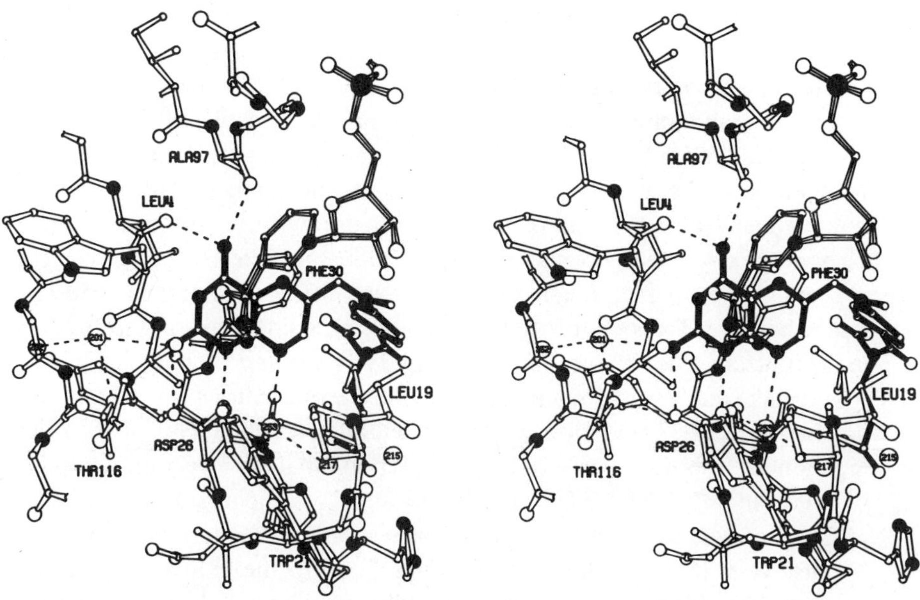

Figure 11. Stereo diagram of the pteridine binding site of *L. casei* dihydrofolate reductase. Methotrexate is indicated by solid bonds, protein by open bonds, and a portion of the NADPH molecule by striped bonds. Carbon atoms are represented by smaller open circles, oxygen atoms by larger open circles, and nitrogen atoms by solid circles. Large numbered circles represent fixed solvent molecules. Hydrogen bonds are indicated by dashed lines. (From Bolin et al., 1982.)

tional van der Waals contacts occur between the pteridine ring and the A side of the nicotinamide ring.

A variety of evidence—namely, spectroscopic (reviewed by Gready, 1980), calorimetric (Subramanian and Kaufman, 1978), and theoretical (Perault and Pullman, 1961)—strongly suggests that N1 is protonated when MTX binds to DHFR. In an elegant series of experiments, Cocco, Blakley, and coworkers have monitored the ^{13}C-NMR signal for MTX enriched at the 2 position of the pteridine ring when bound to DHFR from *L. casei* (Cocco et al., 1981a), *S. faecium* (Cocco et al., 1981b), and bovine liver (Cocco et al., 1983). They find that the ^{13}C chemical shift for enzyme-bound MTX does not change as the pH is raised from 6.0 to 10.0, indicating that the pK_a for N1 must be over 10, that is, 5 pH units above its value in free MTX. By elementary thermodynamic arguments, this suggests that the protonated species of MTX binds to the enzyme (the latter in its COO^- state) with an association constant more than 10^5 times greater than that for the unprotonated species (Cocco et al., 1981b).

The crystallographic studies provide a structural explanation for this observation. When MTX binds to DHFR, a pair of hydrogen bonds is found between the carboxyl side chain of Asp 26,lc on one hand, and the pteridine ring N1 and 2-amino group of the inhibitor on the other. Moreover, the carboxyl group is nearly coplanar with the bound pteridine ring (Fig. 11). It appears very probable, in light of the evidence just cited, that the pteridine ring of bound MTX is protonated at N1 and that the side chain of Asp 26,lc is in the ionized, anionic state. Thus, if bound MTX is deprotonated at N1, it would make one less hydrogen bond with the enzyme in its COO^- state, and no charge-charge interaction, thereby accounting for the factor of 10^5 just mentioned. It should be emphasized at this point that the foregoing analysis of MTX binding indicates nothing directly about a comparison with binding of dihydrofolate in the E·S complex, which is orders of magnitude weaker, because Asp 26,lc is believed to be nonionized in the E·S complex (see Section 7.1). Moreover, the pteridine ring of dihydrofolate is oriented differently in the active site, and has a 4-*oxo* group where MTX has a 4-amino group.

To continue with the description of protein-ligand interactions at the pteridine-binding site, a second hydrogen bond exists between the 2-amino group of MTX and a fixed water molecule that is, in turn, hydrogen bonded to the side chain hydroxyl of Thr 116,lc. Because this threonine is strictly conserved in all DHFR sequences, the bridging water molecule is almost certainly an invariant feature of the enzyme structure (Bolin et al., 1982).

Other than N1, the only endocyclic ring nitrogen involved in hydrogen bonding is N8. A structurally conserved fixed water molecule, Wat 253,lc, lies 3.3 Å from N8 directly in line with its lone pair orbital. In turn, this water molecule can hydrogen bond with two side chains and a second molecule of fixed solvent.

The inhibitor's 4-amino group donates hydrogen bonds to the backbone carbonyl oxygens of Leu 4,lc and Ala 97,lc (Bolin et al., 1982). Corresponding backbone carbonyls exist in geometrically analogous locations in both the *E. coli* and chicken DHFRs. Notwithstanding the apparently fortuitous location of the carbonyl of residue 97,lc for hydrogen bonding to the 4-amino group of MTX, which is of course an artificial inhibitor with no role in biology, it would appear that the observed geometry in this region of the molecule is probably important for enzyme function. Perhaps it is involved in activating the C4 position of the cofactor's nicotinamide ring for hydride transfer (see Section 7.3).

4.3. Trimethoprim

4.3.1. *Binding to* E. coli *DHFR*

A model for trimethoprim (TMP) binding to *L. casei* DHFR in solution was first proposed by Cayley et al. (1979) based on NMR transfer of saturation experiments that permitted identification of certain proton resonances of the bound inhibitor. Subsequent crystallographic studies by Baker et al. (1981) revealed that for the TMP binary complex with *E. coli* DHFR, the inhibitor's two aromatic rings were held in a nearly perpendicular conformation closely resembling one of two possible conformers predicted from the NMR investigation. Shortly thereafter, Matthews and Volz (1982) demonstrated that the geometry of TMP binding to chicken DHFR is significantly different than that reported earlier for the bacterial enzyme. Recently, the crystallographic structures have been refined and analyzed at high resolution in order to facilitate a detailed comparison of TMP binding to a bacterial and vertebrate DHFR (Matthews et al., 1985a, 1985b).

The geometry of TMP binding to *E. coli* DHFR closely resembles that of MTX binding, especially at their respective 2,4-diaminopyrimidine moieties, which interact identically with the enzyme (Fig. 12). As in the case of MTX binding, the side chain of Asp 27,ec closely approaches both N1 and the 2-amino group of TMP in a geometry that is consistent with the existence of a hydrogen-bonded salt bridge. Evidence from NMR experiments indicates that the pK_a of TMP is increased from about 7.5 in solution to over 10.0 upon binding to DHFR (Roberts et al., 1981; Cocco et al, 1983), with the effect presumably occurring at N1. Thus, N1 is probably protonated and Asp 27 is deprotonated in the complex. A further point of similarity between the MTX and TMP complexes is that a tightly bound solvent molecule mediates a second hydrogen bond between the 2-amino group and the side chain hydroxyl of the conserved residue Thr 113,ec. Finally, to complete the comparison between the environment of the pyrimidine rings in bound MTX and TMP, the same pair of backbone carbonyl oxygens (Ile 5 and Ile 94,ec) lie in the plane of the pyrimidine ring, positioned to accept hydrogen bonds from the 4-amino group. The active site region below the pyrimidine ring is open to solvent and contains several bound water molecules. Struc-

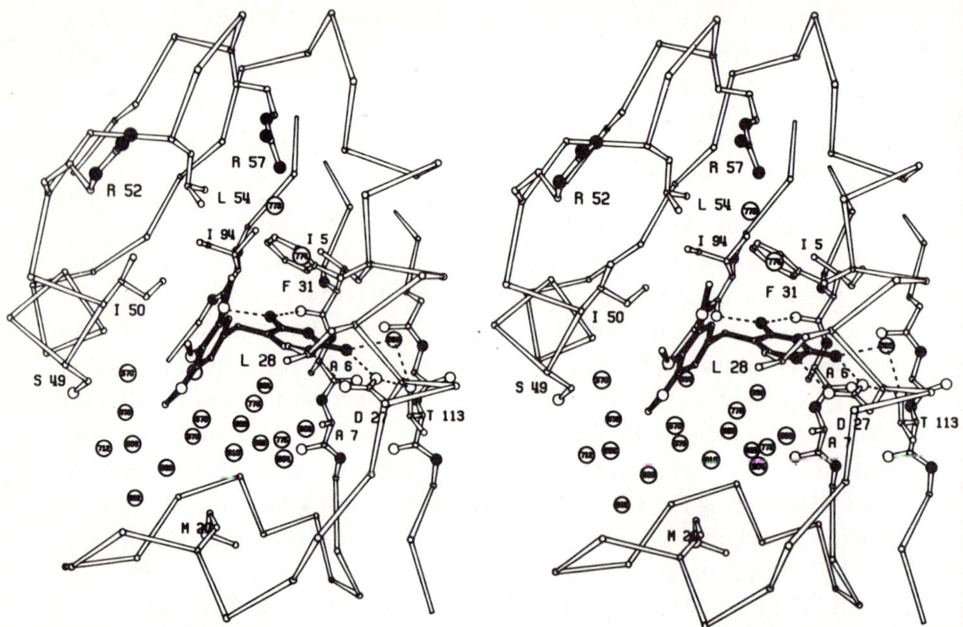

Figure 12. Stereo diagram of the binding of trimethoprim to *E. coli* dihydrofolate reductase. Trimethoprim is indicated by solid bonds and protein by open bonds. Carbon atoms are represented by smaller open circles, oxygen atoms by larger open circles, and nitrogen atoms by solid circles. Large numbered circles represent fixed solvent molecules. Hydrogen bonds are indicated by dashed lines. (From Matthews et al., 1985a.)

tural comparison with the *L. casei* DHFR-NADPH-MTX ternary complex (Filman et al., 1982) suggests that, in the presence of bound cofactor, many of these fixed solvent molecules will be displaced in order to accommodate the nicotinamide ring.

The crystallographic evidence suggests that an important factor underlying the strong affinity of TMP for *E. coli* DHFR is the favorable van der Waals interaction between the trimethoxybenzyl group of TMP and protein residues on two helices, αB and αC, at the entrance to the active site cleft. Ser 49 and Ile 50 are located at the C-terminal end of αC where they form a wall, composed of both main chain and side chain atoms, against which the trimethoxybenzene ring of TMP rests. On the opposite side of the active site cleft, Leu 28, in the middle of αB, provides the major protein contacts with the side chain of TMP.

Before describing TMP binding to chicken DHFR, it is instructive to consider how the structural results presented so far have guided the design of analogs of TMP with enhanced potency against *E. coli* DHFR. All DHFRs of known sequence have a structurally invariant arginine residue (Arg 57,ec) situated at one side of the active site cleft in a position where the guanidinium side chain can interact via charge-mediated hydrogen bonds with the α carboxylate of substrate or inhibitors such as MTX. On the other hand TMP has no such anionic substituent, suggesting that it might be possible to further enhance the binding of TMP-like molecules to *E. coli* DHFR if analogs containing appropriately placed carboxylate groups could be synthesized (Kuyper et al., 1982). At the time this work was begun, only the *E. coli* DHFR-MTX binary complex had been studied crystallographically, so there was a certain amount of guesswork in deciding where to put the anionic substituent and how long to make the arm tethering it to TMP's benzyl side chain. In retrospect, had the structure of the *E. coli* DHFR-TMP complex been known at the time, essentially all guesswork could have been eliminated. Nevertheless, Kuyper et al. (1982) were successful in designing a 3′-carboxypentyloxy analog of TMP that was 55-fold more inhibitory toward *E. coli* DHFR than TMP itself. Subsequent X-ray crystallographic studies of *E. coli* DHFR in binary complexes with two members of this acid-containing series of compounds confirmed that the carboxylate groups do indeed interact in the expected manner with the side chain of Arg 57,ec (Kuyper et al., 1985).

4.3.2. *Weak Binding to Chicken DHFR Results From Loss of a Crucial Hydrogen Bond*

Structural comparison of chicken and *E. coli* DHFR complexes containing TMP reveals several notable differences in inhibitor-binding geometry. The most obvious differences involve: (1) hydrogen bonding at the 4-amino substituent of the pyrimidine ring, (2) enzyme-inhibitor interactions at the trimethoxybenzyl group, (3) inhibitor side chain torsion angles, and (4) a large swinging movement of the side chain of Tyr 31 when TMP binds to the chicken enzyme (Matthews et al., 1985a).

As would be expected, the pyrimidine-binding site in chicken DHFR is analogous to the corresponding site in bacterial DHFRs (Fig. 13). However, an important finding is that there are certain peculiarities about the way TMP binds to the chicken enzyme. One such observation is that the pyrimidine ring is inserted almost 1 Å more deeply into the active site cleft than are the pyrimidines of TMP or MTX bound to the *E. coli* enzyme. Thus,

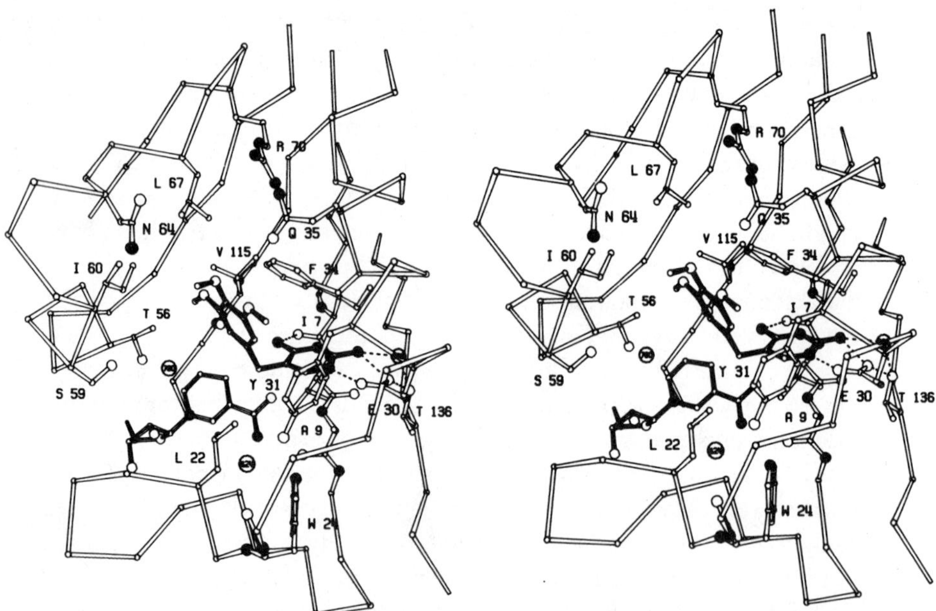

Figure 13. Stereo diagram of the binding of trimethoprim and NADPH to chicken dihydrofolate reductase. Trimethroprim is indicated by solid bonds, protein by open bonds, and a portion of the NADPH molecule by striped bonds. Carbon atoms are represented by smaller open circles, oxygen atoms by larger open circles, and nitrogen atoms by solid circles. Large numbered circles represent fixed solvent molecules. Hydrogen bonds are indicated by dashed lines. (From Matthews et al., 1985a.)

whereas the 4-amino group of TMP participates in two hydrogen bonds with backbone carbonyl oxygens of *E. coli* DHFR, only one of the corresponding hydrogen bonds remains intact in the chicken DHFR-NADPH-TMP ternary complex—that between the 4-amino group of TMP and the backbone carbonyl of Ile 7,cl. Other hydrogen bonds between the pteridine ring and chicken DHFr are analogous to those already discussed for TMP binding to *E. coli* DHFR, including the important hydrogen bond-mediated salt linkage between the active site carboxylate (Glu 30 in chicken DHFR) and TMP's N1 and 2-amino group. Recent NMR data are also consistent with the notion that, at least for TMP and MTX, interaction between the inhibitor's protonated N1 and the active site carboxylate group is very similar for DHFRs from three bacterial and two vertebrate species (Cocco et al., 1983; Birdsall et al., 1983).

For TMP bound to *E. coli* DHFR, the inhibitor's benzyl side chain is positioned low in the active site pocket, pointing down toward the nicotinamide binding site (the "down" conformation which positions the inhibitor side chain in the "lower" cleft), whereas in chicken DHFR the benzyl group is accommodated in a side channel running upward and away from the cofactor (the "up" conformation which positions the inhibitor side chain in the "upper" cleft). Clearly, interactions between the benzyl group and the enzyme are quite different in the two cases. Birdsall et al. (1983) and Cocco et al. (1983) have made similar inferences from their NMR studies, and have gone on to propose that species-specific differences in the way a particular DHFR interacts with the TMP side chain is the major factor modulating inhibitor selectivity.

In a series of difference Fourier analyses of inhibitor binding to chicken DHFR, Matthews et al. (1985b) have shown that this is probably not the major factor influencing TMP selectivity because the trimethoxybenzyl group of TMP binds to chicken DHFR in the same hydrophobic pocket that accommodates the side chains of various other pyrimidine and triazine inhibitors that, unlike TMP, are extremely potent inhibitors of vertebrate DHFRs. Moreover, these authors were able to show that conformational changes accompanying TMP binding to chicken DHFR and torsional differences about the benzyl carbon for TMP bound to *E. coli* and chicken DHFRs are probably of little importance in explaining TMP's poor affinity for vertebrate DHFRs. What was very clear from these crystallographic investigations, however, was that only for inhibitors that are weakly bound to chicken DHFR did Matthews et al. (1985a) observe an altered binding geometry in which the 2,4-diamino heterocycle occupies a position approximately 1 Å closer to αB where it can no longer hydrogen bond to the carbonyl of Val 115 (Ile 94, ec). Thus, the crystallographic evidence strongly suggests that loss of this hydrogen bond is a significant factor in TMP's low affinity for chicken DHFR and, therefore, must be an important determinant of the drug's selectivity.

4.4. Active Site Structural Differences and Inhibitor Selectivity

In attempting to explain why TMP binds differently to *E. coli* and to chicken DHFR, Matthews et al. (1985b) have pointed out that residues on opposite sides of the active site cleft in chicken DHFR are about 1.5–2.0 Å further apart than are structurally equivalent residues in the *E. coli* enzyme. Due to both the increased width across the cavity bounded by αB and αC in chicken DHFR and the presence of Tyr or Phe at position 31 in vertebrate DHFRs, instead of Leu 28 in *E. coli* DHFR, model-building experiments conclusively demonstrate that the TMP side chain cannot be favorably accommodated in the lower cavity of vertebrate DHFR. Specifically, if hydrophobic interactions on the left side of the cleft of chicken DHFR between the side chain of Ile 60 and a hypothetical trimethoxybenzyl group of TMP in the down conformation are required to be similar to the corresponding interactions found for the *E. coli* DHFR-TMP binary complex, then it is not possible to position the side chain of Tyr 31,cl, or the corresponding Phe side chain of other vertebrate DHFRs, to provide favorable hydrophobic contacts on the right side of the cleft analogous to those provided by Leu 28 in the *E. coli* enzyme. In other words, a gap would be left between the benzyl group and the enzyme. The upshot of these considerations is that for chicken DHFR, the trimethoxybenzyl group of TMP can only be adequately accommodated in an alternative binding site, the upper cleft, which simultaneously results in a disruption of hydrogen bonding to the 2,4-diaminopyrimidine, thereby reducing the binding of TMP to vertebrate DHFR.

5.   STRUCTURAL CHANGES RESULTING FROM LIGAND BINDING

5.1.   Definitions and General Considerations

Proteins are complex, dynamic molecules that can presumably exist in a large number of conformational substates of nearly the same free energy. Transition between these substates requires overcoming the associated potential energy barriers, and is thermally driven. Protein molecules in different substates will have similar gross structural features but may differ in local conformation, because at physiological temperature, the energies required for radical structural deformations are large compared with kT. As discussed in Section 6.5, there is good evidence that, for at least some DHFRs, certain conformational substates are in slow equilibrium, and several of these have been characterized based on differences in kinetic and spectroscopic properties.

Such considerations suggest that when ligands bind to a protein such as DHFR, the relative populations of previously accessible substates may be altered and, in certain instances, new conformational substates may appear as well. Changes of this type, collectively referred to as conformational changes, have been inferred from spectroscopic, NMR, X-ray diffraction, and kinetic studies of ligand binding to DHFR. What is not clear is the extent to which these conformational changes accompanying ligand binding result in significant changes in enzyme structure, and to what extent, if any, they may be important in understanding DHFR function.

Part of the problem is one of definition. For example, significant changes in NMR chemical shifts can result from very small changes, on the order of 0.1 Å in atomic position, changes that are within the error limits of even high-resolution structural studies on most proteins. In the few cases where direct comparison of atomic coordinates for the same DHFR with and without bound ligands is possible, the crystallographic data suggests that despite several notable exceptions, these ligand-induced structural changes are localized, few in number, less than $1-2$ Å in magnitude, and may even differ from one species of DHFR to another. Therefore, it would appear that many of the induced conformational changes that have been postulated to occur in DHFR, based on NMR chemical shift data, are more properly viewed as resulting from ligand-mediated redistributions of closely related, rapidly interconverting substates of the apoenzyme. In other words, ligand-binding preferentially stabilizes certain conformers in a spectrum of conformations accessible to the apoenzyme, producing very small changes in the time-averaged atomic coordinates, but resulting in resolvable chemical shift differences.

5.2. NMR Studies

During the last 5 years, considerable success has been achieved in assigning NMR resonances to specific nuclei in several bacterial DHFRs, principally those from *L. casei* and *S. faecium*. When assignments can be made for various nuclei in different parts of the DHFR molecule, these individual NMR probes, in principle, provide a powerful means by which conformational changes accompanying complex formation can be monitored. In practice, it is often difficult to separate indirect effects mediated by conformational change from direct effects on an assigned chemical shift due simply to proximity (Blakley et al., 1978, and references therein.)

The proton magnetic resonance spectrum of *L. casei* DHFR apoenzyme is altered in characteristically specific ways, particularly in the aromatic region, by additions of dihydrofolate, folate, MTX, or cofactor, suggesting that these various enzyme-ligand complexes may exist in conformationally distinct states (Pastore et al., 1974). The ternary complex with $NADP^+$ and TMP exists in solution as a mixture of two slowly interconverting conformational states with apparently widespread conformational differences because six of the seven histidine C2-H resonances have different chemical shifts in these two states (Birdsall et al., 1984). Evidence for cofactor-induced conformational changes in the benzoyl-binding site has been obtained by line-shape analysis of ^{19}F resonances for 3′,5′-difluoro MTX bound to *L. casei* DHFR (Clore et al., 1984). The rate of benzoyl ring flipping increases 2.6 to 2.8-fold upon addition of $NADP^+$ or NADPH to form the ternary complex.

Blakley and coworkers in their NMR studies of *S. faecium* DHFR have made extensive use of biosynthetic incorporation of ^{13}C amino acids in order to provide specific magnetic probes for monitoring ligand binding to the enzyme. Certain methionine (Blakley et al., 1978), arginine (Cocco et al., 1978), and tryptophan (London et al., 1982) side chains containing such labels show changes in chemical shift, upon binding inhibitors and cofactors, that are most readily explained by postulating extensively propagated ligand-induced conformational changes. A similar study by Pastore et al. (1981) monitored resonances from the four ^{13}C-labeled tryptophan side chains of *L. casei* DHFR, and demonstrated that ligand binding leads to a sharpening of the band assigned to Trp 5, indicating an altered conformational mobility.

It has been suggested that the cooperative enhancement of substrate or inhibitor binding in the presence of cofactor is due, at least in part, to ligand-induced conformational changes (Birdsall et al., 1980a, 1980b; Matthews et al., 1978). Comparison of NMR spectra for selectively deuterated and fluorine-labeled *L. casei* DHFR in the presence of MTX with those obtained for the corresponding ternary complexes shows numerous chemical shift differences that could arise from such conformational changes (Feeney et al., 1977; Kimber et al., 1977). More recently, Birdsall et al. (1984) have shown that certain coenzyme analogs, such as thio-$NADP^+$, exhibit cooperativity in binding with TMP that cannot result from direct interaction between the nicotinamide and TMP, because the nicotinamide of thio-$NADP^+$ is believed not to be in contact with that enzyme, but rather must arise from conformational changes induced by the binding of the rest of the coenzyme. Cofactor-binding effects are assigned to proton chemical shifts for three histidine (Gronenborn et al., 1981a) and two tryptophan residues (Feeney et al., 1980a) that are far removed from the coenzyme-binding site. Therefore, these resonances must be monitoring ligand-induced conformational changes. Addition of MTX to *L. casei* DHFR-NADP$^+$ causes His 22, in the loop connection βA to αB, and either Trp 21 or Trp 133 to become more exposed, explicitly demonstrating in this case that the induced changes involve residues in the vicinity of the active site (Feeney et al., 1980b). On the other hand, negative cooperativity is observed for binding of folinic acid to *L. casei* DHFR, where the association constant is reduced 600-fold in the presence of NADPH compared to enzyme alone (Birdsall et al., 1981). Proton chemical shift data indicate that this pronounced effect is apparently accompanied by conformational changes in the adenine-binding site and that the changes are different for NADPH and NADP$^+$ (Birdsall et al., 1981).

5.3. Kinetic Studies

Additional evidence suggestive of induced conformational changes arises from kinetic studies of ligand binding to DHFR. Dunn and King (1980), using the stopped-flow fluorescence technique, found that the rate constants for inhibitor binding to the *L. casei* enzymes-NADPH complex differ by only a factor of 4 for 10 inhibitors of widely varying structure. Due to the high observed activation energies, the authors argue that the initial association is not diffusion controlled, and suggest that the initial ternary complex must undergo rapid rearrangement. Williams et al. (1980) have analyzed the progress of reaction curves for inhibition of *S. faecium* DHFR by 19 folate analogs, and find evidence for rapid formation of the enzyme-NADPH-inhibitor complex followed by a slow isomerization, with a forward rate constant of 2–6 min^{-1}. Clearly, this postulated rearrangement cannot be the same as the much faster process revealed in the experiments of Dunn and King. In neither case is it clear whether the isomerizations are protein conformational changes or simple geometrical repositionings of the inhibitor. Further discussion of related matters is given in Section 6.5.2.

5.4. X-ray Structural Results

5.4.1. Comparison of E. coli DHFR-MTX and L. casei DHFR-NADPH-MTX

Although spectroscopic and kinetic studies demonstrate that ligand-mediated conformational changes occur for several DHFRs, they have been much less successful in describing these rearrangements in structural terms. Comparison of molecular coordinates derived from diffraction measurements on *E. coli* DHFR-MTX and *L. casei* DHFR-NADPH-MTX suggested to Matthews et al. (1979) that cofactor binding to the enzyme-MTX binary complex was accompanied by concomitant changes of $1-3$ Å in the loop connecting βA to αB, and that any geometrical rearrangements elsewhere in the structure must be even smaller. Such a comparison can never be definitive, because the enzymes derive from different species and, moreover, the conformation of loop βA to αB in the *E. coli* binary complex is affected to an undetermined extent by intermolecular contacts in the crystallographic asymmetric unit and by the presence of a bound Ca^{2+} ion.

5.4.2. Cofactor Binding to Chicken DHFR

In order to unambiguously characterize geometrical changes occurring on ligand binding, it is necessary to compare structures for ligated and unligated states of the same enzyme. It became possible to make such comparisons only recently, when diffraction studies of chicken DHFR were broadened to include structural investigations of the apoenzyme (D.A. Matthews and D.J. Filman, unpublished results) and a variety of enzyme-NADPH-inhibitor ternary complexes (Matthews et al., 1985a, 1985b) in addition to the holoenzyme structure reported previously (Volz et al., 1982).

Crystals of apo chicken DHFR are not isomorphous with those for the holoenzyme, suggesting in itself that conformational changes occur on binding to NADPH. As a practical matter, the crystallographic nonisomorphism precludes direct determination of the conformational differences by difference Fourier methods, so that the structures had to be solved independently. Whereas the holoenzyme structure has been determined at high resolution with an R factor of 0.21 for 1.6 Å data (Matthews et al., 1985a), the apoenzyme structure is only at an intermediate resolution with an R factor of 0.24 for 3 Å data (D.A. Matthews and D.J. Filman, unpublished results). Therefore, apparent geometrical differences of less than 1 Å for these two enzyme states may simply be due to errors in the current apoenzyme coordinates.

The most striking structural consequence of NADPH binding to chicken DHFR is a major rearrangement of polypeptide backbone and side chain atoms in the neighborhood of the adenine-binding site, particularly for residues $91-96$ in the loop connecting βD to αE. In apo chicken DHFR, residues 91 through 96 form an irregular loop connecting βD with a short, five-residue helix αE. As a result of cofactor binding, this loop moves over to cap the adenine-binding pocket, providing extensive van der Waals contacts all along the exterior edge of the pyrimidine portion of the ring and positioning five additional residues, 92−96, in α helical conformation at the N-terminus of αE. Backbone movements for residues 91 and 92 are about 6 Å. Even larger rearrangements occur for some of the side chains; for example, the side chain nitrogen of Lys 91 is displaced 15 Å on binding to NADPH. The crystallographic evidence suggests that cofactor binding does not induce analogous conformational changes in *E. coli* or *L. casei* DHFR. Comparison of coordinates for the *E. coli* DHFR-MTX binary and the *L. casei* DHFR-NAPDH-MTX ternary complex reveals no significant structural differences in the adenine-binding region, which suggests that any structural adjustments occurring as a result of cofactor binding must be very small. In this context, it is interesting to note that helix αE in the *E. coli* DHFR binary complex, without NADPH, is fully formed and shortened by only a single residue in comparison with the corresponding helix in the chicken DHFR-NADPH binary complex. The single, additional α helical residue in the chicken enzyme results from amino acid insertions, characteristic of vertebrate DHFRs, that follow αE. These results provide direct structural evidence supporting the hypothesis that conformational changes accompanying ligand binding can be species specific, and may not be especially important for understanding structure-function relationships for DHFRs in general.

Binding of the AMN portion of NADPH to chicken DHFR causes smaller backbone adjustments (1−2 Å) for residues 76 and 77 and 3 Å displacement of the Arg 77 side chain, which in its new location forms one side of the adenine-binding cleft. The largest remaining differences occur at Lys 54 and Lys 55, which in the holoenzyme interact, respectively, with the 2′-phosphate and pyrophosphate groups of the bound cofactor. The side chain of Lys 54 is apparently disordered in apo chicken DHFR, while in the holoenzyme, it is hydrogen bonded to the 2′-phosphate. Lys 55 in the apoenzyme is positioned so that in the absence of conformational readjustments, its side chain would sterically interfere with cofactor binding. In the holoenzyme, the Lys 55 side chain is repositioned (5.8 Å movement for $N\zeta$) to form a charge-mediated hydrogen bond with the pyrophosphate moiety. The structurally analogous residue in *E. coli* DHFR is His 45, and model building experiments suggest that its side chain position in the binary complex with MTX would also have to be altered to permit binding of NADPH (Matthews et al., 1985a).

Surprisingly, there are no clear indications of conformational differences between apo and holo chicken DHFR in the nicotinamide-binding site or adjacent regions of the active site. Some small structural variation, $1-1.5$ Å, occurs in the loop connecting βA to αB, particularly near to Asn 21. If these differences are real and not a consequence of errors in the incompletely refined apoenzyme structure, they may be reporting NADPH-induced changes similar to those described previously by Matthews et al. (1978) in their comparison of the *E. coli* DHFR-MTX binary complex and the *L. casei* DHFR-NADPH-MTX ternary complex.

Crystallographic studies of chicken apoenzyme have clarified one point concerning the structural reason for enhanced inhibitor binding in the presence of NADPH. On the basis of earlier structural comparisons, Matthews et al. (1978) proposed that in the case of *E. coli* DHFR, ligand cooperativity could be explained, in part, by cofactor-induced movement of Met 19 from its location in the enzyme-MTX binary complex, where it does not interact with MTX, into a new position where it would be in direct contact with the inhibitor. The structurally analogous residues in *L. casei* and chicken DHFRs are Leu 20,lc and Leu 22,cl and, in both cases, the respective enzyme-NADPH complexes have the leucine side chains positioned so they can provide favorable van der Waals interactions with bound inhibitors. However, the conformations of Leu 22 in apo and holo chicken DHFR are indistinguishable at the current level of model refinement, suggesting that, at least for the chicken enzyme, cofactor binding does not influence the orientation of this side chain and that the altered position for the structurally analogous Met 19 in *E. coli* DHFR is therefore probably unrelated to the absence of bound NADPH.

5.4.3. Ligand Binding to Chicken DHFR-NADPH

Ligand binding to the chicken DHFR-NADPH binary complex has been studied crystallographically by difference Fourier methods for a wide variety of inhibitors and substrates, including pyrimidines, triazines, quinazolines, pyridopyrimidines, and pterins (Matthews et al. 1985a, 1985b, and unpublished results). The observed conformational changes accompanying formation of the ternary complexes are small, on the order of 1 Å or less, and localized to the active site. An exception occurs for certain $3',4',5'$-trisubstituted benzyl pyrimidines, which, upon binding, cause a significant repositioning of the Tyr 31 side chain with an average movement of 5 Å for all atoms beyond Cβ. In the holoenzyme complex, the carboxylate group of Glu 30 is apparently quite mobile, because a $2F_o - F_c$ map shows only very weak density beyond Cγ. When inhibitors bind at the active site, the side chain of Glu 30 is repositioned slightly, less than 1 Å, and locked in place by interactions with the inhibitor's N1 and 2-amino groups. It is also clear that TMP binding to the chicken holoenzyme causes small structural perturbations at the carboxyl end of αC, which moves 0.1−0.2 Å in order to accommodate binding at the inhibitor's trimethoxybenzyl group (Matthews et al., 1985a). Phenyltriazine binding causes small adjustments, 0.1−0.2 Å, in protein conformation for backbone atoms in αB on the opposite side of the active site cleft. The side chain of Tyr 31 is also affected by phenyltriazine binding, but the induced conformational change is much smaller, 0.8 Å, and in the opposite direction to that caused by TMP (Matthews et al., 1985b).

6. ENZYME-KINETIC PROPERTIES AND SUBSTRATE/COFACTOR BINDING

6.1. Some Questions and Problems

Given the enormous amount of detailed stereochemical information about the DHFR molecule contained in the known crystal structures, one may now ask if the structures explain any enzymic properties. Ultimately, one would like to understand how this particular geometrical arrangement of polypeptide backbone and side chains contrives so specifically and efficiently to accelerate the overall redox reaction summarized by the equation given in the first paragraph of this review. But, in fact, the overall reaction as written is simply a net summation over several individual steps, classically referred to as the "kinetic mechanism," but which perhaps ought to be called a "reaction pathway" or a "kinetic sequence." The problem, then, is to resolve the overall reaction into discrete individual steps, and to determine how each one may be controlled and accelerated by the stereochemical machinery of the DHFR molecule. Phrased more succinctly, a high-resolution reaction pathway is needed to accompany the high-resolution structure, and one is not yet available.

In this section, a summary of what is known about the DHFR reaction pathway is presented. Unfortunately, it has not yet been as thoroughly studied as a few other enzymic reactions, and DHFR presents several intrinsic difficulties requiring extra effort to overcome. One is that the overall reaction itself is fairly complicated, involving three reactants (including the H^+ ion) and two products. Moreover, equilibrium for the overall reaction is far to the right, that is, in favor of fully reduced tetrahydrofolate and oxidized $NADP^+$, with an equilibrium constant variously estimated to be 5.6×10^{11} at pH 7.5 (Mathews and Huennekens, 1963) and 0.84×10^{11} at pH 8.0 (Nixon and Blakley,1968). Thus, straightforward kinetic measurements on the reverse reaction that could, in principle, supply valuable mechanistic information are not feasible in this system.

Another stumbling block to kinetic studies on DHFR is that ground-state binding of both substrate and cofactor is unusually strong; most species of DHFR have K_m values for NADPH and dihydrofolate between 0.5 and $5\ \mu M$, making ordinary spectrophotometric assay methods impractical at concentrations below K_m (Hitchings and Baccanari, 1984). Radiometric assay methods have been developed, however, that may make it feasible to extend kinetic measurements much further down into the lower ranges of enzyme and substrate concentrations (Hayman et al., 1978; Rothenberg et al., 1980).

Further complexities arise in the study of DHFR kinetics owing to the existence of a variety of conformations of the enzyme molecule. Several lines of evidence, including spectroscopic and X-ray crystallographic as well as kinetic data, indicate the occurrence of a number of distinct conformational transitions upon binding of substrate, cofactor, and inhibitors to DHFR. The nature of these conformational changes is still largely obscure, but the X-ray crystallographic data thus far imply that they must be structurally quite minor (see Section 5). Nevertheless they cause some interesting complications when trying to interpret kinetic data.

6.2. Hysteretic Phenomena

One class of kinetic complications that are probably of this type are the hysteresis effects, wherein rates of reaction change slowly in response to a rapid change in substrate, cofactor, or inhibitor concentrations (Frieden, 1970). Thus, Baccanari and Joyner (1981) observed that the shapes of reaction velocity curves obtained with *E. coli* DHFR depend on the order of substrate addition. In particular, although a linear initial velocity curve results when the reaction is started by addition of NADPH, quite different behavior is seen when the reaction is started by addition of dihydrofolate after preincubation with cofactor. In the latter case, the initial velocity is low by a factor of up to 2, and then increases over a period of 12−14 minutes to a final velocity of only 70% of that observed for the NADPH-initiated reaction (Baccanari and Joyner, 1981). The simplest explanation for this behavior invokes the existence of a low-activity, slowly interconverting enzyme-cofactor complex, presumably with a conformation that differs somehow from the high-activity form.

A different hysteretic phenomenon, with a half-time of 9 seconds, in contrast to 10 minutes, and at enzyme concentrations of about 50 nM, in contrast to 0.2 nM, has been observed by Penner and Frieden (1985). In stopped flow studies, these investigators find that rapid addition of substrate and cofactor to the untreated enzyme results in a rate that is initially low by a factor of 2.3 and that slowly increases but fails to attain its expected maximum. However, this lag is abolished by preincubation of the enzyme with either substrate or cofactor. Penner and Frieden considered their results to be reasonably consistent with the findings of Dunn, King, and coworkers (Dunn et al., 1978; Dunn and King, 1980; Cayley et al., 1981) on the kinetics of ligand binding to *L. casei* and *E. coli* DHFRs (see Section 6.5). They also make the interesting observation that hysteretic behavior of this type is usually observed in multisubunit allosteric enzymes, but not in presumably simple enzymes such as DHFR, which are composed of a single polypeptide chain.

6.3.   Specificity for Substrate and Cofactor

The kinetic parameters for different species of DHFR vary considerably, but, for purposes of comparison, the value for k_{cat} is 16.5 sec^{-1}; K_m for dihydrofolate is 0.47 μM; and K_m for NADPH is 2.5 μM (Stone and Morrison, 1982), all for the *E. coli* enzyme. Although all DHFRs show strong specificity for dihydrofolate, its poly-γ-glutamyl derivatives, and its close synthetic homologs, as compared with any other potential substrate, some DHFRs—notably those derived from vertebrate sources—will slowly reduce folate, dihydrobiopterin, and biopterin (Kisliuk, 1984, pp. 27–28). The rates of folate reduction are generally about one-third to 1/10th those for dihydrofolate in the case of some vertebrate enzymes, but no more than 1/1000th in the case of DHFR from *E. coli*. Isozyme I isolated from *S. faecium* is not measurably active toward folate at all (Blakley, 1984). Additionally, the optimum pH for folate reduction is approximately 4.5, whereas for dihydrofolate, it is in the neighborhood of 7.0 (see Blakley, 1984, Tables 5.2–5.4). It should be noted, however, that in very few instances has substrate specificity been carefully examined with purified enzyme preparations.

Specificity for the nicotinamide nucleotide cofactor is not absolute, but NADPH is always strongly preferred over NADH. For example, in the case of *E. coli* DHFR, the K_m value for NADH is 268 μM as compared with 3.2 μM for NADPH (Baccanari et al., 1975). Transfer of the hydride ion from C4 of the nicotinamide to C6 of dihydrofolate takes place from the A side (the 4-pro-R hydrogen) of the nicotinamide ring (Pastore and Friedkin, 1962; Blakley et al., 1963) to the *si* face of the dihydropteridine ring, and results in a product tetrahydrofolate containing an asymmetric center at C6. The absolute configuration at C6 is such that the hydrogen points down into the plane of the paper when dihydrofolate is depicted as in Figure 1 (Fontecilla-Camps et al., 1979; Armarego et al., 1980). This configuration is designated as S in tetrahydrofolate, but would be correctly designated as R in tetrehydrofolate derivatives where the hydrogen on N10 is substituted by a one-carbon moiety (Kisliuk, 1984, p. 6). An important conclusion derived from knowledge of the absolute configuration of enzymatically produced tetrahydrofolate, when combined with earlier information from X-ray crystallography on the binding of MTX at the active site, is that the pteridine ring of enzyme-bound substrate must be turned over by 180° with respect to the pteridine ring of bound MTX.

When the substrate is folate, hydride transfer to C7 likewise occurs to the same *si* face of the pteridine ring as with dihydrofolate (Charlton et al., 1979; Poe, 1980). According to Kisliuk (1984, p. 27), it is possible that, in fact, the 5,6-double bond of folate is reduced initially, with subsequent rearrangement to 7,8-dihydrofolate.

6.4. Activation

A curious and unusual kinetic property of some DHFRs is their activation by salts, urea, and sulfhydryl reagents. The review by Blakley (1984) presents a thorough compilation of examples with comprehensive references, but here it will suffice to summarize what is known about the phenomenon. A few bacterial and vertebrate DHFRs are moderately activated by salts such as KCl or NaCl at 0.3–0.5 M, while others are either unaffected or are actually inhibited. Moreover, all the vertebrate DHFRs activated by salt are also activated by urea at approximately 2 M as well. In the case of chicken DHFR, for which urea activation has been studied in some detail (Kaufman,

1968), urea and thiourea increase k_{cat} by fourfold to 10-fold and substantially increase K_m for both dihydrofolate and NADPH. Finally, most vertebrate DHFRs are similarly activated by reagents that react with sulfhydryl groups, such as organic mercurials or tetrathionate. For the chicken enzyme, this activation can be as much as 12-fold, and is accompanied by increased sensitivity to heat denaturation (Barbehenn and Kaufman, 1980, 1982). It is likely that the reactive cysteine residue is Cys 11 in chicken DHFR or Cys 6 in other vertebrate DHFRs, because cysteine is not present in the N-terminal portion of the polypeptide chain in the bacterial enzymes, which are not activated by mercurials. An open question is whether this phenomenon suggests a biologically significant mechanism for controlling the level of DHFR activity in vertebrate tissues.

A structural explanation for the activation phenomenon would no doubt shed light on the stereochemical mechanism of the DHFR molecule. But in the interim, an attractive speculation can be formulated, based on the well-known property of urea as a denaturant and the thermal sensitivity of the cysteine-modified enzyme. Perhaps the common thread is a general loosening of the enzyme's native folded structure by these activating agents. Such loosening could increase the rate of a conformational isomerization that is rate limiting, for instance, one required for transfer of a proton to the substrate (see Section 7.1). This theme of conformational equilibrium, it will be seen, recurs constantly in discussions of the molecular properties of DHFR (see Section 5).

6.5. Cofactor Binding

6.5.1. Dissociation Constants

Cofactor dissociation constants, as measured directly on binary complexes of NADPH with various species of DHFR at neutral pH, are in the range of $0.02-3.9\ \mu M$. Binding of the oxidized cofactor is somewhat weaker for all species studied, with $NADP^+$ exhibiting dissociation constants that are greater than those for NADPH by factors of $3.2-18$ (Blakley, 1984, Table 5.9). Based on analysis of kinetics measurements, Stone and Morrison (1982) calculate dissociation constants of $0.51\ \mu M$ and $18\ \mu M$ for the binary complexes of *E. coli* DHFR with NADPH and $NADP^+$, respectively, a differential factor of 35. The largest difference between binding affinity for oxidized and reduced cofactor is seen for *L. casei* DHFR at pH 6.0, where binding of the oxidized cofactor is weaker by a factor of 1200; dissociation constants are 0.01 μM and 12 μM, respectively (Dunn et al., 1978). According to Fersht (1985, p. 404), it is usually the case for other nicotinamide nucleotide-dependent enzymes that NADH binds more strongly than NAD^+. This differential binding effect is often attributed to a hydrophobic-binding pocket discriminating against the positively charged nicotinamide ring of the oxidized cofactor, although the nicotinamide-binding pocket in DHFR, as mentioned in Sections 3.3 and 7.3, actually contains several polar oxygen atoms.

Binding data obtained with the cofactor analogs, such as 2′-phospho-adenosine-5′-diphosphoribose and ATP-ribose which lack a nicotinamide ring but bind nearly as well as $NADP^+$, suggest that the nicotinamide ring itself contributes relatively little to cofactor binding energy (Stone et al., 1984).

6.5.2. Kinetic Evidence for Conformational Multiplicity

Several lines of evidence suggest that cofactor binding is not a simple, single-step process, but rather involves at least two conformational states of the enzyme and/or cofactor. Dunn, King, and coworkers (Dunn et al., 1978; Dunn and King, 1980; Cayley et al., 1981) have made stopped-flow kinetic measurements of both the association and dissociation rates for substrate, cofactor, and inhibitors bound to *L. casei* DHFR and *E. coli* DHFR isolated from the strain RT500 (see Section 8.2). Both quenching of the enzymes' fluorescence and enhancement of the NADPH fluorescence upon binding were monitored as a function of time in these studies, which led to the conclusion that formation of all the binary complexes occurs rapidly and reversibly by way of a mechanism that may involve up to three distinguishable steps. Both NADPH and the substrates, dihydrofolate and folate, appear to bind to only one of two equilibrium conformers of the enzyme, although the inhibitors bind to both. A rapid bimolecular assocation is found to be governed by rate constants of about 10^6 to 10^7 M^{-1} sec^{-1} depending on the particular ligand and the species of DHFR. A slower unimolecular step with a pH-dependent rate constant of $0.06-0.012$ sec^{-1} is also observed, and is thought to represent interconversion between conformers of the free enzyme, only one of which can bind NADPH. The major difference between the *E. coli* and *L. casei* enzymes is the pH dependence of the proportion of fast-binding conformer, suggesting that the interconversion is not controlled by the same structural mechanism in the two species and may not be a fundamental property of DHFRs in general.

The kinetics of ligand binding to form the ternary enzyme-inhibitor-coenzyme complex from a binary complex are simpler. Only a single fast phase is seen, with apparent bimolecular rate constants in the range 2×10^6 to $3 \times 10^7 \, M^{-1} \, sec^{-1}$ (Dunn and King, 1980). This simplicity is consistent with the binary complexes existing in essentially a single conformation.

Overall dissociation is evidently governed by an equilibrium between two or more conformers of the enzyme-ligand complex. It is not possible to distinguish from such experiments, however, what the nature of the conformational differences are, for example, whether the enzyme structure itself moves or whether the ligand shifts its position. Ligands are tightly bound in the ternary complex, with dissociation rate constants of less than 1 sec^{-1}.

6.5.3. NMR Evidence for Conformational Multiplicity

The most direct noncrystallographic evidence for the possible involvement of more than one conformational state in the binding of cofactor is provided by NMR studies carried out principally by the Mill Hill group on the *L. casei* DHFR and by Blakley and coworkers on *S. faecium* DHFR.

The Mill Hill group has examined the spectra given by a number of magnetic nuclei in a variety of complexes of *L. casei* DHFR, and shown that two states are observed for the enzyme-$NADP^+$-TMP ternary complex (Birdsall et al., 1984), while three conformational states can be distinguished for the enzyme-$NADP^+$-folate ternary complex (Birdsall et al., 1982). On the other hand, only a single conformation is observed for the binary enzyme-$NADP^+$ or the ternary enzyme-$NADP^+$-MTX complex (Hyde et al., 1980a, 1980b).

Examination of the ternary complex of *L. casei* DHFR with TMP and $NADP^+$ by multinuclear NMR techniques (Birdsall et al., 1984) indicates the existence of two conformations, I and II, that interconvert slowly, about 6 sec^{-1} at 30°, and are present in comparable proportions. In conformation I, the nicotinamide ring of the cofactor is bound to the enzyme, whereas in conformation II, the nicotinamide ring has swung away from the enzyme surface into solution. Conformational flexibility at the pyrophosphate bridge of the dinucleotide is thought to make this motion possible. The ratio of conformation I to conformation II is 1:2 for the TMP ternary complex, but varies from less than 0.1 to 2.3 for complexes with various TMP analogs or analogs of $NADP^+$ (Gronenborn et al., 1981b). Moreover, conformation I appears to be very similar to that of the binary enzyme-$NADP^+$ and ternary enzyme-$NADP^+$-MTX complexes (Hyde et al.,

1980a, 1980b). Apparently the equilibrium between conformations I and II is quite sensitive to details of nonbonded interactions, which is a conclusion also consistent with the finding that the nicotinamide ring itself, particularly in its oxidized state, contributes little to the binding energy of the cofactor. At this point, it should also be noted that only a single structure, with the nicotinamide ring bound to the enzyme, presumably conformation I, was observed in the X-ray structures of the ternary complexes *L. casei* DHFR-NADPH-MTX (Matthews et al., 1978; Filman et al., 1982) and chicken DHFR-NADPH-TMP (Matthews et al., 1985a).

A more complicated array of apparent conformational states was seen by Birdsall et al. (1982) for the *L. casei* DHFR ternary complex with folate and $NADP^+$, which is a dead-end complex. Three conformers were observed with the differences between them localized to the neighborhood of the pteridine and nicotinamide rings. The possible significance of this conformational equilibrium for the molecular mechanism of DHFR is discussed in Section 7.1.4.

A different situation with regard to the effect of cofactor binding on conformational multiplicity is reported for *S. faecium* DHFR by London et al. (1982). In this case, ^{13}C NMR spectra were obtained for enzyme into which ^{13}C-labeled tryptophan had been incorporated by enrichment of the growth medium. Although rigorous identification of each of the four resonances was not possible, a significant observation for present purposes was that resonance 3, probably Trp 22, exhibited a double structure in the apoenzyme that was not observed in any of the nucleotide complexes examined. Thus, this particular conformational multiplicity appears to occur in the cofactor-free enzyme, but not in the enzyme-cofactor complex, which is the opposite of the situation just described for the *L. casei* enzyme.

6.5.4. *Cooperativity with Inhibitors*

Another interesting aspect of the DHFR-cofactor interaction is that the cofactor binds considerably more tightly in some ternary complexes of DHFR than in binary complexes, that is, in complexes containing inhibitors such as MTX or TMP in addition to cofactor. This phenomenon is usually termed cooperativity. Thermodynamics requires, of course, that the effect be reciprocal, producing equivalently enhanced binding of the substrate or inhibitor to the ternary complex.

Cooperativity factors of 2−680 have been observed for the cofactor itself, and still more dramatic effects are seen with some cofactor analogs (Blakley, 1984, Table 5.10). Blakley (1984) has pointed out that the degree of cooperativity is much greater for NADPH and its analogs than for the corresponding nucleotides with oxidized nicotinamide rings, 680 v. 12.5 for the MTX complex with *L. casei* DHFR. This observation correlates with the stronger binding of the reduced cofactor mentioned previously.

In contrast to the marked cooperativity observed between binding of the cofactor and cofactor analogs, on the one hand, and binding of inhibitors containing 2,4-diaminopyrimidine and 2,4-diaminopteridine rings on the other, there appears to be little cooperativity between the cofactor and the substrate, dihydrofolate, itself (Stone et al., 1984). Moreover, the opposite effect, negative cooperativity, is reported with 5-formyltetrahydrofolate, that is, folinic acid (Birdsall et al., 1981). This behavior may reflect some kind of conformational difference between enzyme-substrate and enzyme-inhibitor complexes that goes beyond simply turning over the heterocyclic ring, perhaps involving the local structure of the enzyme itself.

From examination of the crystal structures, cooperativity was proposed to result, at least in part, from direct interaction between the nicotinamide nucleoside portion of the cofactor and the bound inhibitor molecule (Bolin et al., 1982; Matthews et al., 1985a). Quite a different conclusion was drawn, on the basis of structural interpretations of NMR spectroscopic data, by Birdsall et al. (1984). They point out that cofactor analogs that are bound predominantly in conformation II, that is, the nicotinamide ring not in contact with the enzyme, or those lacking a nicotinamide ring entirely, nevertheless show 4.7- to 6.1-fold cooperativity in binding with TMP to the *L. casei* enzyme. However, the magnitudes of these effects are relatively small. Detailed computer simulations carried out by Hagler et al. (personal communication) for *E. coli* DHFR, on the other hand, suggest that favorable van der Waals interactions between the nicotinamide nucleoside and the trimethoxyphenyl ring of TMP are primarily responsible for the high cooperative specificity of that inhibitor. It should be kept in mind, of course, that the DHFRs of *L. casei* and *E. coli* are not identical structures, so the two cases are not precisely comparable. In any event, achieving a more detailed understanding of these phenomenon would require increasingly sophisticated energy calculations coupled with information on the exact structures of the various complexes.

7. STEREOCHEMISTRY OF THE CATALYTIC MECHANISM

7.1. The Kinetic Mechanism

7.1.1. *Overview*

Overall, DHFR kinetics appear to be most nearly consistent with a rapid equilibrium random mechanism. That is, substrate and cofactor can bind in any order, and when bound are essentially in equilibrium with free substrate and cofactor. The slow, rate-determining events in the reaction sequence must then occur at one or more of the subsequent steps, for example, during isomerization, if any, protonic rearrangement, catalytic hydride transfer, or product release. In addition, there is some degree of formation of several nonproductive or dead-end complexes. Earlier studies gave conflicting results, in certain cases seeming to favor an ordered mechanism in which NADPH is required to bind first before productive binding of substrate can take place. However, a considerable body of spectroscopic and kinetic evidence now indicates that either substrate or cofactor can bind to the apoenzyme in the absence of the other.

In recent years, Stone and Morrison (1982, 1983, 1984) have reported on their extensive investigations of the kinetic mechanism of *E. coli* DHFR. One of their conclusions is that the reaction is of the rapid equilibrium random type involving the two dead-end complexes, enzyme-DHF-THF and enzyme-NADP-DHF. Additionally, the data of Cha et al. (1981), obtained from studies of tight binding by MTX and initial velocity analysis, are also interpreted as supporting a rapid equilibrium random mechanism for *L. casei* DHFR. However, it should be kept in mind that most enzymic mechanisms are elusive, and indeed, in a more recent paper, Stone et al. (1984) are careful to qualify their conclusions by noting that the DHFR "only approximates to a rapid equilibrium, random mechanism."

If substrate binding is not slow, then, could hydride transfer or product release be rate limiting? Hydride transfer seems to be eliminated as a possibiility by the finding that there is little or no deuterium isotope effect with NADPD, the deuterated cofactor, at neutral pH (Stone and Morrison, 1982).

7.1.2. *Protonation State of Enzyme and Substrate*

The kinetic mechanism of DHFR catalysis was further illuminated by pH studies on the *E. coli* enzyme (Stone and Morrison, 1984). In this investigation, the pH dependence of kinetic parameters was examined over the pH range 6.0–9.9, and in the presence of a saturating NADPH concentration of 100 μM. Under these conditions, both the maximum velocity, V, and the effective second-order rate constant for interaction between enzyme and substrate, V/K, decreased with increasing pH as if controlled by an ionization with pK_a of about 8. Stone and Morrison proposed on the basis of this and other evidence that a catalytically productive complex results only when unprotonated dihydrofolate binds to the protonated holoenzyme, and that binding of dihydrofolate to the unprotonated holoenzyme results in a dead-end complex. They further proposed that the ionizing group with apparent pK_a of 8 is the carboxyl group of Asp 27 and, following the suggestion originally made by Matthews et al. (1978) on the basis of the X-ray structure of the *L. casei* DHFR-NADPH-MTX complex, that this group is the proton source responsible for protonation of the substrate at N5. Protonation at N5 is generally believed to be a necessary preliminary step to hydride transfer.

A similar conclusion was drawn by Baccanari et al. (1981) from comparative kinetic and binding studies with the wild type v. a mutant *E. coli* DHFR and their inhibition by a Ba^{2+} ion (see Section 8.2). These authors interpreted their results as showing that the protonated state of Asp 27 is involved in the catalytic reaction while the ionized state is important for binding of TMP and MRX.

It should be emphasized at this point that a mechanistic role for protonation of N5 is entirely hypothetical; protonation of dihydrofolate is not observed in the ground state enzyme-substrate complex, but is believed to occur in the activated transition state (Subramanian and Kaufman, 1978; Hood and Roberts, 1978; Cocco et al., 1981b; Stone and Morrison, 1983). However, because, by definition, it corresponds to the highest energy on the reaction coordinate, the transition state itself cannot be directly observed.

7.1.3. Role of Asp 27,ec

Plausible though it might seem that the role of Asp 27 is to donate a proton to N5, at least two puzzling questions are raised by this suggestion. One pertains to the relative pK_a's, and the other to the geometrical positioning of the Asp 27 side chain. The first is considered here, and the second in Section 7.2.2.

With respect to the pK_a question, one should note that the pK_a of N5 in dihydrofolate in solution has been determined to be 3.8 (Poe, 1977). A carboxylic acid group in solution with a pK_a of 4 should be able to transfer a proton to N5 with reasonable facility, but, in fact, the carboxyl group of Asp 27 is thought to have a pK_a of 6.3 as determined from the pH dependence of binding of the inhibitor 2,4-diamino-6,7-dimethylpteridine by *E. coli* DHFR (Stone and Morrison, 1983). Moreover, as noted above, a kinetically observable pK_a of 8 has been associated with Asp 27, which would make proton transfer still more difficult. In fact, if one accepts the argument that the elevated pK_a of Asp 27 is due to its being buried in the hydrophobic interior of the DHFR molecule, then by the same argument, the pK_a of N5 would also have to be much lower than its solution value of 3.8 because it becomes positively charged when protonated. In short, it is unlikely that the enzyme stabilizes the required charge separation

$$-COOH + -N= \longrightarrow -COO^- + -NH^+=$$

simply by placing the groups in a nonpolar environment.

7.1.4.   *A Conformational Change Involved in Proton Transfer?*

Stone and Morrison (1984) conclude by suggesting that the rate-limiting step in the overall reaction sequence is probably a conformational change associated with this protonation. A particularly interesting conformational equilibrium that may be relevant to the protonation step has been described by Birdsall et al. (1982). According to these investigators, multinuclear NMR data obtained for the *L. casei* DHFR-NADP$^+$-folate dead-end complex show the existence of three interconverting states, I, IIa, and IIb, with an upper limit at 11°C of 19 sec^{-1} estimated for the rate of interconversion. This value would be consistent with the observed values of k_{cat} of other DHFRs. Conformational differences between these states appeared to be confined to the neighborhood of the nicotinamide ring. As an explanation for the pH dependence of the relative populations of the three states, it was suggested that the group responsible is the active site aspartate side chain, Asp 26 in the *L. casei* enzyme, with a normal pK_a of < 5 in states IIa and IIb, but an elevated pK_a of > 7 in state I. Such a conformational switch, if it also occurs in the reactive enzyme-NADPH-dihydrofolate complex, may be just the molecular mechanism required for proton transfer in the transition state (Roberts, 1983). In particular, state (IIa + IIb) may represent the conformation required for proton transfer from the active site aspartic acid to the substrate; and state I may represent the conformation in which the aspartic acid is reprotonated by the solvent. A structural explanation for these observations remains to be postulated; in fact, Birdsall et al. (1982) noted that an alternative interpretation is possible, though less likely. It would be interesting to determine whether other species of DHFR show the same multiplicity of states in their ternary complexes with folate.

7.1.5. Product Release

One problem, kinetically speaking, must be mentioned before going to consider some further mechanistic implications of the X-ray structure. In particular, it is not clear from the data presented in the reports just discussed that product release can be eliminated as a rate-determining step. In fact, a subsequent paper by Stone et al. (1984) cites unpublished data on substrate activation by NADPH, implying that release of tetrahydrofolate from the binary enzyme-product complex is slow relative to the other steps of the reaction sequence. Unfortunately, others have been unable to observe this phenomenon (E. Howell, personal communication) so the matter must be left for further investigation.

7.2. Stereochemistry of Dihydrofolate Binding

7.2.1. A Model for Productive Binding

It is now appropriate to examine the likely stereochemistry of dihydrofolate binding to DHFR as deduced from the accumulated X-ray crystallographic data, and the implications of this stereochemistry for the enzymic mechanism. It should be emphasized that as yet, owing to technical difficulties in obtaining suitable crystals, no crystal structure determination has been carried out on a DHFR complexed with a substrate or product, that is, on a DHFR with a bound folate, dihydrofolate, or tetrahydrofolate molecule. However, a very informative, partially refined structure has been determined for the ternary complex of chicken DHFR containing both $NADP^+$ and biopterin (Matthews et al., in preparation), which is the fully oxidized form of the cofactor involved in hydroxylation of aromatic amino acids. Because biopterin contains the same pteridine ring as folate, and because reduction of biopterin is catalyzed by chicken DHFR (Kaufman, 1967), while dihydrobiopterin is slowly reduced in the presence of rabbit brain DHFR (Pollock and Kaufman, 1978), it is likely that the stereochemistry with which biopterin binds at the substrate site is a reasonable approximation to the binding of the pteridine ring of folate or dihydrofolate. The important point to be emphasized is that this structure essentially confirms the conclusion from model-building experiments, to be discussed, that the pteridine ring of substrate is rotated by 180° relative to the observed orientation of 2,4-diamino-heterocycle rings of inhibitors without appreciably altering the geometry of the surrounding enzyme. It should also be mentioned, parenthetically, that in the absence of evidence to the contrary, the pteridine ring of folate is assumed to bind to the enzyme in the same way as the dihyropteridine ring of dihydrofolate.

Essential features of the interaction between DHFR and the pteridine rings of both MTX and dihydrofolate are schematized in Figure 14, adapted from Bolin et al. (1982). The nicotinamide ring of the NADPH cofactor is not shown in Figure 14, but it is behind the plane of the paper. Figure 14a is based on crystal structures of two binary complexes between bacterial DHFRs and MTX, which are described in Section 4.2, whereas Figure 14b is based on a hypothetical model for the binding of dihydrofolate. The latter was derived by rotating the pteridine ring of MTX by 180° about an axis roughly coincident with the C6−C9 bond and the C2−NH$_2$ bond. Additional small adjustments to the rest of the ligand molecule were permitted, but the geometry of the enzyme and bound waters were kept invariant while maximizing hydrogen bonding to the ligand molecule. All components of Figure 14a, including the water molecules, have the same geometry in the two bacterial DHFR crystal structures. In the chicken DHFR structure, the only noteworthy difference in this area is that the side chain of Asp 27,ec or Asp 26,lc is replaced by the side chain of Glu 30,cl.

The geometry of the pteridine ring in the chicken DHFR-NADP$^+$-biopterin complex confirms all the essential features of the foregoing model. A perhaps significant difference, however, is that the pteridine ring of biopterin is translated downward, as seen in Figure 14b, by approximately 1.5 Å. Consequently, one carboxylate oxygen of Glu 30 appears to make simultaneous hydrogen bonds with both N3 and the 4-oxo oxygen atom of the pteridine ring, while the other carboxylate oxygen seems to interact simultaneously with both N3 and the 2-amino nitrogen atom. Whether biopterin bound to chicken DHFR represents the binding geometry of the actual substrate more accurately than does the above model, however, remains to be seen.

Before considering the implications of this model for productive substrate binding, an NMR observation by Blakley and his colleagues (Cocco et al., 1981b) has to be reconciled with the proposed model (Roberts, 1983). Cocco et al. examined the ^{13}C NMR spectra of C2-enriched MTX,

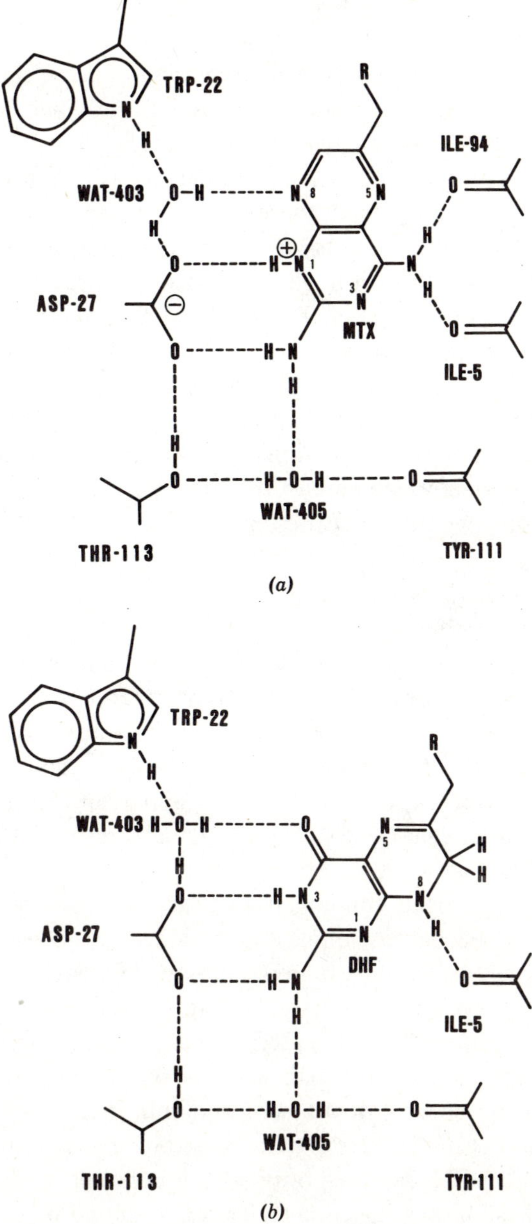

Figure 14. Schematic diagram of (*a*) methotrexate and (*b*) dihydrofolate binding to DHFR. Sequence numbers are for the *E. coli* enzyme. (Adapted from Bolin et al., 1982.)

aminopterin, and folate, both free in solution and bound to *S. faecium* DHFR, as a function of pH. A particularly relevant observation was that binding of folate to the enzyme produced no change in the chemical shift for C2 of folate or in its pH dependence. Such a result would be unexpected, assuming that the proposed model for folate binding is correct, if the carboxyl side chain of Asp 27 were in its negatively charged anionic state. However, as already discussed, there is some evidence favoring a protonated, electrically neutral state for the Asp 27 side chain when substrate is productively bound. Thus, it might be plausibly argued that the chemical shift at C2 and its pH dependence would be almost the same as in aqueous solution, and that is exactly what Cocco et al. (1981a, 1981b) report. Of course, it is also possible that the model is wrong, that the inhibitor and biopterin complexes have been misleading, and, as has often been suggested, that a large conformational change in the enzyme structure accompanies productive binding of the substrate. However, there is little evidence for this hypothesis at present.

7.2.2. Mechanistic Implications

Three features of the hypothetical substrate-binding model depicted in Figure 14*b* deserve special emphasis: (1) the side chain of Asp 27 is not near N5 of the pteridine ring, but rather it is hydrogen bonded to N3 and the 2-amino group. (2) No hydrogen bond donor or acceptor groups of the enzyme are near N5. (3) Two water molecules are included in the hydrogen-bond network involving the pteridine ring and Asp 27. It should also be noted that the hydrogen-bond network originally proposed in Figure 9*a* of Bolin et al. (1982) has been slightly modified. Other schemes with different orientations of hydrogen-bonding groups can also be envisioned.

What are the mechanistic implications of these structural observations? One can imagine two distinct, but not mutually exclusive, functions for the side chain carboxyl group of Asp 27. The carboxyl group might be the ultimate donor of a proton to the pteridine ring of the substrate; and, in the ionized state, the carboxylate might serve to stabilize a positively charged, protonated pteridine ring in the activated transition state complex. The paradoxical aspect of this idea is that the carboxyl group, in fact, does not interact directly with N5 where protonation is expected to occur. To further pursue the question of the geometry at N5, it should be pointed out that stereoelectronic theory (Deslongchamps, 1983) predicts in detail how a protonated N5 should be pyramidalized to facilitate approach of an attacking hydride ion at C6. Because it is already known that the direction of approach is from behind the plane of the pteridine ring, as depicted in Figure 14*b*, theory predicts that the lone-pair orbital on N5 should be antiperiplanar to the bond being formed at C6, that is, it should be pointed toward the viewer. Thus, the proton attached to N5 should be behind the plane of the pteridine ring in Figure 14*b*.

On this basis, one might have expected the enzyme to position a hydrogen bond donor next to N5, but above the plane, and a hydrogen bond acceptor below the plane in order to stabilize the desired geometry. It should be remarked, parenthetically, that most of the dehydrogenases of known three-dimensional structure have a catalytic BH^+ group forming a hydrogen bond with the carbonyl or alcohol oxygen atom of the substrate, which is thought to stabilize the transition state by this type of mechanism (Fersht, 1985). In fact, as just mentioned, there are no hydrogen bond donors or acceptors near N5 in DHFR; the side chain of Asp 27, the obvious candidate, interacts instead with the ring nitrogen N3 and the 2-amino group. What then is the function of a conserved carboxyl group at this position? Perhaps it stabilizes the transiently protonated, positively charged pteridine ring in a more subtle way, exerting a more remote influence. Perhaps it acts as a proton donor to the ring, either directly or via the fixed water molecule Wat 403. However, as will be discussed in Section 8.3, evidence from mutagenesis experiments seems to rule out a role for the carboxyl group in stabilization of the protonated transition state, and to favor a role as proton donor.

7.3.   Stereochemistry of Cofactor Binding

In this section, the binding of the NADPH cofactor by DHFR is discussed from the perspective of a possible stereochemical mechanism whereby the C4 position of the nicotinamide ring is activated with respect to hydride transfer. Much more definitive structural information can be brought to bear on this aspect of the mechanism than for dihydrofolate activation, because two of the high-resolution DHFR crystal structures pertain to complexes containing bound NADPH: those of the *L. casei* and of the chicken enzyme (Filman et al., 1982; Volz et al., 1982; Matthews et al., 1985a). No DHFR structures are yet available that contain bound folate in either its reduced or oxidized form.

In contrast to the mystification raised by the stereochemical environment of the pteridine ring of a bound MTX molecule, the geometry of the enzyme neighboring the nicotinamide ring immediately suggests an attractive hypothesis (Filman et al., 1982). In short, it is observed that the nicotinamide binding pocket contains an array of polar or hydrogen-bonding oxygen atoms that seem to be strategically positioned to stabilize just those canonical resonance structures expected to contribute partial carbonium ion character at C4. The argument can be readily followed with the aid of Figures 15 and 16. Figure 15 shows the resonance structures available to the nicotinamide ring, omitting for simplicity those involving charge separation at the carboxamide group. Evidently, the transition state should most closely resemble the central canonical structure on the second line, with a positive charge localized at C4. Note that if this particular electronic isomer could be uniquely stabilized and isolated, the ring nitrogen would possess an unshared electron pair and tend toward a pyramidal geometry. The electronic isomer in question might be stabilized, for example, by donating a hydrogen bond to the unshared pair. But also note that the two flanking structures have the same feature, and therefore would resonate with the central structure. Thus, it is reasonable to suppose that a combination of all three of these three canonical resonance structures taken together best represents the desired transition state.

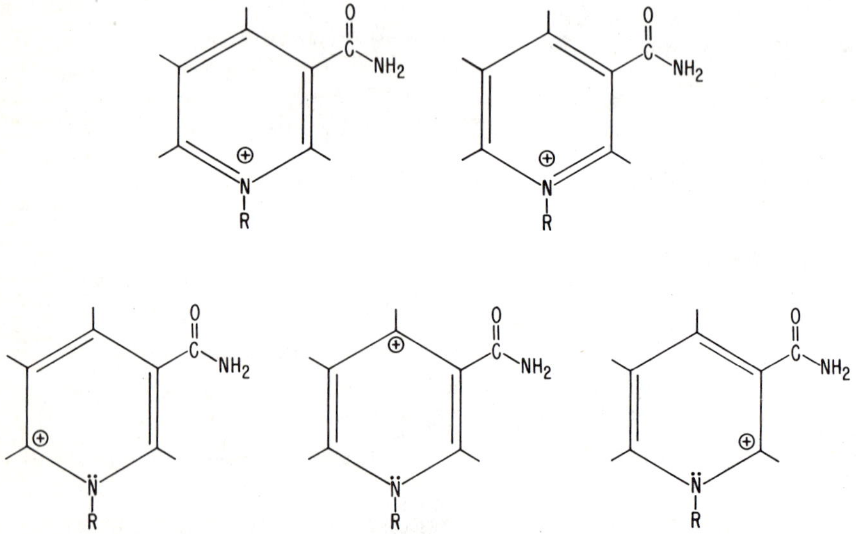

Figure 15. Canonical resonance structures for the nicotinamide ring of $NADP^+$.

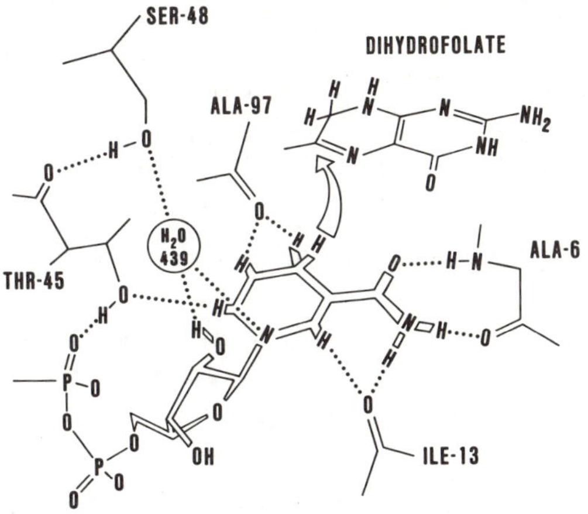

Figure 16. Environment of the nicotinamide ring of bound cofactor. Sequence numbers are for *L. casei* DHFR. (From Filman et al., 1982.)

From such considerations, it is apparent that the required transition state stabilization could be achieved by suitably positioning some group capable of donating a hydrogen bond to a lone pair on the ring nitrogen, together with an arrangement of partial negative charges on polar groups close to the 2, 4, and 6 positions of the nicotinamide ring. The environment of the nicotinamide ring of an NADPH molecule bound to *L. casei* DHFR is shown in Figure 16 (Filman et al., 1982). The same geometry is observed for NADPH bound to the chicken enzyme, and is inferred from model-building studies on the *E. coli* enzyme. Evidently, the fixed water molecule (Wat 439,lc) located above the ring nitrogen could act as the required hydrogen bond donor; and the carbonyl oxygen of Ile 13,lc, the carbonyl oxygen of Ala 97,lc and the side chain hydroxyl of Thr 45,lc could act as the polar, perhaps hydrogen bond-accepting, groups interacting with positions 2, 4, and 6. It is also interesting, but of uncertain significance, that the carbonyl oxygens of Ala 97,lc and of the carboxamide group are symmetrically disposed about the C4 position, that the same carbonyl group immediately precedes the conserved gly-gly connected by a *cis* peptide bond, and that the 4-amino group of 2,4-diamino-heterocyclic inhibitors hydrogen bonds to this carbonyl oxygen.

Two lines of evidence lend these ideas some credence independently of the X-ray structures. First, based on their study of the liver alcohol dehydrogenase reaction and reasoning from secondary kinetic isotope effects observed with ^{15}N-labeled NAD^+, Cook et al. (1981) concluded that N1 must be pyramidalized in some intermediate or transition state. Secondly, Sigman et al. (1978) found that N-substituted dihydronicotinamides, in which an oxygen function could approach the pyridine ring, showed substantially enhanced rates of oxidation in aprotic solvents.

Is the environment of the nicotinamide ring in other $NAD(P)^+$-dependent oxido-reductases similar to that just described? It is probably too early to know because none of the structures with which comparisons might be made have yet been taken to high resolution and refined to convergence. However, a recent report describing the 2.9 Å partially refined structure of horse liver alcohol dehydrogenase (Eklund et al., 1984) notes that a back-bone carbonyl oxygen and a threonine Oγl are in van der Waals contact with C2 and C4, respectively. The locations of fixed water molecules were not given.

If the molecular mechanism suggested by Filman et al. (1982) for activation of the nicotinamide ring is general, comparisons among the various structures should answer some interesting questions. For example, given no other constraints, one would suppose that N1 in the putative transition state may be pyramidalized toward either the A face or the B face of the nicotinamide ring. Does stereoelectronic theory require that the lone-pair orbital projects from the same side of the ring as that to/from which hydride transfer takes place, that is, the A or 4-pro-R side for DHFR? If so, a hydrogen bond donor, such as Wat 439 in the above *L. casei* DHFR model, should also be found on that side in each enzyme structure, that is, on the A side for A side-specific oxido-reductases, or on the B side for enzymes with the opposite specificity.

Another open question that might be illuminated by comparing structures of nonhomologous oxido-reductases concerns the possible role of conformational equilibria in activation of the nicotinamide ring. Inspection of the DHFR crystal structures has so far yielded no obvious clues as to what, if any, molecular motions may be involved.

8.　MUTANTS AND MUTAGENESIS

8.1.　Perspectives

The past 1 or 2 years have witnessed a surge of interest among structural enzymologists in the application of a powerful new tool adopted from the field of molecular genetics. The technology is referred to as "site-directed mutagenesis" or "protein engineering" or some similar appelation. Because it lies well beyond the scope of this review, the technology will not be described in detail here. The interested reader will find an overview of the methodology and a compilation of references in the review by Craik (1985). Suffice it to say that it is now possible, by the application of recombinant DNA and mutagenesis methods, deliberately to construct virtually any desired mutation in a cloned gene, including not only single or multiple base changes, but insertions and deletions as well. In a suitable expression system, large quantities of the mutant gene product can be obtained for intensive study by the usual enzymological and crystallographic methods, and for comparison with the wild-type enzyme molecule. The subject is germane to this, review because *E. coli* DHFR has been among the first of the enzymes to which these techniques have been applied (Villafranca et al., 1983), and in Section 8.3 some of the findings are summarized.

8.2. Spontaneously Occurring Mutants

First, it is relevant to mention the few mutant DHFRs that have already been observed in *E. coli* by classical methods and at least partially characterized. Many reported examples of spontaneous mutation in the *E. coli fol* gene giving rise to regulatory mutants or altered enzymes with decreased inhibitor binding will not be included in this brief survey because they have not been further characterized at the molecular level.

Poe et al. (1972) isolated a mutant DHFR from an MTX-resistant strain of *E. coli* B dubbed MB 1428. It was this mutant for which the first DHFR crystal structure was determined. When the wild-type sequence became available, the MB 1428 enzyme turned out to contain the mutation Glu 154 to Lys; it may also contain Asn/Asp interchanges at positions 37 and 87, but contrary to the published sequence, Asp 142 is known from the refined X-ray structure (Bolin, 1982) to be unchanged from the wild type. Kinetic constants for the MB 1428 enzyme (Poe et al., 1972; Williams et al., 1973) are approximately the same as for the wild-type enzyme (Stone and Morrison, 1982; E. Howell, personal communication) within the rather large experimental error typically characteristic of such determinations. However, it appears that the mutant enzyme binds MTX much less tightly than the wild type in the binary complex. Williams et al. (1973) report K_d values ranging from $1-27$ nM for MB 1428, depending on the ionic strength, whereas the corresponding figure for the wild type is given as 0.12 nM by Stone and Morrison (1983) and 0.07 nM by Howell (personal communication). If the comparison is valid and the usual caveat concerning variability of experimental conditions applies, it is difficult to rationalize the difference in structural terms because the mutation Glu 154 to Lys is far from the inhibitor-binding site.

An interesting mutant *E. coli* DHFR described by Baccanari and coworkers (Baccanari et al., 1975; 1977) has already been mentioned briefly in Section 7.1. The mutant enzyme was obtained from a TMP-resistant strain (RT500) that expressed two distinct forms of DHFR, designated forms 1 and 2. Form 1 may be identical with the wild-type enzyme, but is reported as having Glu 118 substituted by a glutamine residue, while form 2 contains, in addition, the mutation Leu 28 to Arg (Baccanari et al., 1981). The two forms differ considerably in their kinetic and binding properties. Form 2 is approximately 10-fold less active, with a turnover number of 126 min^{-1} at pH 7.0, while form 1 has about the same activity as the wild-type enzyme with a turnover number of 1800 min^{-1}. The affinity of form 2 for inhibitors is also much reduced, especially in the case of TMP, for which the dissociation constant is 600 nM as compared to 15 nM for form 1. Similarly, the kinetic inhibition constants K_i for TMP are 106 nM for form 2 and 1.3 nM for form 1 (Baccanari et al., 1982). The pH activity profile for form 2 is also unusual, showing a continuously increasing activity as pH decreases, such that its activity at pH 5.0 is fivefold greater than at pH 7.0. Interestingly, the properties of form 1 in the presence of 10 mM Ba^{2+} resemble those of form 2. Baccanari et al. (1981) suggested that the positively charged guanidinium side chain of Arg 28 in the mutant, or alternatively the Ba^{2+} ion in the case of the wild-type enzyme, interacts directly with an ionized Asp 27. It seems unlikely, however, that the explanation can be quite so simple because the geometry would be difficult to rationalize; dihydrofolate itself actually binds more tightly by a factor of 10 to the mutant, and Ba^{2+} has no effect at 10 mM on the DHFRs from *Neisseria* gonorrhoeae or rat liver (Baccanari et al., 1981), even though these species of the enzyme also contain a homologous aspartic or glutamic acid at the equivalent of position 27. A perhaps more plausible interpretation is that the guanidinium or Ba^{2+} ion perturbs a conformational equilibrium that is involved in the enzyme's activity, as discussed in Section 7.1. In this context, it is interesting to note that Cayley et al. (1981) found that the proportions of their two postulated conformers differed for forms 1 and 2.

Two other apparently mutant *E. coli* DHFRs have been reported by Poe et al. (1979), but are not yet fully sequenced. They were isolated from TMP-resistant strains MB 3746 and MB 3747. The MB 3746 enzyme seems to be kinetically indistinguishable from the wild type, but the DHFR derived from MB 3747 displays significantly altered properties: (1) a Michaelis constant of 7.3 μM for dihydrofolate that is considerably larger than the $0.4-0.7$ μM found for the wild type (Stone and Morrison, 1982; Howell, unpublished); (2) an inhibition constant for TMP of 11 nM, which is about 10 times greater than for the wild type; and (3) a pH optimum below 6.4. A partial amino acid sequence shows the substitution Met 20 to Ile. A structural interpretation of these differences has not been proposed.

8.3. Preview of Findings From Mutagenesis Studies

There are a growing number of mutant *E. coli* DHFRs that have been deliberately constructed by site-specific mutagenesis techniques. Because much of the work is at a preliminary stage, it will be presented more as a preview rather than a review.

Kraut and coworkers (Villafranca et al., 1983; Howell et al., 1986) have generated several mutant *E. coli* DHFRs designed to address specific questions based on the high-resolution structure of the wild-type enzyme. Mutant DHFRs on hand at this time include Asp 27 to Asn, Ser, Glu, and Leu; Pro 39 to Cys; Gly 95 to Ala; Trp 22 to His; Trp 30 to Ser; Arg 44 to Thr; Ser 63 to Glu; and the double mutation containing both Thr 44 and Glu 63. At present, the most thoroughly characterized of these are the Asp 27 to Asn or Ser and the Pro 39 to Cys mutants.

At pH 7.0, the catalytic constant k_{cat} for the Asn 27 mutant is about 1/300th of that for wild-type DHFR, but catalysis increases rapidly with decreasing pH, so that k_{cat} at pH 5.0 for the mutant has reached 1/40th of the wild-type, pH 7.0 value. This behavior may be consistent with extensive nonproductive binding of substrate to the mutant DHFR at neutral pH, which diminishes rapidly at low pH. A ready explanation is that protonated dihydrofolate is the productive substrate species for the mutant DHFR, while unprotonated dihydrofolate is the nonproductive species. It can be argued from these observations that the catalytic role of Asp 27 is donation of a proton to the substrate rather than stabilization of the positively charged, protonated transition state by means of a charge-charge interaction. Similar behavior is observed for the Asn 27 to Ser mutant, but the two mutants differ in their pH dependence of K_m, and the mechanistic details must be more complicated than this simple intrepretation suggests.

A Cys 39 mutation was designed to introduce a disulfide bridge with Cys 85 into the DHFR molecule. The disulfide bond is not formed, however, in the cytosol of *E. coli*, but can be made in vitro quite readily by oxidation with disulfide-exchange reagents such as DTNB. Both the reduced and disulfide-bridged form of the mutant enzyme have activities that are close to the wild type. However, the disulfide-bridged form of the mutant DHFR shows enhanced resistance to unfolding by guanidine hydrochloride and decreased cooperativity of folding.

A set of three mutants has been constructed to investigate the preference of DHFR for the cofactor NADPH as opposed to NADH. One of these, a double mutant of Arg 44 to Thr and Ser 63 to Glu, exhibits a greatly enhanced relative tolerance for NADH through decreased activity as compared with the wild-type enzyme.

Gly 95 has been changed to alanine in a mutation designed to investigate the structural and functional implications of an invariant Gly 95-Gly 96 linked by a *cis* peptide bond. This peculiar architectural feature of the DHFR molecule immediately follows a variable residue, the backbone carbonyl of which, however, always points toward the nicotinamide C4 of the bound cofactor. Moreover, this same carbonyl accepts a hydrogen bond from the 4-amino group of MTX, TMP, and other inhibitors. The mutant has very low but detectable activity.

It is very likely that further studies of this kind will contribute much to a deeper understanding of the DHFR molecule in particular, and enzymes in general. This will be especially true when mutagenesis experiments are supported by X-ray crystallographic structure determinations to ascertain exactly what geometrical transformations have been wrought by each mutation.

REFERENCES

Antonjuk, D. J., Birdsall, B., Cheung, H. T. A., Clore, G. M., Feeney, J., Gronenborn, A., Roberts, G. C. K., and Tran, T. Q. (1984). *Br. J. Pharmacol.* **81**, 309–315.

Armarego, W. L., Waring, P., and Williams, J. W. (1980). *J. Chem. Soc. Chem. Commun.* **8**, 334–336.

Arnott, S., and Hukins, D. L. (1969). *Nature (London)* **224**, 886–888.

Baccanari, D. P., and Joyner, S. S. (1981). *Biochemistry* **20**, 1710–1716.

Baccanari, D. P., Phillips, A., Smith., S., Sinski, D., and Burchall, J. (1975). *Biochemistry* **14**, 5267–5273.

Baccanari, D. P., Avarett, D., Briggs, C., and Burchall, J. (1977). *Biochemistry* **16**, 3566–3572.

Baccanari, D. P., Stone, D., and Kuyper, L. (1981). *J. Biol. Chem.* **256**, 1738–1747.

Baccanari, D. P., Daluge, S., and King, R. W. (1982). *Biochemistry* **21**, 5068–5075.

Baccanari, D. P., Tansik, R. L., Paterson, S. J., and Stone, D. (1984). *J. Biol. Chem.* **259**, 12291–12298.

Baker, D. J., Beddell, C. R., Champness, J. N., Goodford, P. J., Norrington, F. E. A., Smith, D. R., and Stammers, D. K. (1981). *FEBS Lett.* **126**, 49–52.

Barbehenn, E. K., and Kaufman, B. T. (1980). *J. Biol. Chem.* **255**, 1978–1984.

Barbehenn, E. K., and Kaufman, B. T. (1982). *Arch. Biochem. Biophys.* **219**, 236–247.

Bennett, C. D., Rodkey, J. A., Sondey, J. M., and Hirschmann, R. (1978). *Biochemistry* **17**, 1328–1377.

Birdsall, B., Burgen, A. S. V., and Roberts, G. C. K. (1980a). *Biochemistry* **19**, 3723–3731.

Birdsall, B., Burgen, A. S. V., and Roberts, G. C. K. (1980b). *Biochemistry* **19**, 3732–3737.

Birdsall, B., Burgen, A. S. V., Hyde, E. I., Roberts, G. C. K., and Feeney, J. (1981). *Biochemistry* **20**, 7186–7195.

Birdsall, B., Gronenborn, A., Hyde, E. I., Clore, G. M., Roberts, G. C. K., Feeney, J., and Burgen, A. S. V. (1982). *Biochemistry* **21**, 5831–5838.

Birdsall, B., Roberts, G. C. K., Feeney, J., Dann, J. G., and Burgen, A. S. V. (1983). *Biochemistry* **22**, 5597–5604.

Birdsall, B., Bevan, A. W., Pascual, C., Roberts, G. C. K., Feeney, J., Gronenborn, A., and Clore, G. M. (1984). *Biochemistry* **23**, 4733–4742.

Blakley, R. L. (1969). *The Biochemistry of Folic Acid and Related Pteridines.* North-Holland, Amsterdam.

Blakley, R. L. (1984). In *Folates and Pterins*, Vol. 1, *Chemistry and Biochemistry of Folates*, R. L. Blakley, and S. J. Benkovic, Eds., Chap. 5. John Wiley & Sons, New York, pp. 191–253.

Blakley, R. L., Ramasastri, B. U., and McDougall, B. M. (1963). *J. Biol. Chem.* **238**, 3075–3079.

Blakley, R. L., Cocco, L., London, R. E., Walker, T. E., and Matwiyoff, N. A. (1978). *Biochemistry* **17**, 2284–2293.

Bolin, J. T. (1982). Ph.D. Dissertation. University of California, San Diego.

Bolin, J. T., Filman, D. J., Matthews, D. A., Hamlin, R. C., and Kraut, J. (1982). *J. Biol. Chem.* **257**, 13650−13662.

Brändén, C. -I. (1980). *Quart. Rev. Biophys.* **13**, 317−338.

Brändén, C. -I., and Eklund, H. (1980). In *Dehydrogenases Requiring Nicotinamide Coenzyme, Experimentia Supplementum*, Vol. 36, J. Jeffrey, Ed. Birkhauser Verlag, Basel, pp. 40−84.

Burchall, J. (1984). In *Folate Antagonists as Therapeutic Agents*, F. M. Sirotnak, W. D. Ensminger, J. J. Burchall, and J. A. Montgomery, Eds., Vol. 1. Academic Press, New York, pp. 133−150.

Burchall, J. J. (1983). In *Inhibition of Folate Metabolism in Chemotherapy*, G. H. Hitchings, Ed. Springer-Verlag, New York, pp. 55−74.

Burchall, J. J., and Hitchings, G. H. (1965). *Mol. Pharmacol.* **1**, 126−136.

Cayley, P. J., Albrand, J. P., Feeney, J., Roberts, G. C. K., Piper, E. A., and Burgen, A. S. V. (1979). *Biochemistry*, **18**, 3886−3894.

Cayley, P. J., Feeney, J. and Kimber, B. J. (1980). *Int. J. Biol. Macromol.* **2**, 251−255.

Cayley, P. J., Dunn, S. M. J., and King, R. W. (1981). *Biochemistry* **20**, 874−879.

Cha, S., Kim, S. Y. R., Kornstein, S. G., Kantoff, P. W., Kim, K. H., and Naguib, F. N. M. (1981). *Biochem. Pharmacol.* **30**, 1507−1515.

Charlton, P. A., Young, D. W., Birdsall, B., Feeney, J., and Roberts, G. C. K. (1979). *J. Chem. Soc. Chem. Commun.* **20**, 922−924.

Chothia, C. (1973). *J. Mol. Biol.* **75**, 295−302.

Clore, G. M., Gronenborn, A. M., Birdsall, B., Feeney, J., and Roberts, G. C. K. (1984). *Biochem. J.* **217**, 659−666.

Cocco, L., Blakley, R. L., Walker, T. E., London, R. E., and Matwiyoff, N. A. (1978). *Biochemistry* **17**, 4285−4290.

Cocco, L., Temple, C., Jr., Montgomery, J. A., London, R. E., and Blakley, R. L. (1981a). *Biochem. Biophys. Res. Commun.* **100**, 413−419.

Cocco, L., Groff, J. P., Temple, C., Jr., Montgomery, J. A., London, R. E., Matwiyoff, N. A., and Blakley, R. L. (1981b). *Biochemistry* **20**, 3972−3978.

Cocco, L., Roth, B., Temple, C., Jr., Montgomery, J. A., London, R. E., and Blakley, R. L. (1983). *Arch. Biochem. Biophys.* **226**, 567−577.

Cook, P. F., Oppenheimer, N. J., and Cleland, W. W. (1981). *Biochemistry* **20**, 1817−1825.

Craik, C. S. (1985). *BioTechniques* **3**, 12−19.

Deslongchamps, P. (1983). *Stereoelectronic Effects in Organic Chemistry*. Pergamon Press, Oxford.

Dunn, S. M. J., and King, R. W. (1980). *Biochemistry* **19**, 766−773.

Dunn, S. M. J., Batchelor, J. G., and King, R. W. (1978). *Biochemistry* **17**, 2356−2364.

Eklund, H., Samama, J., and Jones, T. A. (1984) *Biochemistry* **23**, 5982–5996.

Feeney, J., Birdsall, B., Roberts, G. C. K., and Burgen, A. S. V. (1975). *Nature (London)* **257**, 564–566.

Feeney, J., Roberts, G. C. K., Birdsall, B., Griffiths, D. V., King, R. W., Scudder, P., and Burgen, A. (1977). *Proc. R. Soc. Lond, B.* **196**, 267–290.

Feeney, J., Roberts, G. C. K., Thomson, J. W., King, R. W., Griffiths, D. V., and Burgen, A. S. V. (1980a). *Biochemistry* **19**, 2316–2321.

Feeney, J., Roberts, G. C. K., Kaptein, R., Birdsall, B., Gronenborn, A., and Burgen, A. S. V. (1980b). *Biochemistry* **19**, 2466–2472.

Ferone, R., Burchall, J. J., and Hitchings, G. H. (1969). *Mol. Pharmacol.* **5**, 49–59.

Fersht, A. (1985). *Enzyme Structure and Mechanism*, 2nd Ed. W. H. Freeman & Co., New York, pp. 109–111, 404.

Filman, D. J., Bolin, J. T., Matthews, D.A., and Kraut, J. (1982). *J. Biol. Chem.* **257**, 13663–13672.

Fleming, M. P., Datta, N., and Grüneberg, R. N. (1972). *Br. Med. J.* **1**, 726–728.

Fontecilla-Camps, J. C., Bugg, C. E., Temple, C., Jr., Rose, J. D., Montgomery, J. A., and Kisliuk, R. L. (1979). *J. Am. Chem. Soc.* **101**, 6114–6115.

Freisheim, J. A., and Matthews, D. A. (1984). In *Folate Antagonists as Therapeutic Agents*, F. M. Sirotnak, W. D. Ensminger, J. J. Burchall, and J. A. Montgomery, Eds., Vol. 1. Academic Press, Orlando, pp. 69–131.

Frieden, C. (1970). *J. Biol. Chem.* **245**, 5788–5799.

Futterman, S. (1957). *J. Biol. Chem.* **228**, 1031–1038.

Futterman, S., and Silverman, M. (1957). *J. Biol. Chem.* **224**, 31–40.

Gready, J. E. (1980). *Adv. Pharmacol. Chemother.* **17**, 37–102.

Greenberg, G. R. (1954). *Fed. Proc.* **13**, 745.

Gronenborn, A., Birdsall, B., Hyde, E. I., Roberts, G. C. K., Feeney, J., and Burgen, A. S. V. (1981a). *Biochemistry* **20**, 1717–1722.

Gronenborn, A., Birdsall, B., Hyde, E. I., Roberts, G. C. K., Feeney, J., and Burgen, A. S. V. (1981b). *Mol. Pharmacol.* **20**, 145–153.

Hansch, C., Hathaway, B. A., Guo, Z., Selassie, C. D., Dietrich, S. W., Blaney, J. M., Langridge, R., Volz, K. W., and Kaufman, B. T. (1984). *J. Med. Chem.* **27**, 129–143.

Hathaway, B. A., Guo, Z., Hansch, C., Delcamp, T. J., Susten, S. S., and Freisheim, J. (1984). *J. Med. Chem.* **27**, 144–149.

Hayman, R., McGready, R., and Van der Weyden, M. B. (1978). *Anal. Biochem.* **87**, 460–465.

Hitchings, G. H. (1983). In *Inhibition of Folate Metabolism in Chemotherapy*, G. H. Hitchings, Ed. Springer-Verlag, New York, pp. 11–23.

Hitchings, G. H., and Smith, S. L. (1980). *Adv. Enzym.*, **18**, 349–371.

Hitchings, G. H., and Baccanari, D. P. (1984). In *Folate Antagonists as Therapeutic*

Agents, F. M. Sirotnak, J. J. Burchall, W. D. Ensminger, and J. A. Montgomery. Eds., Vol. 1. Academic Press, Orlando, pp. 151–173.

Hitchings, G. H., Elion, G. B., Vanderwerff, H., and Falco, E. A. (1948). *J. Biol. Chem.* **174**, 765–766.

Hol, W. G. J., Van Duÿnen, P. T., and Berendsen, H. J. C. (1978). *Nature (London)* **273**, 443–446.

Hood, K., and Roberts, G. C. K. (1978). *Biochem. J.* **171**, 357–366.

Howell, E. E., Villafranca, J. E., Warren, M. S., Oatley, S. J., and Kraut, J. (1986). *Science* **231**, 1123–1128.

Hyde, E. I., Birdsall, B., Roberts, G. C. K., Feeney, J., and Burgen, A. S. V. (1980a). *Biochemistry* **19**, 3738–3746.

Hyde, E. I., Birdsall, B., Roberts, G. C. K., Feeney, J., and Burgen, A. S. V. (1980b). *Biochemistry* **19**, 3746–3754.

Jaffe, J. J. (1972). In *Comparative Biochemistry of Parasites*, H. V. Borsche, Ed. Academic Press, New York, pp. 219–233.

Jolivet, J., Cowan, K. H., Curt, G. A., Clendeninn, N. J., and Chabner, B. A. (1983). *N. Engl. J. Med.* **309**, 1094–1104.

Joyner, S. S., Fling, M. E., Stone, D., and Baccanari, D. P. (1984). *J. Biol. Chem.* **259**, 5851–5856.

Jolivet, J., Cowan, K. H., Curt, G. A., Clendeninn, N. J., and Chabner, B. A. (1983). *N. Engl. J. Med.* **309**, 1094–1104.

Kaufman, B. T. (1968). *J. Biol. Chem.* **243**, 6001–6008.

Kaufman, S. (1967). *J. Biol. Chem.* **242**, 3934–3943.

Kimber, B. J., Griffiths, D. V., Birdsall, B., King, R. W., Scudder, P., Feeney, J., Roberts, G. C. K., and Burgen, A. S. V. (1977). *Biochemistry* **16**, 3492–3500.

Kisliuk, R. L. (1984). In *Folate Antagonists as Therapeutic Agents*, F. M. Sirotnak, J. J. Burchall, W. B. Ensminger, and J. A. Montgomery, Eds. Vol. 1. Academic Press, Orlando, pp. 1–68.

Kozloff, L. M. (1983). In *Bacteriophage T4*, C. K. Mathews, E. M. Kutter, G. Morig, and P. B. Berget, Eds. American Society for Microbiology, Washington, D.C., pp. 25–31.

Kuyper, L. F., Roth, B., Baccanari, D. P., Ferone, R., Beddell, C. R., Champness, J. N., Stammers, D. K., Dann, J. G., Norrington, F. E. A., Baker, D. J., and Goodford, P. J. (1982). *J. Med. Chem.* **25**, 1120–1122.

Kuyper, L. F., Roth, B., Baccanari, D. P., Ferone, R., Beddell, C. R., Champness, J. N., Stammers, D. K., Dann, J. G., Norrington, F. E. A., Baker, D. J., and Goodford, P. J. (1985). *J. Med. Chem.* **28**, 303–311.

London, R. E., Groff, J. P., Cocco, L., and Blakley, R. L. (1982). *Biochemistry* **21**, 4450–4458.

Luzzati, V. (1952). *Acta Crystallogr.* **5**, 802–810.

Masters, J. N., and Attardi, G. (1983). *Gene* **21**, 59−63.

Mathews, C. K., and Huennekens, F. M. (1963). *J. Biol. Chem.* **238**, 3436−3442.

Matthews, D. A., and Volz, K. W. (1982). In *Molecular Structure and Biological Activity*, J. Griffin, and W. Duax, Eds. Elsevier Science Publications, New York, pp. 13−25.

Matthews, D. A., Alden, R. A., Bolin, J. T., Freer, S. T., Hamlin, R., Xuong, N., Kraut, J., Poe, M., Williams, M., and Hoogsteen, K. (1977). *Science* **197**, 452−455.

Matthews, D. A., Alden, R. A., Bolin, J. T., Filman, D. J., Freer, S. T., Hamlin, R., Hol, W. G. J., Kisliuk, R. L., Pastore, E. J., Plante, L. T., Xuong, N., and Kraut, J. (1978). *J. Biol. Chem.* **253**, 6946−6954.

Matthews, D. A., Alden, R. A., Freer, S. T., Xuong, N. and Kraut, J. (1979). *J. Biol. Chem.* **254**, 4144−4151.

Matthews, D. A., Bolin, J. T., Burridge, J. M., Filman, D. J., Volz, K. W., Kaufman, B. T., Beddell, C. R., Champness, J. N., Stammers, D. K., and Kraut, J. (1985a). *J. Biol. Chem.* **260**, 381−391.

Matthews, D. A., Bolin, J. T., Burridge, J. M., Filman, D. J., Volz, K. W., and Kraut, J. (1985b). *J. Biol. Chem.* **260**, 392−399.

Matthews, D. A., Smith, S. L., Baccanari, D. P., Burchall, J. J., Oatley, S. J., and Kraut, J. (1986). *Biochemistry* **25**, 4194−4204.

Mosher, R. A., DiRenzo, A. B., and Mathews, C. K. (1977). *J. Virol.* **23**, 645−658.

Nixon, P. F., and Blakley, R. L. (1968). *J. Biol. Chem.* **243**, 4722−4731.

Osborn, M. J., and Huennekens, F. M. (1958). *J. Biol. Chem.* **233**, 969−974.

Pastore, E. J., and Friedkin, M. (1962). *J. Biol. Chem.* **237**, 3802−3810.

Pastore, E. J., Kisliuk, R. L., Plante, L. T., Wright, J. M., and Kaplan, N. O. (1974). *Proc. Natl. Acad. Sci. U.S.A.* **71**, 3849−3853.

Pastore, E. J., Plante, L. T., Kisliuk, R. L., Wright, J. M., Strumpf, D., and Kaplan, N. O. (1981). *Fed. Proc. Fed. Am. Soc. Exp. Bio..* **40**, 1871.

Pattishall, K. H., Acar, J., Burchall, J. J., Goldstein, F. W., and Harvey, R. J. (1977). *J. Biol. Chem.* **252**, 2319−2323.

Penner, M. H., and Frieden, C. (1985). *J. Biol. Chem.* **260**, 5366−5369.

Perault, A. -M., and Pullman, B. (1961). *Biochim. Biophys. Acta* **52**, 266−280.

Poe, M. (1977). *J. Biol. Chem.* **252**, 3724−3728.

Poe, M. (1980). *Fed. Proc. Fed. Am. Soc. Exp. Biol.* **39**, 1856.

Poe, M., Greenfield, N. J., Hirshfield, J. M., Williams, M. N., and Hoogsteen, K. (1972). *Biochemistry* **11**, 1023−1030.

Poe, M., Breeze, A. S., Wu, J. K., Short, C. R., and Hoogsteen, K. (1979). *J. Biol. Chem.* **254**, 1799−1805.

Pollock, R. J., and Kaufman, S. (1978). *J. Neurochem.* **30**, 253−256.

Reddy, B. S., Saenger, W., Mühlegger, K., and Weimann, G. (1981). *J. Am. Chem. Soc.* **103**, 907–914.

Richardson, J. S. (1976) *Proc. Nat. Acad. Sci. USA* **73**, 2619–2623.

Roberts, G. C. K. (1983). In *Chemistry and Biology of Pteridines*, J. A. Blair, Ed. Walter de Gruyter & Co., Berlin, pp. 197–214.

Roberts, G. C. K., Feeney, J., Burgen, A. S. V., and Daluge, S. (1981). *FEBS Lett.* **131**, 85–88.

Rossmann, M. G., Liljas, A., Brändén, C. -I., and Banaszak, L. J. (1975). In *The Enzymes*, P. D. Boyer, Ed., 3rd ed., Vol. 11. Academic Press, New York, pp. 62–102.

Roth, B., and Cheng, C. C. (1982). In *Progress in Medicinal Chemistry*, G. P. Ellis, and G. B. West, Eds., Vol. 19. Elsevier Biomedical Press, Amsterdam, pp. 269–331.

Rothenberg, S. P., Iqbal, N. P., and DaCosta, N. (1980). *Anal. Biochem.* **103**, 152–156.

Sigman, D. S., Hajdu, J., and Creighton, D. J. (1978). In *Bioorganic Chemistry*, E. E. van Tamelen, Ed., Vol. IV. Academic Press, New York, pp. 385–407.

Sirotnak, F. M., Burchall, J. J., Ensminger, W. B. and Montgomery, J. A., Eds. (1984). *Folate Antagonists as Therapeutic Agents*, Vol. 1. Academic Press, Orlando.

Smith, S. L., Stone, D., Novak, P., Baccanari, D. P., and Burchall, J. J. (1979). *J. Biol. Chem.* **254**, 6222–6225.

Stammers, D. K., Champness, J. N., Dann, J. G., and Beddell, C. R. (1983). In *Chemistry and Biology of Pteridines*, J. A. Blair, Ed. Walter de Gruyter, Berlin, pp. 567–571.

Stone, D., and Smith, S. (1979). *J. Biol. Chem.* **254**, 10857–10861.

Stone, S. R., and Morrison, J. F. (1982). *Biochemistry* **21**, 3757–3765.

Stone, S. R., and Morrison, J. F. (1983). *Biochim. Biophys. Acta* **745**, 247–258.

Stone, S. R., and Morrison, J. F. (1984). *Biochemistry* **23**, 2753–2758.

Stone, S. R., Mark, A., and Morrison, J. F. (1984). *Biochemistry* **23**, 4340–4346.

Stura, E. A., Zanotti, G., Babu, Y. S., Sansom, M. S. P., Stuart, D. I., Wilson, K. S., Johnson, L. N., and Van de Werve, G. (1983). *Mol. Biol.* **170**, 529–565.

Subramanian, S., and Kaufman, B. T. (1978). *Proc. Natl. Acad. Sci. U.S.A.* **75**, 3201–3205.

Van Beelen, P., Van Neck, J. W., de Cock, R. M., Vogels, G. D., Guijt, W., and Haasnoot, C. A. G. (1984). *Biochemistry* **23**, 4448–4454.

Velick, S. F. (1961). In *Light and Life*, W. D. McElroy, and B. Glass, Eds. Johns Hopkins Press, Baltimore, 108–143.

Villafranca, J. E., Howell, E. E., Voet, D. H., Strobel, M. S., Ogden, R. C., Abelson, J. N., and Kraut, J. (1983). *Science* **222**, 782–788.

Volz, K. W., Matthews, D. A., Alden, R. A., Freer, S. T., Hansch, G., Kaufman, B. T., and Kraut, J. (1982). *J. Biol. Chem.* **257**, 2528−2536.

Wang, C. C., Verham, R., Tzeng, S. F., Aldritt, S., and Cheng, H. W. (1983). *Proc. Natl. Acad. Sci. U.S.A.* **80**, 2564−2568.

Weber, G. (1957). *Nature (London)* **180**, 1409.

Williams, J. W., Duggleby, R. G., Cutler, R., and Morrison, J. F. (1980). *Biochem. Pharmacol.* **29**, 589−595.

Williams, M. N., Poe, M., Greenfield, N. J., Hirschfield, J. M., and Hoogsteen, K. (1973). *J. Biol. Chem.* **248**, 6375−6379.

Yoshimoto, M., and Hansch, C. (1976). *J. Med. Chem.* **19**, 71−98.

Zakrzewski, S. F., and Nichol, C. A. (1958). *Biochem. Biophys. Acta* **27**, 425−426.

2

Alcohol Dehydrogenase

HANS EKLUND
CARL-IVAR BRÄNDÉN
Department of Chemistry and Molecular Biology
Swedish University of Agricultural Sciences
Uppsala Biomedical Center
Uppsala, Sweden

CONTENTS

1. INTRODUCTION

2. METHODS

3. DESCRIPTION OF THE LADH MOLECULE AND ITS RELATION TO INTRON-EXON ARRANGEMENTS
  3.1.   The Coenzyme-Binding Domain
  3.2.   The Catalytic Domain
  3.3.   The Zinc Atoms

4. CONFORMATIONAL DIFFERENCES BETWEEN APO- AND HOLOENZYME

5. COENZYME BINDING
  5.1.   Adenine Binding
  5.2.   Adenosine Ribose Binding
  5.3.   Pyrophosphate Binding
  5.4.   NMN-Ribose Binding
  5.5.   Nicotinamide Binding

6. SUBSTRATE AND INHIBITOR BINDING
  6.1.   The Substrate-Binding Pocket
  6.2.   Substrate Binding
  6.3.   Inhibitor Binding

7. ASPECTS OF THE CATALYTIC MECHANISM
  7.1.   The Role of Water
  7.2.   Proton Abstraction
  7.3.   The Role of Zinc
  7.4.   Hydride Transfer
  7.5.   Stereochemical Aspects of the Catalytic Reaction
  7.6.   Comparison with other Dehydrogenases
  7.7.   The Ordered Mechanism of the Enzyme

8. MODEL BUILDING OF HOMOLOGOUS ALCOHOL AND POLYOL DEHYDROGENASES
  8.1.   Human LADH
  8.2.   Maize Alcohol Dehydrogenase
  8.3.   Yeast Alcohol Dehydrogenase

8.4.   Sorbitol Dehydrogenase

8.5.   Structural Implications of Comparisons of Homologous
       Dehydrogenses

9.   FUTURE PROSPECTS

REFERENCES

1.   INTRODUCTION

Horse liver alcohol dehydrogenase, LADH (1.1.1.1.), is an NAD^+-dependent enzyme that catalyzes the oxidation of various primary and secondary alcohols to the corresponding aldehydes and ketones. The enzyme has a molecular weight of 80,000 and is a dimer of two identical subunits. Each subunit has one coenzyme-binding site and two firmly bound zinc atoms, one of which is essential for catalysis. The sequence of the 374 amino acids in the single polypeptide chain of one subunit (Jörnvall, 1970) and the X-ray structure of the apo- and holoenzymes have been determined (Eklund et al., 1976, 1981). Coenzyme binding induces a conformational change of the enzyme from an open to a closed form (Eklund and Brändén, 1979). The structural results have been correlated to other physical and chemical studies in a number of reviews (Brändén et al., 1975; Brändén and Eklund, 1978, 1980; Klinman, 1981; Plapp, 1982; Jeffery, 1982; Eklund and Brändén, 1983).

The main reaction catalyzed by alcohol dehydrogenase is, in principle, a very simple reaction. An alcohol group is oxidized by the removal of a proton from the hydroxyl group and by the transfer of a hydride ion from the adjacent carbon atom to NAD^+. An essential point for a discussion of the structural requirements of this reaction is that the hydride ion is directly transferred from the substrate to the C4 atom of the nicotinamide moiety of NAD^+ (Klinman, 1981).

Alcohol dehydrogenase, thus, must recognize and bind NAD in such a way that the nicotinamide is correctly positioned in the active site. The binding energy of NAD to an enzyme is potentially very high. Fersht (1977) estimates that a dissociation constant of 10^{-20} M would occur if all the potential binding energies were realized. Because reasonably high K_m values are an important component of efficient catalysis, it is essential that some of the possible interactions between enzyme and coenzyme do not occur. Thus, there is a conflict between the requirements of recognizing NAD and positioning the nicotinamide, on the one hand, and sufficiently weak binding on the other. This conflict may be one important reason why LADH is a fairly slow enzyme, because dissociation of reduced coenzyme is the rate-limiting step in the oxidation of alcohols (Dalziel, 1975).

Furthermore, the enzyme must recognize and bind the substrate in such a way that the hydride to be transferred is correctly positioned in the active site. However, the substrate molecule and the nicotinamide moiety may not be rigidly fixed relative to each other. The electronic rearrangement of carbon from sp^2 hybridization in the aldehyde to sp^3 in the alcohol requires different preferred pathways for hydride transfer, and thus different relative orientations.

Direct transfer of a hydride ion is facilitated in a hydrophobic environment where water is excluded. Thus, there is also a conflict between this requirement and the fact that the active site must be open and available for positioning of both nicotinamide and the substrate. This apparent conflict is resolved by induced conformational changes involving domain rotations, such that the active site is shielded from solution after binding of coenzyme and substrate (Eklund and Brändén, 1979; Eklund et al., 1981, 1984).

2.   METHODS

All experimental X-ray data from LADH crystals have been measured at 4°C by diffractometers equipped with conventional single-counter detectors. Four different crystal forms have been studied (Table 1). Heavy atom derivatives have been prepared for all crystal forms. These were used for phase angle determination for the orthorhombic and triclinic forms, which were solved to high resolution by multiple iomorphous replacement methods. These two crystal forms correspond to two different conformations of the enzyme; an open form for the apoenzyme and a closed form for the holoenzyme. For the two monoclinic forms, the heavy atom derivatives were used for phasing only at low resolution. From the corresponding electron density maps, it was established that one of these monoclinic crystal

Table 1. Different Crystal Forms of Alcohol Dehydrogenase Investigated with X-ray Diffraction

Type of Complex	Crystal Form	Resolution Å	Content of Asymmetric Unit (D)	Crystallographic R factor	References
Apoenzyme	Orthorhombic C222$_1$	2.4	40,000	0.20	Eklund et al., 1976 Jones, unpublished
Holoenzyme	Triclinic P1	2.9	2 × 40,000 +2 NADH	0.22	Eklund et al., 1981, 1984
	Monoclinic P2$_1$	2.9	2 × 40,000 +2 NADH	0.29	Unpublished
Lysine-modified enzyme	Monoclinic C2	3.2	42,000	0.29	Plapp et al., 1983

forms contained the enzyme in its apo conformation and the other in the holo conformation. Molecular replacement methods were then used to obtain the details of these structures to high resolution.

All four crystal forms have been refined crystallographically using different methods (Deisenhofer and Steigemann, 1975; Sussmann et al., 1977; Jack and Levitt, 1978; Hendrickson and Konnert, 1980) in combination with computer graphics (Jones, 1978, 1982). The orthorhombic crystals for the apoenzyme have been refined to a lower R factor at higher resolution than the other forms.

A number of different complexes with coenzyme, substrate, and inhibitors and with different active site metal atoms which crystallize in these different forms have been studied by difference Fourier techniques. These studies are summarized in Table 2.

The tracing of the polypeptide chain was unambiguous in all structures, with the exception of the first three residues that are flexible at the surface of the enzyme molecule. Most side chains and carbonyl oxygen atoms are also confidently located in most structures. The main exceptions are a few

Table 2. Crystallographic Investigation of Alcohol Dehydrogenase Complexes

Enzyme Complex	Resolution (Å)	Reference
1. Orthorhombic Crystals		
1. Apoenzyme	2.4	Eklund et al., 1976
2. ADP-ribose	2.9	Nordström & Brändén, 1975
		Eklund et al., 1984
3. 8-Br-ADP-ribose	4.5	Nordström & Brändén, 1975
4. ADP	3.7	Zeppezauer et al., 1975
5. PyAD	4.5	Samama et al., 1977
6. 3-I-Py-AD	4.5	Samama et al., 1977
7. 5-Methyl-PyAD	3.7	Samama et al., 1981
8. NADH-imidazole	2.4	Cedergren-Zeppezauer, 1983
9. Imidazole	2.9	Boiwe & Brändén, 1977
		Eklund et al., 1982b
10. Pyrazole	3.2	Eklund et al., 1982b
11. 1,10-phenanthroline	4.5	Boiwe & Brändén, 1977
12. Carboxymethylated LADH	4.5	Zeppezauer et al., 1975
13. Carboxymethylated LADH + NADH	3.2	Cedergren-Zeppezauer et al., 1985
14. H$_2$NADH	2.9	Cedergren-Zeppezauer et al., 1982

(continued)

Table 2. *(continued)*

Enzyme Complex	Resolution (Å)	Reference
15. Cibachron blue	3.7	Biellmann et al., 1979
16. 5-I-Salicylate	4.5	Einarsson et al., 1974
17. ANS	2.9	Brändén et al., 1978
18. $-$Zn at active site	2.7	Schneider et al., 1983a
19. $-$Zn at active site + Co	2.4	Schneider et al., 1983a
20. $-$Zn at active site + Cd	2.4	Schneider et al., 1985

2. Triclinic Crystals

Enzyme Complex	Resolution (Å)	Reference
1. NADH + DMSO	2.9	Eklund et al., 1981
2. NAD$^+$/NADH + bromobenzylalcohol/ aldehyde	2.9	Eklund et al., 1982a
3. NAD$^+$ + F$_3$C-CH$_2$OH	4.5	Plapp et al., 1978
4. H$_2$NADH + DACA	2.9	Cedergren-Zeppezauer et al., 1982
5. NAD$^+$ + Pyrazole	2.9	Eklund et al., 1982b
6. NAD$^+$ + 4-I-Pyrazole	2.9	Eklund et al., 1982b
7. NAD$^+$ + 2-4(4-pyrazolyl) butyl	2.9	Horjales 1985
8. NAD$^+$ $-$ Zn at active site + pyrazole	2.9	Schneider et al., unpublished
9. NAD$^+$ $-$ Zn at active site + 4-I-pyrazole	2.9	Schneider et al., unpublished
10. NADH-Zn (at active site)	2.9	Schneider et al., 1983b
11. NADH	4.5	Plapp et al., 1978
12. NADH + Cd $-$ Zn (at active site)	2.7	Schneider et al,. 1985
13. NAD$^+$/NADH + bromobenzyl-alcohol/aldehyde $-$ Zn (at active site) + Cd		Schneider et al., 1985

3. Monoclinic Crystals (P2$_1$)

Enzyme Complex	Resolution (Å)	Reference
1. NADH + DMSO	2.4	Unpublished
2. NAD$^+$ + F$_3$C-CH$_2$OH	3.2	Unpublished
3. NADH + Co $-$ Zn (at active site)	2.9	Unpublished

4. Monoclinic Crystals (C2)

Enzyme Complex	Resolution (Å)	Reference
1. The enzyme with lysine residues modified by isonicotinamidylation	3.2	Plapp et al., 1983
2. -''- + NADH	3.2	Plapp et al., 1983

charged side chains at the surface of the molecule. The resolution, however, is, in no case better than 2.4 Å. This resolution does not allow interpretation of the results with an accuracy better than a few tenths of an Ångström. This is most critical for the accuracy of bond angles for the metal coordination. A difference of a few tenths of an Ångstrom in ligand position can result in differences of 5°−10° in the bond angles.

3. DESCRIPTION OF THE LADH MOLECULE AND ITS RELATION TO INTRON-EXON ARRANGEMENTS

The general structural features of the open and closed forms of the LADH molecule are similar. The dimeric molecule is shaped approximately like a prolate ellipsoid with dimensions of 40 × 55 × 100 Å. The molecule is more bulky at the ends of the long axis than in the central region. A drawing of the dimeric molecule is shown in Figure 1.

The subunits are divided into two domains that are separated by a cleft containing a deep pocket. The pocket accommodates the substrate and the

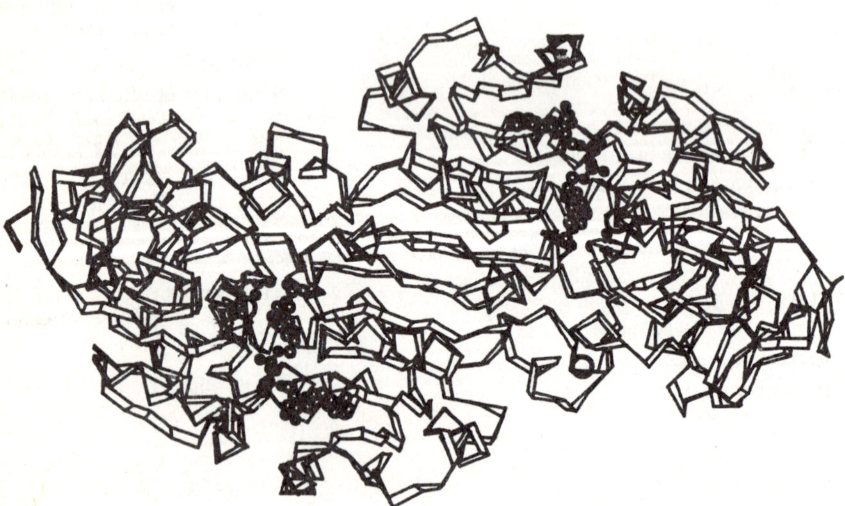

Figure 1. Drawing of the dimeric alcohol dehydrogenase molecule. The coenzyme NADH, DMSO, and the two zinc atoms of each subunit are drawn in thick lines and the main chains as a ribbon in thin lines. (We are indebted to A. Lesk for supplying the plot program.)

nicotinamide moiety of the coenzyme. One domain binds the coenzyme and the other provides ligands to the catalytic zinc atom, as well as to most of the groups that control substrate specificity. The two domains are unequal in size; the catalytic domain is larger and comprises 231 residues, whereas the coenzyme-binding domain is built from 143 residues. The dimer molecule is formed such that the two subunits are joined together into a central compact core through their coenzyme-binding domains. The catalytic domains are at the two ends of the dumbbell-shaped molecule, and the catalytic sites are in the junctions between these domains and the core. There are two covalent connections between the domains that are close together in space.

The subunits of other NAD^+-dependent dehydrogenases are similarly divided into two structurally and functionally different domains (Rossmann et al., 1975), one of which binds the coenzyme while the other provides the residues necessary for substrate binding and catalysis. The NAD^+-binding domains of lactate dehydrogenase (Adams et al., 1970), alcohol dehydrogenase (Brändén et al., 1973), and glyceraldehyde-3-phosphate dehydrogenase (Buehner et al., 1973) are all similar in structure and bind the coenzyme in a similar way. The catalytic domains, on the other hand, have quite different structures. Despite the similarities in structure of the coenzyme-binding domain, there is no sequence homology between these enzymes.

When these relationships were discovered approximately 10 years ago, it was suggested (Ohlsson et al., 1974; Rossmann et al., 1974) that they reflected evolutionary relationships, and that these and other complex enzymes have evolved by reassorting and joining genes that coded for simple ancestral polypeptide chains. At that time, no plausible genetic mechanism was known that could easily account for such rearrangements. Some years later, after the discovery of intron-exon arrangements in eukaryotic genes, Gilbert (1978) proposed that exons might correspond to units of protein function and that this feature would be a mechanism for increasing the rate of evolution (Brändén et al., 1984). The intron-exon arrangement in the recently determined gene sequence of maize alcohol dehydrogenase (Dennis et al., 1984) has been compared with the three-dimensional structure of the liver enzyme (Brändén et al., 1984) to examine if there is any clear relation between the nine intron positions in the gene and structural or functional modules in the protein.

Both liver and maize alcohol dehydrogenase are dimeric enzymes, each composed of two polypeptide chains of 374 and 379 amino acids, respectively. They exhibit 52% sequence identity using the alignment shown in

Table 3. The amino acid sequence of the maize subunit has been deduced by Dennis et al. (1984) from the nucleotide sequence of the maize Adh 1 gene. The high sequence identity is a strong indication that the three-dimensional structures of the subunits are very similar. Furthermore, an analysis of the sequence differences has shown that all structurally important residues are

Table 3. Comparison Between Liver and Maize Alcohol Dehydrogenase Sequences

```
                                    ↓1
  1                 10  ↓                20                      30
  S  T  A  G  K  V  I  K  C  K  A  A  V  L  W  E  E  K  K  P  F  S  I  E  E  V  E  V  A  P
M A  T  A  G  K  V  I  K  C  K  A  A  V  A  W  E  A  G  K  P  L  S  I  E  E  V  E  V  A  P

31                  40                      50           ↓2         60
  P  K  A  H  E  V  R  I  K  M  V  A  T  G  I  C  R  S  D  D  H  V  V  S  G  T  L  V  T  P
  P  Q  A  M  E  V  R  V  K  I  L  F  T  S  L  C  H  T  D  V  Y  F  W  E  A  K  G  Q  T  P  V

61                  70  ↓3                  80                      90
  L  P  V  I  A  G  H  E  A  A  G  I  V  E  S  I  G  E  G  V  T  T  V  R  P  G  D  K  V  I
  F  P  R  I  F  G  H  E  A  G  G  I  I  E  S  V  G  E  G  V  T  D  V  A  P  G  D  H  V  L

91                  100                     110                    120
  P  I  F  T  P  Q  C  G  K  C  R  V  C  K  H  P  E  G  N  F  C  L  K  N  D  L  S  M  P  R
  P  V  F  T  G  E  C  K  E  C  A  H  C  K  S  A  E  S  N  M  C  D  L  L  R  I  N  T  D  R

121                 130                     140                       150
  G  T  M     Q  D  G  T  S  R  F  T  C  R  G  K  P  I  H  H  F  L  G  T  S  T  F  S  Q  Y  T
  G  V  M  I  A  D  G  K  S  R  F  S  I  N  G  K  P  I  Y  H  F  V  G  T  S  T  F  S  E  Y  T

                                                              4↓
151                 160                     170                ↓180
  V  V  D  E  I  S  V  A  K  I  D  A  A  S  P  L  E  K  V  C  L  I  G  C  G  F  S  T  G  Y
  V  M  H  V  G  C  V  A  K  I  N  P  Q  A  P  L  D  K  V  C  V  L  S  C  G  Y  S  T  G  L

                                                         ↓5
181                 190                     200                     210
  G  S  A  V  K  V  A  K  V  T  Q  G  S  T  C  A  V  F  G  L  G  G  V  G  L  S  V  I  M  G
  G  A  S  I  N  V  A  K  P  P  K  G  S  T  V  A  V  F  G  L  G  A  V  G  L  A  A  A  G  G

                                                  ↓6
211                 220                     230                     240
  C  K  A  A  G  A  A  R  I  I  G  V  D  I  N  K  D  K  F  A  K  A  K  E  V  G  A  T  E  C
  A  R  I  A  G  A  S  R  I  I  G  V  D  L  N  P  S  R  F  E  A  R  K  F  G  C  T  E  F

241                 250  ↓7                 260                     270
  V  N  P  Q  D  Y  K  K  P  I  Q  E  V  L  T  E  M  S  N  G  G  V  D  F  S  F  E  V  I  G
  V  N  P  K  D  H  N  K  P  V  Q  E  V  L  A  E  M  T  N  G  G  V  D  R  S  V  E  C  T  G

                                       ↓8
271                 280                     290                     300
  R  L  D  T  M  V  T  A  L  S  C  C  Q  E  A  Y  G  V  S  V  I  V  G  V  P  P  D  S  Q  N
  N  I  N  A  M  I  Q  A  F  E  C  V  H  D  G  W  G  V  A  V  L  V  G  V  P  H  K  D  A  E

301                 310                     320                     330
  L  S  M  N  P  M  L  L  L  S  G  R  T  W  K  G  A  I  F  G  G  F  K  S  K  D  S  V  P  K
  F  K  T  H  P  M  N  F  L  N  E  R  T  L  K  G  T  F  F  G  N  Y  K  P  R  T  D  L  P  N

                       ↓9
331                    340                  350                     360
  L  V  A  D  F  M  A  K  K  F  A  L  D  P  L  I  T  H  V  L  P  F  E  K  I  N  E  G  F  D
  V  V  E  L  Y  M  K  K  E  L  E  V  E  K  F  I  T  H  S  V  P  F  A  E  I  N  K  A  F  D

361                 370
  L  L  R  S  G  E  S  I  R  T  I  L  T  F
  L  M  A  K  G  E  G  I  R  C  I  I  R  M  E  N
```

Arrows show the positions of introns in the maize gene.
Line 1: horse liver alcohol dehydrogenase; line 2: maize alcohol dehydrogenase.

either identical or conservatively substituted. One can safely assume that the polypeptide fold and subunit arrangement of maize alcohol dehydrogenase are the same as in the liver enzyme. A schematic diagram of the polypeptide fold with the intron positions is shown in Figure 2.

The intron positions of the maize gene occur within the triplets coding for residues 11, 71, 179, and 232 and after the triplets coding for residues 56, 206, 252, 284, and 338 (Dennis et al., 1984). All numbers refer to the horse LADH sequence. The positions along the polypeptide chain of the two domains and of the secondary structural elements in the coenzyme-binding domain are shown in Figure 3, as well as their relation to the intron-exon arrangement of the maize gene. The coenzyme-binding domain is separated from the main part of the catalytic domain by intron 4 and from the extra C-terminal β-α-β module of the catalytic domain by intron 9. Intron 4 at position 179 is close to the first covalent connection between the domains, whereas intron 9 at position 338 is 20 residues away from the second connection. These 20 residues form a helix that connects the last parallel strand of the coenzyme-binding domain, βF, with the C-terminal structural unit of the catalytic domain.

3.1.　The Coenzyme Binding Domain

The coenzyme-binding domain is a typical α-β domain (Levitt and Chothia, 1976) with a central β sheet of six parallel strands surrounded by α helices on both sides. The arrangement of strands observed in such domains provides an excellent framework for ligand binding (Brändén, 1980). The coenzyme-binding domain in alcohol dehydrogenase is very regularly built from six α-β units (Fig. 4).

Rao and Rossmann (1973) showed that domains with this topology could be divided into two approximately identical halves, each associated with a mononucleotide-binding area. Intron 7 after position 252 divides the domain into two halves, each of which consists of three α-β units (Eklund et al., 1976). Intron 4 is positioned after one turn in the first helix of the first of these units. Intron 7 is positioned in the corresponding position in the second half. Thus, both exon 5 and exon 8 start at structurally equivalent positions in these units. The mononucleotide-binding domain is a structural module that has been observed in a number of different proteins (Brändén, 1980). The fact that these modules are here so clearly separated by an intron lends

Figure 2. Schematic diagram of the polypeptide fold of the alcohol dehydrogenase dimer. One subunit is shaded. The intron positions are marked and numbered from the amino end. (We are indebted to Bo Furugren for this drawing.)

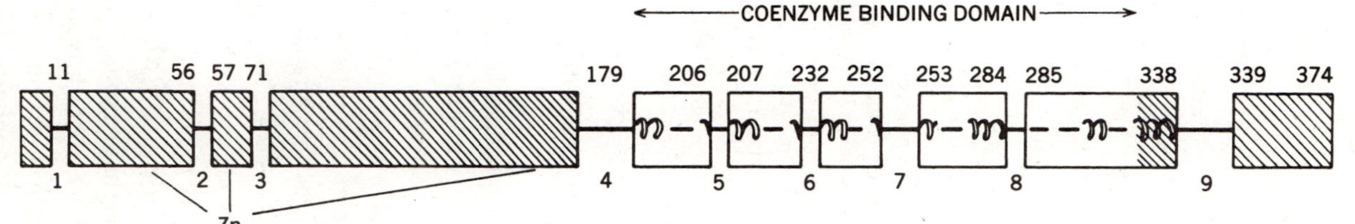

Figure 3. Schematic representation of the polypeptide chain of alcohol dehydrogenase. Residue numbers refer to the sequence of the liver enzyme. Gaps represent intron positions in the maize gene. Residues belonging to the catalytic domain are shaded. Residues belonging to the helices and strands of pleated sheet in the coenzyme binding domain are marked as well as the zinc ligands.

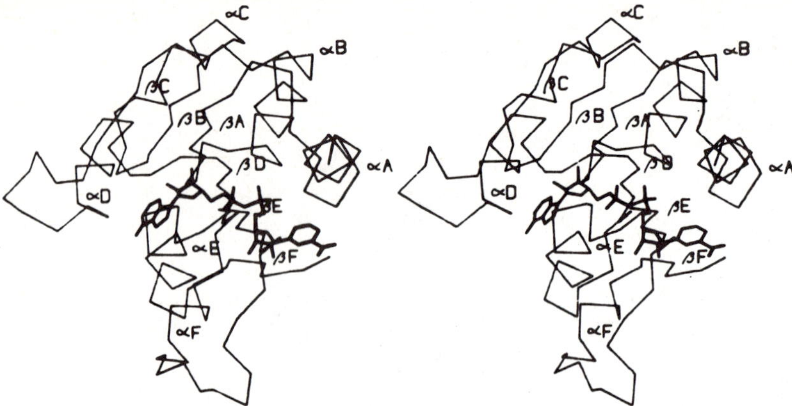

Figure 4. Coenzyme binds at the carboxyl end of the coenzyme binding domain. In the stereo diagram, the Cα atoms of the coenzyme binding domain are connected with thin lines and the coenzyme atoms with thick lines. The nomenclature is adopted to the regular character of the domain in alcohol dehydrogenase and is slightly different from the one used for LDH and in comparisons (Rossmann et al., 1975). The secondary structure elements, which are assigned on the basis of main chain dihedral angles, contain the following residues; βA, 193−198; βB, 218−223; βC, 239−242; βD, 264−267; βE, 287−291; βF, 312−318; αA, 174−187; αB, 202−214; αC, 226−235; αD, 250−258; αE, 272−280; αF, 305−310. The only deviation from this regular α-β pattern is a strand (βS) between βE and αF involved in the subunit interactions.

considerable strength to the theory that introns may separate structural modules in proteins.

Blake (1983) has attempted to produce a comprehensive view on the origin and role of exons in protein evolution. He assumes that fundamental early coding units of 20−40 residues might have been potential supersecondary structures such as β-β, α-α, or α-β units. These could then be readily fused to form larger peptides representing early protein molecules. One such early molecule must have been a nucleotide-binding protein. Granted that the present intron-exon arrangement in proteins, to some extent, reflects early stages of gene arrangement, the intron positions found in the first mononucleotide-binding unit strongly support this view on protein evolution. The DNA region for the first mononucleotide-binding domain, residues 179−252, contains three exons. Intron positions 5 and 6 are after residue 206 and at residue 232. Structurally, these positions are approximately after the first turn of helices αβ and αC. The three exons each comprise an α-β unit with some additional residues after the β strand. Exons

5, 6, and 7 approximately correspond to αA-βA, αB-βB, and αC-βC, respectively.

The second mononucleotide-binding domain in alcohol dehydrogenase contains only one intron, which is positioned after residue 284 in the loop before strand βE. There is no similar correlation between exons and structural repeat of α-β units in the latter half of the coenzyme-binding domain.

3.2.   The Catalytic Domain

The catalytic domain is comprised of two regions of the polypeptide chain: residues 1−174 and 319−374. Both ends of the polypeptide chain are found within this domain. The two zinc atoms of the subunit are bound to ligands from this domain, and are located in the molecule as shown in Figure 1. There are only five helical segments in the catalytic domain. In contrast, a large number of the residues are in β-sheet regions, with the polypeptide folded into an intricate network of similarly long antiparallel β strands. The five helices present in the catalytic domain are comprised of residues 46−56, 100−105, 166−174, 324−338, and 355−364. These are at the surface of the molecule outside the interior β-sheet regions. In general, one side of each helix faces the solution while the other side provides nonpolar residues for one of two hydrophobic cores. The hydrophobic regions form areas of interaction between helices and β-sheet regions.

The major portion of the catalytic domain, residues 1−174, does not have as extensive an intron-exon pattern in the maize gene as the coenzyme-binding domain. There are four exons in this region of the chain comprised of residues 1−11, 11−56, 57−71, and 71−179. The last is very long and dominates the domain. The intron between residues 56 and 57 is located between α1 and a short extended chain in a region that is of variable length among species. Two of the introns are in β-sheet strands, at residues 11 and 71. The first intron at residue 11 is in the first β-strand. Exon 1 is located at the surface of the domain, and this part is of variable length in different species. In alcohol dehydrogenase from *Bacillus stearothermophilus* (Bridgen et al., 1973), the domain starts at position 9 just before the observed intron position in the maize gene. The intron at position 71 divides the domain into two large units, where the first half lies in a doughnut on the surface of exon 4. The central feature of the catalytic domain is an orthogonal β-sheet structure as shown in Figure 5 (Chothia and

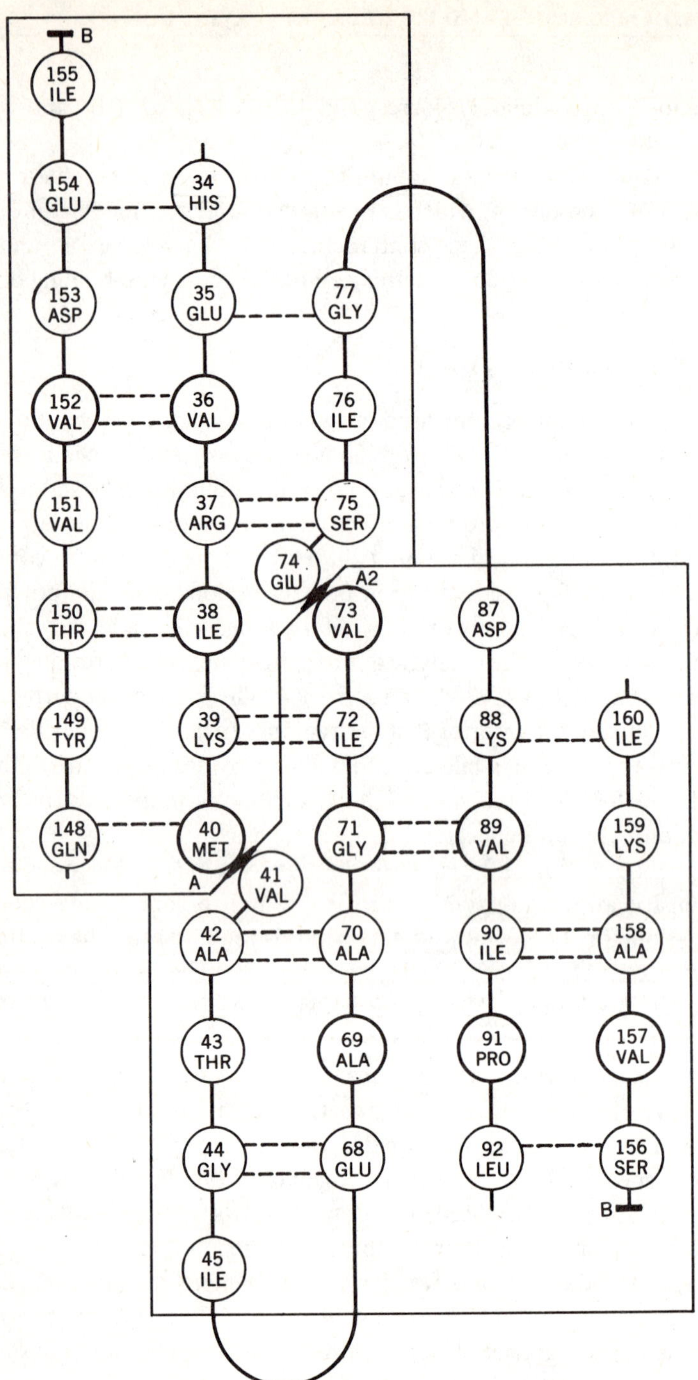
B
155
ILE
154
GLU
34
HIS
153
ASP
35
GLU
77
GLY
152
VAL
36
VAL
76
ILE
151
VAL
37
ARG
75
SER
74
GLU
A2
73
VAL
87
ASP
150
THR
38
ILE
149
TYR
39
LYS
72
ILE
88
LYS
160
ILE
148
GLN
40
MET
71
GLY
89
VAL
159
LYS
A
41
VAL
42
ALA
70
ALA
90
ILE
158
ALA
43
THR
69
ALA
91
PRO
157
VAL
44
GLY
68
GLU
92
LEU
156
SER
B
45
ILE

Janin, 1982). These structures were shown to have two closed corners and two splayed corners. The intron at position 71 is close to one of these corners, and the intron divides the orthogonal sheet into its two halves.

3.3. The Zinc Atoms

The catalytic domain provides the ligands to the catalytic zinc atom: Cys 46, His 67, and Cys 174 (Fig. 6). The peptide units of all three zinc ligands are firmly positioned in secondary structural elements, and none is part of a flexible loop region. Cys 46 is the first residue of a short helix comprising residues 46−55 immediately after the last residue of a β strand. His 67 is the first residue of a central β strand of a pleated sheet, and Cys 174 is in the middle of the helix that joins the two domains. Each ligand to this zinc atom is in a separate exon (Fig. 3).

There is also a second zinc atom in each subunit. This zinc is liganded in a tetrahedral arrangement by four sulfur atoms from cysteine residues 97, 100, 103, and 111 (Fig. 7). These cysteine residues are part of a lobe that projects out of the catalytic domain having only a few side chain interactions with the remaining part of the subunit. Accessibility calculations show that this zinc is completely buried by its ligands and is not accessible to the solution.

The function of the second zinc atom and the protein lobe that surrounds it is unknown, although a structural role has been suggested (Drum et al., 1969). The fact that all ligands to this zinc are close together along the polypeptide chain supports such a role. It is usually found that metal atoms having a catalytic role in proteins have their ligands from different parts of the polypeptide chain, whereas ligands to atoms with a structural role are close together. Examples of the latter type include the calcium binding in thermolysin (Matthews et al., 1974) and the zinc binding in aspartate transcarbamylase (Monaco et al., 1978).

Figure 5. Plan showing the residues and hydrogen bonds that form the orthogonal β-sheet packing in LADH. Dotted lines indicate hydrogen bonds. Residues marked with an asterisk have Ramachandran angles outside the β region. Residues surrounded by thick circles form the β sheet to β-sheet contacts. The separate β sheets are boxed by a thin continous line. The orthogonal packing involves bending the strands at the point labeled A1 and A2 and joining them at B. The figures are drawn so that the near β sheet on the top left folds forward over the far β sheet on the bottom right. Note the diagonal pattern formed by contact residues and the staggered pattern formed by the strands. (From Chothia and Janin, 1982.)

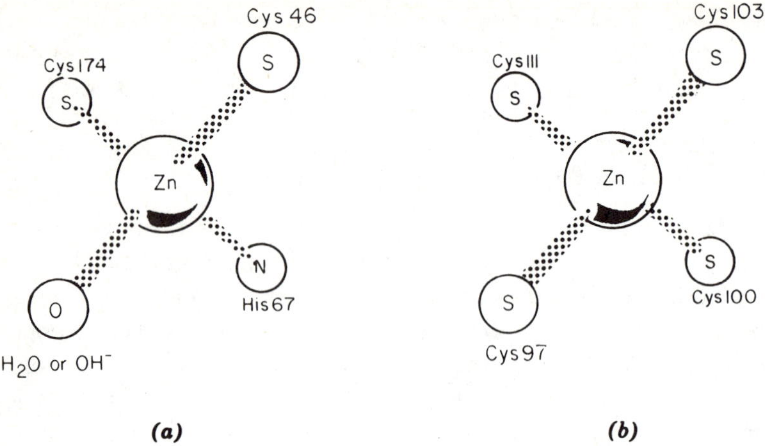

Figure 6. Schematic diagram showing the zinc ligands in LADH. (*a*) Ligands to the catalytic zinc atoms; (*b*) Ligands to the structural zinc atom.

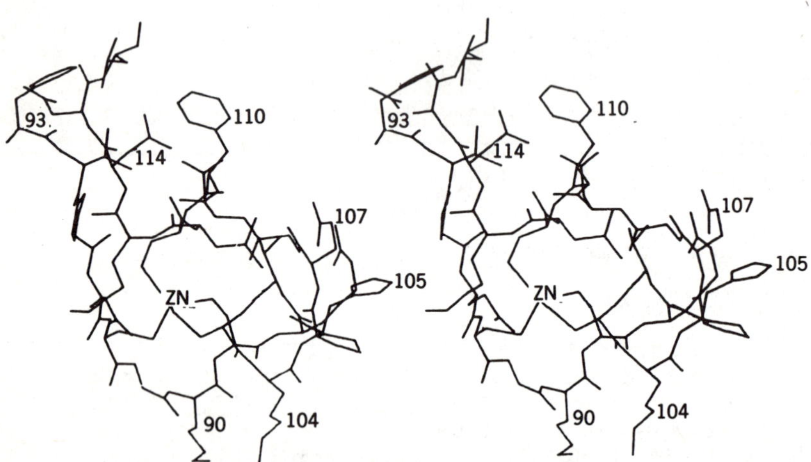

Figure 7. The structural zinc atom is bound in a protrusion from the catalytic domain formed by residues 94–113. The zinc atom is bound by cysteine residues 97, 100, 103, and 111.

4.   CONFORMATIONAL DIFFERENCES BETWEEN APO- AND HOLOENZYME

When the coenzyme binds, it triggers a large change of conformation of the enzyme (Brändén and Eklund, 1978; Eklund and Brändén, 1979; Eklund et al., 1981). A substantial rearrangement of the relation between the three main parts of the dimeric molecule takes place. A comparison of vectors between pairs of chemically identical atoms in the two subunits in the apoenzyme and ternary complex was made to objectively determine the differences (Eklund et al., 1981). It was apparent that the relative orientations between atoms in the coenzyme-binding domains are very similar in the two structures, whereas many atoms in the second domain differ. It was also obvious that residues 294−298 in the coenzyme-binding domain are subject to a local conformational change. The two coenzyme-binding domains have similar relative orientations, whereas the two catalytic domains are rotated relative to each other. The main conformational differences are best described as rigid body rotations of each catalytic domain with respect to the central core of the dimer. This is schematically shown in Figure 8.

The extent of the conformational change is demonstrated when the

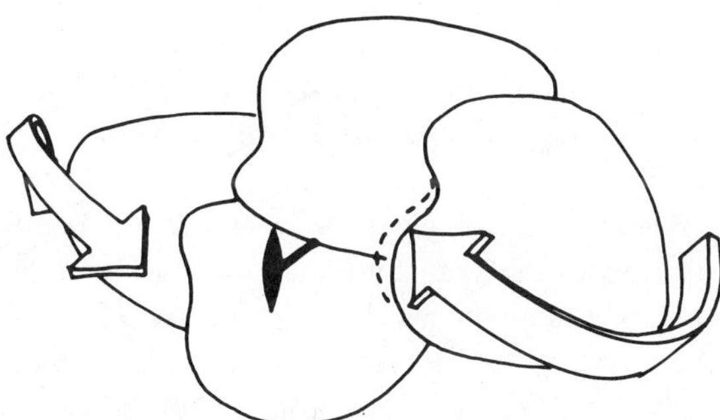

Figure 8 A schematic drawing of the main conformational changes of the alcohol dehydrogenase molecule in the transition from apoenzyme to ternary complex. The arrows describe the movements of the catalytic domains at the ends of the central core of the two coenzyme binding domains. The molecular twofold axis, which is crystallographic for the apoenzyme but not for the holoenzyme, is shown at the center of the molecule. The front side, which faces the viewer, contains the area that binds the two coenzyme molecules. (We are indebted to Bo Furugren for this drawing.)

subunits are compared by superimposing the α-carbon atoms of the coenzyme-binding domains of the two forms of the enzyme. Some parts of the catalytic domain are then very differently located in space with differences of up to 7 Å. There are several side chains at the surface that differ by more than 10 Å. On the other hand, if the catalytic domains of the two forms are separately compared, the differences are small. The internal structures of the catalytic domains, therefore, are very similar, indicating that the observed movement of the catalytic domain is essentially a rigid body rotation. Similar observations have been made for immunoglobulins (Huber et al., 1976), hexokinase (Bennett and Steitz, 1978), citrate synthase (Remington et al., 1982), and glyceraldehyde-3-phosphate dehydrogenase (Leslie and Wonacott, 1984).

The regions of minimum change occur near residues 44−46, 67, 68, 93, and 110, where the differences between α-carbons of the two forms are only of the order of 1 Å. These residues are centered in one region of the molecule, and thus define a hinge region for the rotation. This hinge region is essentially the hydrophobic core between helices of residues 166−174 and 324−338. It is significant that these helices also form the covalent connections between the two domains. A structurally similar hinge region has been observed in other enzymes (Lesk and Chothia, 1984). Hydrophobic side chains are packed in this core. The rotation of the catalytic domain can be accomplished by a small sliding movement of these two helices without any large structural changes within the domains. The catalytic domains rotate 10° around an axis through the center of the domain. By this rotation, the domains move 3−4 Å closer to the coenzyme-binding domain. The catalytic and coenzyme-binding domains have two covalent connections on the back side of the cleft, while there are only a few weak van der Waals contacts on the front side. These van der Waals interactions must change to allow the rotation of the catalytic domains as shown in Figure 9.

Large movements occur distant from the hinge region in the interaction area between the two domains. Residues 294−297 of the coenzyme-binding domain must be moved away in order to make room for residues 51−58 of the catalytic domain. Stereo views of this region in the two structures are given in Figure 10. The α-carbon atom of Pro 296 is changed 5 Å and the γ-carbon of this residue is changed as much as 8−9 Å. A remarkable feature of the conformational change of this loop is that the side chain of Val 294 points in opposite directions in the two structures. In the apoenzyme structure, it points away from the active site cleft, whereas in the ternary complex, it

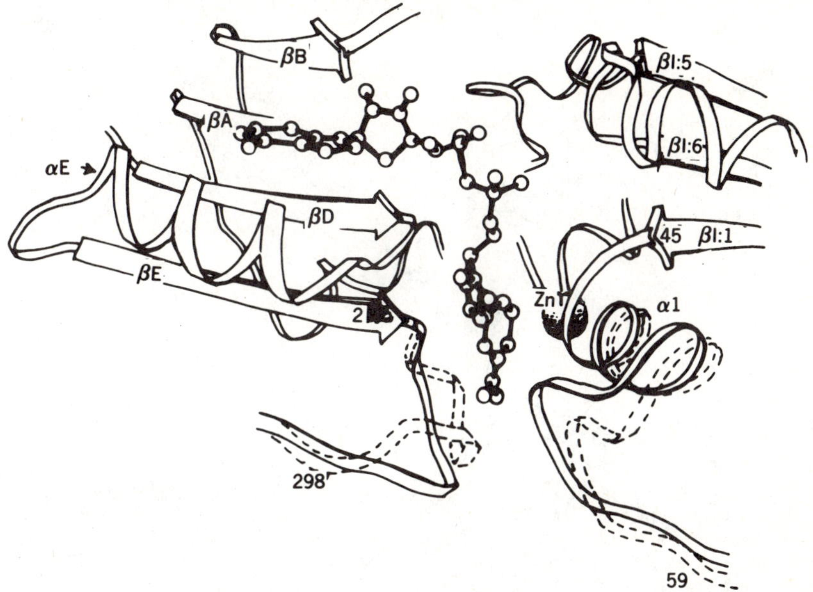

Figure 9. The domain-domain interaction in the triclinic structure with the active site and the coenzyme. The parts of the corresponding orthorhombic structure that have different conformations are shown in broken lines. (We are indebted to Bo Furugren for this drawing.)

is directed into the cleft and is in contact with the nicotinamide ribose. As a result of these movements, the interaction areas between the domains change considerably. In the apoenzyme structure, there are only weak contacts between a few residues from each domain in this region (see Fig. 10). Residues 55−56 are in contact with residues 296−297. In the ternary complex, these parts have moved so that the contact area is shifted by about 5 Å and involve other residues in the same region. An example of the drastic change in this interface region between the domains is that the side chains of Leu 57 and Val 294 are about 15 Å away from each other in the apoenzyme, but only 4−5 Å in the ternary complex.

In addition to the large, rigid body rotation of the catalytic domain, there are small but significant changes in the relative orientation of the coenzyme-binding domain (Eklund et al., 1984). This can be described as a rotation of the domain by 1.5° around an axis across the strands of the β sheet. The coenzyme-binding domains move in opposite directions toward the catalytic

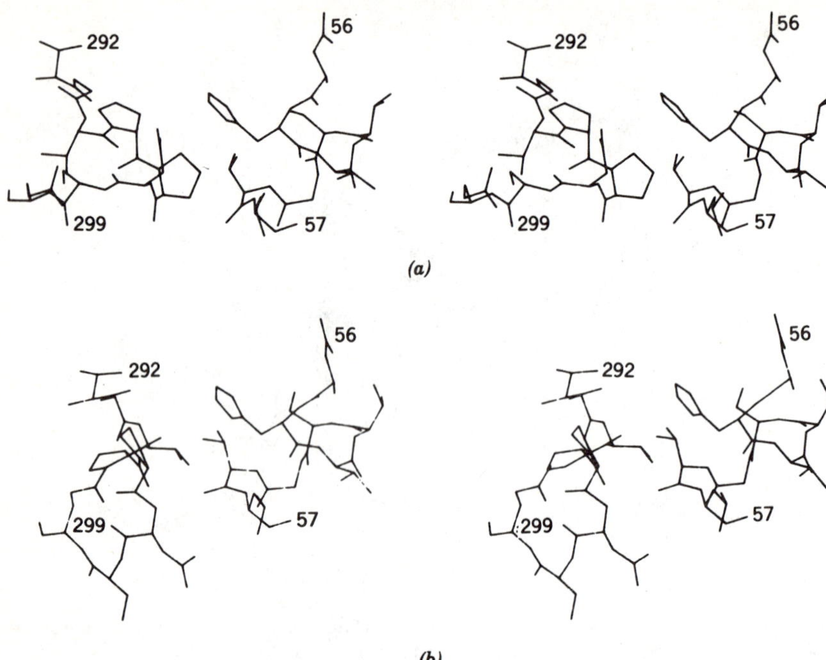

Figure 10. Stereo diagram of the domain-domain contact area in (*a*) the apoenzyme and (*b*) the ternary complex.

domain in each subunit (Fig. 11). NAD rotates together with the coenzyme-binding domain, maintaining similar interactions. The net result is that NAD is pushed deeper into the enzyme.

The complete coenzyme is essential for inducing the conformational change of LADH. The closed conformation of LADH has never been observed in the absence of coenzyme. Furthermore, the ADP-ribose portion of the coenzyme is not sufficient to produce the large conformational change (Nordström and Brändén, 1975). A requirement for the rearrangement of the domains during the conformational change is the large movement observed of the loop region of residues 292–298 of the coenzyme-binding domain (Eklund et al., 1981). This part of the structure acts as a wedge between the domains; when it moves, the catalytic domain can rotate closer to the coenzyme-binding domain. Therefore, the change of conformation of residues 292–298 must be an early step in the transition from

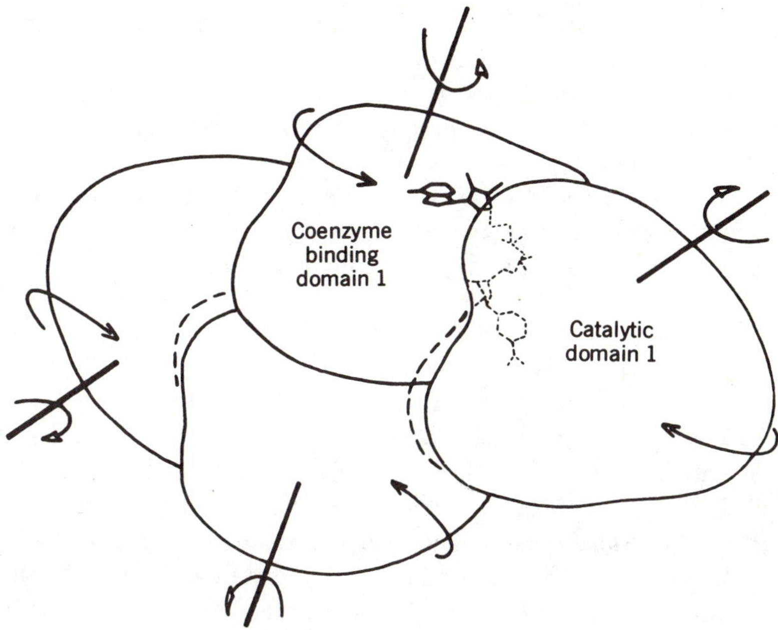

Figure 11. Schematic drawing showing the conformational change of the enzyme. The domains in each subunit rotate against each other. The catalytic domain rotates in 10° around an axis through the domain, while the coenzyme binding domain rotates 1.5° through an axis approximately across the pleated sheet strands. The coenzyme parts covered by the domain are drawn with dashed lines.

apoenzyme conformation to holoenzyme conformation. There are, however, few contacts between region 292−298 and the coenzyme. No obvious mechanisms as to how coenzyme binding triggers this loop movement have been deduced based on the formation of charge interactions or hydrogen bond patterns. The hydrogen bond formed between the carbonyl oxygen of residue 292 and the carboxamide group may be one minor component. In addition to this hydrogen bond, there are only a few van der Waals contacts between the coenzyme and Val 292 and Val 294. Consequently, there is no strong driving force in the interactions between the protein loop and the coenzyme.

An equilibrium mechanism is probably operating. The enzyme may exist in both conformations, but the equilibrium favors the open conformation in the absence of coenzyme. When coenzyme is bound, it stabilizes the closed

conformation and the equilibrium is shifted. The reduced coenzyme stabilizes the closed conformation to such an extent that the dissociation of
reduced coenzyme, which by necessity is coupled to the opening of the coenzyme binding cleft, is rate limiting for turnover with alcohol. Conformational transition in this process seems, in fact, to be the rate-determining
step of coenzyme dissociation (Czeisler and Hollis, 1973). The oxidized
coenzyme cannot stabilize the closed conformation to the same extent because of the positive charge on the nicotinamide ring.

5. COENZYME BINDING

The adenosine moiety of the coenzyme binds in a cleft in the coenzyme-
binding domain. The remaining atoms of the coenzyme molecule bind in the
large cleft region between the two domains of the subunit (Fig. 1) and
interact with residues from both domains (Eklund et al., 1984). The clefts
are approximately 30 Å long, 7 Å in diameter, and extend in both directions
from the coenzyme-binding site. At the nicotinamide site, the cleft has a
natural continuation into the substrate-binding site. The adenosine-binding
site is easily accessible from solution, while the nicotinamide-binding site is
in the center of the molecule, buried deep inside the protein. The coenzyme
molecule binds with its pyrophosphate moiety at the center of the carboxyl
end of the parallel β sheet of the coenzyme-binding domain. This site is
created by the two $(\alpha\text{-}\beta)_3$ units of the coenzyme-binding domain (Fig. 4).

5.1. Adenine Binding

The adenine ring is sandwiched, in van der Waals contact, between the
isoleucine side chains of residues 224 and 269, which line the cleft on both
sides (Fig. 12). N6A of the adenine ring points toward the solution and is in
contact with the side chain of Arg 271. The combination of a primarily
nonpolar adenine-binding site with the positive charge of Arg 271 has been
suggested as the reason for the perturbation of the adenine spectra in
ADP-ribose and coenzyme bound to LADH (Subramanian et al., 1981).
Only four atoms of the adenine ring have parts of their surfaces accessible to
solvent (Eklund et al., 1984). The final electron density map shows peaks
that can be interpreted as water molecules forming hydrogen bonds to N1A,

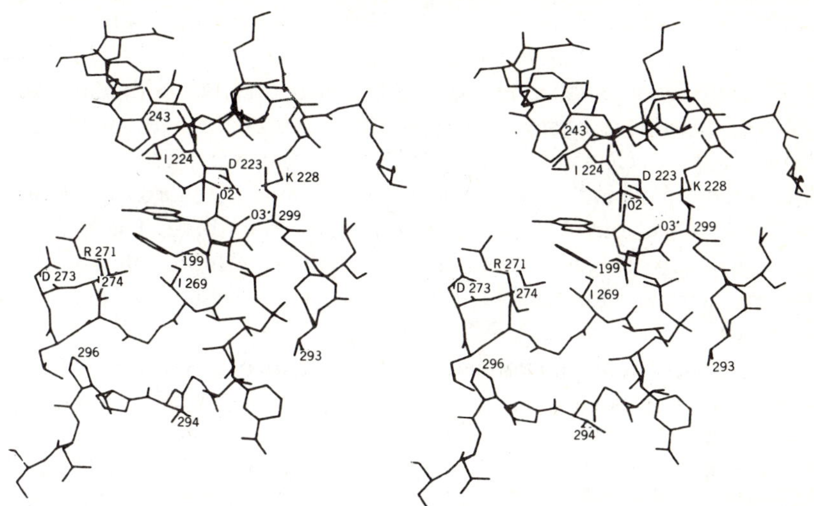

Figure 12. Stereo diagram showing the adenosine binding site of the coenzyme binding domain with the bound NADH molecule. The adenine ring fits into the slit between Ile 224 and Ile 269. Arg 271 covers part of one surface of the ring from the surrounding solvent. The adenosine binding site has Asp 223 in a central position. Hydrogen bonds are drawn as dotted lines.

N6A, and N7A, respectively. No hydrogen bond is formed between the adenine ring and the protein.

The adenine-binding site in alcohol dehydrogenase is not specific for adenine and can accommodate many other hydrophobic ligands (Einarsson et al., 1974; Biellmann et al., 1979; Samama and Eklund, 1985). This is in contrast, for example, to citrate synthase, which has a specific hydrogen bond pattern for adenine (Remington et al., 1982).

5.2. Adenosine Ribose Binding

The adenosine ribose-binding site is at the surface of the coenzyme-binding domain. Glycine residues at positions 199, 201, 202 allow the coenzyme to come close to the main chain. The first two of these residues are conserved in all alcohol dehydrogenases (Jörnvall et al., 1978; Dennis et al., 1984) as well as in other dehydrogenases (Rossmann et al., 1975). In a mutant of alcohol dehydrogenase from *Drosophila*, the position corresponding to residue 199 is substituted by an Asp (Thatcher, 1980), with the consequence that the enzyme has lost its ability to bind NAD (Thatcher and Retzios, 1980).

The site is dominated by Asp 223, which together with the glycine residues are the only structurally conserved residues between different dehydrogenases (Rossmann et al., 1975). This aspartic residue is crucial for discrimination in binding NAD and NADP because the side chain forms hydrogen bonds to the O2'A and O3'A atoms of the ribose. The hydrogen-bonding pattern is shown schematically in Figure 13. Glu 50 in glutathione reductase interacts with the FAD coenzyme in a very similar manner (Schulz et al., 1982).

O3'A is also hydrogen bonded to the side chain of Lys 228. Modification of this residue can enhance the activity of the enzyme (Plapp, 1970;

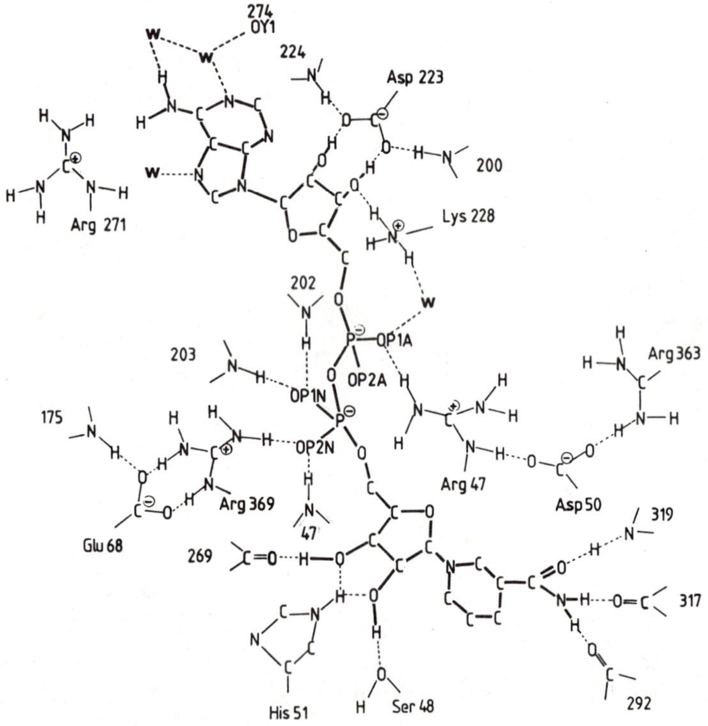

Figure 13. Schematic drawings showing hydrogen bonds between the enzyme and NADH. Hydrogen bonds are drawn as dashed lines and are assigned if the distance between hydrogen bond acceptor and donor is less than 3.3 Å. Water molecules are labeled W.

Zoltobrocki et al., 1974). Amidination of most of the amino groups of horse LADH with methyl isonicotinimidate increases the activity of the enzyme about 10-fold. The modified protein crystallizes readily, and the structure has been solved (Plapp et al., 1983). Of the 30 lysine residues per subunit, 23 were found to be modified, 2 are probably buried and were not modified, and 5 were accessible but had no visible substituents. The overall conformation of the isonicotinimidylated enzyme is very similar to that of native apoenzyme, despite the extensive chemical modification and the different packing arrangement. Enzyme molecules were loosely packed in the crystal lattice, with interactions involving six isonicotinimidylated lysines and 11 other amino acid residues. The isonicotinimidyl groups at the end of the molecule apparently stabilize a different mode of packing than is found with native apoenzyme.

When NADH binds to the modified enzyme, the isonicotinimidyl group on lysine residue 228 interferes with the interactions that normally exist between the unmodified amino group and the ribophosphate of the AMP portion of NAD. The absence of this interaction might activate the enzyme by causing the coenzyme to dissociate more rapidly (Plapp et al., 1983).

5.3. Pyrophosphate Binding

The pyrophosphate moiety of the coenzyme binds in the central part of the cleft between the domains. It is almost completely buried within the protein (Eklund et al., 1984), which is possible due to the polar character of the binding site. This site is situated between the amino ends of two helices: αB of the coenzyme-binding domain and α1 of the catalytic domain (Fig. 14). The pyrophosphate oxygen atoms form hydrogen bonds to main chain nitrogen atoms. Furthermore, the side chains of Arg 47 and Arg 369 hydrogen bond to each phosphate of the pyrophosphate bridge (Fig. 13). The complex charge distribution around the coenzyme is shown in Figure 15.

The dipole moments of the helices make the binding site more positive (Hol et al., 1978). Similarly, Lys 228, which is the only lysine close to the coenzyme, provides an additional positive charge in the coenzyme-binding site between the negative charges of Asp 223 and the pyrophosphate moiety of the coenzyme. A lysine residue, Lys 58, is present at an equivalent position in lactate dehydrogenase (Eventoff et al., 1977) and is conserved in all species (Eventoff et al., 1977; Hensel et al., 1983). Lys 44 has a similar position in malate dehydrogenase (Birktoft et al., 1982).

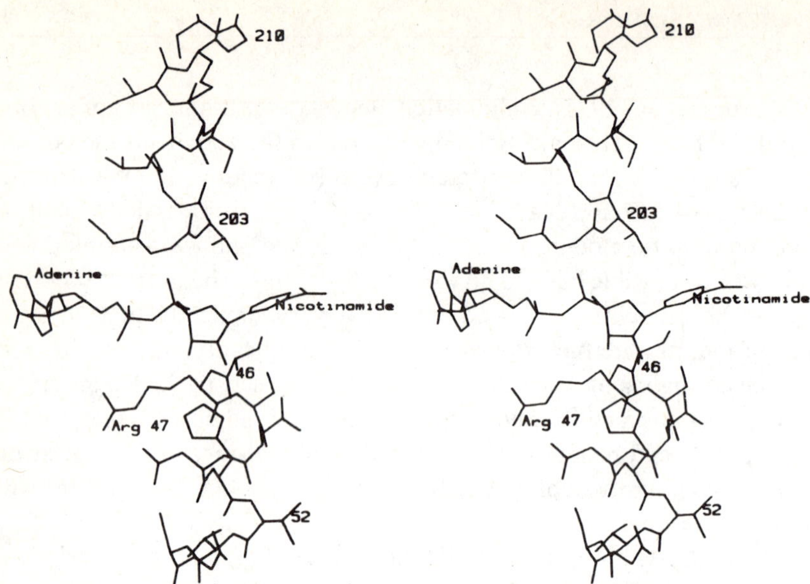

Figure 14. Coenzyme is bound at the amino end of two helices: α1 from the catalytic domain (residues 47–54) and αB from the coenzyme binding domain (residues 202–214).

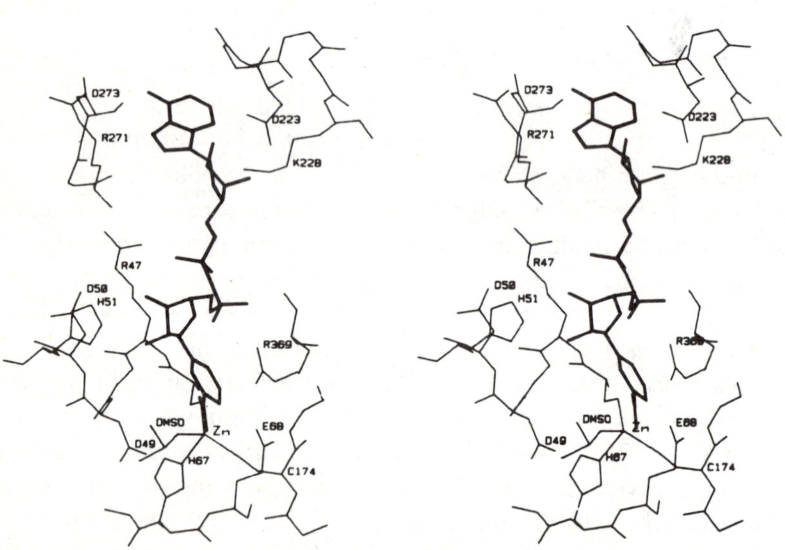

Figure 15. Binding site for NAD in liver alcohol dehydrogenase is surrounded by a number of charged residues. In the stereo diagram, we show these together with the residues that interact with them. The active site zinc atom with its protein ligands and the bound inhibitor DMSO are also plotted. The coenzyme molecule is drawn in thick lines.

5.4. NMN–Ribose Binding

The NMN-ribose binds in a narrow cleft between the two domains where hydrogen bonds are formed with the side chains of Ser 48 and His 51, and with a main chain carbonyl oxygen atom (Fig. 13). Possible roles of Ser 48 and His 51 in the proton transfer from the active site to the solution have been suggested (Eklund et al. 1974, 1982a).

In the ternary complexes, the NMN-ribose part is almost completely buried, while in complexes with the open form of the enzyme, ADP-ribose (Eklund et al., 1984), and H_2NADH (Cedergren-Zeppezauer et al., 1982), the site is much more open. Calorimetric measurements for AMP, ADP, and ADP-ribose binding suggest that the additional ribose moiety in the latter molecule does not contribute to the binding enthalpy (Schmid et al., 1978).

5.5. Nicotinamide Binding

The nicotinamide ring binds in a cleft in the interior of the protein, close to the center of the molecule, about 15 Å from surrounding solvent. The ring interacts on one side with Thr 178, Val 203, and Val 294. The other face is directed toward the active site, and is close to the catalytic zinc atom and the sulfur ligands of Cys 46 and Cys 174 (Fig. 16). In ternary complexes, there are interactions between the nicotinamide ring and substrates or inhibitors (Eklund et al., 1981, 1982a, 1982b; Cedergren-Zeppezauer et al,. 1982). The oxygen atom of the carboxamide is hydrogen bonded to the main chain nitrogen atom of residue 319. The nitrogen atom of the carboxamide group is hydrogen bonded to the carbonyl oxygens of residues 292 and 317 (Fig. 16).

In the structure of dihydrofolate reductase, there are three protein oxygen atoms that are close to and in the plane of the nicotinamide ring (see Chapter 1). Filman et al. (1982) have suggested that these oxygen atoms stabilize nicotinamide ring intermediates during catalysis. Alcohol dehydrogenase also has three oxygen atoms close to the nicotinamide ring: the two carbonyl oxygen atoms that hydrogen bonds to the carboxamide nitrogen atom and the side chain oxygen of Thr 178. The threonine side chain is positioned in van der Waals contact to the C4 atom of the nicotinamide ring. The oxygen atom lies in the direction of the hydrogen atom of C4 and the B side of NADH. Thr 178 is conserved in all known homologous alcohol dehydrogenase (Jörnvall et al., 1978; Dennis et al., 1984).

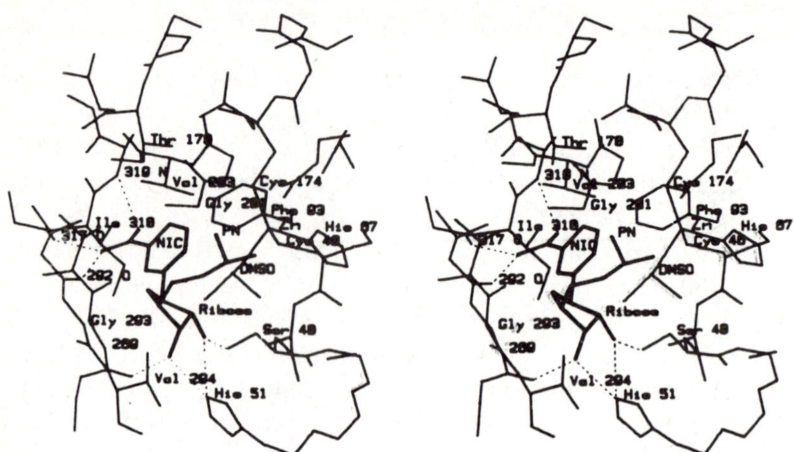

Figure 16. Nicotinamide binding site is located deep in the enzyme molecule. The NMN half of the coenzyme is shown in thick lines and the neighboring protein residues, Zn, and the inhibitor DMSO are displayed with thin lines. Hydrogen bonds are plotted with dotted lines.

The role of the active site zinc atom in NADH binding has been investigated by crystallographic methods (Schneider et al., 1983b). The active site metal-depleted enzyme undergoes the same conformational changes and has the same conformation as the native enzyme. An obvious conclusion of this result is that the zinc and its ligands do not determine the conformation of the coenzyme, not even its NMN moiety. The removal of the zinc atom and the subsequent small rearrangement of the former zinc ligands have minor effects on the nicotinamide position. The tight fit of the ribose and the nicotinamide with the enzyme gives the ring little flexibility. It seems probable, therefore, that the substrate, rather than the nicotinamide ring, has the required positional flexibility to accept a hydrogen from a stationary nicotinamide C4 atom when its C1 atom has sp^2 configurations, and to donate a hydrogen in sp^3 configuration along the same relative reaction path.

6. SUBSTRATE AND INHIBITOR BINDING

6.1. The Substrate–Binding Pocket

The substrate-binding cleft is much larger and more hydrophobic in LADH than in the other known dehydrogenases. This is reflected in the substrate specificity of the enzymes. LADH has a broad specificity toward alcohols with uncharged substituents, whereas both LDH and GPDH have strict specificity; each towards a specific group of small charged substrates.

The catalytic zinc atom in LADH is positioned deep inside the subunit, close to the covalent junction between the two domains. It is not accessible to solution from the backside nor from the top side of the molecule as seen in Figure 1. In the open conformation, it is accessible from the front side through the crevice between the domains. In the closed conformation, this crevice is closed. From the bottom side, there is a deep pocket 5–10 Å wide and approximately 20 Å long from the protein surface to the zinc atom. This is the substrate-binding pocket. This zinc atom is accessible from solution through this pocket both in the open and closed forms. The bottom of this pocket is formed by the zinc atom and its ligands.

The three protein ligands are arranged at three corners of a distorted tetrahedron. The fourth corner of this distorted tetrahedron points into the substrate-binding pocket. This coordination position is occupied by a water molecule in the open form of the enzyme. In the closed form, it is occupied by water, substrate or inhibitor molecules.Direct binding to zinc of these ligands has been observed in all crystallographic studies of the closed form. Deviations from tetrahedral symmetry in bond angles is observed, especially in the angle between the two sulfur atoms of Cys 46 and Cys 174. Bond angles involving the ligand also deviate significantly from tetrahedral coordination due to geometrical constraints at the active site, depending principally on interactions with Ser 48.

The only polar groups in the pocket close to zinc are the zinc ligands, the nicotinamide moiety of the coenzyme, and the side chain of Ser 48. The remaining part of the inside wall of the pocket is lined with hydrophobic side chains from residues Leu 57, Phe 93, Phe 110, Leu 116, Phe 140, Leu 141, Leu 294, and Ile 318, which are from the same subunit as the zinc ligands, as well as Met 306 and Leu 309 from the other subunit. The substrate-binding site near the zinc is narrow, because Phe 93 and Ser 48 limit access. Further away, the binding site is more open and forms a wide barrel (Fig. 17). In this region, it is limited in one direction by the nicotinamide and in the other directions by protein side chains. In the open form, there are several firmly fixed water molecules in the inner part of the pocket outside the coordination sphere of zinc (Eklund et al., 1984). In ternary complexes in the closed form, there are no such water molecules because the coenzyme and substrate occupy that space.

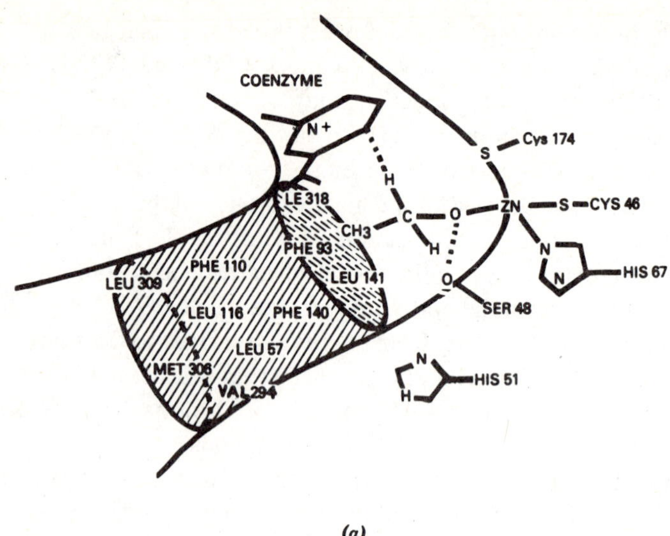

(a)

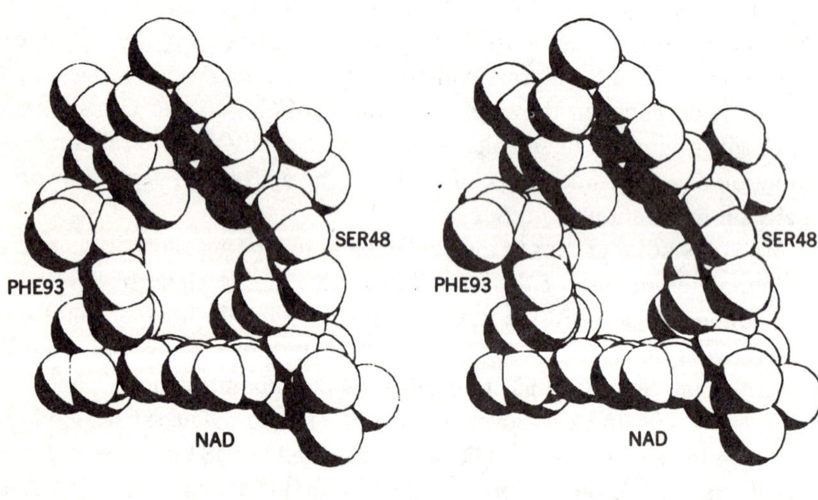

(b)

Figure 17. (*a*) Schematic representation of the substrate binding pocket in LADH. (*b–d*) Stereo diagrams of the substrate binding pocket. The view is from the zinc atom through the open channel toward the solution: (*b*) Space-filling drawing of the empty pocket; (*c*) bromobenzyl alcohol bound in hydrophobic pocket; and (*d*) space-filling drawing of the pocket with bound bromobenzyl alcohol in its observed orientation.

104

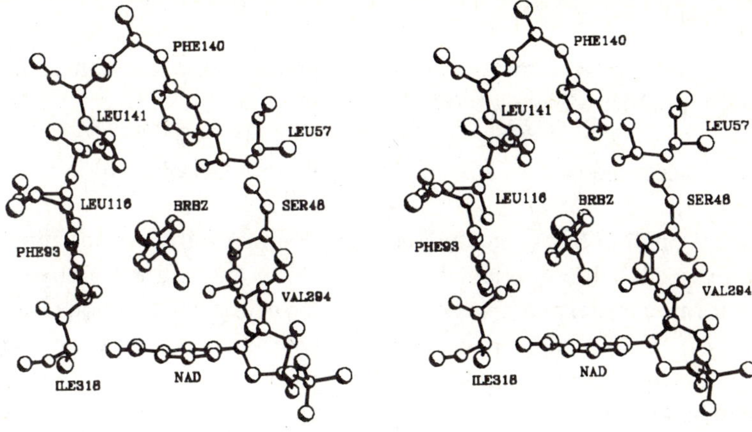

(c)

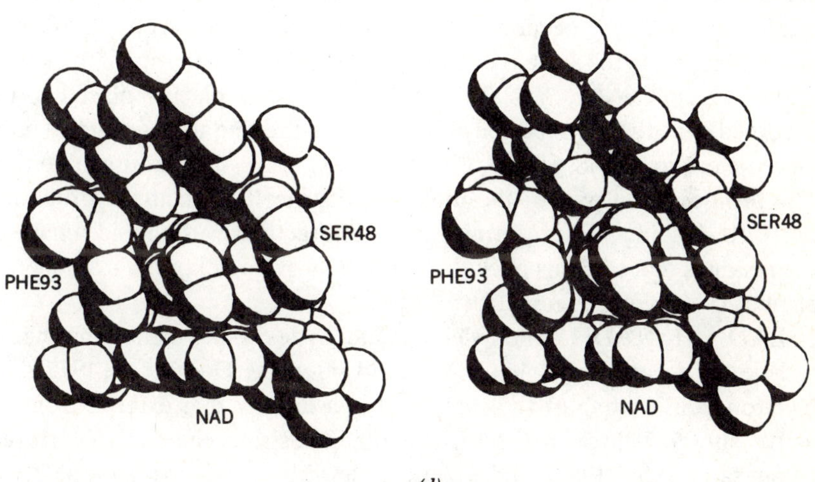

(d)

Figure 17. (*continued*)

105

6.2. Substrate Binding

The reaction catalyzed by alcohol dehydrogenase is the interconversion of alcohol and aldehyde via a ternary complex:

$$\text{enzyme-NAD}^+\text{-alcohol} \rightleftharpoons \text{enyzme-NADH-aldehyde}$$

Substrate binding has been studied crystallographically in three different complexes: (1) trifluoroethanol, which is not converted to aldehyde and which forms a ternary complex on the ethanol side of the reaction (Plapp et al., 1978); (2) dimethylaminocinnamaldehyde (Cedergren-Zeppezauer et al., 1982), which forms a stable ternary complex with a coenzyme analog on the aldehyde side; and (3) an equilibrium complex between bromobenzyl alcohol and bromobenzyl aldehyde (Eklund et al., 1982a). In solution, the equilibrium favors alcohol and the oxidized coenzyme. Of these three complexes, the first two yield information about the ternary alcohol and aldehyde complexes, and the third is similar to an active complex.

The substrates and the substrate analog bind to the active site zinc atom in all three complexes. The position and conformation of bound bromobenzyl alcohol can be very accurately and distinctly deduced from its electron density maps due to the heavy bromine atom and the planarity of the benzyl group (Eklund et al., 1982a). A benzyl alcohol molecule placed in the observed density has its oxygen atom bound to the zinc atom and its C1 atom placed close to Ser 48, His 67, Phe 93, Cys 174, and the nicotinamide ring. The bromophenyl part of the molecule fits nicely in to the hydrophobic substrate-binding pocket, forming many interactions with the side chains of the residues that line this pocket. A stereo diagram of bound bromobenzyl alcohol is shown in Figure 18.

LADH-H_2-NADH-dimethylcinnamaldehyde crystals exhibit the characteristic 468 nm absorption band for bound aldehyde (Dunn et al., 1975). The electron density map of the complex has a density that extends from the active site zinc through the hydrophobic substrate channel (Cedergren-Zeppezauer et al., 1982), and is compatible with a planar aromatic aldehyde bound to the active site zinc atom.

The size of the substrate-binding cleft is partly reconciled to the nature of the substrate by changes in side chain conformations. The side chain of Leu 116, in particular, changes its conformation in the bromobenzyl alcohol complex, as well as in the complex with dimethylaminocinnamaldehyde

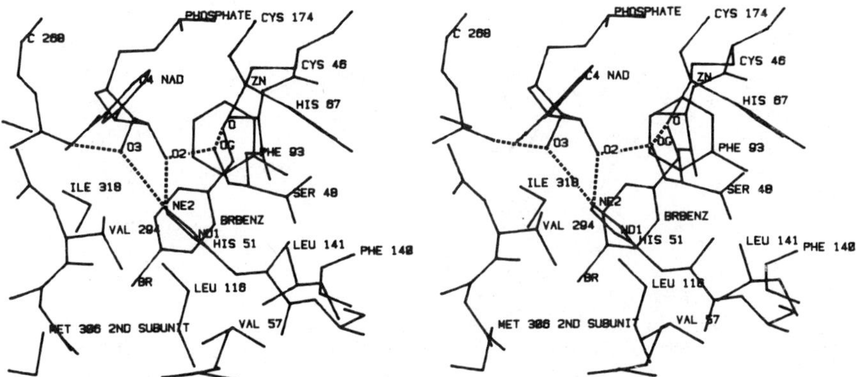

Figure 18. Stereo drawing of bromobenzyl alcohol bound to LADH. The hydrogen bond system in the active site involves bromobenzyl alcohol, nicotinamide ribose, Ser 48, His 51, and Ile 269.

compared to the complex with dimethylsulfoxide. Met 306 moves in the aldehyde complex to avoid unfavorable contacts with the dimethylamino group.

The binding mode of bromobenzyl alcohol and dimethylaminocinna-maldehyde found in the crystallographic study represents the most stable form of the complex. This form differs from the complex that should be present at the hydride transfer step, because the hydrogen to be transferred does not point towards C4 of the nicotinamide ring. Furthermore, the distance between these atoms are longer than expected for an active complex. Model-building experiments show, however, that the position of the bromobenzyl alcohol molecule could be easily rotated around an axis through the oxygen and bromine atoms of the molecule, such that it is in an ideal position for hydride transfer (Eklund et al., 1982a). This position is illustrated in Figure 19.

6.3. Inhibitor Binding

The crystallographic studies of inhibitor binding to LADH have shown that there are two main classes of inhibitors. One type binds in the coenzyme-binding crevice, preferably in the region where the AMP portion of the coenzyme binds. Inhibitors belonging to this class are the dye cibachrone

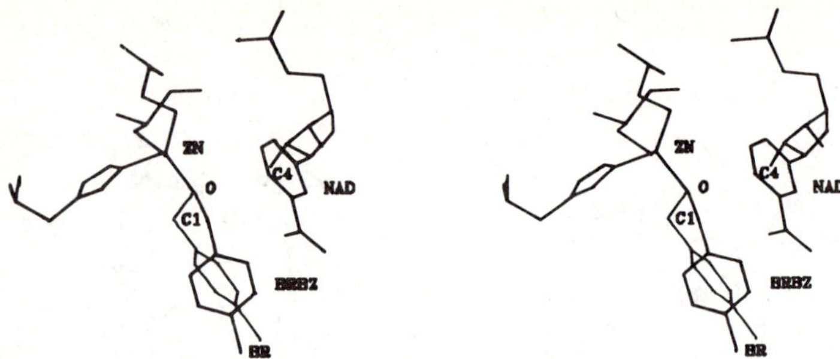

Figure 19. Proposed productive binding mode for bromobenzyl alcohol. The observed mode is shown in thin lines and the productive orientation by thick lines. The position of the oxygen atom of the bromobenzyl alcohol in the active orientation was fixed, and the other atoms were moved essentially by a rotation around the Zn-oxygen bond.

blue (Biellmann et al., 1979), the spectroscopic probe anilino-napthalene-sulfoxide, ANS, (Einarsson et al., 1974; Brändén et al., 1979), and salicylate derivatives that have been shown (Einarsson et al., 1974) to simulate binding of the adenosine moiety of nucleotides.

The second type of inhibitors bind in the substrate-binding pocket and compete with the substrate for ligation to the active site zinc atom. The mode of binding of dimethylsulfoxide (Eklund et al., 1981), pyrazole, and some of its derivatives (Eklund et al., 1982b; Horjales 1985), as well as imidazole and orthophenanthroline (Boiwe and Brändén, 1977), have been studied in detail. They all bind directly to the active site zinc atom, and act as inhibitors by competing with substrate binding.

The most thoroughly studied class of inhibitors are heterocyclic zinc-binding compounds such as pyrazole, which normally are substrate competitive. In the absence of coenzyme, imidazole and pyrazole bind in a similar way (Bowie and Brändén, 1977; Eklund et al., 1982b). Both of these rings bind in the narrow slit that exists between Ser 48 and Phe 93, with one nitrogen coordinated to the active site zinc atom (Fig. 20).

Pyrazole forms a very strong complex with enzyme and oxidized coenzyme (Theorell and Yonetani, 1963), and has been extensively studied both in vivo and in vitro as an inhibitor for alcohol dehydrogenase. Crystallographic studies of this ternary complex show that pyrazole binds in the same slit as in the binary complex, and has one of its nitrogen atoms bound to the active site zinc atom. When the pyrazole molecule is positioned in the

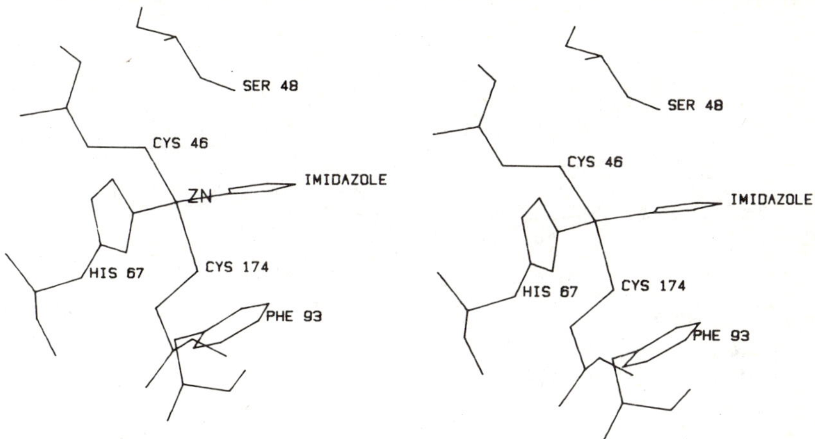

Figure 20. Stereo diagram of bound imidazole to LADH, illustrating the slot between the side chains of Ser 48 and Phe 96 where inhibitors (here imidazole) and substrates bind.

electron density, the second nitrogen atom is 2 Å from the C4 atom of the nicotinamide ring. The free electron pair of this nitrogen is directed towards the C4 atom perpendicular to the plane of the nicotinamide ring. Pyrazole, can therefore, be regarded as a transition state analog with one nitrogen simulating the oxygen of the alcohol group bound to zinc. The other nitrogen simulates the hydride ion that is transferred to C4 during catalysis.

Substituted pyrazole molecules as inhibitors for alcohol dehydrogenase have been studied extensively (Theorell et al., 1969; Dahlbom et al., 1974; Tolf, 1981). It is obvious from the binding of the pyrazole molecule that substitutions in the 3 position interfere with the protein, especially the zinc ligand His 67; substitutions in the 5 position interfere with the coenzyme (Fig. 21). This explains the reduced inhibitory power for 3- and/or 5-substituted pyrazole molecules. In contrast, the 4 position of the pyrazole molecule points into the hydrophobic substrate-binding channel. As a basis for the study of 4-substituted pyrazole molecules, the structure of the 4-iodo-pyrazole complex was determined (Eklund et al., 1982b). The position of the pyrazole ring is nearly the same as in the unsubstituted pyrazole complex. The iodine atom is located in the hydrophobic substrate cleft.

Tolf, Dahlbom, and others have shown that 4-alkyl chains on pyrazole are especially potent inhibitors (Dahlbom et al., 1974; Tolf et al., 1979). The 4-iodo-pyrazole molecule was used as a starting point for model building

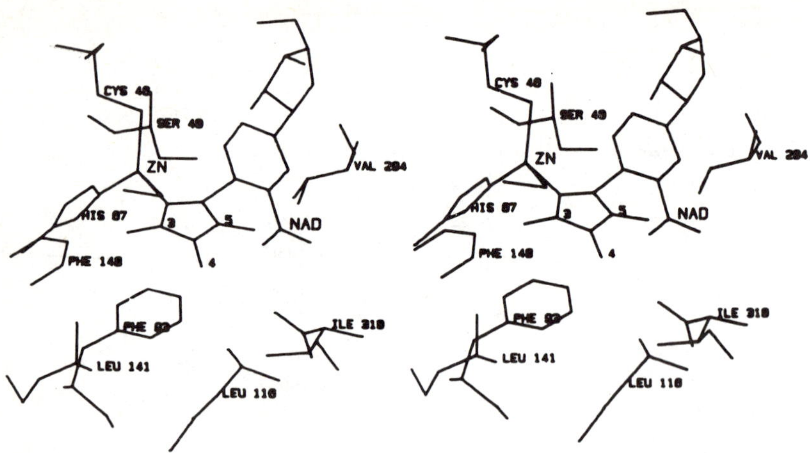

Figure 21. Model of 3,4,5-trimethylpyrazole placed at the active site of the enzyme. The model of the derivative is constructed from a 4-iodopyrazole molecule as bound to the enzyme in the ternary complex.

using computer graphics of 4-substituted alkyl chains (Eklund et al., 1982b). A long alkyl chain was attached in the 4 position in the most extended conformation possible, and only small torsions had to be applied at two places to avoid close contact with the protein.

The large increase in inhibitory power of pyrazole when there are substitutions at the 4 position is a consequence of the hydrophobic substrate cleft (Fig. 22). Hydrophobic substituents in this position increase the number of interactions with the hydrophobic residues of the substrate cleft, and thereby, also increase the stability of the enzyme-NAD$^+$-inhibitor complex. Positively or negatively charged groups, on the other hand, decrease the affinity unless they are remote from the pyrazole ring. In these latter cases, they reach polar residues at the bottom of the substrate-binding site, close to the outer solution. Branched and bulky substituents sterically interfere with protein residues along the narrow hydrophobic channel.

The active site zinc atom is not necessary for the formation of a pyrazole-NAD$^+$-enzyme complex. Crystallographic studies of active site metal-depleted enzyme show nearly identical binding of the 4-iodo-pyrazole (Schneider et al., unpublished results).

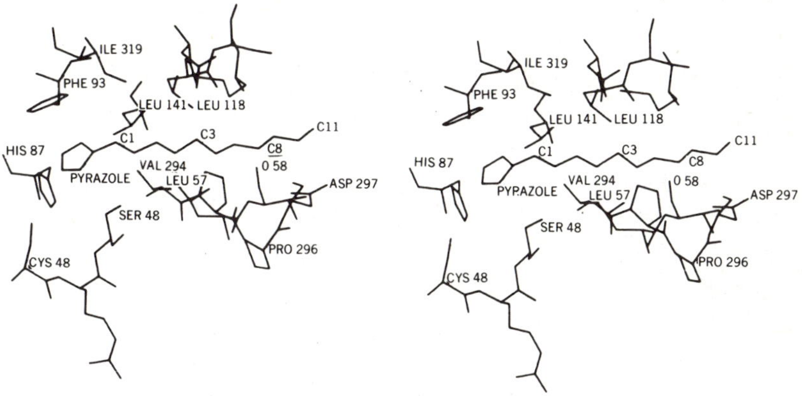

Figure 22. Hydrophobic channel in alcohol dehydrogenase surrounding the 4-alkylpyrazole chain that was positioned by model building. The view is perpendicular to the alkyl chain.

7. ASPECTS OF THE CATALYTIC MECHANISM

The mechanism for oxidation of alcohols has been suggested (Theorell, 1967; Brändén et al., 1975) to contain at least seven different steps (Fig. 23). The proposed mechanism is essentially electrophilic catalysis mediated by the active site zinc atom. The mechanism is valid in the pH region 6.0–10.0. The pH optimum for alcohol oxidation is approximately pH 8.0 (Sund and Theorell, 1963).

7.1. The Role of Water

LADH differs in at least one important aspect from other zinc-containing enzymes for which crystallographic data are known. Water is not a component of the reaction. There is no need for LADH to activate a water molecule as there is in many other zinc enzymes. This is also reflected in the total charge of the zinc-ligand complex.

The protein ligands around zinc, two cysteine sulfur atoms and one histidine nitrogen atom, are preserved in all known zinc-containing alcohol dehydrogenases. This particular ligand arrangement, therefore, must be important either for substrate binding or to provide a proper electronic structure for zinc in electrophilic catalysis. This ligand arrangement is quite

(a) **Overall reaction**

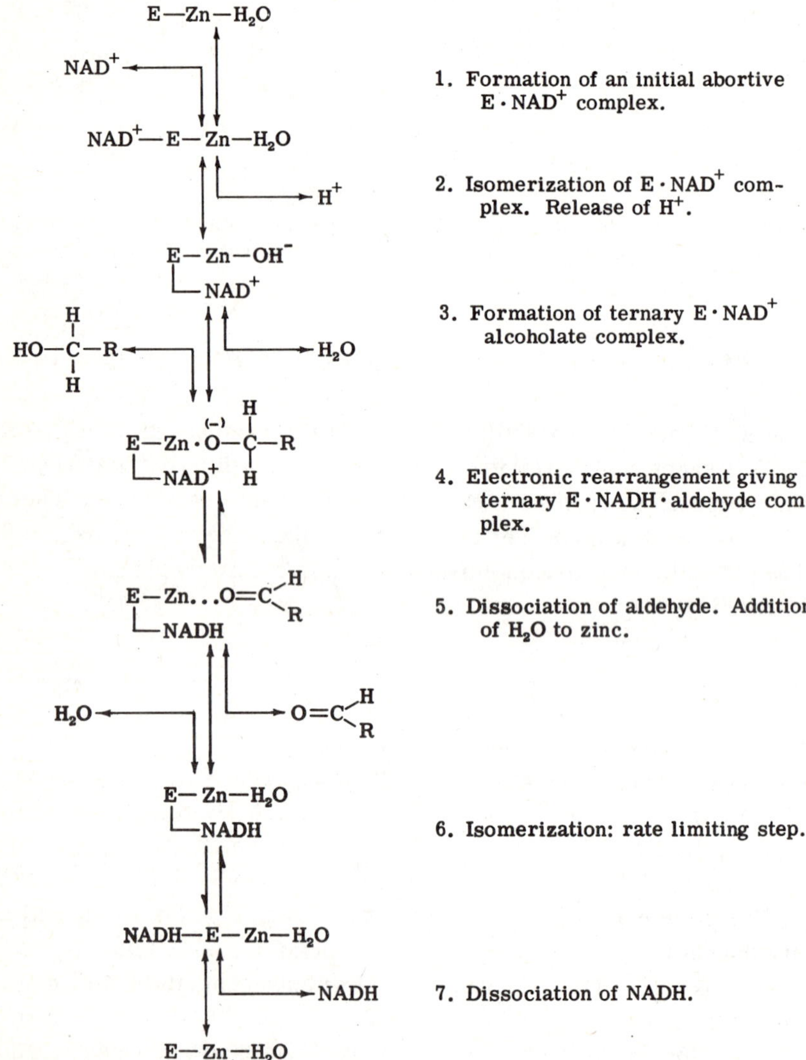

(b) **Reaction steps**

1. Formation of an initial abortive $E \cdot NAD^+$ complex.

2. Isomerization of $E \cdot NAD^+$ complex. Release of H^+.

3. Formation of ternary $E \cdot NAD^+$ alcoholate complex.

4. Electronic rearrangement giving ternary $E \cdot NADH \cdot$ aldehyde complex.

5. Dissociation of aldehyde. Addition of H_2O to zinc.

6. Isomerization: rate limiting step.

7. Dissociation of NADH.

Figure 23. Schematic reaction mechanism for the oxidation of alcohols by LADH.

112

different from that in carboxypeptidase or carbonic anhydrase. The total charge of zinc and its ligands is zero in LADH, whereas it is +2 in carbonic anhydrase (Liljas et al., 1972) and +1 in carboxypeptidase (Rees et al., 1981) and thermolysin (Holmes and Matthews, 1981). This affects the pK_a of a water bound to the zinc atom. The pK_a is 7.0 in carbonic anhydrase, 8.9 in carboxypeptidase, and 11.2 in LADH (Andersson, et al., 1981b). NAD$^+$ binding to LADH is associated with a pK_a shift to 7.6 for a group that has been assumed to be the zinc-bound water hydroxyl (Taniguchi et al., 1967).

Because water is not involved directly in the catalytic reaction of LADH, that is, no hydrolysis or hydration, there is no necessity to suggest a role for a water molecule at the active site of LADH.

Sloan et al. (1975). suggested the presence of a zinc-bound water molecule between zinc and the oxygen atom of the substrate in order to explain calculated distances from NMR measurements. Crystallographic investigations and solution studies of substrates and inhibitors have shown these interpretations to be incorrect. In several cases, such as binding of trifluoroethanol, imidazole, and pyrazole, the same complexes have been studied by both X-ray and NMR methods (Young and Mildvan, 1977; Drysdale and Hollis, 1980). In all of these complexes, crystallographic studies have unambiguously demonstrated direct binding of the ligand to zinc (Fig. 24).

Water was also suggested to bind simultaneously with the substrate to the zinc atom, and to act as a proton acceptor/donor (Dworschack and Plapp, 1977; Makinen and Yim, 1981; Makinen et al., 1983). In the structures that

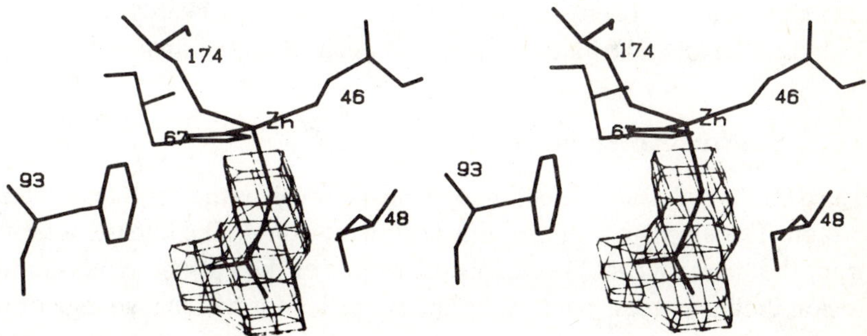

Figure 24. A trifluoroethanol molecule fitted to a difference electron density map at the active site of LADH. The map demonstrates unambiguously the direct binding of trifluoroethanol to the active site zinc atom.

are observed for holoenzyme complexes, zinc is tetracoordinated, and there is no water molecule bound to zinc when the substrate or inhibitor is bound (Eklund et al., 1981, 1982a, 1982b; Cedergren-Zeppezauer et al., 1982).

The free space around the zinc atom is limited by the zinc ligands as well as the side chains of Ser 48 and Phe 93. It is not possible to have two oxygen atoms at a distance of 2 Å from zinc and separated from one another by 2.6 Å. Nor was it possible to build a model of pentacoordinate zinc in the apoenzyme structure with these restrictions while preserving the ligand geometry (Eklund et al,. 1982a). Pentacoordination is even less likely to occur in the ternary complexes, where the nicotinamide ring prevents a ligand from binding over a large part of the remaining free space around the zinc atom. Furthermore, it has been pointed out (Kvassmann et al., 1981) that an extra water molecule would be catalytically disadvantageous because it would tend to increase the pK_a of a zinc-bound alcohol substrate.

It is important to realize that hydride transfer occurs in a completely water-free environment. The inner part of the active site, where the nicotinamide ring as well as the oxygen and C1 atoms of the substrate are situated, is inaccessible to water. The effect of the positive charges of zinc and NAD^+ to polarize the alcohol and, thus, facilitate hydride transfer is enhanced in a water-free environment. Furthermore, if water were present, there would be a high probability for side reactions between water and the hydride ion. An alcohol such as ethanol has an apparent pK_a of 15.7 and, consequently, no tendency to dissociate a proton under normal conditions. The positive charges of the zinc atom and the nicotinamide ring in a water-free environment gives the alcohol a much stronger tendency to dissociate a proton (Kvassman and Pettersson, 1980). The pK_a value associated with ethanol in this situation has been measured to be 6.4 and to be 5.4 and 4.3 for 2-chloroethanol and trifluoroethanol, respectively (Kvassman et al., 1981).

7.2. Proton Abstraction

The overall oxidation of alcohol to aldehyde involves the net release of one proton. The ultimate source of this proton is alcohol. The release of proton from the bound alcohol occurs in the center of the enzyme molecule in a region that is inaccessible to solution by the conformational change of the enzyme. The proton could be picked up by a group at the active site, where it is kept to the end of the reaction and released in the open form during NADH dissociation. Alternatively, the proton could be transferred by groups in the closed form of the enzyme to the surrounding solution. Because there is no proton release associated with NADH dissociation, the last mechanism seems more likely.

A second model for the proton transfer step was originally suggested by Theorell (1967), modified by Shore et al. (1974), and combined with the existing three-dimensional apoenzyme structural data (Eklund et al., 1974; Brändén et al., 1975). In this model, the zinc-bound hydroxyl ion then acts as a base that abstracts a proton from the alcohol. The resulting water molecule is then substituted by the alcoholate ion as a ligand to zinc. In this mechanism, an internal proton transfer occurs from the alcohol to the hydroxyl ion before binding of alcoholate to the zinc atom.

The difference between this proton release scheme and the one first suggested by Kvassman and Pettersson (1980), is that water dissociation precedes alcohol binding in the latter case. The structure of the active site in alcohol dehydrogenase is so narrow that a prior dissociation of water seems more plausible from a structural point of view. Both schemes involve the formation of an alcoholate ion bound to the zinc atom.

7.3. The Role of Zinc

From what can be deduced from our present knowledge, the important role for the active site zinc atom in LADH is to bind the substrate in a suitable position relative to the coenzyme in order for direct hydride transfer to occur. The substrate binds by direct coordination of its oxygen atom to the zinc atom in the fourth corner of a distorted tetrahedral coordination around zinc.

For an alcohol substrate, the creation of an alcoholate ion greatly facilitates hydride transfer. The important role of the zinc atom in alcohol oxidation, therefore is to stabilize the alcoholate ion for the hydride transfer step (Kvassman et al., 1981). Alcoholate ion formation has also been an important component in most mechanisms described for the enzyme (Theorell, 1967; Shore et al., 1974; Brändén et al., 1975; Kvassman and Pettersson, 1980). In the reverse reaction, zinc plays the role of electron attractor, which gives rise to an increased electrophilic character of the aldehyde, consequently facilitating the transfer of a hydride ion to the aldehyde.

A useful tool for further evaluating the role of catalytic zinc atoms in zinc metalloenzymes is to substitute the zinc atom with different metal atoms. In alcohol dehydrogenase, such studies have been complicated by the fact that the enzyme has two different metal sites. Selective replacement of the active site zinc atom has been accomplished by M. Zeppezauer and his coworkers (Maret et al., 1979), and are now being used for a large number of different physicochemical studies.

The crystallographic investigations have helped to clarify important points concerning the properties and function of the catalytic metal ion and its environment in LADH (Schneider et al., 1983a, 1983b, 1985). The assignment of native and exchanged metal ions to their respective binding sites was definitely established. X-ray diffraction studies of active site zinc-depleted enzyme show that small movements of the side chains of Cys 46 and His 67 are the only structural differences between this enzyme and the native protein. The activity of the enzyme can be fully reconstituted to that of the native enzyme by addition of zinc (Maret et al., 1979; Schneider and Zeppezauer, 1983). Crystals lacking the active site zinc atom can be specifically reconstituted with Co, Cd, Cu, etc. (Maret at al., 1979; Andersson et al., 1981; Dietrich et al., 1981).

Investigation of the metal-depleted species also sheds some light on the importance of the protein structure for the coordination geometry required by the catalytic metal ion in its binding site. The three protein ligands, Cys 46, His 67 and Cys 174, originate from rather distant loci along the polypeptide chain. The backbones of these three amino acids are incorporated in well-ordered helices and β sheets. This may explain why removal of this zinc ion does not lead to gross structural changes of tertiary structure. Crystallographic studies of the Co(II)- and Cd(II)-substituted enzyme show that the coordination geometry is preserved. The rigidity of the protein structure, including its metal-binding site, is obviously responsible for the fact that tetrahedral coordination is observed in LADH even with metals that tend to prefer planar coordination geometries, such as Cu(II) and Ni(II) ions (Maret et al., 1979; Dietrich et al., 1981). Thus, LADH represents a macromolecular ligand system that imposes tetrahedral coordination geometries on those metal ions that form stable complexes with the ligand.

Coenzyme binding and the associated conformational transition from the open to the closed form of the enzyme are important aspects of the catalytic mechanism. Zinc does not seem to play any role in these processes because they occur in the same way in enzyme species where the catalytically active zinc atom has been removed (Schneider et al., 1983b).

7.4. Hydride Transfer

No crystallographic study of alcohol dehydrogenase has demonstrated a complex where substrate and nicotinamide are positioned such that hydride ions are transferred back and forth between the two counterparts. However, model building has shown that only slight rearrangement of the observed position of bromobenzyl alcohol in the closed form of the enzyme is needed to position the alcohol so that direct hydride transfer can occur (Eklund et al., 1982a). A schematic diagram of this hypothetical transition state for ethanol oxidation is shown in Figure 25. The model can be built without any change of the protein structure, which indicates that there is no large energy barrier to overcome. This repositioning of the substrate was done without changing the position of the substrate oxygen atom.

This model has been compared to the structure obtained by model building for productive cyclohexanol binding. Atoms C1 and O of these substrates superimpose within 0.3 Å. Atom C2, on the other hand, differs by about 1.3 Å. The plane through O, C1, and C2 is approximately parallel to the plane of the nicotinamide ring (Figure 26).

Although transfer of hydrogen from the alcohol to the nicotinamide is generally described as a hydride transfer, it may occur stepwise with the transfer of a hydrogen atom and an electron or a proton plus two electrons. Such intermediate stages have been discussed generally (Hamilton, 1971) and in connection with alcohol dehydrogenase (Klinman, 1976, 1981; Dworschak and Plapp, 1977; Welsh et al., 1980; Cook et al., 1981).

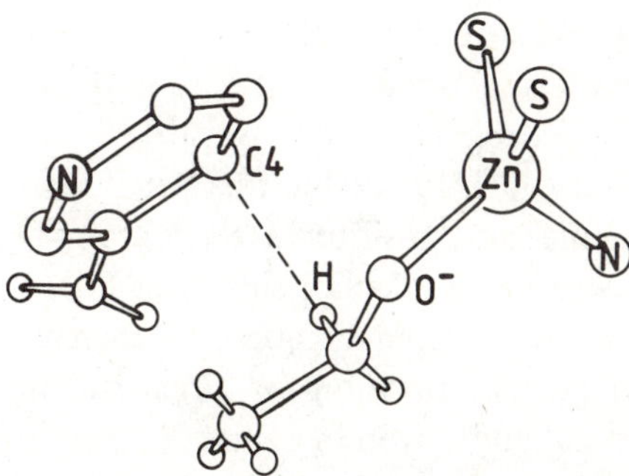

Figure 25. Schematic drawing of the geometry for hydride transfer between ethanol and NAD$^+$

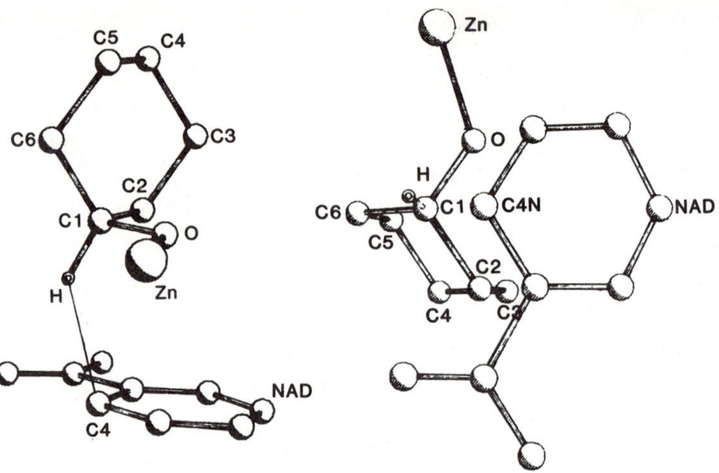

Figure 26. The geometry for hydride transfer for the secondary alcohol cyclohexanol. Two perpendicular views of Zn, substrate and nicotinamide.

7.5.   Stereochemical Aspects of the Catalytic Reaction

In the suggested productive complex of bromobenzyl alcohol (Figure 19), the pro-R hydrogen should point toward C4 of the nicotinamide ring. In this mode of binding, the pro-S hydrogen points toward the benzene ring of Phe 93. It is much more difficult to form a complex with the pro-S hydrogen pointing toward C4, because this would cause unacceptable contacts between the alcohol and the protein. The side chains of Ser 48 and Phe 93 effectively fix the positioning of the carbons in the alcohol. As an illustration of how the structure of the enzyme determines that the pro-R hydrogen is to be trans-ferred, models of ethanol were built into the active site with either the pro-R or pro-S hydrogen pointing toward C4 of the nicotinamide ring (Figure 27) (Eklund et al., 1982a). In this structure, the oxygen and C1 of the ethanol were in the same positions as these atoms in the productive mode of binding for bromobenzyl alcohol shown in Figure 19. Rotation of the methyl group around the oxygen-C1 bond was then used to position the pro-S hydrogen in a favorable position for direct hydrogen transfer to C4 of NAD^+. However, as can be seen in Figure 27, this forces the methyl group of ethanol too close to the benzene ring of Phe 93 at a distance of about 2 Å from several atoms in the ring. In contrast, when the pro-R hydrogen atom is in a suitable position for direct hydride transfer, there are favorable inter-

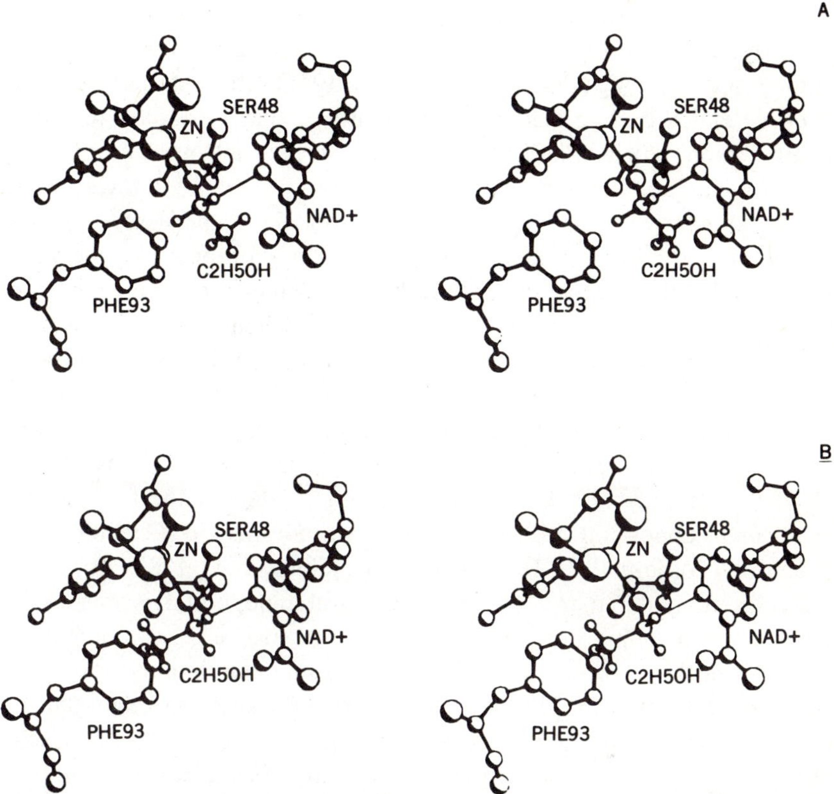

Figure 27. Possible structural basis for stereospecificity of hydrogen transfer. Ethanol was built into the structure with the requirements that the oxygen be ligated to the zinc atom and the bond connecting C1 and the *pro-R* or *pro-S* hydrogen point toward C4 of the nicotinamide ring. (*a*) When the *pro-R* hydrogen points toward the nicotinamide ring, the methyl group of ethanol has favorable interactions with the protein. (*b*) When the *pro-S* hydrogen points toward C4, however, the methyl group of ethanol is too close to the ring of Phe 93.

actions between the methyl group and the nicotinamide ring and Ser-48. In this orientation, the pro-S hydrogen points toward the ring of Phe 93.

Although the stereospecificity of LADH for primary alcohols can be explained in this way, the enzyme is also active on both isomers of secondary alcohols, such as secondary butanol (Dickinson and Dalziel, 1969) and cyclohexanol (Dalziel and Dickinson, 1966). Dutler (1977) has made an extensive and careful kinetic study of all 10 possible methyl-substituted cyclohexanol derivatives. A qualitatively good correlation between these kinetic results and the X-ray structure of the apoenzyme was obtained by crude model building (Dutler and Brändén, 1981). These studies have later been extended and refined using computer graphics coupled to energy minimization (Horjales and Brändén, 1985). A plausible model for productive substrate binding of a secondary alcohol was obtained that gave excellent agreement with the kinetic data. This model was then used to extend Prelog's diamond lattice description (Prelog, 1964) of the active site and to compare the actual active site of LADH with the indirect mapping obtained by kinetic studies (Jones and Beck, 1976). Excellent agreement was obtained demonstrating the power of indirect kinetic mapping of enzyme active sites.

One of the classical aspects of coenzyme binding to alcohol dehydrogenase is the A stereospecificity of the coenzyme. This was first demonstrated for the yeast enzyme (Fisher et al., 1953) and later for the liver enzyme (Levy and Vennesland, 1957). Various reasons for this specificity have been proposed (Popjak, 1970; Benner, 1982; Nambiar et al., 1983; You, 1985, and references cited therein). From a structural standpoint, it is obvious that there is a steric hindrance that selects this stereospecificity. The enzyme positions both the nicotinamide ring and the substrate in a narrow pocket deep inside the protein. There is no space in the pocket to accommodate the nicotinamide ring if it is rotated 180° around the glycosidic bond. Such a change holds the C4 of the nicotinamide ring in the same position, but turns the B side toward the substrate. However, the carboxamide group would then collide with protein residues 178 and 203 and the zinc sphere, especially with the sulfur atoms of Cys 46 and Cys 174. Binding of the coenzyme to LADH with its B-specific side directed toward the active site zinc atom is possible, however, for the open apoenzyme conformation of the enzyme. This has been shown in a nonproductive complex where a zinc-bound imidazole molecule occupies part of the normal nicotinamide-binding site (Cedergren-Zeppezauer, 1983).

7.6. Comparison with Other Dehydrogenases

The chemical reactions catalyzed by NAD^+-dependent dehydrogenases are, in principle, very simple reactions. An alcohol group of a metabolite is oxidized by the removal of a proton from the hydroxyl group and by the transfer of a hydride ion from the reacting carbon to the oxidized form of the coenzyme. Most dehydrogenases catalyze this reaction without involving a zinc atom. In fact, even alcohol dehydrogenase from *Drosophila* (Thatcher, 1980), which is unrelated to the liver and yeast enzymes, requires no zinc atom to catalyze ethanol oxidation efficiently.

One aspect of enzymatic alcohol oxidation is that the enzyme facilitates the removal of the proton from the alcohol molecule. In lactate dehydrogenase (LDH), hydride transfer and proton abstraction from the alcohol group of the substrate occur simultaneously according to the "oil-water-histidine" mechanism proposed by Parker and Holbrook (1977). LDH, as well as other dehydrogenases, uses a histidine residue as a proton abstractor. The reaction must therefore involve a short-lived intermediate stage, because a negatively charged alcoholate substrate is unstable and no system is available to stabilize this intermediate.For alcohol dehydrogenase, the situation is different because the zinc atom at the active site stabilizes the alcoholate ion by forming a metal-alcoholate complex. Thus, the proton abstraction can precede hydride transfer for this enzyme and need not be simultaneous.

7.7. The Ordered Mechanism of the Enzyme

Binding of the coenzyme affects the binding of alcohol and aldehyde drastically, and the binding of these substrates is generally kinetically negligible in the absence of coenzyme. This was the basis for the Theorell and Chance (1951) mechanism proposed originally for horse LADH. This property has later been described as an ordered mechanism, which requires binding of the coenzyme before any binding of substrate (Dalziel, 1975).

The transition from the open structure of the enzyme to the closed form changes properties of the active site, which can be related to the ordered mechanism. First, the positioning of the nicotinamide ring at the active site forms one side of the substrate-binding channel. This hydrophobic barrel-shaped channel, illustrated in Figure 17, is in contrast to the open site in the apoenzyme form. The tighter binding site in the closed form of LADH can accommodate more van der Waals contact interactions with the substrate. Another factor, which could influence the increased binding of substrate to the closed form of the enzyme, is the changed topography of the substrate channel where the positioning of Val 294 into the substrate cleft makes the cleft smaller and allows more van der Waals contacts to the substrate.

The positive charge on the nicotinamide ring is crucial for the enhanced binding of alcohol to the enzyme. The insertion of the positive charge in this largely hydrophobic environment facilitates formation of the negatively charged alcoholate ion. Not only does this compensate for the positive charge at the nicotinamide ring, it also facilitates the binding of alcohol to the active site zinc atom. An alcoholate ion binds much tighter than an uncharged alcohol molecule.

8. MODEL BUILDING OF HOMOLOGOUS ALCOHOL AND POLYOL DEHYDROGENASES

Among mammalian alcohol dehydrogenases, the enzyme from horse liver has been most extensively studied (Brändén et al., 1975). Models of the structure of some homologous alcohol and polyol dehydrogenases have been built (Jörnvall et al., 1978; Brändén et al.,1984; Eklund et al., 1985) from the three-dimensional structure of horse LADH by computer graphics using the program FRODO (Jones, 1978, 1982). Each residue of the horse liver enzyme was replaced with the corresponding residue in the homologous enzymes, with main chain atoms at identical positions and common atoms of the side chains as close as possible to those in the original residue. The main chain of the protein was changed as little as possible. Substitutions were generally modeled by small changes of side chain torsion angles. All changed regions were regularized to conform to standard bond angles and distances. Some of the differences in substrate specificity between the homologous enzymes have been correlated with substrate-docking experiments using the interactive-fitting program TOM (Cambillau et al., 1984). The final three-dimensional models of the subunit structures have good bonded and nonbonded interactions. The internal core structures are highly hydrophobic and contain no internal charged residues that are not present in the horse LADH structure.

8.1. Human LADH

Human LADH consists of a complex isoenzyme system coded, at least, at five gene loci. The amino acid sequences of the two predominant liver isoenzyme subunits, β and γ have recently been determined (Hempel et al., 1984, Bühler et al., 1984) and are 87% and 89% homologous, respectively, with the horse liver amino acid sequence (Table 4). Most substitutions are at the surface of the molecules.

Few substitutions between horse and human ADH isozymes occur at positions at the substrate- or coenzyme-binding sites. One substitution in the β isoenzyme affects the inner part of the substrate pocket: Thr at position 48 instead of Ser in the horse and γ enzymes. The other difference in the substrate pocket is Val 318 for Ile between horse and β1 and Val 141 for Leu between horse and γ1. Both make the substrate pocket wider at some distance from the zinc atom. The extra methyl group of Thr 48 makes the binding site smaller, but not so that it hinders binding of cyclohexanol. The best position obtained for cyclohexanol in the active site of β1 has a few short distances and is, thus, less favorable than for γ1. Furthermore, it differs from that obtained in γ1 in such a way that the hydride to be transferred is pointing in a direction between the nicotinamide ring and Phe 93. Thus, it seems that the lower activity of cyclohexanol for β1 is due to a combination of the smaller binding site and the extra methyl group that limits the movements of cyclohexanol and therefore its possibilities for hydride transfer.

The greater affinity of β1 for small substrates, such as methanol and ethylene glycol as it is expressed in lower K_m values (Wagner et al., 1983), is a consequence of the smaller substrate-binding site due to Thr in position 48.

Residues involved in coenzyme binding are almost completely identical in horse and human ADH. The only exception is residue 48, which is Thr in β1. However, Thr can serve the same function in coenzyme binding as Ser; namely, to form a hydrogen bond to O2′ of the NMN-ribose of NAD.

In the β2 isoenzyme, Arg 47 is substituted by His. A hydrogen bond can be formed from its ND1 atom to OP1A of the pyrophosphates in NAD similar to the Arg-phosphate interactions in the β1 enzyme. The substitution to His shifts the pH optimum of alcohol oxidation and weakens the binding strength between the coenzyme and the enzyme (Jörnvall et al., 1984a).

The subunit interactions in the horse liver structure are dominated by interactions between the large hydrophobic surface of the coenzyme-binding domain and the same area in the second subunit (Eklund et al., 1976). This area is shielded from the surface of the molecule by an antiparallel β strand on each side. This area is largely conserved in the human isoenzymes. Of the internal residues, only residue 303 is changed from a Met in horses to Ile in humans. There are no differences in interaction between the subunits in the two human isoenzymes and only very small differences between the human and the horse enzyme subunits. This is compatible with the occur-

Table 4. Comparison of Horse Liver Alcohol Dehydrogenase with β1 and γ1 Isoenzymes of Human Liver Alcohol Dehydrogenase

	1									10										20										30
Horse	AcS	T	A	G	K	V	I	K	C	K	A	A	V	L	W	E	E	K	K	P	F	S	I	E	E	V	E	V	A	P
β1																V									D					
γ1																L									E					

	31									40										50										60
Horse	P	K	A	H	E	V	R	I	K	M	V	A	T	G	I	C	R	S	D	D	H	V	V	S	G	T	L	V	T	P
β1				Y									V					T	D							N				
γ1				H									A					S	E							N				

	61									70										80										90
Horse	L	P	V	I	A	G	H	E	A	A	G	I	V	E	S	I	G	E	G	V	T	T	V	R	P	G	D	K	V	I
β1					L											V								K						
γ1					L											V								K						

	91									100										110										120
Horse	P	L	F	T	P	Q	C	G	K	C	R	V	C	K	H	P	E	G	N	F	C	L	K	N	D	L	S	M	P	R
β1												V			N				S	Y							G	N		
γ1												I			N				S	Y							G	N		

	121									130										140										150
Horse	G	T	M	Q	D	G	T	S	R	F	T	C	R	G	K	P	I	H	H	F	L	G	T	S	T	F	S	Q	Y	T
β1			L					R					R								L		T							
γ1			L					R					S								V		V							

	151									161										170										180
Horse	V	V	D	E	I	S	V	A	K	I	D	A	A	S	P	L	E	K	V	C	L	I	G	C	G	F	S	T	G	Y
β1						N		A																						
γ1						N		A																						

	181									190										200										210
Horse	G	S	A	V	K	V	A	K	V	T	Q	G	S	T	C	A	V	F	G	L	G	G	V	G	L	S	V	I	M	G
β1					N						P																	A	V	
γ1					K						P																	V	V	

124

```
211               220                 230                 240
C K A A G A A R I I G V D I N K D K F A K A K E V G A T E C
                    A                               L
                    A                               L

241               250                 260                 270
V N P Q D Y K K P I Q E V L T E M S N G G V D F S F E V I G
I                           K     T D
I                           K     T D

271               280                 290                 300
R L D T M V T A L S C C Q E A Y G V S V I V G V P P D S Q N
          M A S   L     H     C   T                 A
          M A S   L     H     C   T                 D

301               310                 320                 330
L S M N P M L L L S G R T W K G A I F G G F K S K D S V P K
    I             T               V Y             E G I
    I             T               I F             E S V

331               340                 350                 360
L V A D F M A K K F A L D P L I T H V L ? F E K I N E G F D
                    S     A       H V
                    S     A       N I

361               370
L L R S G E S I R T I L T F
    H     K         V
    R     K         V
```

Line 1 Horse liver alcohol dehydrogenase (Jörnvall, 1970); Line 2 Human liver alcohol dehydrogenase isoenzyme β1 (Hempel et al., 1984); Line 3 Human liver alcohol dehydrogenase isoenzyme γ1 (Buhler et al., 1984).

rence of heterodimeric isoenzymes between the human isoenzymes, and corresponding mixture in vitro between the human and horse forms.

8.2.   Maize Alcohol Dehydrogenase

The amino acid sequence of maize alcohol dehydrogenase (MADH) has been deduced by Dennis et al. (1984, 1985) from the gene sequence of the maize ADH1 and ADH2 genes. At the amino acid level, they have been shown to be highly homologous, approximately 50%, to the horse LADH and less homologous, approximately 20%, to yeast alcohol dehydrogenase. The MADH model could be constructed without changing the general structure of the protein. For the main chain, only minor differences exist, and all side chain substitutions are compatible with the same fold of the molecule. No new charged side chain is found in internal regions.

The residues binding the zinc atom at the active site are conserved, although approximately half of the residues involved in substrate and coenzyme binding differ between maize and horse. Differences in substrate specificity between the two species can be related primarily to the difference between Ser and Thr at position 48. There are also substitutions of residues further away from the zinc atom in the hydrophobic substrate pocket. Ser 48 and His 51 have been thought to play a role in the proton release from the active center. In the MADH, the corresponding residues are Thr and Tyr, which cannot serve the same function.

The only substitution in the substrate- and coenzyme-binding sites between the maize ADH1 and ADH2 enzymes are located in the adenine-binding area. Arg at position 228 in ADH1 may be correlated to the higher activity for this enzyme.

8.3.   Yeast Alcohol Dehydrogenase

Yeast and mammalian alcohol dehydrogenases are distantly homologous with many insertions/deletions but have only 25% of all residues conserved (Jörnvall, 1977). A model of the yeast subunit has been built (Jörnvall et al., 1978) that conserves the ligands to both zinc atoms as well as to a number of structurally important glycine residues (Table 5). A large deletion is found for residues 119−139 which are present in the horse liver enzyme but are absent in the yeast enzyme. This segment forms a surface loop in the horse liver subunit far from the active site.

Table 5. Liver, Maize, Yeast and *B. steareothermophilus* Alcohol Dehydrogenase and Sorbitol Dehydrogenase Amino Acid Sequences

	1									10										20										30
	S	T	A	G	K	V	I	K	C	K	A	A	V	L	W	E	E	K	K	P	F	S	I	E	E	V	E	V	A	P
M	A	T	A	G	K	V	I	K	C	K	A	A	V	A	W	E	A	G	K	P	L	S	I	E	E	V	E	V	A	P
M	T	I	P	D	K	Q	L	A	A	V	F	H	T	H	G	G	P	E	N	V	K	F	E	E	V	P	V	A	E	
				S	I	P	E	T	Q	K	G	V	I	F	Y	E	S	H	G	K	L	E	Y	K	D	I	P	V	P	K
							M	K	A	A	V	V	E	Q	F	K	K	P	L	Q	V	K	E	V	E	K	P	K		
		A	K	P	A	A	E	N	L	S	L	V	V	H	G	P	G	D	L	R	L	E	N	Y	P	I	P	E		

31									40										50													60
P	K	A	H	E	V	R	I	K	M	V	A	T	G	I	C	R	S	D	D	H	V	V	S	G	T				L	V	T	P
P	Q	A	M	E	V	R	V	K	I	L	F	T	S	L	C	H	T	D	V	Y	F	W	E	A	K			G	Q	T	P	V
P	G	Q	D	E	V	L	V	N	I	K	Y	T	G	V	C	H	T	D	L	H	A	L	Q	G	D		W	P	L	P	A	K
P	K	A	N	E	L	L	I	N	V	K	Y	S	G	V	C	H	T	D	L	H	A	W	H	G	D		W	P	L	P	T	K
I	S	Y	G	E	V	L	V	R	I	K	A	C	G	V	C	H	T	B	L	H	A	A	H	G	B		(W)	P	V	P	(K, K)	
P	G	P	N	E	V	L	L	K	M	H	S	V	G	I	C	G	S	D	V	H	Y	W	Q	G	R	I	G	D	F	V	V	K

61									70										80										90
L	P	V	I	A	G	H	E	A	A	G	I	V	E	S	I	G	E	G	V	T	T	V	R	P	G	D	K	V	I
F	P	R	I	F	G	H	E	A	G	G	I	I	E	S	V	G	E	G	V	T	D	V	A	P	G	D	H	V	L
M	P	L	I	G	G	H	E	G	A	G	V	V	V	K	V	G	A	G	V	T	R	L	K	I	G	D	R	V	G
L	P	L	V	G	G	H	E	G	A	G	V	V	V		G	M	E	N	V	K	G	W	K	I	G	D	Y	A	G
K	P	M	V	L	G	H	E	A	S	G	T	V	V	K	V	G	S	L	V	R	H	L	Q	P	G	D	R	V	A

91										100										110										120
P	L	F	T	P		Q	C	G	K	C	R	V	C	K	H	P	E	G	N	F	C	L	K	N	D	L	S	M	P	R
P	V	F	T	G		E	C	K	E	C	A	H	C	K	S	A	E	S	N	M	C	D	L	L	R	I	N	T	D	R
V	K	W	M	N	S	S	C	G	N	C	E	Y	C	M	K	A	E	E	T	I	C	P	H	I	Q	L	S	G		
I	K	W	L	N	G	S	C	M	A	C	E	Y	C	E	L	G	N	E	S	N	C	P	H	A	D	L	S	G		
I	Q	P	G	A		P	R	Q	T	D	E	F	C	K	I	G	R	Y	N	L	S	P	T	I	F	F	C	A		

(continued)

Table 5. (continued)

```
121                     130                     140                     150
 G T M     Q D G T S R F T C R G K P I H H F L G T S T F S Q Y T
 G V M I A D G K S R F S I N G K P I Y H F V G T S T F S E Y T
                                     Y T V D G T F Q H Y C
                                     Y T H D G S F Q Q Y A
                                     T P P D D G N L C R F Y

151                     160                     170                     180
 V V D E I S V A K I D A A S P L E K V C L I G C G F S T G Y
 V M H V G C V A K I N P Q A P L D K V C V L S C G I S T G L
 I A N A T H A T I I P E S V P L E V A A P I M C A G I T C Y
 T A D A V Q A A H I P Q G T D L A E V A P V L C A G I T V Y
 K H N A N F C K Y L P D N V T F E E G A L I   E P L S V G I

181                     190                     200                     210
 G S A V K V A K V T Q G S T C A V F G L   G G V G L S V I M G
 G A S I N V A K P P K G S T V A V F G L   G A V G L A A A G G
 R A L K E   S K V G P G E W I C I P G A G G G L G H L A V Q Y
 K A L K S   A N L M A G H W V A I S G A A G G L G S L A V Q Y
 H A C R R   A G V T L G N K V L V C G A   G P I G L V N L L A

211                     220                     230                     240
 C K A A G A A R I I G V D I N K D K F A K A K E V G A T E C
 A R I A G A S R I I G V D L N P S R F E E A R K F G C T E F
 A K A M A M   R V V A I D T G D D K A E L V K S F G A E V F
 A K A M G Y   R V L G I D G G E G K E E L F R S I G G E V F
 A K A M G A A Q V V V T D L S A S R L S K A K E V G A D F I

241                     250                     260                     270
 V N P Q D Y K K P I Q E V L T E M S N G G   V D F S F E V I G
 V N P K D H N K P V Q E V L A E M T N G G   V D R S V E C T G
```

```
L D F K K E A D M I E A V K A     C T N G G     A H G T L V L S T
I D F T K E K D I V G A V L K     A T N G G     A H G V I N V S V
L E I S N E S P E E I A K K V E G L L G S K P E V T I E C T G

271               280                 290                 300
R L D T M V T A L S C C Q E A Y G V S V I V G V P P D S Q N
N I N A M I Q A F E C V H D G W G V A V L V G V P H K D A E
S P K S Y E Q A A G F A R     P G S T M V T V S M P A G A K L
S E A A I E A S T R Y V R     A N G T T V L V G M P A G A K C
V E T S I Q A G I Y A T H     S G G T L V L V G L G S E M T

301               310                 320                 330
L S M N P M L L L S G R T W K G A I F G G F K S K D S V P K
F K T H P M N F L N E R T L K G T F F G N Y K P R T D L P N
G A D I F W L T V K         M L K I C G S H V G N R I D S I
C S D V F N Q V V K         S I S I V G S Y V G N R A D T R
S     V P     L V H A A     T R E V D I K G V F R Y C N T W P M

331               340                 350                 360
L V A D F M A K K F A L D P L I T H V L P F E K I N E G F D
V V E L Y M K K E L E V E K F I T H S V P F A E I N K A F D
E A L E Y V S R         G L V K P Y Y K V Q P F S T L P D V Y R
E A L D F F A R         G L I K S P I K V V G L S T L P E I Y E
A I S M L A S K S V N V K P L V T H R F P L E K A L E A F E

361               370
L L R S G E     S I R T I L T F
L M A K G E     G I R C I I R M E N
L M H E N K I A G R I V L D L S K
K M E K G Q V V G R Y V V D T S K
T S K K G L     G L K V M I K C D P S D Q N P
```

Line 1 Horse liver ADH (Jörnvall, 1970); Line 2 Maize ADH (Dennis et al., 1984); Line 3 Yeast *Schizosaccharomyces pombe* (Russel and Hall, 1983); Line 4 Yeast (Jörnvall, 1977); Line 5 *Bacillus stearothermophilus* ADH (Only until 60) (Bridgen et al., 1973); Line 6 Sorbitol dehydrogenase (Jeffery et al., 1984).

Residues that participate in coenzyme binding, especially the AMP group, are not conserved with the exception of Gly 199 and Asp 223, which are essential for proper recognition and binding of the adenosine ribose. Catalytically important residues, such as the zinc ligands Ser 48 and His 51, are identical or conservatively substituted. In the substrate-binding pocket, exchanges have occurred to more bulky residues: Leu 57 to Trp, Phe 93 to Trp, and Phe 140 to Tyr. These differences narrow the space available for substrates and can be correlated to the stricter substrate specificity of yeast alcohol dehydrogenase (YADH) for ethanol and similar small substrates.

YADH is tetrameric, whereas LADH is dimeric. Residues in the hydrophobic dimer interaction area of LADH are completely different in YADH, which strongly indicates that this subunit interaction is not preserved in the tetramer. From the subunit model of YADH, it was not possible to deduce the subunit arrangement in the tetramer.

8.4. Sorbitol Dehydrogenase

The amino acid sequence of sorbitol dehydrogenase (SDH) was aligned to horse, yeast, plant, and bacterial alcohol dehydrogenase (Eklund et al., 1985). There are two stretches of the amino acid sequence of SDH that clearly differ from that of horse LADH. The first is at residues 300−315, which constitute the main subunit-subunit interaction area of LADH. Here, the five-residue shorter sequence of the tetrameric SDH is analogous to what is found for tetrameric YADH, and is probably related to the quaternary structure.

The other region of clear differences between SDH and LADH is the sequence corresponding to residues 110−186. By introducing a similar gap in the region 120−140 as in tetrameric YADH, an alignment was obtained that minimizes deviations from the subunit conformation of LADH. A drastic consequence of this alignment is that the third zinc ligand, at position 174, is not conserved (Eklund et al, 1985). Instead of a cysteine residue as in all alcohol dehydrogenases (Table 5), the alignment gives a glutamic acid residue in SDH. This enzyme contains one zinc atom per subunit (Jeffrey et al., 1981; Jörnvall et al., 1984b), but nothing is experimentally known about the zinc coordination (Jeffrey et al., 1984). A glutamic acid can act as a zinc ligand in enzymes, and has been found to do so for carboxypeptidase (Rees et al., 1981) and thermolysin (Holmes and Matthews, 1981). Both a cysteine and a glutamic acid residue should be negatively charged as zinc ligands, so the formal net charge of zero for the zinc ligand complex would be the same in either case.

The substrate specificity of SDH and LADH differ in several respects. LADH acts on primary alcohols, but SDH oxidizes the second alcohol group of sorbitol. This requires that the active site region close to the zinc atom be different in the two enzymes. The principal difference, with respect to the size of the active site cleft between the model of SDH and the observed structure of LADH, is caused by having a proline residue in SDH corresponding to Phe 93 in LADH. Proline at this position could allow the substrate to enter deeper into the protein and could provide space for direct binding of the second alcohol oxygen to the zinc atom. In LADH, it is not possible to position the sorbitol molecule with the second alcohol group bound to zinc. The primary alcohol group would then come too close to the side chain of Phe 93. The replacement of this Phe by a Pro also affects the accessibility of residue 319, such that the lysine side chain at this position in the SDH model can reach the substrate cleft. When sorbitol is thus positioned for hydride transfer, the primary alcohol would be in a proper position to form hydrogen bonds to Glu 174 (in the numbering system of LADH; see Table 5) as well as to Lys 319 (Fig. 28). A glutamic acid residue as a zinc ligand should, in this structure, be favorable compared with a

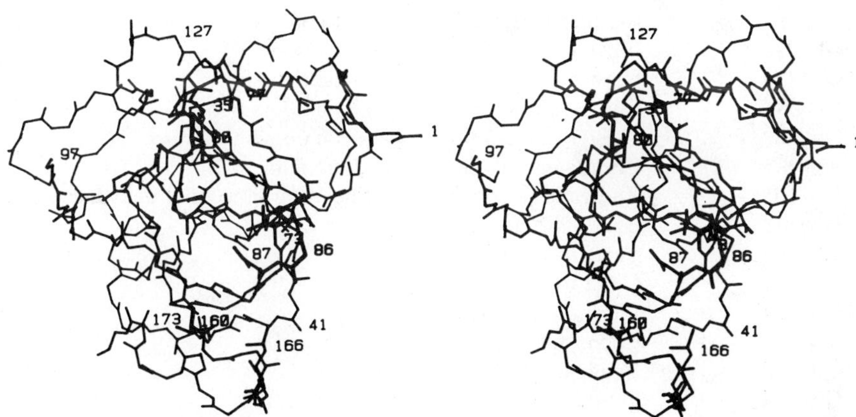

Figure 28. One area at the backside of the catalytic domain contains many conserved residues. This stereo drawing shows residues that are conserved in all species and in most species with thick lines. The main chain is drawn with thin lines.

cysteine residue for oxidation of the secondary alcohol group at C2 rather than the primary alcohol group at C1.

8.5. Structural Implications of Comparisons of Homologous Dehydrogenases

The recent publication of DNA sequences of YADH and MADH (Bennetzen and Hall, 1982; Russel and Hall, 1983; Russel et al., 1983; Dennis et al., 1984, 1985) and the amino acid sequence of the distantly related enzyme, SDH (Jörnvall et al., 1984b), has presented a unique possibility to analyze which residues are of importance for a catalytically functional enzyme. These enzymes are so distantly related that very few residues are common to all enzymes. Residues conserved in all species of alcohol dehydrogenase and in SDH are listed in Table 6.

Table 6. Residues Conserved in All Mammalian Maize and Yeast Alcohol Dehydrogenase and Liver Sorbitol Dehydrogenase

Glu 35	Cluster on backside of catalytic domain
Cys 46	Zn-ligand
Asp 49	Hydrogen bond to the side chain of Zn-ligand His 67
Pro 62	Only cis-proline in the LADH structure
Gly 66	Internal, little space
His 67	Zn-ligand
Glu 68	Ion pair and hydrogen bond to 369
Gly 71	Cluster on backside of catalytic domain
Gly 77	Cluster on backside of catalytic domain
Val 80	Cluster on backside of catalytic domain
Gly 86	Cluster on backside of catalytic domain
Asp 87	Cluster on backside of catalytic domain
Cys 103	Zn-ligand to noncatalytic zinc atom in alcohol dehydrogenase
Gly 192	Bend
Gly 199	Close to coenzyme
Gly 201	Close to coenzyme
Gly 204	Internal, little space
Asp 223	Coenzyme binding
Gly 236	Bend
Gly 260	Bend
Val 290	Internal
Val 292	Internal
Gly 320	Bend

Important residues at the active site are strictly conserved. These include the zinc ligands, Cys 46 and His 67, and other residues in the environment of the zinc, such as Asp 49, and Glu 68. Most of the remaining conserved residues are glycines. Glycine has the possibility to adopt unusual conformational angles not allowed for residues having a side chain, and consequently are important for the folding of proteins. They are commonly found in loop regions between secondary structural elements.

One of the important features for the folding of a domain are residues that favor a switch from one secondary structural element to the next. Analysis of the variation among the residues that comprise the domains of the alcohol dehydrogenase family show that a certain pattern is present for such loop regions.

The coenzyme-binding domain, is built by alternating β strands and α helices. Glycine 192 between αA and βA is conserved in all species. The loop between βA and αB contains two glycine residues or one glycine plus one proline at positions 201–202. Region 211–217, which makes a long bend between αB and βB, often has glycine in position 215, but this is not an absolute requirement because 215 is Ala in *Schizosacchoromyces pombe* ADH. Furthermore, this region is very rich in alanines. Glycine 236 between αC and βC is conserved in all species, and residues 237 is either a glycine or an alanine. Glycine 260 between αD and βD is conserved, and residue 261 is a glycine in many species. There is also a glycine residue either at position 286 or 287 between αE and βE in all species. The catalytic domain requires a glycine at positions 77, 86 and 320. A glycine residue is also found at either position 365 or 368 between the last helix and last strand and at position 55.

The results indicate that there is a strong tendency for a glycine residue in the connection from a helix to a strand, at position 192, 236–237, 260–261, 286–287, and 365–368, for the α-β structure of the coenzyme-binding domain. A similar pattern has been observed for the first domain of phosphoglycerate kinase (Blake and Rice, 1981). Glycines are also quite common in the loop regions at the other side of the domain between the β strands and α helices, but the requirements are less strict. One exception is between βA and αB, where glycines are required at position 199 and 201 for the binding of coenzyme, and at position 204, for the tight packing of αB to βA.

The pattern is somewhat different for the antiparallel strands of the catalytic domain, where prolines are more frequent and seem to be a requirement in the regions around 30 and 62. Residue 62, which is the only *cis* proline in the LADH enzyme is conserved. Prolines are, in general, unevenly distributed and are clustered in certain regions in the catalytic domain, at positions 6, 18–20, 27–33, 58–62, 85, 91–96. The only resemblance to this in the coenzyme-binding domain is at 295–296 in the loop that changes conformation in LADH.

Residues involved in packing of internal regions of both domains are less strictly conserved than Gly and Pro in the loop regions between secondary structural elements. In most cases, compensating changes in size can keep the proper packing interactions in these hydrophobic cores. Three valines are conserved at positions 80, 290, and 292. At a few positions, glycine is conserved due to lack of space of side chains, such as at positions 69, 213, 215, and 237. Ile, Leu, Met, and Phe are never completely conserved in any of these cores.

In addition to the active site, there is a second area with many conserved residues. Surprisingly, this area is at the backside of the catalytic domain far from the active site. This area includes residues 35, 77, 80, 86, and 87, which are shown in Figure 29. The area is not involved in subunit interactions, at least not in the dimeric enzymes. It is located at one edge of the orthogonal β-sheet structure (Chotia and Janin, 1982), which constitutes the main part of the catalytic domain. However, the conserved residues do not contribute side chains important for the internal packing of the structure, which would

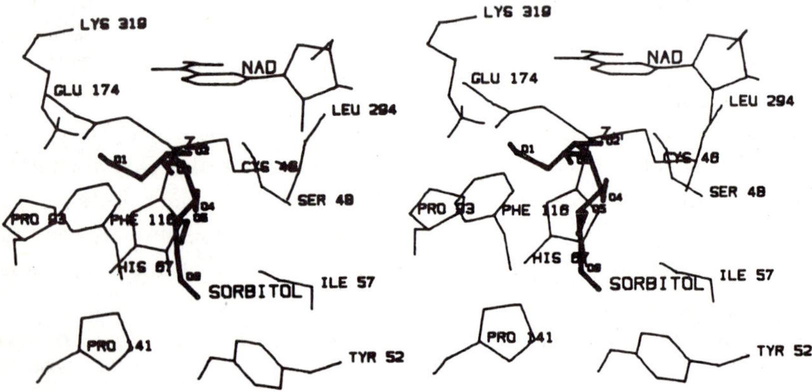

Figure 29. Stereo diagram of a sorbitol molecule positioned in the active site of the sorbitol dehydrogenase model.

have been a plausible explanation for the conservation. One reason for the conservation of these residues might be that they are of vital importance for the folding of the catalytic domain. They might, however, also have some unknown function, such as forming an interaction surface for a larger cytoplasmic complex.

Residues that bind the coenzyme are not conserved to any great extent. One exception is Asp 223, which is important for ribose binding by forming two hydrogen bonds to ribose oxygens (Eklund et al., 1984). The glycine residues at the end of βA of the coenzyme-binding domain are conserved, which is common for dehydrogenase (Rossmann et al., 1975). The residues around the adenine rings form an unspecific slit for this moiety and are not conserved. These residues, Ile 224 and Ile 269 in the LADH molecule, vary considerably among the species. Arg 47 forms a hydrogen bond to the phosphate in the AMP half of the coenzyme. The corresponding residue is histidine in the yeast enzymes and in the human β-1 isoenzyme, which reduces the strength of coenzyme binding. A mutant with arginine in this position in YADH increases coenzyme binding to that of the liver enzyme (Wills and Jörnvall, 1979). A similar substitution is observed within the human isoenzymes (Jörnvall et al., 1984a). The corresponding residue in SDH is a glycine, and no positively charged residues are present in this part of the catalytic domain.

9. FUTURE PROSPECTS

From the combined knowledge of the X-ray structure, kinetic studies, and solution chemistry of LADH, there are a number of specific questions that can be asked regarding the precise role of certain residues in the catalytic mechanism. To date, the scientific tools to answer these questions have been lacking. However, with the possibility of using site-directed mutagenesis (Winter and Fersht, 1984), these questions can now be answered provided the cDNA of the mammalian gene can be cloned and expressed. We will discuss some of these questions as examples of the type of information this powerful new technique might give.

Ser 48 is located at a strategic position in the substrate-binding pocket close to C1 of the substrate. In all known species, this residue is either identical or conservatively substituted to Thr. It thus seems very likely that this residue is important for catalytic efficiency. However, it has not yet been possible to assign a specific role for this residue in the catalytic mechanism. By kinetic studies on mutants, where this residue is substituted with Ala or Cys, it should be possible to deduce at which step this group is important and what its function is.

The side chains of Asp 49 and Glu 68 are close to the zinc atom on the opposite side compared to the substrate (Fig. 15). These residues are also invariant in all species. Asp 49 is completely buried and forms two hydrogen bonds, one to the zinc ligand, His 67, and a second to an internal water molecule. What is the function, if any, of these residues? Are these negative charges necessary to enhance the electrophilic character of zinc?

Which residues are important for the conformational transition between the apo and holo forms? The loop region 294−298 of the coenzyme-binding domain undergoes a local conformational change that must be coupled to the domain rotation. This region has the sequence Val-Pro-Pro-Asp-Ser. Are the proline residues required for this conformational change, and if so, why? From this specific case, one might learn more general aspects of local conformational changes within a protein framework.

Because of its broad substrate specificity, LADH is currently being used as a tool for stereospecific reduction of aldehydes in organic synthesis (Jones and Beck, 1976). However, some of the side chain in the substrate-binding pocket, such as Leu 57, Phe 93, Leu 116, and Leu 141, place severe restrictions on which aldehydes can bind in the pocket. By model building with computer graphics, it is easy to design substitutions of these residues that would allow a much wider range of substrates for stereospecific reductions and increase the usefulness of this enzyme in organic synthetic reactions.

ACKNOWLEDGMENT

This work was supported by a grant from the Swedish Natural Science Research Council. We are grateful to Mrs. A. Rogelius for assistance in the preparation of the manuscript, and to U. Uhlin for help with the drawings.

The abbreviations used are: ADH, alcohol dehydrogenase; LADH, liver alcohol dehydrogenase; YADH, yeast alcohol dehydrogenase; MADH, maize alcohol dehydrogenase; SDH, sorbitol dehydrogenase; H_2NADH, tetrahydro NAD; ANS, 8-anilino 10-naphthalene-sulphonate BrPhCH$_2$OH, BRBZ,BRBENZ-p-bromobenzyl alcohol; and DMSO, dimethyl sulfoxide.

REFERENCES

Adams, M. J., Ford, G. C., Koekoek, R., Lentz, P. J., Jr., McPherson, A., Jr., Rossmann, M. G., Smiley, I. E., Schevitz, R. W., and Wonacott, A. J. (1970). *Nature* **227**, 1098−1103.

Andersson, I., Maret, W., Zeppezauer, M., Brown, R. D. III, and Koenig, S. H. (1981a). *Biochemistry* **20**, 3424−3432.

Andersson, P., Kvassmann, J., Oldén, B., and Pettersson, G. (1981b). *Eur. J. Biochem.* **118**, 119−123.

Benner, S. A. (1982). *Experientia* **38**, 633−637.

Bennet, W. S., Jr., and Steitz, T. A. (1978). *Proc. Natl. Acad. Sci. USA* **75**, 4848−4852.

Bennetzen, J. L. and Hall, B. D. (1982). *J. Biol Chem.* **257**, 3018−3025.

Biellmann, J. -F., Samama, J. -P., Brändén, C. -I., and Eklund, H. (1979). *Eur. J. Biochem.* **102** 107−110.

Birktoft, J. J., Fernley, R. T., Bradshaw, R. A., and Banaszak, L. J. (1982). In *Molecular Structure and Biological Activity*, J. F. Griffin, and W. L. Duax, Eds. Elsevier, Amsterdam, pp. 37−55.

Blake, C. F. (1983). *Nature* **306**, 535−537.

Blake, C. F., and Rice, D. W. (1981). *Phil. Trans. R. Soc. Lond. A.* **293** 23−104.

Boiwe, T. and Brändén, C. -I. (1977). *Eur. J. Biochem.* **77**, 173−179.

Brändén, C. -I. (1980). *Quart. Rev. Biophys.* **13**, 317−338.

Brändén, C. -I., and Eklund, H. (1978). In *Molecular Interactions and Activity in Proteins, Ciba Foundation Symposium 60* (new series), R. Porter and D. W. Fitzsimons, eds. Excerpta Medica, Amsterdam, pp 63−80.

Brändén, C. -I., and Eklund H. (1980). In *Dehydrogenase Requiring Nicotinamide Coenzymes*, J. Jeffrey, Ed. Birkhäuser Verlag, Basel pp. 41−84.

Brändén, C. -I., Eklund H., Nordström, B., Boiwe, T., Söderlund, G., Zeppezauer, E., Ohlsson, I., and Åkeson, Å. (1973). *Proc. Natl. Acad. Sci. USA* **70**, 2439−2442.

Bränden, C. -I., Jörnvall, H., Eklund H., and Furugren, B. (1975). In *The Enzymes*, 3rd. ed., Vol XI. Academic Press, New York, San Francisco, London, pp. 104−190.

Bränden, C. -I., Eklund, H., Zeppezauer, E., Samama, J. -P., and Plapp, B. V. (1978). In *Proceedings of FEBS 12th meeting*, Vol 52, Dresden, pp. 15−23.

Brändén, C. -I., Eklund, H., Cambillau, C., and Pryor, A. J. (1984). *EMBO J.* **3**, 1307−1310.

Bridgen, J., Kolb, E., and Harris, J. I. (1973). *FEBS Lett.* **33** 1−3.

Buehner, M., Ford, G. C., Moras, D., Olsen, K. W., and Rossmann, M. G. (1973). *Proc. Natl. Acad. Sci. USA* **70**, 3025−3054.

Bühler, R., Hempel, J., Kaiser, R., DeZalenski, C., von Wartburg, J. -P., and Jörnvall, H. (1984). *Eur. J. Biochem.* **145**, 447−453.

Cambillau, C., Horjales, E., and Jones, T. A. (1984). *J. Mol. Graphics* **2** 53−54.

Cedergren-Zeppezauer, E. (1983). *Biochemistry* **22**, 5761−5772.

Cedergren-Zeppezauer, E., Samama, J. -P., and Eklund, H. (1982). *Biochemistry* **21**, 4895−4908.

Cedergren-Zeppezauer, E. S., Andersson, I., Ottonello, S., and Bignetti, E. (1985). *Biochemistry* **24**, 4000−4010.

Chothia, C., and Janin, J. (1982). *Biochemistry* **21**, 3955−3965.

Cook, P. F., Oppenheimer, N. J., and Cleland, W. W. (1981). *Biochemistry* **20**, 1817−1825.

Czeisler, L. and Hollis, D. (1973). *Biochemistry* **9**, 1683−1689.

Dahlbom, R., Tolf., B. R., Åkeson, Å., Lundquist, G. and Theorell, H. (1974). *Biochem. Biophys. Res. Comm.* **57**, 549−553.

Dalziel, K. (1975). In *The Enzymes*, P. D. Boyer, Ed., 3rd. ed., Vol XI. Academic Press, New York, San Francisco, London, pp. 1−60.

Dalziel, K., and Dickinson, F. M. (1966). *Biochem. J.* **100**, 491−500.

Deisenhofer, J., and Steigemann, W. (1975). *Acta Cryst.* **B31**, 238−250

Dennis, E. S., Gerlach, W. L., Pryor, A. J., Bennetzen, J. L., Inglis, A., Llewellyn, D., Sachs, M. M., and Peacock, W. J. (1984). *Nucl. Acid Res.* **12**, 3983−3999.

Dennis, E. S., Sachs, M. M., Gerlach, W. L., Finnegan, E. J., and Peacock, W. J. (1985). *Nucl. Acid. Res.*, **13**, 727−743.

Dickinson, F. M., and Dalziel, K. (1967). *Biochem J.* **104**, 165−172.

Dietrich, H., Maret, W., Koslowski, H., and Zeppezauer, M. (1981). *J. Inorg. Biochem.* **14**, 297−311.

Drum, D. E., Li, T. -K., and Vallee, B. L. (1969). *Biochemistry* **8**, 3783−3797.

Drysdale, B. -E., and Hollis, D. P. (1980). *Arch. Biochem. Biophys.* **205**, 267−279.

Dunn, M. F., Biellmann, J. -F., and Branlant, G. (1975). *Biochemistry* **14**, 3176−3182.

Dutler, H. (1977). *Biochem. Soc. Trans.* **15**, 617−620.

Dworschack, R. T., and Plapp, B. V. (1977). *Biochemistry* **16**, 2716−2725.

Einarsson, R., Eklund, H., Zeppezauer, E., Boiwe, T., and Brändén, C. -I. (1974). *Eur. J. Biochem.* **49**, 41−47.

Eklund, H. and Brändén, C. -I. (1979). *J. Biol. Chem.* **254**, 3458−3461.

Eklund, H. and Brändén, C. -I. (1983). In *Zinc Enzymes*, T. Spiro, Ed. J. Wiley & Sons, New York, pp. 124−152.

Eklund, H., Nordström, B., Zeppezauer, E., Söderlund, G., Ohlsson, I., Boiwe, T., and Brändén, C. -I. (1974). *FEBS Lett.* **44**, 200−204.

Eklund, H., Nordström, B., Zeppezauer, E., Söderlund, G., Ohlsson, I., Boiwe, T., Söderberg, B. -O., Tapia, O., Brändén, C. -I., and Åkeson, Å. (1976). *J. Mol. Biol.* **102**, 27−59.

Eklund, H., Samama, J. -P., Wallén, L., Brändén, C.-I. Åkeson, Å., and Jones, T. A. (1981). *J. Mol. Biol.* **46**, 561−587.

Eklund, H., Plapp, B., Samama, J. -P., and Brändén, C. -I. (1982a). *J. Biol. Chem.* **257**, 14349−14358.

Eklund, H., Samama, J. -P., and Wallén, L. (1982b). *Biochemistry* **21**, 4858−4866.

Eklund, H., Samama, J. -P., and Jones, T. A. (1984b). *Biochemistry* **23**, 5982−5996.

Eklund, H., Horjales, E., Jörnvall, H., Brändén, C. -I., and Jeffery, J. (1985). *Biochemistry* **24**, 8005−8012.

Eventoff, W., Rossmann, M. G., Taylor, S. S., Torff, H. -J., Meyer, H., Keil, W., and Klitz, H. -H. (1977). *Proc. Natl. Acad. Sci. USA* **74**, 2677−2681.

Fersht, A. (1977). *Enzyme Structure and Mechanism*. Freeman, Reading and San Francisco.

Filman, D. J., Bolin, J. T., Matthews, D. A., and Kraut, J. (1982). *J. Biol. Chem.* **257**, 13663−13672.

Fisher, H. F., Conn, E. E., Vennesland, B., and Westheimer, F. H. (1953). *J. Biol. Chem.* **202**, 687−697.

Gilbert, W. (1978). *Nature* **271**, 501.

Hamilton, G. A. (1971). In *Progress in Bioinorganic Chemistry*, E. T. Kaiser, F. J. Kzedy, Eds., Vol. 1. Wiley-Interscience, New York, pp. 83.

Hempel, J., Bühler, R., Kaiser, R., Holmquist, B., DeZalenski, C., von Wartburg, J. -P., Vallee, B. L., and Jörnvall, H. (1984). *Eur. J. Biochem.* **145**, 437−445.

Hendrickson, W. A., and Konnert, J. H. (1980). In *Computing in Crystallography*, R., Diamond, S. Ramaseshan, and K. Venkatesan, Eds. The Indian Academy of Science, Bangalore, pp. 13.1−13.26.

Hensel, R., Mayr, U. and Yang, C. -Y. (1983). *Eur. J. Biochem.* **134**, 503−511.

Hol, W. G. J., van Duijnen, P. T., and Berendsen, H. J. C. (1978). *Nature (London)* **273**, 443−446.

Holmes, M. A., and Matthews, B. W. (1981). *Biochemistry* **20**. 6912−6920.

Horjales, E., and Brändén, C. -I. (1985). *J. Biol. Chem.* **260**, 15445−15451.

Horjales, E., (1985). Thesis at Swedish University of Agricultural Sciences.

Huber, R., Deisenhofer, J., Colman, P. M., Matsushima, M., and Palm, W. (1976). *Nature (London)* **264**, 415−420.

Jack, A., and Levitt, M. (1978). *Acta Cryst.* **A34**, 931−935.

Jeffery, J. (1982). In *Stereochemistry*, Ch. Tamm, Ed. Elsevier, Amsterdam, pp. 113−160.

Jeffery, J., Cummins, L., Carlquist, M., and Jörnvall, H. (1981). *Eur. J. Biochem.* **120**, 229−234.

Jeffery, J., Chesters, J., Mills, C., Sadler, P. J., and Jörnvall, H. (1984). *EMBO J.* **3**, 357−360.

Jones, J. B., and Beck, J. F. (1976). In *Applications of Biochemical Systems in Organic Chemistry*, J. B. Jones, C. J. Sih, and D. Perlman, Eds., Part 1. John Wiley & Sons, New York, pp. 107−401.

Jones, T. A. (1978). *J. Appl. Crystallogr.* **11**, 268–272.

Jones, T. A. (1982). In *Computational Crystallography*, D. Sayre, Ed. Oxford University Press, Oxford, pp. 303–317.

Jörnvall, H. (1970). *Eur. J. Biochem.* **16**, 25–40.

Jörnvall, H. (1977). *Eur. J. Biochem.* **72**, 443–452.

Jörnvall, H., Eklund, H., and Brändén, C. -I. (1978). *J. Biol. Chem.* **253**, 8414–8419.

Jörnvall, H., Hempel, J., Vallee, B. L., Bosron, W. F., and Li, T. -K. (1984a). *Proc. Natl. Acad. Sci. USA* **81**, 3024–3028.

Jörnvall, H., von Bahr-Lindström, H. and Jeffrey, J. (1984b). *Eur. J. Biochem.* **140**, 17–23.

Klinmann, J. P. (1976). *Biochemistry* **15**, 2018–2026.

Klinmann, J. P. (1981). *CRC Crit. Rev. Biochem.* **10**, 39–78.

Kvassman, J. and Pettersson, G. (1980). *Eur. J. Biochem.* **87**, 417–427.

Kvassman, J., Larsson, A., and Pettersson, G. (1981). *Eur. J. Biochem.* **114**, 555–563.

Lesk, A., and Chothia, C. (1984). *J. Mol. Biol.* **174**, 175–191.

Leslie, A. G. W., and Wonacott, A. J. (1984). *J. Mol. Biol.* **178**, 743–772.

Levitt, M., and Chothia, C. (1976). *Nature* **261**, 552–558.

Levy, H. R., and Vennesland, B. (1957). *J. Biol. Chem.* **228**, 85–96.

Liljas, A., Kannan, K. K., Bergsten, P. -C., Waara, I., Friedborg, K., Strandberg, B., Carlborn, U., Jarup, L., Lövgren, S., and Petef, M. (1972). *Nature New Biol.* **235**, 131–137.

Makinen, M. W., and Yim, M. B. (1981). *Proc. Natl. Acad. Sci. USA* **78**, 6221–6225.

Makinen, M. W., Maret, W., and Yim, M. B. (1983). *Proc. Natl. Acad. Sci. USA* **80**, 2584–2588.

Maret, W., Andersson, I., Dietrich, H., Schneider-Bernlöhr, H., Einarsson, R., and Zeppezauer, M. (1979). *Eur. J. Biochem.* **98**, 501–512.

Matthews, B. W., Weaver, L. H., and Kester, W. R. (1974). *J. Biol. Chem.* **249**, 8030.

Monaco, H. L., Crawford, J. L., and Lipscomb, W. N. (1978). *Proc. Natl. Acad. Sci. USA* **75**, 5276–5280.

Nambiar, K., Stauffer, D. M., Kolodziej, and Benner, S. A. (1983) *J. Am. Chem. Soc.* **105**, 5886–5890.

Nordström, B., and Brändén, C. -I. (1975). *Structure and Conformation of Nucleic Acids and Protein-Nucleic Acid Interactions*, M. Sundaralingam and S. T. Rao, Eds. University Park Press, Baltimore, pp. 387–395.

Ohlsson, I., Nordström, B., and Brändén, C. -I. (1974). *J. Mol. Biol.* **89**, 339–354.

Parker, D. M., and Holbrook, J. J. (1977). In *Pyridine Dependent Dehydrogenases*, H. Sund, Ed. Walter De Gruyter, Berlin, pp. 485–495.

Plapp, B. V. (1970). *J. Biol. Chem.* **245**, 1727−1735.

Plapp, B. V. (1982). In *Perspectives on Evolution*, R. Milkman, Ed. Sinauer, Sunderland, MA, pp. 129−147.

Plapp, B. V., Eklund, H., and Brändén, C. -I. (1978). *J. Mol. Biol.* **122**, 23−32.

Plapp, B. V., Eklund, H., Jones, T. A., and Brändén, C. -I. (1983). *J. Biol. Chem.* **258**, 5537−5547.

Popjak, G. (1970). In *The Enzymes*, P.D. Boyer, Ed., 3rd ed., Vol. 2. Academic Press, New York, 115−215.

Prelog, V. (1964). *Pure Appl. Chem.* **9**, 119−130.

Rao, S. T., and Rossmann, M. G. (1973). *J. Mol. Biol.* **76**, 241−256.

Rees, D. C., Lewis, M., Honzatko, R. B., Lipscomb, W. N., and Hardman, K. D. (1981). *Proc. Natl. Acad. Sci. USA* **78**, 3408−3412.

Remington, S., Wiegand, G., and Huber, R. (1982), *J. Mol. Biol.* **158**, 111−152.

Rossmann, M. G., Moras, D., and Olsen, K. W. (1974). *Nature* **250**, 194−199.

Rossmann, M. G., Liljas, A., Brändén, C. -I., and Banaszak, L. J. (1975). In *The Enzymes*, P. D. Boyer, Ed. 3rd ed., Vol. 11. Academic Press, New York, pp. 61−102.

Russel, D. W., Smith, M., Williamson, V. M., and Young, E. T. (1983). *J. Biol. Chem.* **258**, 2674−2682.

Russel, P. R., and Hall, B. D. (1983). *J. Biol.Chem.* **258**, 143−149.

Samama, J. -P., and Eklund, H. (1985). In *Pyridine Nucleotide*, Dolphin et al., Eds. John Wiley & Sons, New York, (in press).

Samama, J. -P., Zeppezauer, E., Biellmann, J. -F., and Brändén, C. −I. (1977). *Eur. J. Biochem.* **81**, 403−409.

Samama, J. -P., Wrixon, A. -D., and Biellmann, J. -F. (1981). *Eur. J. Biochem.* **118**, 479−486.

Schmid, F., Hinz, H. -J., and Jaenicke, R. (1978). *FEBS Lett.* **87**, 80−82.

Schneider, G., and Zeppezauer, M. (1983). *J. Inorg. Biochem.* **18**, 59−69.

Schneider, G. Eklund, H., Cedergren-Zeppezauer, E. and Zeppezauer, M. (1983a). *Proc. Natl. Acad. Sci. USA* **80**, 5289−5293.

Schneider, G., Eklund, H., Cedergren-Zeppezauer, E., and Zeppezauer, M. (1983b). *EMBO J*, **2**, 685−689.

Schneider, G., Cedergren-Zeppezauer, E., Knight, S., Eklund, H., and Zeppezauer, M. (1985). *Biochemistry* **24**, 7503−7510.

Schulz, G. E., Schirmer, R. H., and Pai, E. F. (1982). *J. Mol. Biol.* **160**, 287−308.

Shore, J. D., Gutfreund, H., Brooks, R. L., Santiago, D., and Santiago, P. (1974). *Biochemistry* **13**, 4185−4191.

Sloan, D. L., Young, J. M., and Mildvan, A. S. (1975). *Biochemistry* **14**, 1998−2008.

Subramanian, S., Ross, J. B. A., Ross, P. D., and Brand, L. (1981). *Biochemistry* **20**, 4068−4093.

Sund, H., and Theorell, H. (1963). *The Enzymes*, 2nd ed., Vol. 7, Boyer, P. D., ed. Academic Press, New York and London, pp. 25−83.

Sussman, J. L., Holbrock, S. R., Church, G. M., and Kim, S. H. (1977). *Acta Cryst.* **A33**, 800−804.

Taniguchi, S., Theorell, H., and Åkesson, Å. (1967). *Acta Chem Scand.* **21**, 1903−1920.

Thatcher, D. R. (1980). *Biochem. J.* **187**, 875−886.

Thatcher, D. R., and Retzios, A. (1980). *Protides Biol. Fluids* **28**, 157−160.

Theorell, H. (1967). *The Harvey Lectures, Series 61.* Academic Press, New York, pp. 17−41.

Theorell, H., and Chance, B. (1951). *Acta Chem. Scand.* **5**, 1127−1144.

Theorell, H., and Yonetani, T. (1963). *Biochem. Z.* **338**, 537−553.

Theorell, H., Yonetani, T., and Sjöberg, B. (1969). *Acta Chem. Scand.* **23**, 255−260.

Tolf, B. -R. (1981). Thesis at Faculty of Pharmacy, Acta Unversitatis Upsaliensis.

Tolf, B. -R., Piechaczek, J., Dahlbom, R., Theorell, H., Åkeson, Å., and Lundquist, G. (1979). *Acta Chem. Scand.* **B33**, 483−487.

Wagner, F. W., Burger, A. R., and Vallee, B. L. (1983). *Biochemistry* **22**, 1857−1863.

Welsh, K. M., Creighton, D. J., and Klinman, J. P. (1980). *Biochemistry* **18**, 2005.

Wills, C., and Jörnvall, H. (1979). *Nature* **279**, 734−736.

Winter, G., and Fersht, A. R. (1984). *Trends in Biotechnology* **2**, 115−119.

You, K. -S. (1985). *CRC Cri. Rev.* **17**, 313−451.

Young, J. M., and Mildvan, A. S. (1977). In *Alcohol and Aldehyde Metabolizing Systems*, R. G. Thurman, J. R. Williamson, H. R. Drott, and B. Chance, Eds., Vol. 2. Academic Press, New York, pp. 109−117.

Zoltobroci, M., Kim, J. C., and Plapp, B. V. (1974). *Biochemistry* **13**, 899−903.

3

Pyruvate Kinase

HILARY MUIRHEAD
Department of Biochemistry
University of Bristol
Bristol, England

CONTENTS

1. **INTRODUCTION**
    1.1. Biochemical and Functional Properties of Pyruvate
          Kinase
    1.2. Catalytic Activity
    1.3. Possible Catalytic Residues
    1.4. NMR Measurements

2. **STRUCTURE OF PYRUVATE KINASE**
    2.1. Amino Acid Sequence
    2.2. Crystal Structure of Type M Pyruvate Kinase
    2.3. Enzyme-bound Cations
    2.4. Substrate Binding
    2.5. Sequence Homologies

3. **COMPARISON OF PYRUVATE KINASE WITH OTHER
    STRUCTURES**
4. **PROPOSED CATALYTIC ACTIVITY OF MUSCLE PYRUVATE
    KINASE**
REFERENCES

1. INTRODUCTION

1.1. Biochemical and Functional Properties of Pyruvate Kinase

Pyruvate kinase (PK) (ec 2.7.1.40) is a key regulatory enzyme found in all cells and tissues during glycolysis. The properties of the enzyme have been reviewed by Kayne (1973). The overall structure of PK is the same in bacteria, plants and animals; it is a tetramer composed of identical 50−60 kilodalton (kd) subunits. The regulatory properties of PK are particularly important in tissues carrying out glugoneogenesis, where inhibition of the PK-catalyzed reaction is necessary in order to avoid a futile cycle via the pyruvate carboxylase- and phosphoenolpyruvate carboxykinase-mediated steps. In tissues such as liver, where both processes may occur simultaneously, the metabolic regulation of this enzyme is subject to fine control by allosteric effectors. PK catalyzes the second of the two ATP-forming reactions of the glycolytic pathway; namely, the conversion of phosphoenolpyruvate into pyruvate by the addition of a proton and the loss of a phosphoryl group that is transferred to ADP. The product, pyruvate, is the first nonphosphorylated intermediate in the pathway and occupies a central role in metabolism because it takes part in several different biochemical reactions. The major fate of pyruvate is oxidation to acetyl coenzyme A for use in the tricarboxylic acid cycle. However, under anaerobic conditions, pyruvate can undergo reversible reduction to lactate, while with aerobic conditions, it can be converted back to carbohydrate, which is used for the formation of oxaloacetate or malate or for the transamination to alanine. The reaction catalyzed by pyruvate kinase is essentially irreversible in favor of pyruvate and ATP formation, and requires both bivalent and monovalent cations.

$$\text{Phosphoenolpyruvate} + \text{ADP} + \text{H}^+ \underset{\text{Mg}^{2+},\ \text{K}^+}{\rightleftharpoons} \text{Pyruvate} + \text{ATP}$$

$$(K_{eq} \text{ is } 6.45 \times 10^3 \text{ at pH } 7.4)$$

In most cells, the reactants are far removed from the equilibrium position and the reaction is one of the control points of glycolysis. Phosphoenolpyruvate has a high free energy of hydrolysis because the initial reaction product, enolpyruvate, isomerizes almost completely to the more stable keto form. This isomerization involves a large free energy change, such that the overall standard free energy of hydrolysis of phosphoenolpyruvate is greater than that of ATP. The failure of PK to reach equilibrium means that the intermediates of glycolysis are maintained at a reasonable concentration even when glycolysis is occurring at a slow rate. This allows the cells to respond rapidly to any increased requirements for glycolysis, and to provide substrate for synthetic reactions requiring glycolytic intermediates. The enzyme has a low nucleotide specificity, and the 5′-diphosphates of guanosine, inosine, uridine, and cytidine can all serve as phosphate acceptors.

At least four different isoenzymes have been described in mammals. Type M is found in muscle or brain, type L is a major component in liver and a minor component in kidney, type A (alternatively referred to as K or M2) is found in kidney, adipose tissue, and lung, and type R is found in red blood cells (Ibsen, 1977). Genetic studies suggest that two structural genes exist: one coding for the M and A types and one coding for the L and R types (Peters, et al., 1981; Moore and Bulfield, 1981). The enzyme found in yeast has properties rather similar to those of type A (Hunsley and Suelter, 1969). Types A and L are allosterically regulated, and the protein activity is modulated by many positive and negative effectors showing sigmoidal kinetics with respect to the substrate phosphoenolpyruvate. The more active conformation is favored by phosphoenolpyruvate, fructose-1,6-bisphosphate, and low pH. The less active conformation is favored by ATP, alkaline pH, alanine, and other gluconeogenic amino acids. The addition of micromolar quantities of fructose biophosphate enhances the affinity for phosphoenolpyruvate. In contrast, the type M enzyme has been considered not to possess allosteric properties, although conditions have been described under which the type M enzyme displays kinetic properties typical of a regulatory enzyme (Phillips and Ainsworth, 1977). High concentrations of phenylalanine at high pH inhibit the type M enzyme allosterically (Kayne and Price, 1972). The control of gluconeogenesis has been reviewed by Hers and Hue (1983). In order to prevent a futile loop during gluconeogenesis, the mammalian liver enzyme is regulated by a cyclic-AMP-dependent protein kinase. Phosphorylation of a specific serine residue results in the incorporation of two to four phosphates per tetramer, and leads to changes in the kinetic properties of the enzyme. The apparent affinity for phosphoenolpyruvate and fructose bisphosphate decreases, whereas the affinity for the allosteric effectors ATP and alanine is increased while V_{max} is unaffected. The changes are reversed by dephosphorylation. Ligands protect the type L enzyme from phosphorylation and subsequent inactivation (Ekman et al., 1976; Bergstrom et al., 1976). Treatment of perfused rat livers or hepatocytes with glucagon causes stimulation of gluconeogenesis and phosphorylation of PK. The inactive proportion is higher during starvation and diabetes than under control conditions. The amount of PK in rat liver is increased up to fivefold by a high carbohydrate diet and decreased threefold during starvation. This seems to be related to the amount of functional mRNA coding for the enzyme.

Most studies of PK have used the type M enzyme from rabbit muscle or the allosteric enzyme from yeast or liver. The amino acid sequences of the enzymes from yeast, chicken muscle, and cat muscle and the partial sequence of the liver enzyme are known (Burke et al., 1983; Lonberg and

Gilbert, 1983; Lorimer et al., 1985; Hoar et al., 1985). It seems unlikely that there are large structural differences between the different types of enzyme. The experimental results suggest that all forms of PK have essentially the same structure, which can take up one of several closely related conformations. Presumably, the relative stability of the different conformations varies between the different types of enzyme, so that the proportion in each conformational state varies. In vivo, the type M enzyme will usually be in the fully active conformation, while the exact conformation of the other iso-enzymes will depend upon the concentration of substrates and the various allosteric activators and inhibitors. Depending upon the precise nature of the intersubunit contacts, the tetrameric quatenary structure may well vary.

1.2. Catalytic Activity

In general, kinases transfer the γ-phosphoryl group of ATP to some nucleophile acceptor molecule. Other purine and pyrimidine nucleoside triphosphates can replace ATP as the phosphoryl donor. All kinases require a bivalent metal cation, usually magnesium (Mg), for catalysis. The Mg^{2+} is used to chelate with the ATP^{4-} substrate, and this Mg-ATP chelate is the actual substrate taking part in the reaction. Mildvan (1970) has pointed out that ternary complexes between an enzyme, E, metal ion, M, and substrate, S, may be enzyme bridge complexes (M-E-S), substrate bridge complexes

$$(\text{E-S-M}), \text{ or metal bridge complexes } (\text{E-M-S or } E\underset{\diagdown}{\overset{\diagup}{}}\genfrac{}{}{0pt}{}{M}{|}{S})$$

Type I kinases do not form binary metal complexes, but have ternary metal-substrate-enzyme complexes. PK differs from the type I kinases in that it forms a binary complex with the bivalent cation Mg^{2+} and ternary complexes of the type enzyme-metal-substrate. The Mg, which can form six coordinate bonds to produce octahedral complexes, forms a bridge between enzyme and substrate. In addition to the enzyme-bound bivalent cation, the nucleotide-bound bivalent cation is required. A subgroup of kinases, which includes PK, requires, in addition to the bivalent cation, a monovalent cation (K^+ or NH_4^+) that may be necessary for activity or may lead to enhanced activity. In the case of PK, there is an absolute requirement for a monovalent cation, which is known to move on the binding of phosphoenolpyruvate, binding close to the bivalent cation at the active site (Kayne, 1973). Metal ions such as those associated with PK may function in several ways in enzymatic reactions. The bivalent cation can function as a proton with a positive charge >1, that is, the metal acting as a superacid in electrophilic catalysis. Other roles may involve inducing electronic distortion in the substrate by coordination with the metal, which may act as an electron sink alternately removing and returning electrons from the substrates in order to stabilize the transition state, or which may neutralize the net charge on anions such as ADP and ATP, reducing electrostatic repulsions and consequently increasing the rate of attack of a nucleophile on the electrophilic phosphorus. Another possible role is the stabilization of leaving groups. The general role proposed for monovalent cations is to induce a conformational change in the enzyme yielding the active form.

A phosphoryl transfer reaction ($R'OPO_3^{2-} + ROH \rightarrow ROPO_3^{2-} + R'OH$) can take place using one of two general types of mechanisms. In a dissociative mechanism (SN1), the leaving group is first eliminated to give an unstable metaphosphate intermediate followed by the rapid addition of a nucleophile. In an associative mechanism (SN2), the nucleophile adds first to give a pentacovalent intermediate followed by the elimination of the leaving group. Associative mechanisms may be subdivided into in-line and adjacent mechanisms, depending on which side the attacking nucleophile enters. The in-line associative mechanism leads to inversion of configuration at the phosphorus atom, while the adjacent associative mechanism leads to retention of configuration (Lowe et al., 1981). For PK, in common with other kinases, experimental data suggest an in-line associative mechanism with a nucleophilic attack on the phosphorus to generate a trigonal bipyrimidal intermediate and inversion of configuration (Mildvan et al., 1976, Hassett et al., 1982). Isotope studies show that proton addition occurs stereospecifically at the 2-*si* face of phosphoenolpyruvate, which may be accessible to solvent (Rose, 1970). In the reverse reaction, the enolization of pyruvate may be decoupled from phosphorylation (Rose, 1960). Tritium exchange experiments show that in the presence of inorganic phosphate and pyruvate, PK will catalyze the enolization of pyruvate. The enolate form of pyruvate does not exist in significant quantities free in solution, but is generated from pyruvate at the active site by general base catalysis. Thus, a base on the enzyme is necessary. In the normal reaction, this nucleophile then attacks the γ-P_i of ATP. These results suggest that enolpyruvate is an intermediate in the reaction. Thus, the reaction catalyzed by PK may be described as the interaction of PK and phosphoenolpyruvate with phosphoryl transfer from the oxygen of the enolate anion of pyruvate to ADP, producing ATP (Rose, 1981).

Kinetic studies reveal an equilibrium random-order mechanism for the type M enzyme, in which the binding of one substrate affects the binding of the others. Pyruvate can form dead-end complexes with other substrates apart from phosphoenolpyruvate, emphasizing the requirement for the correct substrates to cooperate in efficient binding. ATP competitively inhibits the binding of both ADP and phosphoenolpyruvate, suggesting overlap of the ATP and phosphoenolpyruvate-binding sites (Ainsworth and Macfarlane, 1973; Dann and Britton, 1978; Reynard et al., 1961). This indicates that direct phosphoryl transfer occurs between the substrates, with the phosphoryl groups being coordinated to the bivalent cation. There is also evidence that P_i can bind at either the phosphoenolpyruvate or ATP sites.

In addition to the normal physiologically important reaction and the enolization of pyruvate, PK is able to catalyze a number of other reactions, including phosphoryl transfer from ATP to an acceptor such as the fluoride ion or hydroxylamine (with bicarbonate as a cofactor), decarboxylation of oxaloacetate (Creighton and Rose, 1976), and detritiation of pyruvate (Robinson and Rose, 1972).

1. Enolization of pyruvate

$$H_3C - \underset{\displaystyle \overset{\displaystyle O}{\|}}{C} - COO^- \ \rightleftharpoons \ H_2C = \underset{\displaystyle \overset{\displaystyle O^-}{|}}{C} - COO^- + H^+$$

2. Oxaloacetate decarboxylase

$$^-OOC - \underset{\displaystyle \overset{\displaystyle O}{\|}}{C} - CH_2 - COO^- + H^+ \ \rightleftharpoons \ ^-OOC - \underset{\displaystyle \overset{\displaystyle O}{\|}}{C} - CH_3 + CO_2$$

3. Fluorokinase

$$F^- + HCO_3^- + ATP^{4-} \ \rightleftharpoons \ F - \underset{\displaystyle \underset{\displaystyle O^-}{|}}{\overset{\displaystyle \overset{\displaystyle O}{\|}}{P}} - O^- + HCO_3^- + ADP^{3-}$$

4. Hydroxylamine kinase activity

$$NH_2OH + HCO_3^- + ATP^{4-} \rightleftharpoons H_2N - O - \overset{\displaystyle O}{\underset{\displaystyle O^-}{\overset{\displaystyle \|}{P}}} - O^- + HCO_3^- + ADP^{3-}$$

In reactions 3 and 4, HCO_3^- is required in catalytic amounts, presumably to bind at the sites normally occupied by the pyruvate carboxylate ion, and induces a protein conformational change allowing F^- or NH_2OH close approach to the γ-P_i of ATP. The cationic cofactors are required in all the reactions. Rose (1970) studied the enolization reaction using tritiated pyruvate. This work suggested that P_i or P_i analogs are required as well as the physiological cofactors. A dianion is necessary at the γ-P_i-binding site, indicating that such a dianion is necessary for pyruvate to bind in the correct orientation. Thus, at the phosphoenolpyruvate-binding site, there are highly specific groups governing binding. The nucleotide-binding site is relatively nonspecific. In the enolization reaction, other nucleoside triphosphates may replace ATP and several nucleoside diphosphates can act as phosphoryl acceptors. The binding of a fluorescent analog of ADP suggests loose binding because the base has high mobility (Barrio et al., 1973). Several bivalent cations other than Mg^{2+} will bind to the enzyme. pH profiles suggest that Mn^{2+} will bind to the enzyme when a cationic group of pK_a 6.9 is deprotonated in solution. Good solvent access to the active site is likely because of the high turnover number of the enzyme (about 250 sec^{-1}), the stereospecific addition of protons, and the tritium exchange reaction. Because no phosphatase activity is found, the water molecules must be restricted in some way. The lack of evidence for the direct involvement of other groups suggests that catalysis is mainly effected by metal ions and the attacking base.

1.3. Possible Catalytic Residues

Various techniques have been used to identify possible groups at the active site of PK. The rabbit muscle enzyme loses its ability to catalyze phosphoryl transfer when an essential lysyl ε-amino group is converted to the *N*-trinitrophenyl derivative (Flashner et al., 1973; Johnson et al., 1979; Russell et al.,

1979) or when an essential sulfhydryl group is converted to an intramolecular disulfhide bridge (Flashner et al., 1973). It has been suggested that both of these residues are involved in the binding of nucleotides. In the yeast enzyme, substitution of three sulfhydryl groups per subunit leads to total inactivation of the enzyme (Wieker and Hess, 1972). The most reactive sulfhydryl group is neither part of the active nor the allosteric site. Reaction of the second group blocks substrate binding and the third forms part of the allosteric site. Other chemical modification experiments implicate a critical tyrosine residue (Colman, 1983). From the variation of Mn^{2+} binding with pH, Mildvan and Cohn (1965) calculated apparent pK's of 6.9 and 7.8, and suggested histidine and an amino group or an atypical sulfhydryl group as binding groups for Mn^{2+}. NMR studies also indicated an imidazole ligand (Meshitsuka et al., 1980). More recent work by Dougherty and Cleland (1985) identified these pK's as belonging to the metal ligand (6.9) and water in the inner coordination sphere of the metal (7.8).

1.4. NMR Measurements

Measurements of the perturbation of longitudinal relaxation rates of magnetic nuclei of substrates by an enzyme-bound paramagnetic probe which are proportional to the inverse sixth power of the paramagnetic-magnetic nuclei distances may be used to determine the distances. Mildvan and coworkers have studied the conformations of various substrates bound to PK using paramagnetic probes bound to two sites on the rabbit muscle enzyme; namely, Mn^{2+} or Co^{2+} at the bivalent cation site and Cr^{3+} bound to ATP (Mildvan et al., 1976). Tridentate Cr-ATP is not a substrate for the phosphoryl transfer reaction of PK, although it replaces ATP in promoting the enzyme-catalyzed enolization of pyruvate in the presence of K^+ and Mg^{2+}. These results demonstrated that rabbit muscle PK requires two bivalent cations per active site for the catalysis of the enolization of pyruvate in the presence of ATP. One bivalent cation (site 1) is tightly bound directly to the enzyme; the second, more weakly bound bivalent cation (site 2) is directly coordinated to the phosphoryl groups of ATP and does not interact directly with the enzyme. The essential role of the bivalent cation at site 1 is shown by the requirement for Mg^{2+} or Mn^{2+} for the enolization of pyruvate in the presence of the substitution inert Cr^{3+}-ATP complex. The essential role of the bivalent cation at site 2 is shown by the sigmoidal dependence of the rate of detritiation of pyruvate on Mn^{2+} concentration in the presence of high concentrations of enzyme and ATP. The electrophilicity of the nucleotide-bound metal (site 2) determines the rate of enolization of pyruvate. Both bivalent cations are necessary for the enolization of pyruvate as well as for the phosphoryl transfer reaction (Gupta et al., 1976; Dunaway-Mariano et al., 1979). The NMR results indicate direct coordination of the site 2 bivalent cation to the α-, β-, and γ-phosphoryl groups of enzyme-bound ATP and to the α- and β-phosphoryl groups of enzyme-bound ADP. Thus, enzyme-bound Mg-ATP is a tridentate complex with coordination of all three phosphoryl groups by the metal (Gupta and Mildvan, 1977). However chemical studies indicate that the Δ isomer of the β, γ bidentate chelate of Cr^{3+}-ATP is active in a single phosphoryl transfer of PK, and is the most effective metal in promoting the enolization of pyruvate (Dunaway-Mariano et al., 1979). This may result from a difference in conformation of the Mg^{2+} and Cr^{3+} complexes of ATP.

As shown above, the overall reaction catalyzed by PK may be divided into the two half reactions.

$$\text{Phosphoenolpyruvate} + \text{ADP} \xrightleftharpoons{M^{2+}, M^+} \text{Pyruvate enolate} + \text{ATP}$$

and

$$\text{Pyruvate enolate} + H^+ \xrightleftharpoons{ATP, M^{2+}, M^+} \text{Pyruvate}$$

In the direction of the reaction described above, the nucleotide-bound Mg migrates from α, β-ADP coordination to β-ADP phosphoenolpyruvate coordination. This is followed by the transfer of the phosphoryl group giving Mg-ATP and the enolate of pyruvate. Finally, the enolate is protonated to give the keto form of pyruvate.

The sets of distances from the paramagnetic reference points were used to construct an active complex (enzyme-K^+-Mn^{2+}-Cr-ATP-pyruvate) and a hypothetical complex (enzyme-K^+-Mn^{2+}-Cr-ATP-phosphoenolpyruvate). The results suggest that all three of the ATP phosphorous atoms are about 5 Å from the site 1 bivalent cation, indicating coordination to the metal, but as a second sphere complex with another ligand (possibly water) intervening.

The NMR results for binding of substrate could also be explained by equilibrium between inner and outer sphere pyruvate complexes with a preponderance of a catalytically inactive outer sphere complex. Also, the phosphorus of phosphoenolpyruvate is about 1 Å from the position of the γ-phosphorus of ATP and about 5 Å from the Mn^{2+}. This would allow the transition state to form, and the metal would act either through a water molecule or through an enzyme-donated ligand to activate the transferred phosphoryl group. The conformation of the enzyme-bound ATP is extended, with an average dihedral angle at the glycosidic bond ($X = 30° \pm 10°$) that differs little from that found for such nucleotides when free in solution (Granot et al., 1979). This result is consistent with the high mobility of the base and the low nucleotide specificity of the enzyme. The distance of the enzyme-bound K^+ from the enzyme-bound Mn^{2+} is 4.9 Å in the phosphoenolpyruvate complex and 8.2 Å in the absence of phosphoenolpyruvate (Reuben and Kayne, 1971). The former distance allows for the direct coordination of the monovalent cation by the carboxyl group of phosphoenolpyruvate. Upon binding to the enzyme, the activating monovalent cation, K^+, would coordinate to the carboxylate group and change the conformation of the complex to the catalytically active form. Other evidence for small conformational changes comes, for example, from the quenching of tryptophyl fluorescence by the addition of the activating cations (Kuczenski and Suelter, 1971).

2. STRUCTURE OF PYRUVATE KINASE

2.1. Amino Acid Sequence

The amino acid sequences (Table 1) of the enzymes from yeast (*Saccharomyces cerevisiae*) and chicken muscle have been determined from corresponding nucleotide sequences (Burke et al., 1983; Lonberg and Gilbert, 1983), and the determination of the sequence of the cat muscle enzyme has been completed recently (Muirhead et al., 1986). Partial sequences from liver PK are also available (Hoar et al., 1985), and short sequences for peptides from other muscle enzymes have been published (Anderson and Randall, 1975; Johnson et al., 1979; Harkins et al., 1983). These peptides can be located in the complete sequence. One peptide contains a highly reactive cysteine residue (Cys 316 in the cat muscle sequence) and another

Table 1. Amino Acid Sequence Comparison of Pyruvate Kinase

```
            -10              *   1                10                  20
Pig liver   E G P A G Y L R R A S V A Q L T Q E L G T A F F Q R Q Q L P A A
Rat liver   E G P A G Y L R R A S V A Q L T Q E L G T A F F Q Q Q Q L P A A
Rat M1                        P K P D S E A G T A F I Q T Q Q L H A A
Cat M1                      Ac-S K P H S D V G T A F I Q T Q Q L H A A
Chick M1                      S K H H   D A G T A F I Q T Q Q L H A A
Yeast
                          (——— α ———)

                         30                40                50
                                                                   #
Pig L       M
Rat L       M A D T F L E H L C L L D I D S Q P V A A R S T S I I A T I G P
Rat M1      M A D T F L E H M C R L D I D
Cat M1      M A D T F L E H M C R L D I D S P P I T A R N T G I I C T I G P
Chick M1    M A D T F L E H M C R L D I D S E P T I A R N T G I I C T I G P
Yeast         M S R L E R L T S L N V   V A G S D L R R T S I I G T I G P
            (———— N α1 ————)               (——— Aβ1 ———)

                       60                70                80
                                              ***              #
Rat L       A S R S V D R L K E M
Cat M1      A S R S V E I L K E M I K S G M N V A R L N F S H G T H E Y H A E
Chick M1    A S R S V D K L K E M I K S G M N V A R L N F S H G T H E Y H E G
Yeast       K T N N P E T L V A L R K A G L N I V R M N F S H G S Y E Y H K S
            (——— Aα1 ———)       (——— Aβ2 ———)       (——————— A

                     90               100              110
                                                          **************************
Cat M1      T I K N V R A A T E S F A S D P I R Y R P V A V A L D T K G P E I
Chick M1    T I K N V R E A T E S F A S D P I T Y R P V A I A L D T K G P E I
Yeast       V I D N A R K S E E L Y P G       R P L A I A L D T K G P E I
            α2 ————————————————————)       (——— Aβ3 ———)       (———
```

                    120   #   130   140   150
                    ****
Cat M1   R T G L I K G S G T A E V E L K K G A T L K I T L D N A Y M E K C
Chick M1   R T G L I K G S G T A E V E L K K G A A L K V T L D N A F M E N C
Yeast   R T G T T T N D V D Y P I P P N H E M I F   T T D D K Y A K A C
            Bβ1 ————————)   (— Bα1 —)   (— Bβ2 —)   (— Bβ3 —)   (———— Bα2 ———)   (—— B

                                      160   170   180
Cat M1   D E N V L W L D Y K N I C K V V E V G S K V Y V D D G L I S L L V
Chick M1   D E N V L W V D Y K N L I K V I D V G S K I Y V D D G L I S L L V
Yeast   D D K I M Y V D Y K N I T K V I S A G R I I Y V D D G V L S F Q V
            α3 ——)   (———————— Bα4 ————————)   (——— Bβ4 ———)

                          190   200   210
Rat M1   #   G G S L G S K K G V N L P G A A V D
Cat M1   K E K G A D   F L V T E V E N G G S L G S K K G V N L P G A A V D
Chick M1   K E K G K D   F V M T E V E N G G M L G S K K G V N L P G A A V D
Yeast   L E V V D D K T L K V K A L N A G K I C S H K G V N L P G T D V D
            (———— Bβ5 ————)   (— Bβ6 ———)   (— Bβ7)   (——— Bβ8 ———)

                    220   ****************************230   240
                                                                                  ****
Rat M1   L P A V S E K D I Q D L K F G V
Cat M1   L P A V S E K D I Q D L K F G V E Q D V D M V F A S F I R K A S D
Chick M1   L P A V S E K D I Q D L K F G V E Q N V D M V F A S F I R K A A D
Yeast   L P A L S E K D K E D L R F G V K N G V H M V F A S F I R T A N D
                (—— Bβ9)   (————— Aα3 —————)   (— Aβ4 —)   (————

            250   260   270   280
                                                            *******************   #
Cat M1   V H E V R K V L G E K G K N I K I I S K I E N H E G V R R F D E I
Chick M1   V H A V R K V L G E K G K H I K I I S K I E N H E G V R R F D E I
Yeast   V L T I R E V L G E Q G K D V K I I V K I E N Q Q G V N N F D E I
            Aα4 ——————————————)   (——— Aβ5 ———)   (——— Aα5 —)   (———— Aα5

                                                                                                (continued)

Table 1. *(continued)*

```
                        290                           300                           310
RMPK                    *********************                                      I  I  G  R
Cat M1    L  E  A  S  D  G  I  M  V  A  R  G  D  L  G  I  E  I  P  A  E  K  V  F  L  A  Q  K  M  M  I  G  R
Chick M1  M  E  A  S  D  G  I  M  V  A  R  G  D  L  G  I  E  I  P  A  E  K  V  F  L  A  Q  K  M  M  I  G  R
Yeast     L  K  V  T  D  G  V  M  V  A  R  G  D  L  G  I  E  I  P  A  P  E  V  L  A  V  Q  K  K  L  I  A  K
          ——)           (—— Aβ6 ——)      (——  Aα6  ——)    (——————————      Aα6      ——————————

                320                           330                           340
RMPK      C  N  R  A                         #
Bovine M                      *************                                  A  E  G  S  D  V  A
Cat M1    C  N  R  A  G  K  P  V  I  C  A  T  Q  M  L  E  S  M  I  K  K  P  R  P  T  R  A  E  G  S  D  C  A
Chick M1  C  N  R  A  G  K  P  I  I  C  A  T  Q  M  L  E  S  M  I  K  K  P  R  P  T  R  A  E  G  S  D  V  A
Yeast     S  N  L  A  G  K  P  V  I  C  A  T  Q  M  L  E  S  M  T  Y  N  P  R  P  T  R  A  E  V  S  D  V  G
          ——————————)    (——— Aβ7 ———)                              (——— Aα7 ———————————

          350                           360                           370                           380
                                        *******************************                               #
Bovine    N  A  V  L  D  G  A  D  C  I  M  L  S  G  E  T  A  K  G  D  Y  P  L  E  A  V  R
Cat M1    N  A  V  L  D  G  A  D  C  I  M  L  S  G  E  T  A  K  G  D  Y  P  L  E  A  V  R  M  Q  H  L  I  A
Chick M1  N  A  V  L  D  G  A  D  C  I  M  L  S  G  E  T  A  K  G  D  Y  P  L  E  A  V  R  M  Q  H  A  I  A
Yeast     N  A  I  L  D  G  A  D  C  V  M  L  S  G  E  T  A  K  G  N  Y  P  I  N  A  V  T  T  M  A  E  T  A
          —————————)        (——— Aβ8 ———)                    (——————— Aα8 ———————————

                390                           400                           410
Rat L                                                                     P  T  E  V  T  A  I  G
Cat M1    R  E  A  E  A  A  M  F  H  R  K  L  F  E  E  L  V  R  G  S  S  H  S  T  D  L  M  E  A  M  A  M  G
Chick M1  R  E  A  E  A  A  M  F  H  R  Q  Q  F  E  E  I  L  R  H  S  V  H  H  R  E  P  A  D  A  M  A  A  G
Yeast     V  I  A  E  Q  A  I  A  Y  L  P  N  Y  D  D     M  R  N  C  T  P  K  P  T  S  T  T  E  T  S  L  P
          ——————————)        (——————— Cα1 ———————————)        (——————— Cα2
```

                    420   430   #   440
Rat L   A V E A S F K C C A A A I I V L T R P G R V A Q L L S Q Y R P R V
Cat M1   S V E A S Y K C L A A A L I V L T E S G R S A H Q V A R Y R P R A
Chick M1   A V E A S F K C L A A A L I V M T E S G R S A H L V S R Y R P R A
Yeast   R V A A V F E Q K A K A I I V L S T S G T T P R L V S K Y R P N C
           ——————————————————————)   (——— Cβ1 —)   (———————— Cα3 ———)

              450   460   470   480
Rat L   A L I
Cat M1   P I I A V T R N H Q T A R Q A H L Y R G I F P V V C K D P V Q E A
Chick M1   P I I A V T R N D Q T A R Q A H L Y R G V F P V L C K Q P A H D A
Yeast   P I I L V T R C P R A A R F S H L Y R G V F P F V F E K E P V S D
           (— Cβ2 —)   (———————————— Cα4 ——————————)   (— Cβ3 —)

                          490   #   500   510
Cat M1   W A E D V D L R V N L A M N V G K A R G F F K H G D V V I V L T G
Chick M1   W A E D V D L R V N L G M N V G K A R G F F K T G D L V I V L T G
Yeast   W T D D V E A R I N F G I E K A K E F G I L K K G D T Y V S I Q G
           (————————————————— Cα5 ————————————————————)   (———— Cβ4 ——)

                       520   530
Cat M1   W R P G S G F T N T M R V V P V P
Chick M1   W R P G S G Y T N T M R V V P V P
Yeast   F K A G A G H S N T L Q V S T V
           (——— Cβ5 ———)

Cat muscle PK sequence, Muirhead et al. (1986).
Chicken muscle sequence, Lonberg and Gilbert (1983).
Yeast sequence, Burke et al. (1983).
Rabbit muscle sequence (RMPK), Anderson and Randall (1975).
Bovine sequence, Johnson et al. (1979).
Liver sequence, Hoar et al. (1985).
The phosphorylated serine in liver PK and residues close to the active site are marked with an *. Introns (as determined by Lonberg and Gilbert, 1985) are marked #. The intron between residues 379 and 380 corresponds with the end of the TIM structure (Banner et al., 1975, 1976). The positions of the elements of secondary structure are indicated.

a lysine residue (Lys 366). It is the trinitrophenylation of the amino group of this lysine that results in inactivation of the enzyme; the lysine is protected when nucleotide is bound. The yeast gene (Burke et al., 1983) codes for a protein of 499 amino acids in a single open-reading frame giving a monomer with a molecular mass of 54,605 daltons, including the N-terminal methionine. The sequence is specified by a highly restricted set of codons. Only 35 of the 61 possible codons are utilized, and 97% of the amino acids are specified by a subset of only 25 codons. This biased codon utilization has been observed for other highly expressed yeast genes. The chicken muscle gene (Lonberg and Gilbert, 1983) codes for 529 amino acids, giving a molecular weight of 57,865. The N-terminal residues of PK isolated from bovine muscle, rabbit muscle, rat liver, and human kidney have been shown to be modified. The modified residue is an α-amino-acetylated serine for the rabbit and bovine muscle enzymes (Brummel et al., 1976). The residue following the first AUG in the cDNA sequence of PK mRNA is a serine, and this is assumed to be the first amino acid of the mature enzyme.

The chicken muscle sequence may be aligned with the yeast sequence, minimizing the number of insertions and maximizing the number of identities and conserved substitutions. The chicken enzyme is longer than the yeast enzyme by 21 residues at the N-terminus and a single residue at the C-terminus. A total of nine additional residues must be inserted at four sites into the yeast sequence and one residue into the chicken sequence in order to align the two proteins. With this alignment, 42% of the residues are identical in the two structures. The cat muscle sequence may be aligned exactly with the chicken sequence. It has one extra residue near the N-terminus, giving a total of 530 residues. The numbering of residues in the cat muscle enzyme is used in this article. When the alignment is compared with the three-dimensional structure of the enzyme, it is seen that regions of low homology and the positions of insertions and deletions correspond to connecting regions between elements of secondary structure. The sequence following the phosphorylated serine in liver PK may be aligned with that of the N-terminus of the type M enzyme. The fact that the phosphorylated serine precedes the N-terminus, which is disordered in the muscle enzyme, is consistent with the evidence that it is near one end of the polypeptide chain (Bergstrom et al., 1976). It is possible that a better alignment for residues 1–22 of the yeast enzyme would be to the N-terminus of the liver enzyme, with the yeast Ser 22 equivalent to the liver Ser-P_i-2, instead of that shown in the Table 1. This would align the serine at position 45 in yeast (at the beginning of Aβ1) with the phosphorylated serine in liver.

The chicken PK gene is interrupted by at least 10 introns, including nine within the coding region (Lonberg and Gilbert, 1985). The 10 exons are of a more uniform length than would be expected if the introns occurred at random within the coding region, and the introns tend to fall between discrete elements of regular secondary structure. There is a tendency for introns to occur in the connecting regions between the end of strands of β sheet and the beginning of α helices. Lonberg and Gilbert suggest that the location of these introns indicates that the PK gene evolved by exploiting RNA splicing to combine smaller-coding units.

Secondary structure predictions were made using the program developed by McLachlan (1977). These predictions are shown in Table 2 and compared with those found in the actual crystal structure. The worst predictions were in domain B (see Section 2.2) and are consistent with the observation that this part of the structure shows some disorder.

2.2. Crystal Structure of Type M Pyruvate Kinase

The crystallographic studies on PK have been carried out using the type M enzyme, which can be prepared in large quantities and is more stable than the type L enzyme. Microcrystals of yeast PK have been obtained, but have not been used for X-ray analysis. Much of the solution work on PK has used the rabbit muscle enzyme. This enzyme has been crystallized in the form of large hexagonal bipyramids (Tietz and Ochoa, 1958), but these are disordered and unsuitable for structural work (Hollenberg et al., 1971). Human muscle PK gives cuboidal crystals (Kubowitz and Ott, 1944), and these have the orthorhombic space group I222 with one subunit in the crystallographic asymmetric unit. These crystals, although more promising than those of the rabbit enzyme, show disorder in one direction. Crystals from PK of cat muscle have a similar morphology to those of the human enzyme, and have the same space group and similar cell dimensions (Stammers and Muirhead, 1975). They are well ordered and are the only crystals found to be suitable for high-resolution structural studies.

The overall dimensions of the tetrameric molecule are 75 Å, 95 Å, and 125 Å. The subunits pack closely together near the center of the tetramer, but on the outside of the molecule, large areas of each subunit are exposed to

Table 2. Secondary Structure Predictions

		Predictions		
	Model	Cat Muscle	Yeast	Liver
	Q2−L7			Helix
Nα1	M21−M29	Helix	Helix	Helix
Aβ1	T45−G51	Sheet	Sheet	Sheet
Aα1	V57−G67	Helix	Helix	Helix
Aβ2	V70−F75	—	Sheet?	
Aα2	H80−S99	Helix	Helix?	
Aβ3	P106−D112	Sheet	Sheet?	
Bβ1	E117−I123	Sheet	Sheet	
Bα1	G125−T128	—	—	
Bβ2	E130−L133	Helix	—	
Bβ3	K134−A137	Helix	Helix?	
Bα2	L139−N145	Helix	—	
Bα3	E149−N154	Helix	Helix?	
Bα4	K161−V173	Sheet	Sheet	
Bβ4	D176−L182	—	—	
Bβ5	K185−F191	—	—	
Bβ6	S201−S204	—	—	
Bβ7	G207−N209	Sheet	Sheet	
Bβ8	G212−D216	—	—	
Bβ9	S221−K223	Helix	Helix	
Aα3	D227−Q234	Helix?	—	
Aβ4	M238−A241	Helix	Sheet	
Aα4	A247−G258	Helix	Sheet?	
Aβ5	K265−I270	Sheet?	Sheet	
Aα5	H273−R277	Helix	—	
	F279−E284	Helix	Helix	
Aβ6	I289−A292	Sheet	Sheet	
Aα6	D295−E299	—	—	
	E303−A319	Helix	Helix	
Aβ7	P322−T327	Helix?	—	
Aα7	M329−V351	Helix?	Sheet?	
Aβ8	D356−S361	—	—	
Aα8	P370−A387	Helix	Helix	
Cα1	H390−H403	Helix	—	
Cα2	M408−Y420	Helix	Helix	—
Cβ1	A425−V429	Helix	Sheet	Sheet
Cα3	R435−A442	Sheet?	Sheet	Sheet?
Cβ2	I449−V452	Sheet	Sheet	
Cα4	N455−L464	Helix?	—	
Cβ3	G467−V472	Sheet	Sheet	
Cα5	W481−R499	Helix	Helix	
Cβ4	I509−G513	Sheet	Sheet	
Cβ5	G519−M524	Sheet	Sheet	

Residue numbering is from the cat muscle sequence. Strong predictions are given, weak predictions have a ?, and "—" indicates no prediction of regular secondary structure.

solvent (Fig. 1). The overall subunit shape is approximately elliptical, with a circular cross-section 45 Å in diameter and an overall length perpendicular to this cross-section of 75 Å. The electron density in that part of the subunit furthest away from the center of the tetramer (domain B) is weaker than that in the rest of the subunit and shows signs of disorder. In the crystal, these low-density regions in neighboring molecules pack together along one axis, giving rise to the disorder observed in the crystals of human muscle PK.

The electron density map at a resolution of 2.6 Å (Stuart et al., 1979) shows that each subunit comprises four domains: an N-terminal domain (residues 1−42)— and domains A (43−115, 224−387), B (116−223), and C (388−530). It has been possible to fit most of the amino acid sequence to the electron density map with the exception of residues 1−9, which appear to be mobile, and parts of domain B where the structure is disordered. Thus, there is one covalent connection between the N-terminal domain and domain A, two covalent connections between domains A and B, and one covalent connection between domains A and C. There are three types of intersubunit contact, one type involving domains A and C, and the other two, mainly domain C. The small N-terminal domain also forms an integral part of the intersubunit contacts. This model was used as the starting point for a restrained parameter least-squares refinement using the Hendrickson-

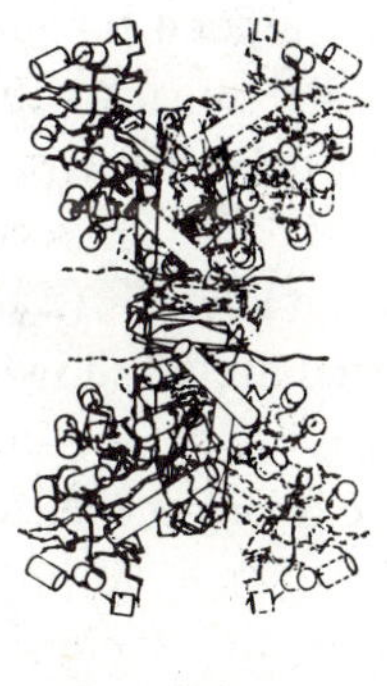

Tetramer   Tetramer

Cat muscle pyruvate kinase   *Cat muscle pyruvate kinase*

Figure 1. Stereo drawing of the tetramer of the crystalline enzyme cat muscle pyruvate kinase. The C domains are at the center of the tetramer. The A domains form the center of each subunit, and the flexible B domains are furthest from the center. The N-terminal domain is between domains A and C and the active site is between domains A and B. (Drawn by computer program of A. M. Lesk.)

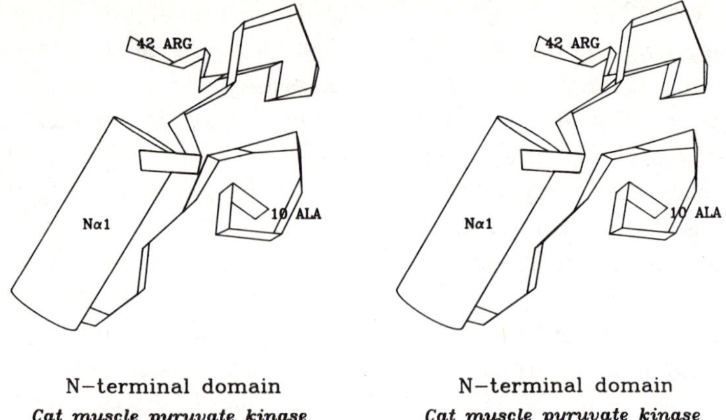

N−terminal domain
Cat muscle pyruvate kinase

N−terminal domain
Cat muscle pyruvate kinase

Figure 2. Stereo drawing of the N-terminal domain. Residues 1−9 are disordered in the muscle enzyme. For this and all other figures, see Table 1 for the correlation between sequence and elements of secondary structure. (Drawn by computer program of A. M. Lesk.)

Konnert programs (Konnert, 1976). The final crystallographic *R* factor was 0.26, and the final structure differed from ideal bond lengths by an overall root-mean-square (r.m.s.) deviation of 0.03 Å.

The N-terminal domain (Fig. 2) is the smallest domain (42 residues). Residues 21−29 form a short α helix and the first nine residues are disordered. The rest of the domain consists of an extended polypeptide chain. The available sequence information for liver PK suggests that the active serine peptide, which is a substrate for AMP-dependent protein kinase, precedes the N-terminus of the muscle enzyme (Table 1). Secondary structure predictions on the 75 residue sequence for rat liver PK (Table 2) suggest that it should have an additional α helix immediately following the active serine residue. This difference in the structure at the N-terminus would affect the intersubunit contacts and help to explain the difference in the regulatory properties.

Domain A (Fig. 3) is a highly symmetrical structure. It consists of an inner cylindrical β sheet of eight parallel strands. Adjacent strands are connected by predominantly helical segments running antiparallel to the strands of sheet, which form an outer cylinder coaxial with the inner one. The fifth and sixth helices both consist of a short, irregular helix followed by a longer, more regular one. The connecting loops between elements of regular sec-

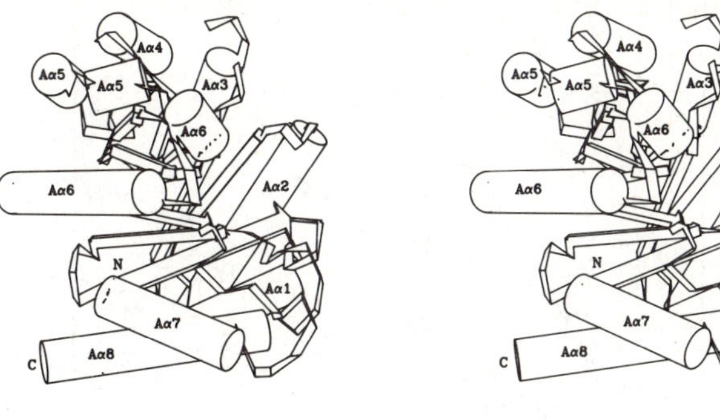

Domain A

Cat muscle pyruvate kinase

Domain A

Cat muscle pyruvate kinase

Figure 3. Stereo drawing of domain A. The 8-stranded α-β barrel structure. The N- and C-termini are labeled (bottom left). Domain B is inserted between strand Aβ3 and helix Aα3 (top right). (Drawn by computer program of A. M. Lesk.)

ondary structure are of variable length. The extent to which the structure possesses eight-fold symmetry has been analyzed (Stuart et al., 1979). By rotating the molecule through multiples of 45° about the axis of the cylinder, six of the eight sheet-helix pairs could be brought approximately into the same plane, which was at an angle of 40° to the axis of rotation. The two elements that deviated most from the eight-fold symmetry were the sheet-helix pairs six and seven. This remarkably symmetrical 8-stranded α-β barrel structure was first found in the glycolytic enzyme triosephosphate isomerase (TIM) isolated from chicken muscle (Banner et al., 1975, 1976; Alber et al., 1981). Subsequently, the same structure has been recognized in the A domain of the type M PK from cat muscle (Levine et al., 1978), bacterial KDPG aldolase from *Pseudomonas putida* (Mavridis et al., 1982), taka-amylase A from the fungus *Aspergillus oryzae* (Matsuura et al., 1980), xylose isomerase (Carrell et al., 1984), and plant glycolate oxidase from spinach (Lindqvist and Branden, 1980). These enzymes come from widely differing sources and have very different structural characteristics and catalytic functions (Table 3; Muirhead, 1983).

The polypeptide chain forms domain B (Fig. 4) between the third strand of β sheet and the third α helix of domain A. The central part of this domain consists of two antiparallel β sheets. Some of these sheets are connected by short, rather disordered helical segments that lie on the surface of the

Table 3. Proteins with 8-Stranded α/β Domain Structures

	TIM		PK	KDPG Aldolase	α-Amylase	Glycolate Oxidase
Reaction	D-Glyceraldehyde-3-phosphate $\Updownarrow$ Dihydroxyacetone phosphate		Phosphoenolpyruvate + ADP + H$^+$ $\Updownarrow$ Pyruvate + ATP	2-keto-3-deoxy-6-phosphogluconate $\Updownarrow$ Pyruvate + glyceraldehyde-3-phosphate	Amylose + H$_2$O $\Updownarrow$ Dextrin + maltose	Glycolate + O$_2$ $\Updownarrow$ Glyoxalate + H$_2$O$_2$
Source	Chicken muscle	yeast	Cat muscle	*Pseudomonas putida*	*Aspergillus oryzae*	Spinach
Resolution (nm)	0.25	0.30	0.26	0.28	0.29	0.55
Subunits (MW)	27,000		55,000	24,000	50,000	35,000
No. of subunits	2		4	3	1	8
No. of domains	1		3	1	2	2
Active site	Intersubunit		Interdomain	Intersubunit	Interdomain	Interdomain
Secondary binding site	No		ADP (Interdomain)	?	Substrate analogous and effectors (on surface)	No
Cofactor	No		Mg^{2+}, K$^+$	No	Ca^{2+}	FMN

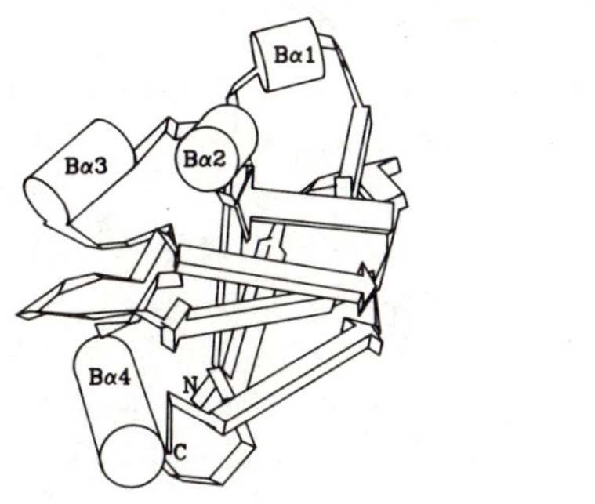
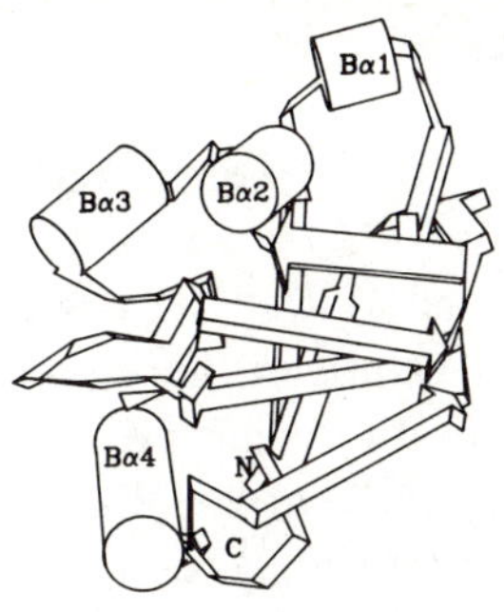

Domain B

Cat muscle pyruvate kinase

Domain B

Cat muscle pyruvate kinase

Figure 4. Stereo drawing of domain B. The antiparallel β structure is flanked on the outside of the tetramer by short helical segments. (Drawn by computer program of A. M. Lesk.)

domain. The position of domain B relative to domain A corresponds to extended loops or domains in the other similar structures.

Domain C (Fig. 5) consists of two long antiparallel α helices followed by a twisted 5-stranded β sheet enclosed by three helices. Strand five of the sheet is antiparallel to the rest. Part of this structure is similar to part of the

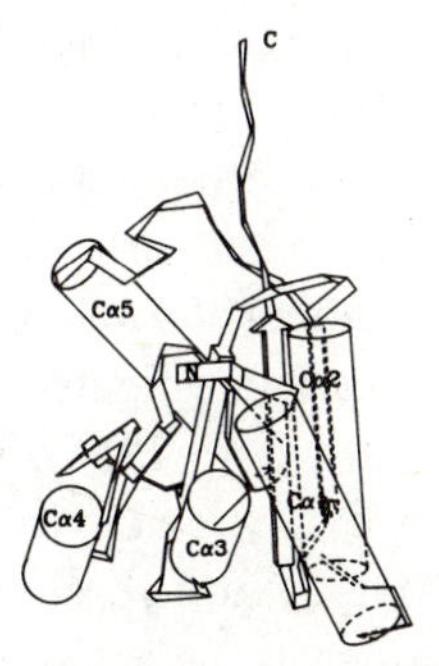
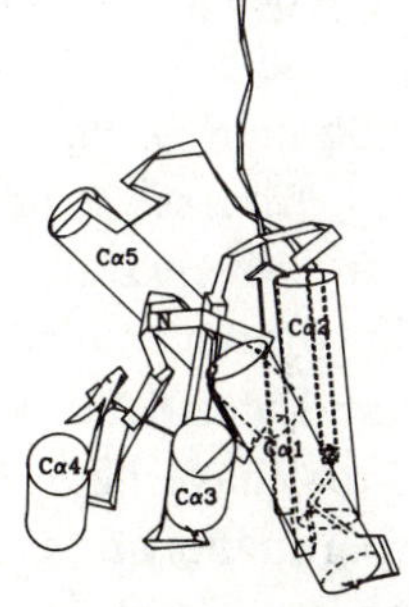

Domain C

Cat muscle pyruvate kinase

Domain C

Cat muscle pyruvate kinase

Figure 5. Stereo drawing of domain C. Two antiparallel helices are followed by a 5-stranded sheet flanked by helices. (Drawn by computer program of A. M. Lesk.)

dehydrogenase fold (Rossmann et al., 1974; Ohlsson et al., 1974). There are extensive intersubunit and interdomain contacts (Stuart et al., 1979). The relationship of the sequence to the elements of secondary structure is shown in Table 1.

Rossmann (1981) has attempted to make some generalizations about domain structure in the glycolytic enzymes. He observes that there is a tendency for those enzymes that require nucleotides as cofactors to possess at least one flat, parallel, β-pleated sheet domain. The nucleotide normally binds to the carboxyl end of this sheet. Second, there is a tendency for those enzymes that require no nucleotide cofactors to have 8-stranded, parallel barrels. The phosphorylated substrate binds to the carboxyl end of this sheet structure. Third, there is a tendency for kinases to contain two flat, parallel, β-pleated sheet domains with the carboxyl ends facing each other and the substrate binds between the domains. Results with hexokinase (Steitz et al., 1981) and phosphoglycerate kinase (Blake and Rice, 1981; Watson et al., 1982) suggest that in these monomeric kinases, the domains move relative to each other. In adenylate kinase (Pai et al., 1977), a phosphoryl-binding loop moves. Last, dehydrogenases invariably contain one flat, parallel, β-pleated sheet domain and one catalytic domain. PK is more similar to the glycolytic enzymes KDPG-aldolase and TIM, which have no nucleotide cofactors, than to other kinases. Phosphofructokinase from *Bacillus stearothermophilus*, another tetrameric kinase, is more similar to the monomeric kinases than to PK (Evans et al., 1981).

Amino acid compositions and numbers of residues conserved in the different sequences are shown in Tables 4, 5, 6, and 7. In PK, the active site (Fig. 6) lies in the pocket between domains A and B in a very similar position relative to domain A as the active sites in the other enzymes with the 8-stranded α-β barrel structure. The substrates and products of the PK reaction are relatively stable in solution; the binding of some of these as well as the binding of the activating cations have been investigated in the crystal (Stammers and Muirhead, 1975; Muirhead et al., 1981, 1984). The positions of residues that have been implicated in the catalytic activity by chemical labeling experiments and conserved in the muscle and yeast sequences are listed in Table 6. There are no cysteine residues in the active site region in either the muscle or yeast sequences. There are only three cysteines in the same position in both sequences. One of these (Cys 151) is at the entrance to the active site, one (Cys 325) is internal, and one (Cys 357) is in the domain interface between domains A and C. The most reactive cysteine in muscle is

Table 4. Amino Acid Compositions

Amino Acid	Yeast	Cat Muscle	Chicken Muscle	No. Conserved in All Three Sequences
Alanine	40	54	58	20
Arginine	25	32	32	16
Aspartate	32	30	32	18
Asparagine	26	17	18	11
Cysteine	7	9	8	3
Glutamate	28	38	36	14
Glutamine	10	12	13	2
Glycine	34	41	42	27
Histidine	7	15	18	3
Isoleucine	37	35	37	17
Leucine	36	39	36	13
Lysine	37	36	36	13
Methionine	11	19	21	5
Phenylalanine	15	16	17	6
Proline	26	21	20	13
Serine	27	30	25	9
Threonine	38	25	25	10
Tryptophan	1	3	3	1
Tyrosine	15	9	8	6
Valine	47	49	44	23
Totals	499	530	529	230

Table 5. The Percentage of Residues Identical in the Different Domains and in the Total Structure of Pyruvate Kinase for the Chicken Muscle, Cat Muscle, and Yeast Enzymes

	N-Terminus	Domain A	Domain B	Domain C	Total
Cat/Chicken	86	96	87	81	88
Cat/Yeast	10	52	39	34	41
Chicken/Yeast	10	51	40	38	42

on the surface, and is replaced by serine in yeast. It would not be possible to form an intramolecular disulfide bridge as suggested by Flashner et al. (1975) without considerable structural changes. Cys 455 in yeast could form part of the allosteric site (Wieker and Hess, 1972). Fourteen lysine residues

Table 6. Cysteine Residues in Cat Muscle and Yeast Sequences

Cat Muscle	Yeast		Position in Structure
Cys 30	Thr	N-domain	Intersubunit contact
Cys 48	Gly	Aβ1	Internal
Cys 151[a]	Cys	Bα3	External, entrance to active site
Cys 164	Thr	Bα4	Internal
Gly 203	Cys	Bβ6	External
Cys 316	Ser	Aα6	External "reactive" cysteine
Cys 325[a]	Cys	Aβ7	Internal
Cys 357[a]	Cys	Aβ8	Domain interface
Ser 401	Cys	Cα1	External
Cys 422	Gln	Cα2/Cβ1	External
Ala 447	Cys	Cα3/Cβ2	Domain interface
Asn 455	Cys	Cα4	Second nucleotide-binding site
Cys 473	Phe	Cβ3	External heavy atom site

Only three cysteines are present in both structures. One of these is at the entrance to the active site and the other two are internal.

[a]Conserved cysteines.

are conserved. Of these, eight are external, three are in the active site, and one is at the entrance to the active site. The "essential" lysine (Lys 366) is at the entrance to the active site and is at least 8 Å away from the phosphate groups of ADP and ATP. There is a network of salt bridges in the vicinity that must be disrupted by substrate binding. The presence of such a large group as trinitrophenol would cause steric hindrance to the binding of a nucleotide, which would explain why the muscle enzyme loses its ability to catalyze phosphoryl transfer (Russell et al., 1979; Johnson et al., 1979). There are no histidines or tyrosines in the active site, although Tyr 147 (Phe in chicken muscle) is near the entrance. There are two conserved arginine residues.

2.3. Enzyme-bound Cations

The crystals of cat muscle PK were grown from ammonium sulfate solution at pH 6.0. Because the binding of the bivalent cation at site 1 is pH dependent with increased binding above pH 7.0, the crystals were transferred to an ammonium sulfate solution at pH 8.0 and soaked with 0.1 M $MnCl_2$ or 0.05 M $MgSO_4$. Difference maps showed a single, large positive

Table 7. Conserved Residues in Cat Muscle and Yeast Sequences

Conserved Lysines (14)

Lys 2	N-terminus, disordered. Gln in liver PK	
Lys 65	Aα1	External
Lys 114[a]	Aβ3/B	Active site (nucleotide-binding site)
Lys 161	Bα4	External
Lys 165	Bα4	External
Lys 206	Bβ7	Subunit interface
Lys 223[a]	B/Aα3	Active site
Lys 262	Aα4/Aβ5	External
Lys 269[a]	Aβ5	Active site, Base?
Lys 310	Aα6	External
Lys 321	Aα6/Aβ7	External
Lys 366[a]	Aβ8/Aα8	Entrance to active site. Labeled Lys
Lys 497	Cα5	External
Lys 503	Cα5/Cβ4	External

Conserved Tyrosines (7)

Tyr 82	Aα2	External
Tyr 147[a]	Bα2/Bα3	Entrance to active site (Phe in chick)
Tyr 160[a]	Bα4	Entrance to active site
Tyr 174	Bα4	External
Tyr 369	Aβ8/Aα8	External
Tyr 444	Cα3/Cβ2	Internal
Tyr 464	Cα4/Cβ3	Internal

Conserved Histidines (3)

His 77	Aβ2/Aα2	Internal
His 83	Aα2	Internal
His 463	Cα4	Interdomain C/A

Conserved Arginines (16)

Arg 42	N/A β1	External
Arg 72 [a]	Aβ2	Active site
Arg 91	Aα2	External
Arg 105	Aβ3	External
Arg 119[a]	Bβ1	Active site
Arg 245	Aβ4/Aα4	External
Arg 254	Aα4	External
Arg 293	Aβ6/Aα6	External
Arg 338	Aα7	External
Arg 341	Aα7	External
Arg 399	Cα1	Intersubunit contact
Arg 444	Cα3/Cβ2	Interdomain
Arg 454	Cβ2/Cα4	External
Arg 460	Cα4	External (interdomain)
Arg 466	Cα4/Cβ3	Interdomain
Arg 488	Cα5	External

[a]Those residues in or at the entrance to the active site.

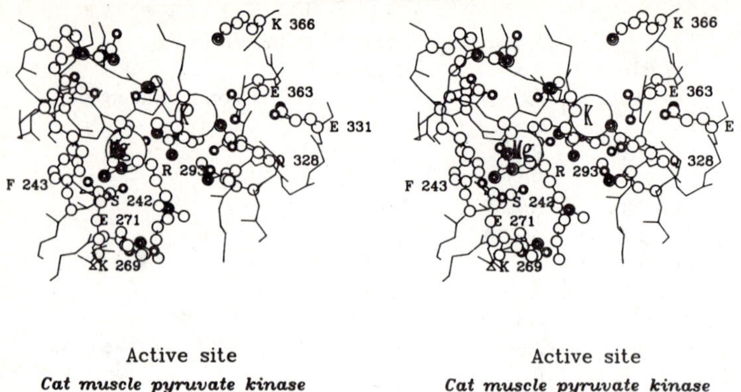

Active site
Cat muscle pyruvate kinase

Active site
Cat muscle pyruvate kinase

Figure 6. Stereo drawing of the active site region of pyruvate kinase. The binding sites for the enzyme-bound bivalent (Mg^{2+}) and monovalent (K^+) cations are shown. (Drawn by computer program of A. M. Lesk.)

peak situated between domains A and B at the carboxyl end of the eight β strands in the α-β barrel of domain A close to its axis. This site for the enzyme-bound bivalent cation was confirmed by the binding of gadolinium at pH 8.0, which was used as a heavy atom in the phase determination. The three probable enzyme ligands are the carboxyl group of Glu 271, which is situated at the end of strand Aβ5, and the main chain carbonyl oxygens of Ala 292 and Arg 293 at the C-terminus of strand Aβ6. This agrees with the suggestion from NMR measurements that the enzyme provides three ligands to the inner sphere of Mn^{2+}, with the remaining three ligands being water producing an octahedral complex (Reuben and Cohn, 1970). Negative density in the difference maps coincides with the side chain of Lys 269. In the native structure, there are two positively charged side chains in the vicinity of this binding site (Lys 269 and Arg 293). Repulsion of the positively charged lysine side chain when the bivalent cation binds would allow the lysine to move closer to the phosphoenolpyruvate-binding site. The movement of the main chain carbonyl groups towards Mn^{2+} would also allow the side chain of Arg 293 to move closer to the phosphate-binding site and away from the cation.

There is no direct crystallographic evidence for the binding site of the monovalent cation. It has not proved possible to bind a monovalent cation to the crystalline enzyme. This can probably be explained by the presence of a high concentration of ammonium ions that would compete for the binding site. The difference maps for the binding of Mn^{2+} and phosphoenolpyruvate both show a large negative peak in the same place; namely, adjacent to their own positive peaks. This must either correspond to movement of the density to another site or to the absence of a group displaced by the substrate. The original 2.6 Å electron density map for the native enzyme, prior to refinement, shows a positive peak that partially overlaps the sites at which Mn^{2+} and phosphoenolpyruvate bind. Its relatively isolated position makes it unlikely to be a side chain. This peak could either represent a bound sulfate, which would be displaced in the binding experiments, or a heavily solvated ammonium ion bound at the monovalent cation site. It is known that the addition of phosphoenolpyruvate to a PK-Mn-thallium complex, where thallium occupies the monovalent site, results in the distance between the monovalent and bivalent cations being reduced from 8.2 Å to 4.9 Å, which is interpreted as the binding of thallium to the substrate carboxyl group. When the substrate binds, the monovalent cation would move from a site corresponding to the isolated peak into the positive density in the difference map. Possible side chains close to this peak are Gln 328 ($A\beta7$) and Glu 363 ($A\beta8$), suggesting that it is more likely to be the monovalent cation site than a sulfate-binding site.

2.4. Substrate Binding

The binding of phosphoenolpyruvate has been studied in the absence of Mn^{2+} at pH 6.0 and in the presence of Mn^{2+} at pH 8.0. The phosphoenolpyruvate density is about 7 Å long, and none of the density is more than 5 Å from the enzyme-bound bivalent cation. This binding site is located in the loops that connect strands $A\beta3$ and $A\beta4-6$ to domain B and $A\alpha4-6$, respectively. The density for Mn-phosphoenolpyruvate is very similar to that of phosphoenolpyruvate by itself, except for the additional density due to the Mn^{2+}. The phosphoenolpyruvate lies too close to the Mn^{2+} to be a second-sphere ligand. It can be oriented to fit the difference density with the phosphate group in the vicinity of Ser 242 ($A\beta4$) and the carboxyl group pointing towards the monovalent cation. As described above, there is a suggestion of conformational changes around Lys 269 ($A\beta5$). These movements could bring the lysine into the correct position to act as the base in catalysis. This position allows protonation from beneath the double bond at the 2-*si* face, as established by Rose (1970). It is possible that for correct binding in the crystal, phosphoenolpyruvate must displace a sulfate ion already bound near the side chain hydroxyl of Ser 242. It seems likely that P_i would be necessary for the correct binding of pyruvate, and this would explain why P_i is necessary for the enolization of pyruvate.

Mn-ATP binds to the crystalline enzyme at pH 8.0. There is no suggestion that domains A and B move relative to each other, as observed for the monomeric kinases, but there are suggestions of considerable main chain movement in domain B and side chain movement in domain A. These movements make interpretation of the difference maps difficult. The phosphates are bound to the loop connecting Aβ3 to domain B and to the enzyme-bound bivalent cation. They bind close to a complex of charged side chains. These groups could undergo a conformational change such that the basic side chains Lys 114 and Arg 119 form salt links to the negatively charged phosphates. Other positively charged side chains in the vicinity are Arg 72 and Arg 293. The adenine ring could make a number of nonspecific hydrophobic contacts with, for example, Ile 118, Val 215, Val 220 and Phe 243. Main chain carbonyl groups are available for hydrogen bonding. This site, which spans a region between the N-terminal and C-terminal portions of domain B and the connections β4-α4 and β5-α5 of domain A, is a large hydrophobic hole that is not particularly specific for adenine. This is compatible with the broad specificity for nucleotides shown by the enzyme. With the exception of Val 220, which is replaced by leucine in yeast, all the side chains mentioned are conserved in the yeast PK sequence. This specific binding of ATP is achieved primarily through its three phosphate groups. Thus, not only is the structure of PK quite different from that of other kinases, but the mode of binding of the nucleotide is also different from that observed in other kinases and in the dehydrogenases (Rossmann et al., 1974; Ohlsson et al., 1974; Stammers and Muirhead, 1975).

At pH 6.0, a second nucleotide-binding site is present: ADP binds in a pocket formed between domains A and C that is accessible from the surface of the tetramer. It interacts with the N-termini of β strands of the cylindrical sheet and the carboxyl end of a helix in domain C. The function of this second nucleotide-binding site is unknown. The demonstration that fructose biphosphate can act as an activator for the rabbit muscle enzyme (Phillips and Ainsworth, 1977) suggests that this could be an allosteric activator-binding site in the muscle enzyme. This binding site is near the N-terminus of the polypeptide chain, which positions it close to the phosphorylated serine in the type L enzyme. A feature of the active site is the number of charged residues in a generally hydrophobic environment consistent with the exclusion of unbound water.

2.5. Sequence Homologies

A comparison of the amino acid sequences for the muscle and yeast enzyme shows the expected high homology between the two muscle enzymes and significant homology between the muscle and yeast enzymes (Table 1). Domain A seems to be more highly conserved than the rest of the structure (Table 5). The residues close to the active site comprise the two strands connecting domain A to domain B: the loops connecting the C-termini of β strands 4−8 of domain A with the N-termini of adjacent α helices plus the side chain of Arg 72 (Aβ2), which makes a salt bridge to Asp 112 (Aβ3) in the native structure. These residues show a remarkable degree of conservation (Table 1). By contrast, inspection of the residues involved in the second nucleotide-binding site between domains A and C shows that only 59% of the side chains are identical in the muscle and yeast sequences. There is also a significant difference between the two main intersubunit contacts. The intersubunit contact that involves helices α6, α7, and α8 of domain A and helix α1 of domain C (Fig. 7) is more highly conserved than the intersubunit contact that involves helices α1 and α2 and strand β4 of domain C (Fig. 8).

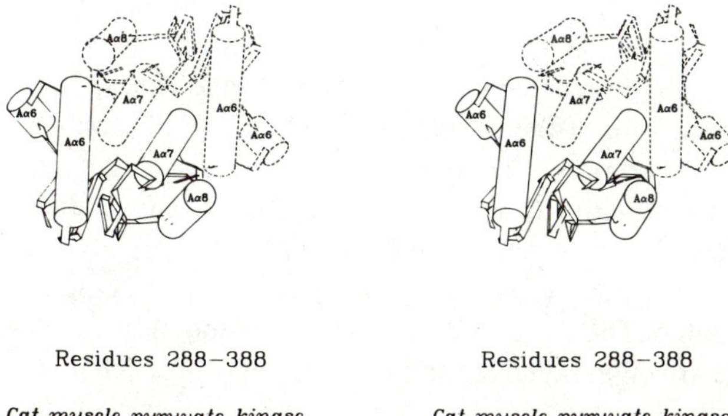

Figure 7. Stereo drawing of the more highly conserved intersubunit contact involving helices Aα6, Aα7, and Aα8. (Drawn by computer program of A. M. Lesk.)

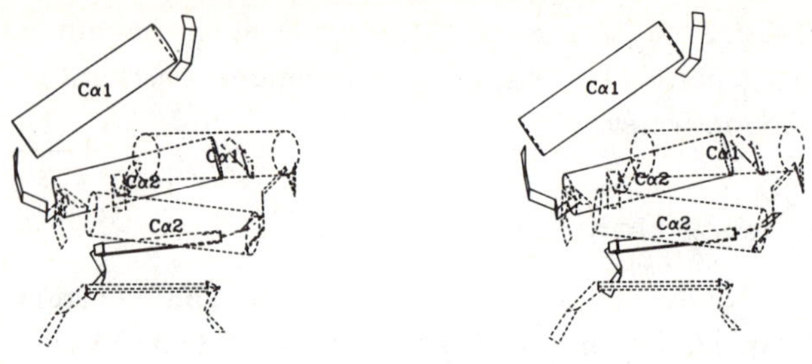

Cat muscle pyruvate kinase   Cat muscle pyruvate kinase

Figure 8. Stereo drawing of the second less highly conserved intersubunit contact involving two helices and a strand of sheet of domain C. (Drawn by computer program of A. M. Lesk.)

The N-terminus is closely involved in the first, more highly conserved intersubunit contact. Other relatively well-conserved regions of the subunit include the mononucleotide-binding fold in Domain C (residues 425−472) and an area on the surface of domain B at the entrance to the active site.

3.   COMPARISON OF PYRUVATE KINASE WITH OTHER STRUCTURES

Qualitatively, PK, TIM, KDPG-aldolase, and other 8-stranded α-β barrel enzymes (Table 3) have a very similar fold (Muirhead, 1983). Prior to the publication of the PK sequences, Tulinsky and his colleagues carried out a detailed comparison of the three structures (Lebioda et al., 1982). A quantitative comparison of the geometry of the main chain folding showed that about 150 α-carbons were equivalent in the three structures, with r.m.s. differences in position of just over 3.0 Å. The nonequivalent regions of the molecules occur principally in the connecting loops between elements of secondary structure. In the active site region, even connecting loops are closely related. The α-β barrel structure has pseudo eight-fold symmetry, so that there are eight possible ways of superimposing each pair of structures; none of these is clearly better than the others. However, the fact that in TIM, a loop that connects the third β strand and the third α helix is in a position corresponding to that of domain B in PK and forms an integral part of the active site and intersubunit contact favors the obvious correlation.

The minimum base change per codon for the superposition is only slightly lower than the random value of 1.54. However, the value would be expected to be slightly lower than random because of the correlation of hydrophobic and hydrophilic residues. The superposition of TIM and KDPG-aldolase indicates equivalence between Lys 144 of aldolase and Glu 165 of TIM. Glu 165 is the putative base in TIM and Lys 144 in aldolase, which has been identified as the residue forming a Schiff base with the carbonyl group of substrates. A superposition of PK and TIM using the comparison program of Rossmann and Argos (1975) showed the superposition of active site residues listed in Table 8. It is clear that the active sites involve residues in similar positions, but are not identical.

The overall α-β barrel structure is very simple, highly symmetrical, and subject to very stringent topological and packing constraints (Sternberg et al., 1981). The lack of any sequence homology and the eightfold symmetry of the superpositions suggest that this particular structure in several quite distinct enzymes may be an example of convergent evolution to a highly symmetrical, highly stable fold. If the existence of this structure in TIM, PK, and aldolase is an example of divergent evolution, then similarities of function would be expected (Levine et al., 1978). Rose (1981) has pointed out that in each case, an essential part of the mechanism is the abstraction of a proton next to a carbonyl group, immobilization of the intermediate in a highly planar form, and a similarity in the position of the functioning base (Fig. 9). In both TIM and PK, a proton is abstracted from C1 and a group

Table 8. Superposition of Pyruvate Kinase and Triosephosphate Isomerase

Structural Element	Function in PK	Residue in PK	Residue in TIM	Function in TIM
Aβ2	Binding of P_i	Arg 72	Gly 42	
Aβ4		Ser 242	His 95	
Aβ5	Putative base	Lys 269	Gly 128	
Aβ5	Binding of bivalent cation	Glu 271	Lys 130	
Aβ7	Binding of monovalent	Gln 328	Ser 211	Phosphate-
Aβ8	cation	Glu 363	Ala 234	binding site

The possible site of the monovalent cation is close to the phosphate-binding site in TIM. Lys 269 is close to the putative base Glu 165 in TIM, but on an adjacent β strand. The other residues are all part of the TIM active site (Alber et al., 1981).

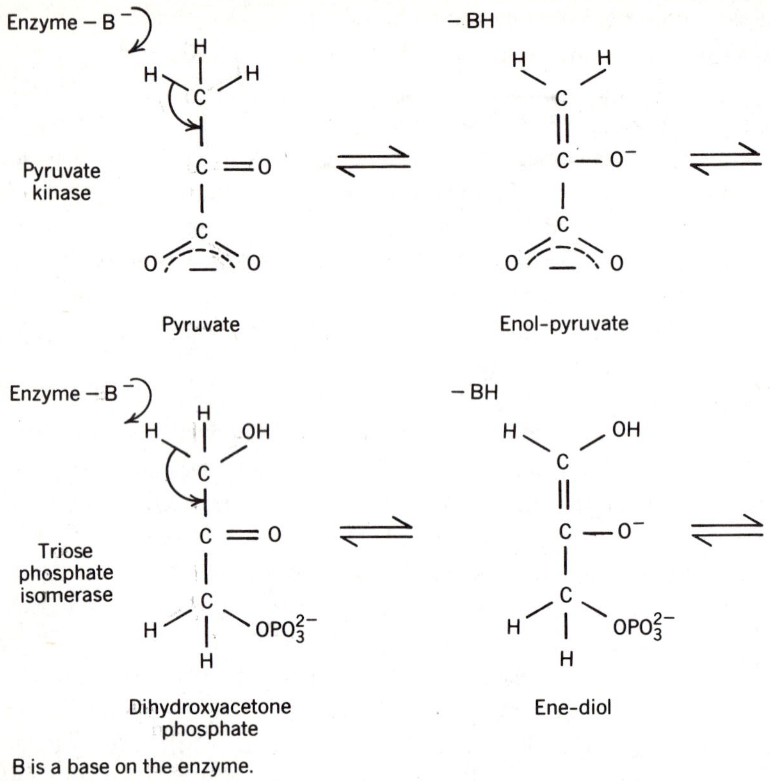

Figure 9. The similar reactions catalyzed by pyruvate kinase (PK) and triosephosphate isomerase (TIM).

attached to C3 has a specific binding site. In the case of TIM, this is the phosphate group that interacts with the N-terminus of a helix; for PK, it is the carboxyl group of phosphoenolpyruvate or pyruvate that interacts with the monovalent cation. In the structural superposition, the proposed binding site for the monovalent cation in PK is very close to that for the sulfate ion in TIM (Table 8). In KDPG-aldolase, the proton is transferred between C3 and C4, and the adjacent carbonyl group forms a Schiff base with lysine.

These comparisons suggest that the α-β barrel structure is just the right size to bind a three-carbon substrate and to stabilize the transition state, with functional groups on the enzyme being supplied by opposite sides of the barrel at the correct distance from the substrate. Aldolase and PK, which have to accommodate a larger or greater number of substrates than TIM, have smaller connecting loops. The positions of the substrate phosphate groups vary. However, in this structure, eight helices are available to bind a substrate phosphate group in whichever position it happens to lie (Hol et al., 1978). Thus, this particular structure combines the advantages of rigidity in its highly regular secondary structure and flexibility in the structure of the connecting loops.

Glycolate oxidase, which catalyzes the oxidation of glycolate to glyoxalate and will also oxidize L-lactate to pyruvate, could fit into the above pattern. The pyruvate might be expected to bind in a similar way to that found in PK. The first step in the reaction is the abstraction of a proton, and a planar *cis* enediol intermediate is a possibility (Lindqvist and Brändén, 1980). It would be necessary to have a binding site for the carboxyl group and a base (or bases) to abstract the two protons that are transferred to flavin mononucleotide (FMN). Similar arguments apply to the reaction catalyzed by xylose isomerase (Carrell et al., 1984). More information from gene structure, for example, may indicate whether such homologies reflect divergent or convergent evolution. In PK, the only intron (Lonberg and Gilbert, 1985) that falls in the middle of an element of secondary structure (between residues 379 and 380 in helix $A\alpha8$) coincides with the C-terminus of TIM.

A better case for divergent evolution is observed for part of domain C and the dehydrogenases. Levine et al. (1978) pointed out that part of the structure of domain C ($\beta1\alpha3\beta2\alpha4\beta3$) formed a mononucleotide-binding fold defined as one-half of the nicotinamide adenine dinucleotide (NAD)-binding domain found in the dehydrogenases and some kinases (Rossmann et al., 1974). Two mononucleotide-binding folds are connected by a dyad axis to form the NAD-binding unit. This section of domain C superimposes well upon the mononucleotide-binding fold of lactate dehydrogenase (Levine et al., 1978). The intron-exon structure of maize alcohol dehydrogenase (MADH) has been determined (Dennis et al., 1984), and the structure and location of the NAD-binding region can be inferred from the known three-dimensional structure of homologous horse liver enzyme (Brändén et al., 1973, 1984). Lonberg and Gilbert (1985) have shown that if the mononucleotide-binding fold of PK is superimposed on the first binding fold of the NAD-binding domain of horse liver alcohol dehydrogenase (LADH) to give an r.m.s. distance between α-carbons of 2.6 Å, then the amino acid sequence alignment produced by the structural superposition reveals some homology (about 20%) between PK and alcohol dehydrogenase. This aligns the chicken PK sequence with the MADH sequence such that one of the PK introns is four nucleotides from an equivalent alcohol dehydrogenase intron. These introns fall approximately in the middle of the region of amino acid sequence homology.

4. PROPOSED CATALYTIC ACTIVITY OF MUSCLE PYRUVATE KINASE

The results of the chemical and crystallographic investigations of the catalytic activity of muscle PK described above may be summarized as follows. PK catalyzes the transfer of the phosphoryl group of phosphoenolpyruvate to Mg-ADP via an enolate intermediate that is protonated stereospecifically to form pyruvate. It is unique among kinases in its absolute requirement for enzyme-bound cations, Mg^{2+} and K^+. One role for the enzyme-bound bivalent cation is to coordinate substrates in the large active site pocket. It is probable that phosphoenolpyruvate only binds correctly in the presence of the bivalent cation, and that the binding of Mg^{2+} enhances the binding of the phosphate group. Thus, Mg^{2+} assists in orienting the phosphates of ATP and ADP to perform phosphoryl transfer between ATP and pyruvate and between ADP and phosphoenolpyruvate, respectively. It would also orient the transferring phosphoryl group. The nucleotide-bound Mg^{2+} serves to reduce the electrostatic repulsion felt between the phosphoryl donor ($\gamma\, P_i$ or phosphoenolpyruvate) and the nucleophile ($\beta\, P_i$ or enolate), as well as to position the α and $\beta\, P_i$'s at the active site. A nucleotide-bound Mg^{2+} performs these functions in other kinases, such as phosphoglyceratekinase (Blake and Rice, 1981; Watson et al., 1982). There are several charged side chains forming an extensive network of salt bridges that could interact with the phosphate or the nucleotide-bound cation. The crystallographic difference maps suggest the presence of extensive conformational changes in the loops connecting elements of secondary structure when nucleotides bind. Various suggestions have been made for the role of the monovalent cation. Nowak and Mildvan (1972) have suggested that it acts both to coordinate the phosphoenolpyruvate carboxyl group and to participate in phosphoryl transfer. The presence of bound K^+ raises the affinity of PK-Mn^{2+} for phosphoenolpyruvate and vice versa. Such an interdependence implies coordination of the substrate by potassium, with the carboxylate being implicated by the lack of this effect in substrates with a blocked or absent carboxyl group. K^+ binding may change the conformation of the E-Mn-phosphoenolpyruvate complex to a catalytically active form. The proposed K^+-binding site is in the loops connecting Aβ7 to Aα7 and Aβ8 to Aα8, so that it would be very easy for the monovalent cation to move closer to the substrate when Mg or phosphoenolpyruvate is bound. Model building (D. Barford, D. A. Clayden, and H. Muirhead, unpublished results) shows that the phosphoryl donor and acceptor bind at sites such that phosphoryl transfer can occur directly from donor to acceptor in agreement with the NMR results (Mildvan et al., 1976). Thus, a conformational change in the enzyme is not required to bring the two substrates close enough together to react. Such a change is characteristic of monomeric kinases (Pai et al., 1977; Steitz et al., 1981; Blake and Rice, 1981; Watson et al., 1982). Because the substrate-binding sites are essentially on the same domain, any major conformational change of one relative to the other would seem unlikely. The mode of binding of the substrates to the crystalline enzyme is compatible with the in-line associative mechanism for the phosphoryl transfer reaction, because the two substrates bind correctly oriented for direct phosphoryl transfer. This agrees with the fact that phosphoryl transfer occurs with inversion of configuration (Lowe et al., 1981). Phosphoenolpyruvate and the phosphoryl groups of ATP are coordinated to the enzyme-bound bivalent cation, forming predominantly inner sphere complexes. The separation and orientation of the substrates are in accord with a direct phosphoryl transfer mechanism, possibly mediated by the bivalent cation. The monovalent cation may also be involved in phosphoryl transfer by polarizing the phosphate group to be transferred. The enolization reaction is brought about by the interaction of a base, possibly Lys 269, with the methyl group, which may be enhanced by the polarizing effect of the monovalent cation coordinated to the carboxyl group of pyruvate.

Recently, Dougherty and Cleland (1985) have carried out pH studies on PK using the nonphysiological substrates, oxaloacetate, glycolate, hydroxyl-amine, and fluoride, as well as the physiological substrates. From various pH profiles, they have identified three catalytically important groups.

(1) There is a group on the enzyme, when protonated, that prevents the binding of Mg^{2+} or Mn^{2+}. Meshitsuka et al. (1980) suggested that a histidine residue was involved in the binding of the bivalent cation. However, there are no histidine residues close to the active site in the cat muscle enzyme, and Glu 271 appears to be the metal ligand that is protonated below pH 6.9.

(2) The binding of Mg^{2+} is stronger when a group with a pK of $9.0-9.2$ is ionized. Cleland suggests that this group is water in the coordination sphere of Mg^{2+}. The kinetic measurements indicate that the carbonyl oxygen of oxaloacetate must replace water in the coordination sphere of Mg^{2+} prior to decarboxylation, rather than remain as an outer sphere complex. The other substrates must also replace the water prior to phosphorylation by Mg-ATP. The enolate intermediate is then protonated to a Mg^{2+}-coordinated keto acid, which leaves the coordination sphere of Mg^{2+} and is released from the enzyme. This is in accordance with EPR measurements (G. H. Reed, personal communication) showing the presence of inner sphere complexes for PK-Mn-oxalate and PK-Mn-lactate, and possibly for PK-Mn-pyruvate. Reed postulates the phosphoenolpyruvate chelate

$$
\begin{array}{c}
\text{O} \\
\text{O}\diagdown\underset{\text{P}}{\big|}\diagup\text{O} \\
\text{PK}-\text{Mn}\ \text{-----}\ \text{O}-\text{C}=\text{CH}_2 \\
\diagdown\!\!\diagdown\ \text{O}-\text{C}=\text{O}
\end{array}
$$

with the bridging oxygen coordinated to the Mn^{2+}, again suggesting that pyruvate binds directly to the enzyme-bound Mn^{2+}. Reed also suggests that the carboxylate group of pyruvate and phosphoenolpyruvate act as a bridging ligand for the bivalent and monovalent cations. The distance of 4.9 Å for the K^+-Mn^{2+} distance in the presence of phosphoenolpyruvate is approximately correct for a carboxylate-bridging ligand. An inner sphere complex would be more effective in the enolization of pyruvate. Binding of Mg^{2+} would make the enolate of pyruvate a reasonable leaving group. This suggests that in the phosphorylation of glycolate, the critical step is the formation of the inner sphere complex, which requires the glycolate hydroxyl to be ionized and the bound water to be protonated. Phosphoenolpyruvate binds tightly as a predominantly inner sphere complex, while pyruvate would form a predominantly second sphere complex, but does occasionally enter the inner coordination sphere with resulting catalysis.

(3) When Mg^{2+} is the bivalent cation, a group with a pK of 8.3 is the acid-base catalyst responsible for the interconversion of pyruvate and enolpyruvate. In the direction of pyruvate phosphorylation, the acid-base catalyst must be nonprotonated so that it can abstract a proton from C3 of pyruvate to form an enolate. In the forward reaction, it must be protonated so that it can donate a proton to the intermediate following phosphoryl transfer. The pK value suggests either lysine or histidine. There are no histidine residues near the active site, although there are several lysine residues. The pK of 9.2 for water coordinated to Mg^{2+} argues for a net positive charge in the vicinity of the pyruvate-binding site (Arg 293). This would lower the pK of the acid-base catalytic group, indicating a lysine rather than a histidine residue (Lys 269). The acid-base catalytic group with a pK of 8.3 must be protonated so that it can donate a proton to the intermediate enolate following phosphoryl transfer. The chemical mechanism with the natural substrates would involve phosphoryl transfer with a Mg^{2+}-bound enolate and Mg-ADP with metal coordination of the enolate serving to make it a good leaving group.

Cleland suggests that phosphoryl transfer is between Mg-ADP and a Mg^{2+}-bound enolate of phosphoenolpyruvate and that the leaving group is transferred to the metal ion. If Mg^{2+} were not present to bind the enolate, the latter would probably be too poor a leaving group to produce a sufficiently rapid reaction between phosphoenolpyruvate and Mg-ADP. The combination of the results from solution studies with evidence from the crystallographic analysis suggests the amino acid residues most likely to be involved in the catalytic activity of PK.

Step 1 Phosphoenolpyruvate and ADP bind.

(a) P_i is essential for the correct binding of pyruvate.

(b) K^+ binds to Glu 363 and Gln 328.

(c) P_i binds near Ser 242, Arg 72, and Arg 293.

(d) Phosphoenolpyruvate can displace a water molecule from the inner co-ordi-nation sphere of Mg^{2+}.

(e) The phosphoryl group is transferred to ADP

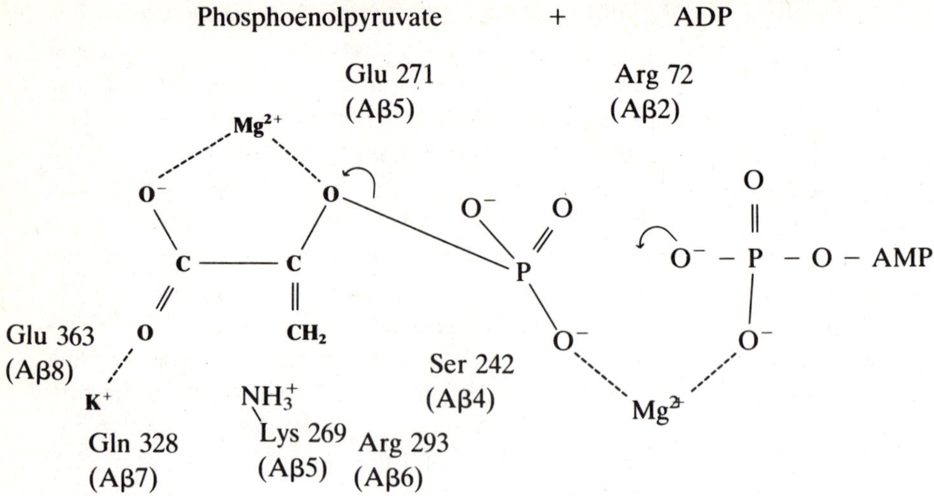

Step 2   ATP-enolate intermediate is formed.
(a) Enolate is stabilized by Mg^{2+}.
(b) Enolate of pyruvate and ATP formed.

Step 3 Lysine supplies a proton leading to the formation of pyruvate and ATP.

Pyruvate   +   ATP

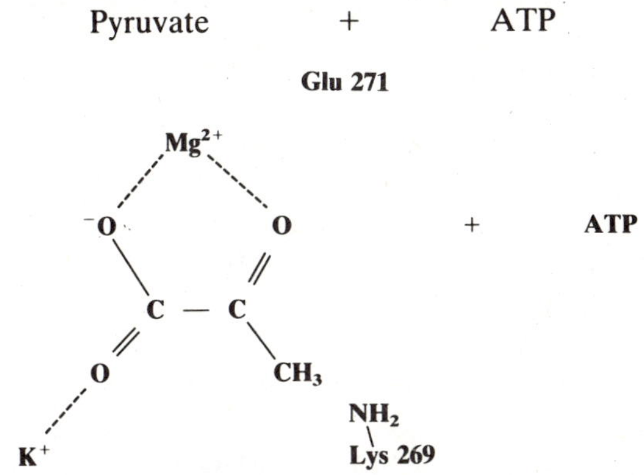

Step 4 Water enters and outer sphere complexes are formed.
Step 5 Pyruvate leaves, giving an ATP complex.

ACKNOWLEDGMENTS

I thank W. W. Cleland, N. Lonberg, and G. C. Reed for communicating results prior to publication; L. A. Fothergill-Gilmore, E. Schiltz, and collaborators for their work on the amino acid sequence of the muscle enzyme; D. Barford and D. A. Clayden for model building; and A. M. Lesk for his assistance in drawing the pictures of PK. The crystallographic work at Bristol was supported by the SERC.

REFERENCES

Ainsworth, S., and MacFarlane, N. (1973). *Biochem. J.* **131**, 223–236.

Alber, T., Banner, D. W., Bloomer, A. C., Petsko, G. A., Phillips, D. C., Rivers, P. S., and Wilson I. A. (1981). *Phil. Trans. Roy. Soc. Lond. B.* **293**, 159–171.

Anderson, P. J., and Randall, R. F. (1975). *Biochem. J.* **145**, 575–579.

Banner, D. W., Bloomer, A. C., Petsko, G. A., Phillips, D. C., Pogson, C. I., and Wilson, I. A. (1975). *Nature* **255**, 609–614.

Banner, D. W., Bloomer, A. C., Petsko, G. A., Phillips, D. C., and Wilson I. A. (1976). *Biochem. Biophys. Res. Comm.* **72**, 146–155.

Barrio, J. R., Secrist, J. A., Chien, Y-H., Taylor, P. J., Robinson, J. L., and Leonard, N. J. (1973). *FEBS Lett.* **29**, 215–218.

Bergstrom, G., Ekman, P., Dahlqvist, U., Humble, E., and Engstrom, L. (1976). *FEBS Lett.* **56**, 288–291.

Blake, C. C. F., and Rice, D. W. (1981). *Phil. Trans. Roy. Soc. Lond. A.*, **293**, 93–104.

Brändén, C-I., Eklund, H., Nordstrom, B., Boiwe, T., Soderlund, G., Zeppezauer, E., Ohlsson, I., and Akeson, I. (1973). *Proc. Natl. Acad. Sci. USA* **70**, 2439–2442.

Brändén, C-I., Eklund, H., Cambillau, C., and Pryor, A. J. (1984). *EMBO J.* **3**, 1307–1310.

Brummel, M. C., Boaz, D. P., Flashner, M., and Stegink, L. D. (1976). *Biochim. Biophys. Acta* **438**, 102–107.

Burke, R. L., Tekamp-Olson, P., and Najarian, R. (1983). *J. Biol. Chem.* **258**, 2193–2201.

Carrell, H. L., Rubin, B. H., Hurley, T. J., and Glusker, J. P. (1984). *J. Biol. Chem.* **259**, 3230–3236.

Colman, R. F., (1983). *Ann. Rev. Biochem.* **52**, 67–91.

Creighton, D. J., and Rose, I. A. (1976). *J. Biol. Chem.* **251**, 61–68.

Dann, L. G., and Britton, H. G. (1978). *Biochem J.* **169**, 39–54.

Dennis, E. S., Gerlach, W. L., Pryor, A. J., Bennetzen, J. L., Inglis, A., Llewellyn, D., Sachs, M. M., Ferl, R. J., and Peacock, W. J. (1984). *Nucleic Acids Res.* **12**, 3983–4000.

Dougherty, T. M., and Cleland, W. W. (1985). *Biochemistry*.

Dunaway-Mariano, D., Benovic, J. L., Cleland, W. W., Gupta, R. K., and Mildvan, A. S. (1979). *Biochemistry* **18**, 4347–4354.

Ekman, P., Dahlqvist, U., Humble, E., and Engstrom, L. (1976). *Biochim. Biophys. Acta* **429**, 374–382.

Evans, P. R., Farrants, G. W., and Hudson, P. J. (1981). *Phil. Trans. R. Soc. Lond. B.* **293**, 53–62.

Flashner, M., Tamir, I., Mildvan, A. S., Meloche, H. P., and Coon, M. J. (1973). *J. Biol. Chem.* **248**, 3419–3425.

Granot, J., Kondo, H., Armstrong, R. N., Mildvan, A. S., and Kaiser, E. T. (1979). *Biochemistry* **18**, 2339–2345.

Gupta, R. K., and Mildvan, A. S. (1977). *J. Biol. Chem.* **252**, 5967–5976.

Gupta, R. K., Oesterling, R. M., and Mildvan A.S. (1976). *Biochemistry* **15**, 2881–2887.

Harkins, R. N., Nocton, J. C., Russell, M. P., Fothergill, L. A., and Muirhead, H. (1983). *Eur. J. Biochem.* **136**, 341–346.

Hassett, A., Blattler, W., and Knowles, J. R. (1982). *Biochemistry* **21**, 6335–6340.

Hers, H. G., and Hue, L. (1983). *Ann. Rev. Biochem.* **52**, 617–653.

Hoar, C. G., Nicoll, G. W., Schiltz, E., Schmitt, W., Bloxham, D. P., Byford, M. F., Dunbar, B., and Fothergill, L. A. (1985). (in preparation).

Hol, W. G. J., van Duijnen, P. T., and Berendsen, H. J. C. (1978). *Nature* **273**, 443–446.

Hollenberg, P. F., Flashner, M., and Coon, M. J. (1971). *J. Biol. Chem.* **246**, 946–953.

Hunsley, J. R., and Suelter, C. H. (1969). *J. Biol. Chem.* **244**, 4815–4822.

Ibsen, K. H. (1977). *Cancer Res.* **37**, 341–353.

Johnson, S. C., Bailey, T., Becker, R. R., and Cardenas, J. M. (1979). *Biochem. Biophys. Res. Commun.* **90**, 525–530.

Kayne, F. J. (1973). In *The Enzymes*, P. D. Boyer, Ed., 3rd ed., Vol. 8. Academic Press, New York, pp. 353–382.

Kayne, F. J., and Price, N. C. (1972). *Biochemistry* **11**, 4415–4420.

Konnert, J. H., (1976). *Acta Cryst.* **A32**, 614–617.

Kubowitz, F., and Ott, P. (1944). *Biochem J.* **317**, 193–203.

Kuczenski, R. T., and Suelter, C. H. (1971). *Biochemistry* **10**, 2862–2872.

Lebioda, L., Hatada, M. H., Tulinsky, A., and Mavridis, I. (1982). *J. Mol. Biol.* **162**, 445–458.

Levine, M., Muirhead, H., Stammers, D. K., and Stuart, D. I. (1978). *Nature* **271**, 626–630.

Lindqvist, Y., and Brändén, C-I. (1980). *J. Mol. Biol.*, **143**, 201–211.

Lonberg, N., and Gilbert, W. (1983). *Proc. Natl. Acad. Sci. USA*, **80**, 3661–3665.

Lonberg, N., and Gilbert W. (1985). *Cell* **40**, 81–90.

Lowe, G., Cullis, P. M., Jarvest, R. L., Potter, B. V. L., and Sproat, B. S. (1981). *Phil. Trans. Roy. Soc. Lond. B.* **293**, 75–92.

Matsuura, Y., Kusunoki, M., Harada, W., Tanaka, N., Iga, Y., Yasuoka, N., Toda, H., Narita, K., and Kakudo, M. (1980). *J. Biochem.* **87**, 1555–1558.

Mavridis, I. M., Hatada, M. H., Tulinsky, A., and Lebioda, L. (1982). *J. Mol. Biol.* **162**, 419–444.

McLachlan, A. D. (1977). *Int. J. Quant. Chem.* **12**, Suppl. i, 371–385.

Meshitsuka, S., Smith, G. M., and Mildvan, A. S. (1980). Abstract, Am. Chem. Soc., National Mtg., Biol-24.

Mildvan, A. S. (1970). In *The Enzymes*, P. Boyer, Ed., 3rd ed., Vol. 2. Academic Press, New York, p. 446.

Mildvan, A. S., and Cohn, M. (1965). *J. Biol. Chem.* **240**, 238–246.

Mildvan, A. S., Sloan, D. L., Fung, C. H., Gupta, R. K., and Melamud, E. (1976). *J. Biol. Chem.* **251**, 2431–2434.

Moore, K. J., and Bulfield, G. (1981). *Biochem. Genet.* **19**, 771–781.

Muirhead, H. (1983). *Trends in Biochem. Sci.* **8**, 326–329.

Muirhead, H., Grant, J. P., Lawton, M. A., Midwinter, C. A., Nocton, J. C., and Stuart, D. I. (1981). *Biochem. Soc. Trans.* **9**, 212–213.

Muirhead, H., Clayden, D. A., and Barford, D. (1984). *13th International Congress of Crystallography* **2**, 1–24.

Muirhead, H., Clayden, D. A., Barford, D., Lorimer, C. G., Fothergill-Gilmore, L. A., Schiltz, E., and Schmitt, W. (1986). EMBO J. **5**, 475–481.

Nowak, T., and Mildvan, A. S. (1972). *Biochemistry* **11**, 2819–2828.

Ohlsson, I., Nordstrom, B., and Brändén, C-I. (1974). *J. Mol. Biol.* **89**, 339–354.

Pai, E. F., Sachsenheimer, W., Schirmer, R. H., and Schultz, G. E. (1977). *J. Mol. Biol.* **114**, 37–45.

Peters, J., Nash, H. R., Eicher, E. M., and Bulfield, G. (1981). *Biochem. Genet.* **19**, 759–769.

Phillips, F. C., and Ainsworth, S. (1977). *Int. J. Biochem.* **8**, 729–738.

Reuben, J., and Cohn, M. (1970). *J. Biol. Chem.* **245**, 6539–6546.

Reuben, J., and Kayne, F. J. (1971). *J. Biol. Chem.* **246**, 6227–6234.

Reynard, A. M., Hass, L. F., Jacobsen, D. D., and Boyer, P. D. (1961). *J. Biol. Chem.* **236**, 2277–2283.

Robinson, J. L., and Rose, I. A. (1972). *J. Biol. Chem.* **247**, 1096–1105.

Rose, I. A. (1960). *J. Biol. Chem.* **235**, 1170–1177.

Rose, I. A. (1970). *J. Biol. Chem.* **245**, 6025–6056.

Rose, I. A. (1981). *Phil. Trans. R. Soc. Lond. B.* **293**, 131–143.

Rossmann, M. G. (1981). *Phil. Trans. R. Soc. Lond. B.* **293**, 191–203.

Rossmann, M. G. and Argos, P. (1975). *J. Biol. Chem.* **250**, 7525–7532.

Rossmann, M. G., Moras, D., and Olsen, K. W. (1974). *Nature* **250**, 194–199.

Russell, M. P., Harkins, R. N., and Fothergill, L. A. (1979). *Biochem. Soc. Trans.* **7**, 714–715.

Stammers, D. K., and Muirhead, H. (1975). *J. Mol. Biol.* **95**, 213–225.

Steitz, T. A., Shoham, M., and Bennett, W. S. (1981). *Phil. Trans. R. Soc. Lond. B.* **293**, 43–52.

Sternberg, M. J. E., Cohen, F. E., Taylor, W. R., and Feldmann, R. J. (1981). *Phil. Trans. R. Soc. Lond. B.* **293**, 177–189.

Stuart, D. I., Levine, M., Muirhead, H., and Stammers, D. K. (1979). *J. Mol. Biol.* **134**, 109–142.

Tietz, A., and Ochoa, S. (1958). *Arch. Biochem. Biophys.* **78**, 477–493.

Watson, H. C., Walker, N. P. C., Shaw, P. J., Bryant, T. N., Wendell, P. L., Fothergill, L. A., Perkins, R. E., Conroy, S. C., Dobson, M. J., Tuite, M. F., Kingsman, A. J., and Kingsman, S. M. (1982). *EMBO J.* **1**, 1635–1640.

Wieker, H. J., and Hess, B. (1972). *Hoppe-Seyler's Z. Physiol. Chem.* **353**, 1877–1893.

4

Structural Basis for Catalysis by Aspartate Aminotransferase

J. N. JANSONIUS
M. G. VINCENT
Department of Structural Biology
Biozentrum, University of Basel
Basel, Switzerland

CONTENTS

1. INTRODUCTION

2. SOLUTION PROPERTIES OF ASPARTATE AMINOTRANSFERASE

3. CRYSTALLOGRAPHIC STUDIES
3.1. Historical Overview
3.2. Experimental
3.3. The Structure of the PLP Form of Mitochondrial Aspartate Aminotransferase
3.4. The "Closed" Structure of the Inhibited PLP Enzyme
3.5. The Structures of Pig and Chicken Cytosolic Aspartate Aminotransferase
3.6. Inhibition Experiments with Crystalline AAT
3.7. Optical and Kinetic Studies of AAT Crystals

4. CORRELATION OF CRYSTALLOGRAPHIC RESULTS WITH CHEMICAL AND PHYSICOCHEMICAL DATA
4.1. Amino Acid Sequences
4.2. The Stereochemistry of Enzymic Transamination
4.3. Binding of Substrates and Inhibitors
4.4. Properties of the Coenzyme in AAT
4.5. Chemical Modifications
4.6. Differences in Behavior of Cytosolic and Mitochondrial Isoenzymes

5. CATALYTIC MECHANISM
5.1. Early Proposals
5.2. Kinetic Studies
5.3. pK Values of Active Site Groups
5.4. Crystallographic Proposals
5.5. Kinetic Isotope Effects

6. FUTURE STUDIES

REFERENCES

1. INTRODUCTION

Pyridoxal-5′-phosphate (PLP) or vitamin B_6 aldehyde (Fig. 1) is an extremely versatile catalyst. In aqueous solution, it is able to catalyze a variety of transformations of amino acids, including transamination, racemization, α decarboxylation, and, depending on the amino acid side chain, β or γ elimination and replacement reactions. All of these reactions are also catalyzed by enzymes, which use PLP as a cofactor. Naturally, the enzymatic reactions are many orders of magnitude faster, and high reaction and substrate specificity are observed.

Braunstein and Kritzmann (1937) were the first to demonstrate reversible enzymic amino group transfer. Snell (1945) discovered a few years later nonenzymic conversion of pyridoxal (PL) into pyridoxamine (PM) in the presence of glutamate, and suggested that this reaction might also occur in vivo, as pyridoxal is related to the natural coenzyme of transaminating enzymes. Subsequently, it was demonstrated that pyridoxal phosphate carries out this function. These findings evoked extensive studies of catalysis of PL in solution, as well as of the corresponding enzymatic reactions that were shown to be largely catalyzed by PLP-dependent enzymes. Independently, Braunstein and Shemyakin (1953) and Metzler et al. (1954) presented a mechanistic scheme for PL catalysis, which has become textbook knowledge (Metzler, 1977) and is believed to apply to both nonenzymic and enzymic PL catalysis.

The first PLP enzyme to be isolated in pure form and in large quantities was aspartate aminotransferase (Jenkins et al., 1959). This paved the way for extensive studies of its properties by a wide range of techniques, and although in subsequent years, many further PLP enzymes have been isolated, aspartate aminotransferase has remained the most extensively studied representative of its class to date. Before we embark on a description of the X-ray crystallographic studies that started independently in three laboratories in the mid-1970s, we will give a brief overview of the properties of aspartate aminotransferase as revealed by solution studies.

2. SOLUTION PROPERTIES OF ASPARTATE AMINOTRANSFERASE

Although partially purified preparations of aspartate aminotransferase had been obtained in 1945 (Lichstein et al., 1945; Green et al., 1945), the isolation, by Jenkins et al. (1959) of the pig heart (cytosolic) enzyme in large quantities and in rather pure form marks the end of the prehistory of this enzyme and the start of extensive investigations of its properties. Over the years, the physicochemical, structural, spectroscopic, and catalytic properties of both cytosolic and mitochondrial (Boyd, 1961) aspartate aminotransferase have been studied in great detail. Several excellent reviews of the results of these studies have appeared. The more recent reviews include Ivanov and Karpeisky (1969), Fasella and Turano (1970), Dunathan (1971), Braunstein (1973), Metzler (1979), and Torchinsky (1985). The most complete and up-to-date source of information on the enzyme is provided by the recent monograph, *Transaminases* (Christen and Metzler, 1985).

Aspartate:2-oxoglutarate aminotransferase (EC 2.6.1.1), usually called aspartate aminotransferase (AAT) or aspartate transaminase (and formerly glutamate, aspartate transaminase (GAT), or glutamate:oxaloacetate transaminase (GOT)), is an α_2 dimeric enzyme of $M_r{\sim}90{,}000$, which contains one molecule of the coenzyme PLP per subunit. It catalyzes the reversible reaction

$$\text{L-aspartate} + \text{2-oxoglutarate} \rightleftharpoons \text{oxaloacetate} + \text{L-glutamate}$$

by means of a ping-pong bi-bi mechanism (Cleland, 1963), in which the substrate amino group temporarily resides on the coenzyme (pyridoxamine phosphate, PMP). In the PLP form of the enzyme, the coenzyme is covalently linked to the ϵ-amino group of a specific lysine residue by means of an aldimine linkage (Hughes et al., 1962). In higher animals, two isoenzymes are found that are encoded on discrete nuclear genes: a cytosolic form (cAAT) and a mitochondrial form (mAAT). The latter is synthesized as a precursor with an N-terminal extension which is imported into the mitochondria and processed to the mature enzyme (Jaussi et al., 1982; Christen et al., 1984).

The complete amino acid sequences of pig cAAT (Ovchinnikov et al., 1973; Doonan et al., 1974, 1975), pig mAAT (Kagamiyama et al., 1977, 1980; Barra et al., 1977, 1980), chicken cAAT (Shlyapnikov et al., 1979), chicken mAAT (Graf-Hausner et al., 1983), rat mAAT (Huynh et al., 1980), and *Escherichia coli* AAT (Kondo et al., 1984) and partial sequences of the enzymes from turkey mitochondria (Barra et al., 1979), horse cytosol (Martini et al., 1984), and mitochondria (Martini et al., 1983) and from human mitochondria (Barra et al., 1984) have been determined.

The cytosolic enzymes consist of 411−412 amino acid residues per chain, whereas the mitochondrial enzymes are comprised of 401 residues, and the *E. coli* enzyme of 396 residues. The sequence identity of corresponding isoenzymes from different species is about 85%, much higher than that between isoenzymes from the same species (46% for chicken and 48% for pig; Christen et al., 1984). The *E. coli* enzyme is equally distant from both vertebrate isoenzymes, having about 40% sequence identity with both the mitochondrial and cytosolic enzymes from pig, while ~30% of the residues are identical in all three enzymes (Kondo et al., 1984). It has become customary to number all sequences relative to the "oldest" and longest sequence of pig cAAT. In this numbering scheme, Lys 258 is the coenzyme-binding residue. Although cooperativity (negative or positive, depending on pH) in coenzyme binding has been observed (Arrio-Dupont and Vergé, 1982), manifold experiments document subunit independence in catalysis (Boettcher and Martinez-Carrion, 1975, 1976; Schlegel et al. 1977; Schlegel and Christen, 1978).

The essential features of the coenzyme for proper binding and for catalysis have been determined through studies with coenzyme analogs (e.g., Furbish et al., 1969; for reviews, see Braunstein, 1973; Metzler and Fonda, 1985). Such studies supported the notion that PLP binds to the enzyme as a dipolar ion, with a protonated N1 and a deprotonated 3′-OH (Fig. 1),

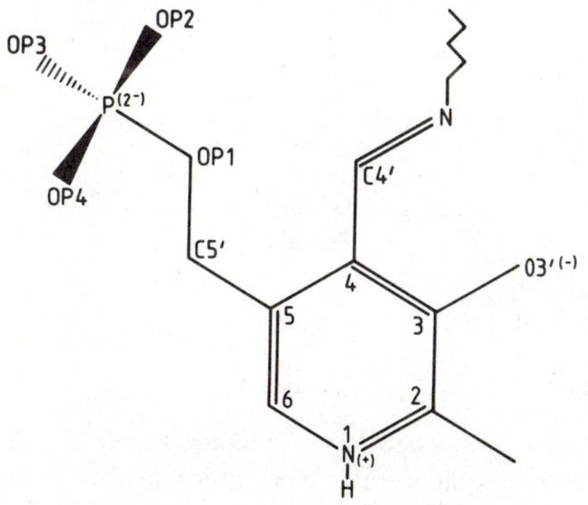

Figure 1. Pyridoxal phosphate, as it is bound in aspartate aminotransferase, with an aldimine double bond to Lys 258, a protonated N1, and a deprotonated O3′. The numbering of the atoms is standard with the exception of the four oxygens binding to the phosphorus atom, which have been labeled as in Kirsch et al. (1984).

corresponding to the assumption in earlier reaction schemes for PLP-assisted catalysis of reactions with amino acids (Metzler et al., 1954; Braunstein and Shemyakin, 1952).

The reaction scheme for a half-transamination by AAT, as it is understood now, is shown in Figure 2. PLP acts as a pH indicator. At low pH, the aldimine nitrogen is protonated, and the enzyme shows a maximum in its absorption spectrum at ~430 nm. At higher pH, this peak gradually disappears with concomitant appearance of a maximum at ~360 nm (Ia, Fig. 2). The pK_a of this process is normally 6.3 (Jenkins et al., 1959), but it depends on the concentration of certain anions. The active form of the enzyme is the high pH form with unprotonated aldimine. When an amino acid substrate enters the active site, it first forms a Michaelis complex with the enzyme. Subsequently, a proton from its charged amino group shifts to the coenzyme aldimine to produce the yellow 430 nm absorbing species (Ib, Fig. 2).

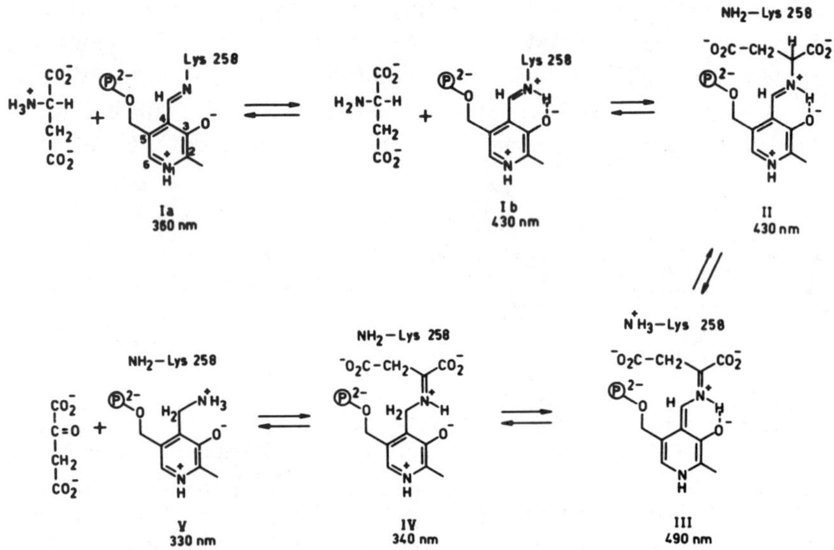

Figure 2. Reaction scheme for a half-transamination reaction of aspartate aminotransferase with aspartate to produce oxaloacetate. The intermediates shown correspond to: I, PLP enzyme + aspartate, where Ia is the unprotonated internal aldimine and Ib the protonated internal aldimine; II, external aldimine; III, quinonoid intermediate; IV, ketimine intermediate; V, PMP enzyme + oxaloacetate. The approximate wavelengths of absorption maxima of the intermediates are given. The role of Lys 258 as proton donor/acceptor (Section 5) is emphasized. For further information, see text.

Nucleophilic attack of the uncharged substrate amino group on the C4' carbon atom results in a transient geminal diamine intermediate absorbing at ~340 nm. In a second step, Lys 258 is released with an uncharged ε-amino group, while an "external" aldimine is formed between coenzyme and substrate (II, Fig. 2). This process is known as "transaldimination." The next step is deprotonation of the α-carbon atom by a proton-attracting group in the active site. The released, unprotonated lysine ε-amino group (of Lys 258) has often been considered an attractive candidate for this function (Snell, 1962; Ivanov and Karpeisky, 1969; Morino et al., 1974). Alternatively, a histidine residue has been implicated on the basis of the observation that enzyme inactivation exactly paralleled photo-oxidation of two histidine residues, while substrate binding was not impaired and slow transamination still occurred, suggesting that the active site was essentially intact (Martinez-Carrion et al., 1967, 1970a; Peterson and Martinez-Carrion, 1970).

The α-deprotonation results in a quinonoid intermediate that is stabilized by resonance of its extensive system of conjugated double bonds, and has an absorption maximum near 490 nm (III, Fig. 2). It is, however, only a transient species with normal substrates. Spectroscopic evidence for the occurrence of this intermediate has been found for the cytosolic isoenzyme, but not for the mitochondrial isoenzyme (Jenkins and Taylor, 1965; Michuda and Martinez-Carrion, 1969a). On the other hand, a very strong and persistent 495 nm band with a shoulder near 465 nm is observed with both enzymes upon addition of D, L-*erythro*-3-hydroxyaspartate, which is a slowly transaminating quasi-substrate. Only the L-isomer interacts with AAT, which is a phenomenon generally observed with substrates and inhibitors of this enzyme (Jenkins et al., 1959).

The rate-determining step in transamination of *erythro*-3-hydroxyaspartate occurs after formation of the ketimine intermediate, which is the product of reprotonation at the C4' carbon atom (Hammes and Haslam, 1969), whereas in the reaction with normal substrates it probably corresponds to the formation of the ketimine intermediate, which absorbs at 330−340 nm (IV, Fig. 2) (Fasella and Hammes, 1967). In this intermediate, the ketimine double bond is no longer conjugated with the aromatic ring. Hydrolysis can occur, leading to the product complex of PMP enzyme (with 330 nm absorbance) and oxoacid product (V, Fig. 2), from which the latter dissociates to make room for the oxoacid substrate in the second half-reaction.

Further studies with the enzyme, together with their interpretation in light of the known three-dimensional structure, will be reviewed after a detailed description of the structure and crystallographic inhibition studies of relevance for the catalytic mechanism will be given in the next sections. Although a large number of references has been included in this review, the list should by no means be considered complete. It is simply impossible to do justice to the incredibly vast amount of literature on the subject that has been accumulated over the years. Therefore, we had to make a choice that is undoubtedly subjective. However, we hope to have covered most studies that are especially relevant to the catalytic mechanism. Less emphasis has been given to chemical and physical studies that produced structural information and to studies that cannot (yet) be related to the spatial structure of the enzyme. We apologize to all those authors whose valuable work has been sacrificed to meet the limitations of this review.

3. CRYSTALLOGRAPHIC STUDIES

3.1. Historical Overview

After several decades of intensive solution studies on AAT, the successful growth of large, well-diffracting crystals of three different enzymes were reported for chicken cAAT (Borisov, et al., 1977), pig cAAT (Arnone et al., 1977), and chicken mAAT (Gehring et al., 1977). In fact, Borisov et al. (1977) also described a low-resolution (5 Å) electron density map together with a possible chain tracing. A more detailed report appeared 1 year later (Borisov et al., 1978). The enzyme was described as a symmetrical dimer with subunits that could be approximated by ellipsoids with axes $40 \times 48 \times 60$ Å. Nine helices were recognized, one of which was 40 Å long. The crystals (see Table 1) were grown in the presence of D, L-2-methylaspartate and displayed the yellow color typical for the 430 nm absorption band of a protonated pyridoxal phosphate aldimine. However, addition of cysteine sulfinate resulted in rapid disappearance of this color due to exchange of methylaspartate in the active site for cysteine sulfinate, followed by transamination to the PMP enzyme. With 2-oxoglutarate, the yellow color reappeared. Thus, the active sites were both accessible in the crystal, and the enzyme was catalytically competent.

The apoenzyme was produced by extensive soaking of crystals of the PMP enzyme, prepared as described above, in a solution containing 0.8 M

Table 1. Crystal Forms of Aspartate Aminotransferase

Enzyme Source	Crystallization Conditions	Space Group	Cell Dimensions Axes (Å); Angles (deg)			Z	V_M	Enzyme Form or Derivative	References
Chicken cytosol	2 M CsCl; ~30% sat. ammonium sulfate; 100 mM potassium phosphate, pH 7.5; 50 mM 2-methylaspartate; 0.5 mM EDTA; 7.5% MPD at 4°C.	P2₁2₁2₁	62.7 90.0	118.1 90.0	124.5 90.0	2	2.5	PLP enzyme + 2-methylaspartate	a,b
Pig cytosol	8% PEG, M_r 6000; 40 mM sodium acetate, pH 5.4 at 4°C.	P2₁2₁2₁	125.1 90.0	130.8 90.0	55.7 90.0	2	2.5	PLP enzyme	c, d
Chicken mito-chondria	16−20% PEG, M_r 4000; 10 mM sodium phosphate, pH 7.5 at room temperature.	P1	55.6 85.2	58.6 109.2	75.9 115.6	2	2.3	PLP enzyme; apoenzyme	e
	22−24% PEG, M_r 4000; 20 mM sodium phosphate, pH 7.5 at room temperature.	P2₁	57.1 90.0	52.2 101.4	136.3 90.0	2	2.2	PLP enzyme + 2-methylaspartate; apoenzyme + PPL-Asp	f
	20−24% PEG, M_r 4000; 10 mM sodium phosphate, pH 7.5 at 4°C.	C222₁	69.7 90.0	91.4 90.0	128.5 90.0	1	2.3	PLP enzyme + 2-methyl-aspartate or maleate; apoenzyme + PPL-Asp or PPL-Glu	g, h
Pig mito-chondria	16−17% PEG, M_r 4000; 10 mM sodium phosphate, pH 7.5 at room temperature.	P1	55.9 85.0	58.5 109.8	77.0 115.9	2	2.4	PLP enzyme	i
Ox mito-chondria	18% PEG, M_r 4000; 10 mM glycine-NaOH, pH 9.1 at room temperature.	P1	55.6 85.5	58.6 109.1	76.0 115.6	2	2.3	PLP enzyme	j

Source: a, Borisov et al. (1977); b, Borisov et al. (1978); c, Arnone et al. (1977); d, Arnone et al. (1982); e, Gehring et al. (1977); f, Thaller et al. (1981); g, D. Picot (unpublished); h, Jansonius et al. (1985); i, Eichele et al. (1979b); j, Capasso et al. (1979).

Abbreviations: Z, number of subunits per asymmetric unit; V_M, unit cell volume (Å) per dalton molecular weight (Matthews, 1968); MPD, 2-methyl-2,4-pentane diol; PEG, polyethylene glycol; the percentages are to be taken as (w/v); and PPL-Asp, N-(5'-phosphopyridoxyl)-L-Aspartate.

195

phosphate at pH 4.95. A subsequent electron density difference map showed two symmetry-related high-density features, about 30 Å apart, which were identified as the PL moieties. These defined the active sites, which were found near the subunit interface. Another interesting feature that these authors observed was the occurrence of a large conformational change in one subunit with respect to the "parent" 2-methylaspartate complex upon formation of the free PLP enzyme, the PMP enzyme, and an isonicotinoyl hydrazon derivative (but *not* of the apoenzyme) at a distance 20–30 Å from the coenzyme.

A 4.4 Å resolution study of chicken mAAT (Eichele et al., 1979a) performed on triclinic crystals of the PLP enzyme with one dimer per unit cell (Table 1) confirmed many of the structural features described for the cytosolic isoenzyme. Not unexpectedly, these isoenzymes appeared to be structurally homologous. Five of the seven helices, identified with certainty in mAAT, could be correlated with a helix found in cAAT. The longest helix in both structures was 50 Å long in mAAT. The molecular dimensions were somewhat larger: 70 × 50 × 40 Å for the subunit and 105 × 60 × 50 Å for the isologous dimer. Directly grown apoenzyme crystals were found to be isomorphous with the PLP holoenzyme. A difference map confirmed the coenzyme positions found by Borisov et al. (1978). The phosphate positions were determined by soaking apoenzyme crystals in an arsenate solution, thus making use of the higher electron density of the latter. The arsenate and, by implication, the phosphate positions in the apoenzyme and holoenzyme were found adjacent to the PL moieties, but closer to the molecular dyad, and they corresponded to high density in the electron density map of the holoenzyme.

Further information came from a derivative, produced by soaking apoenzyme crystals in a solution containing *N*-(5′-phosphopyridoxyl)-3-iodotyrosine (PPL-iodo Tyr), that replaced the inorganic phosphate and produced an analog of covalent intermediates in transamination of 3-iodo Tyr. The latter acts as a substrate for mAAT. A difference map revealed the two iodine positions, which were symmetrically positioned near the active site wall opposite to the neighboring subunit. This low-resolution map showed another interesting feature. Density corresponding to a single stretch of the polypeptide chain, 45 Å long, disconnected itself from one subunit near the entrance of the active site, crossed over to the neighboring subunit, and ended there in a noncovalent interaction. Obviously, this corresponded to either the N- or the C-terminus of the polypeptide chain. A subsequent 3.2

Å electron density map, briefly mentioned by Eichele et al. (1979a), could be interpreted in terms of the complete chain tracing. Side chain density was correlated with the known amino acid sequence of pig mAAT (Kagamiyama et al., 1977) and preliminary sequence information on chicken mAAT from the laboratory of P. Christen. The isolated stretch of polypeptide chain turned out to correspond to the N-terminus. The details of the chain folding, which has been confirmed at higher resolution, will be described later.

Both the apoenzyme and the PPL-iodo Tyr derivative were fully isomorphous with the PLP enzyme, in contrast to the results described by Borisov et al. (1978). However, a subsequent soaking experiment with apoenzyme crystals in a solution containing N-(5′-phosphopyridoxyl)-L-aspartate (PPL-Asp) indicated a large change in conformation, since the crystals tended to shatter. Only under the mildest conditions (0.2 mM PPL-Asp at 4°C) did some crystals survive and could X-ray data be collected. A difference map at 3 Å resolution showed strong alternating positive and negative features localized in one subunit, which were indicative of a bodily movement of a large portion of the structure over 3−4 Å towards the active site (Eichele et al., 1979a). Clearly, this phenomenon corresponded to that observed by Borisov et al. (1978), and the asymmetry had to be ascribed in both cases to inhibition of the movement in one subunit by intermolecular contacts in the crystals.

The chain-folding and active site structure of chicken mAAT, as seen at 2.8 Å resolution, have been described by Ford et al. (1980), and in more detail by Eichele (1980) and Jansonius et al. (1985). Inhibitor-binding studies were carried out by Eichele (1980). The results of these studies, in combination with an accurately built structural model (on a graphics display system) and model-building studies of catalytic intermediates, were used as a basis for a mechanistic proposal (Kirsch et al., 1984).

Cocrystallization of chicken mAAT with D, L-2-methylaspartate gave rise to two additional crystal forms (Table 1). At room temperature, monoclinic crystals could be grown from polyethylene glycol (PEG) solutions (Thaller et al., 1981). They belong to space group P2$_1$ and have one dimeric molecule per asymmetric unit. Its structure has been determined at 4.4 Å resolution (M. G. Vincent, C. Thaller and J. N. Jansonius, unpublished results; Jansonius et al., 1985).

The conformational change, observed for the PPL-Asp derivative, was confirmed; and, as expected, it was expressed identically in both subunits. The same crystal form was produced with a PPL-Asp derivative prepared by combining the apoenzyme with the coenzyme substrate analog in solution (Jansonius et al., 1984a). We refer to the resulting structure as the "closed" conformation.

A third crystal form was likewise first obtained with the 2-methylaspartate derivative. These crystals grew at 4°C from a somewhat more concentrated PEG solution. They are orthorhombic (space group $C222_1$) and are, to date, the only crystal form of AAT from any species with crystallographically expressed twofold symmetry (D. Picot, M. G. Vincent, G. Eichele, and J. N. Jansonius, unpublished results; Kirsch et al., 1984; Jansonius et al., 1984a, 1984b, 1985). At low resolution, the molecular structure of this form is identical to that in the $P2_1$ crystals. Several other enzyme derivatives have produced the same crystal form: PLP enzyme + maleate, PLP enzyme + succinate, and apoenzyme + PPL-Glu. X-ray data to 2.3 Å resolution have been collected from the maleate derivative and the 2-methylaspartate derivative. A least-squares refinement of the former is complete, the final R factor being 15.9%.

X-ray data to 1.92 Å resolution have been collected from the PLP enzyme in space group P1. Refinement of this structure is underway as well. The current R factor is 17.8% at 2.3 Å resolution.

The next step in the study of chicken cAAT was a 3.5 Å electron density map computed with improved phases due to use of the noncrystallographic symmetry (Borisov et al., 1980a, 1980b). However, not all elements of secondary structure were discernible from the map. Further inhibitor-binding studies were described. Finally, at 2.8 Å resolution, the near identity of the structure to that of chicken mAAT (Ford et al., 1980) was recognized. Some small differences were pointed out, mainly arising from insertions in the amino acid sequence. Additional inhibitor-binding studies with arsenate-containing substrate analogs were reported (Borisov et al., 1985).

In the meantime, an independent study of this enzyme based on the same crystals, but with 2-oxoglutarate in place of 2-methylaspartate, was carried out also in Moscow (Harutyunyan et al., 1982, 1984, 1985; Torchinsky et al., 1983). The structural near identity to mAAT was again confirmed, and some additional information on inhibitor binding and conformational changes was given. Lattice-induced asymmetry in inhibitor binding and in the conformational changes induced by these inhibitors were observed again in these studies.

An analogous situation occurred in the structural work on pig cAAT (Arnone et al., 1977, 1982, 1984, 1985b; Hyde et al., 1984). Again, crystals were used with one dimer per asymmetric unit (Table 1), and similar lattice-induced asymmetric conformational changes were observed upon inhibitor binding. In line with this, the β subform of pig cAAT, in which a covalently bound, so far unknown (Arnone et al., 1985b) substrate-like inhibitor inactivates one subunit and induces the "closed" conformation, crystallizes "orderly" with the inhibited subunit always occupying the same crystallographic position corresponding to that of the flexible subunit in the native enzyme.

Since the active α subform of pig cAAT was crystallized in the absence of inhibitors, the conformational changes observed upon inhibitor binding were in the opposite direction to those observed for chicken cAAT.

AAT crystals from two additional species have been described (Table 1). Capasso et al. (1979) crystallized bovine mAAT from PEG and obtained crystals that had essentially identical unit cell dimensions as chicken mAAT. On the other hand, crystals from pig mAAT, grown by macroseeding (Thaller et al., 1981) in PEG with a crystal of chicken mAAT, had slight but significantly different cell dimensions. Nevertheless, a 3.2 Å difference electron density map between the two structures, computed with multiple isomorphous replacement (MIR) phases of the chicken enzyme, showed many features that could be correlated with differences in the amino acid sequences of these two enzymes (Eichele et al., 1979b; Eichele, 1980). Very recently, this structure has been solved by molecular replacement. It is essentially identical to that of the chicken enzyme (M. G. Vincent, unpublished results). Strong similarities in the diffraction patterns allow the same conclusion to be drawn for the bovine and chicken mAATs (Capasso et al., 1979).

The characteristic ultraviolet (UV)-visible absorption spectrum through which PLP, PMP, and their derivatives can be recognized (e.g., Braunstein, 1973) have been used to identify the coenzyme forms in crystals of pig cAAT, chicken cAAT, and chicken mAAT, both in the absence and in the presence of substrates and inhibitors. Moreover, through use of polarized light, unique information has been obtained on the directions of the transition dipole moments corresponding to the π-π^* transitions giving rise to the lowest energy absorption bands of the bound coenzyme derivatives. Since these transition dipole moments are confined to the coenzyme pyridine ring, crystallographic knowledge and polarized spectral data on a given crystalline species can be correlated in favorable cases to determine the direction of a given transition dipole moment in the coenzyme ring plane. Also, reorientations of the coenzyme with respect to the crystallographic axes can be signaled by a change in the relative absorbances of light polarized parallel to these axes.

Crystal spectral studies have been described for pig cAAT by Metzler et al. (1978), Arnone et al., (1984), and Hyde et al. (1984), for chicken cAAT by Makarov et al. (1980, 1981, 1984), for pig mAAT by Mozzarelli et al. (1979), and for chicken mAAT by Eichele et al. (1978), Kirsten (1982), and Vincent et al. (1984). A brief review of this work has been given in Chapter 5c of *Transaminases* (Arnone et al., 1985a).

3.2. Experimental

3.2.1. Crystallization

Until about 1 decade ago, ammonium sulfate, the most common precipitant in protein isolation and purification, was also the precipitant of choice in crystallization. However, it has the disadvantage of providing NH_3 as a ligand for heavy atom-containing complexes, thus preventing in many cases specific binding of such complexes to the protein. It also produces a liquid phase in the crystal of quite high electron density, which is far from being a physiological environment for the protein. Therefore, PEG, which does not have any of these disadvantages and does not penetrate into the crystal, has been used more and more frequently since its introduction by McPherson (1976, 1982).

In fact, of all AATs crystallized to date, only in the case of chicken cAAT has ammonium sulfate been used (and given some trouble by causing transaldimination; see Makarov et al., 1980, 1981). Pig cAAT, chicken mAAT (in three forms), pig mAAT, and bovine mAAT were all crystallized from PEG solutions. A summary of the crystallization conditions and the crystal data is given in Table 1.

3.2.2. Data Collection

In all but one case, the X-ray data have been collected with an automatic X-ray diffractometer (Borisov et al., 1978; Eichele et al., 1979a; Arnone et al., 1982; Harutyunyan et al., 1982). Harutyunyan et al. (1984) have recently used an area detector device. Only the high-resolution data collection on the orthorhombic form of chicken mAAT has been carried out photographically with the oscillation technique (D. Picot, unpublished results). We refer to the original literature for details on data collection.

3.2.3. Heavy Atom Derivatives

These were, as usual, produced by soaking native crystals in solutions containing the heavy atom compounds. The presence of free thiol groups in all of these enzymes was a help in producing mercury-containing derivatives. The noncrystallographic twofold symmetry was strongly expressed in heavy atom binding, and usually gave the first indication of position and direction of the molecular dyad (Borisov et al., 1978; Eichele et al., 1979a; Arnone et al., 1982).

3.2.4. Phase Determination

Standard procedures (Blundell and Johnson, 1976) were used, at least initially, for phase determination. Borisov et al. (1978, 1980b) and Arnone et al. (1982, 1984), who had difficulty in following the polypeptide chain throughout the molecule due to an insufficient number of independent heavy atom sites in their derivatives, successfully used a molecular-averaging and solvent-leveling procedure for improving their phases (Hyde et al., 1984; Borisov et al., 1985); Harutyunyan et al. (1985) also refined their isomorphous replacement phases at 2.8 Å resolution in this way.

3.2.5. Electron Density Maps

Initial electron density maps of a new structure are usually phased by the multiple isomorphous replacement (MIR) method (Blundell and Johnson, 1976). During refinement, calculated phases have to be used. In the structural refinements of chicken mAAT, coefficients $(3F_{obs}-2F_{calc})$ (Steigemann et al., 1976) rather than $(2F_{obs}-F_{calc})$ or F_{obs} were used as structure factor amplitudes in the Fourier syntheses to emphasize the errors in the current model. For modeling inhibitor derivatives, maps computed with amplitudes $(3F_{der}-2F_{nat})$ and calculated native phases—from which the contributions of (possibly) changed parts of the structure had been excluded—were of most use.

3.2.6. Functional Derivatives

Substrate and inhibitor derivative studies are usually carried out by soaking "native" crystals in buffered crystal-stabilizing solutions enriched with a sufficient concentration of the compound of interest. All inhibitor-binding studies with pig and chicken cAAT, and many of those carried out with chicken mAAT, have been done in this way (e.g., Arnone et al., 1985b; Borisov et al., 1985; Harutyunyan et al., 1985; Jansonius et al., 1985). However, all three isoenzymes displayed large conformational changes in one subunit only, although both subunits in solution act independently and identically (Boettcher and Martinez-Carrion, 1975; 1976; Schlegel et al., 1977; Schlegel and Christen, 1978). Thus, an artifactual situation was created that hampered the interpretation of the observations. In addition, it was also possible that even the observed effects in one subunit might be restricted due to lattice constraints and did not, therefore, correspond to the enzyme behavior in solution. Cocrystallization of enzyme and inhibitor is a better alternative. This method was applied to chicken mAAT inhibited with 2-methylaspartate. It gave rise to two new crystal forms, space groups $P2_1$ and $C222_1$ (Table 1). In both cases, independent structure determinations were carried out to 4.4 Å using diffractometer data (D. Picot, C. Thaller and M. G. Vincent, unpublished results). In addition, a maleate derivative structure in $C222_1$ was solved at 2.3 Å resolution with data collected with an oscillation camera. A 2.3 Å data set to the same resolution was collected from the 2-methylaspartate derivative (D. Picot, unpublished results). The crystallographic refinement of the maleate derivative is complete ($R = 15.9\%$), while that of the native enzyme is underway ($R = 17.8\%$) at 2.3 Å resolution. A meaningful comparison of the two structures is thus possible and will be discussed in the next two subsections.

3.3. The Structure of the PLP Form of Mitochondrial Aspartate Aminotransferase

3.3.1. *Subunit and Domain Structure*

Although diffractometer X-ray data of the PLP enzyme form of chicken mAAT have been collected to 1.92 Å nominal resolution (M.G. Vincent, unpublished results) only data to 2.3 Å have so far been used in the refinement of the structure with the method of Hendrickson and Konnert (1980a, 1980b). Individual temperature factors, and 352 water molecules have been included to date. Eichele (1980) and Jansonius et al. (1985) have described the molecule in considerable detail, as it presented itself in a MIR-phased electron density map at 2.8 Å resolution. Refinement has improved the orientation of peptide bonds as well as side chain conformations, but nowhere has the path of the polypeptide chain had to be altered. Thus, the descriptions given earlier suffice. Each of the two subunits, essentially identical in structure (refinement treats them as independent entities), consists of three parts: (1) an extended N-terminal arm (residues 3-14); (2) a small domain, including residues 15−47 and 326−410; and (3) a large domain consisting of residues 48−325. The latter includes Lys 258, to which the coenzyme pyridoxyl phosphate is bound through an aldimine double bond. We will use the amino acid numbering of the 412 residue pig cAAT sequence throughout (Ovchinnikov et al., 1973; Doonan et al., 1974, 1975). With respect to this, the 401 residue chicken mAAT sequence has deletions at positions 1, 2, 65, 127, 128, 131, 132, 153, 407, 411, and 412 (Graf-Hausner et al., 1983). The interface between the twofold-related subunits is made through their "large" or "coenzyme-binding" domains. The regions 47−70, 142−149, 262−266, 290−301, and residue 106 make up the interface, which is relatively small and not very tight. A significant contribution to the stabilization of the dimer is made by interaction of residues 3−11 of each subunit with the "bottom" of the coenzyme-binding domain of its neighbor: residues 122−125, 183, 249−251, and 270−286. The side chains of Trp 5 and Trp 6 are inserted into a hydrophobic pocket, lined with the side chains of Phe 122, Phe 123, Phe 125, Leu 218, Val 251, Phe 270, and Val 273. Figure 3 shows a perspective view of the α-carbon chain of a dimer with one open and one closed subunit. Figures 4−6 are stereo pictures of α-carbon chains of the dimer (closed form), the PLP-binding domain, and the small domain, respectively.

3.3.2.   Secondary Structure

mAAT has a high content of secondary structure, comprising 16 α helices, a large pleated sheet structure of seven strands (six of which run parallel), and two small double-stranded sheets (one parallel, the other antiparallel). In addition, there are numerous bends. Table 2 tabulates the secondary structure in the manner of Ford et al. (1980). In this assignment, the same criteria have been used as in Jansonius et al. (1985) to distinguish helices from bends. With this definition (essentially that of Lewis et al., 1973), bends can overlap with helices. Bend regions can be longer than four residues due to the occurrence of successive and multiple bends, as defined by Isogai et al. (1980). Only those residues taking part in at least one hydrogen bond, or

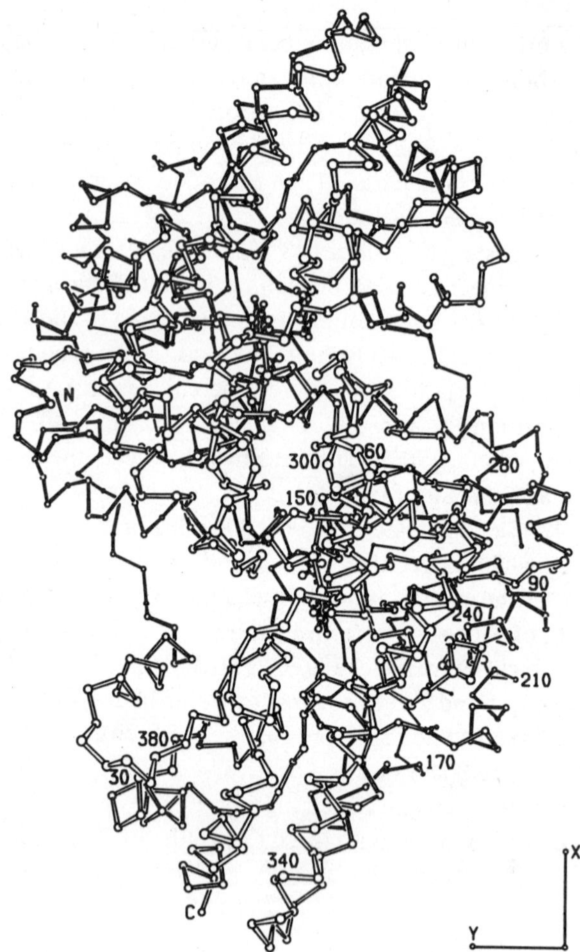

Figure 3. Perspective view along the molecular dyad of the α-carbon chain of the mAAT dimer in a hybrid open/closed structure (lower subunit S open, upper subunit S* closed) as observed upon soaking triclinic crystals of the apoenzyme (Table 1) in PPL-Asp or of the PLP enzyme in maleate (Kirsch et al., 1984). The maleate ion is shown. In the lower subunit, N- and C-termini are indicated and sequence numbers are given for selected residues. Also shown are the directions of the laboratory axes X, Y, and Z defining the molecule. X is roughly parallel to the longest dimension of the dimer, while Z coincides with the molecular dyad.

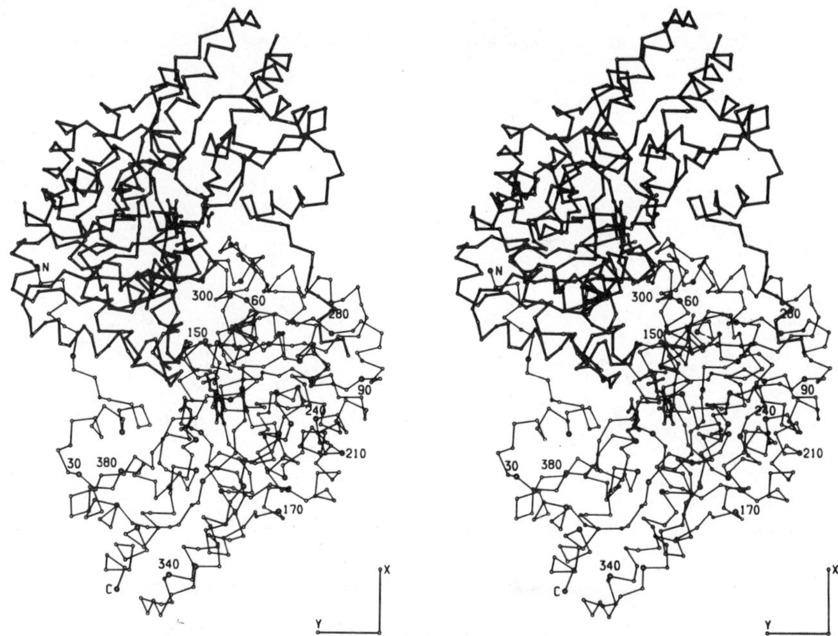

Figure 4. Stereo view along the molecular dyad of the α-carbon model of mAAT in the "closed" structure (space group C222₁) with bound maleate ion. Both PLP and maleate are included. One subunit is drawn with double lines and the other with single lines between atoms. In the latter, N- and C-termini are labeled, and every tenth residue in the (pig cAAT) sequence is drawn with a larger circle, with selected examples of these being numbered. The directions of the laboratory axes X, Y, and Z, of which Z corresponds to the molecular dyad, are indicated.

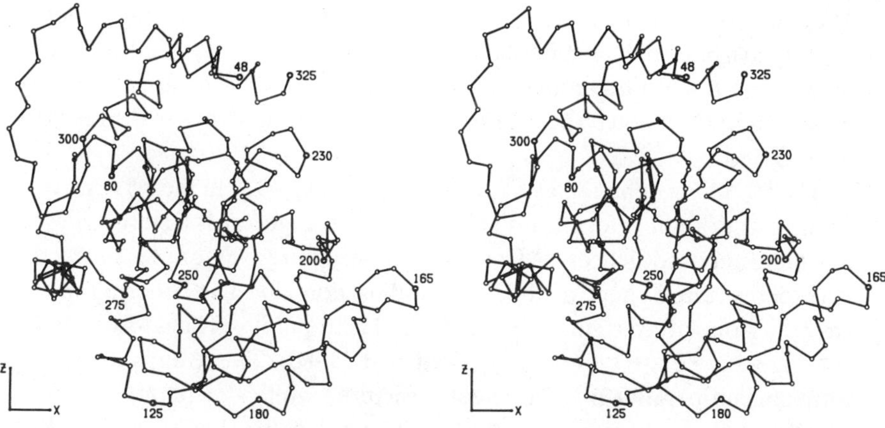

Figure 5. Stereo view, along the laboratory Y axis, of the coenzyme-binding domain, residues 48–325. Only α-carbon atoms are drawn and connected. Pyridoxal phosphate, bound covalently to Lys 258, is shown. Note the position of its phosphate near the N-terminus of helix 5 (residues 108–122), and the extensive α/β structure. Numbered residues are indicated by an enlarged circle.

205

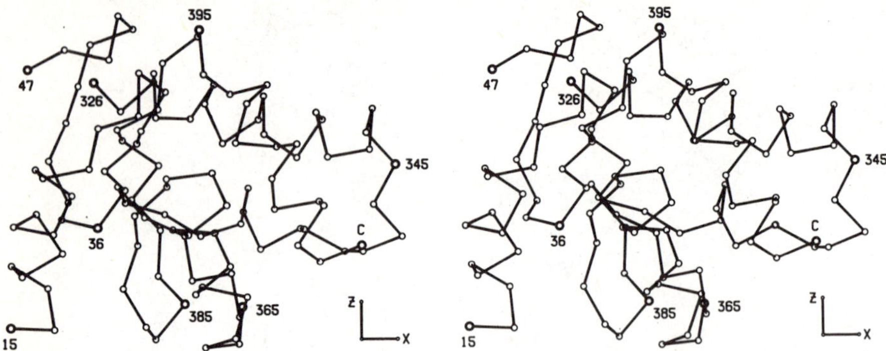

Figure 6. Stereo view, along the laboratory Y axis, of the α-carbon model of the small domain (residues 15–47 and 326–410). Numbered residues are indicated by enlarged circles. A comparison with Figures 4 and 5 demonstrates that the domain definitions are structurally logical.

having neighbors on both sides involved in hydrogen bonding, have been counted as being in pleated sheets. The assignments in Table 2 are given for both crystallographically independent subunits (S and S*) of the dimer in the P1 crystal form, as well as for the single independent subunit in the "closed" structure of the maleate derivative in space group C222₁ (see Section 3.4.). They are essentially identical. In Table 2, residues that belong simultaneously to a helix or a pleated sheet and to a bend have been listed as H or B, rather than t. With these assignments, we find 50% helical structure, 22% bends, and 14% pleated sheet structure, leaving another 14% of the residues unclassified.

The 50 Å long helix 13 has a kink near residue Gly 325 in the P1 structure. In the orthorhombic form, this is less apparent. Figure 7 is a stereo view of the superimposed helices 13 from the two structures. Figure 8*a* shows the extended pleated sheet of the PLP-binding domain. The ordering of the strands, from left to right is *a g f e d b c*. The topology (Richardson, 1981) is +5x, +1x, −2x, −1x, −1x, −1 (Ford et al., 1980). The strands *a, g, f* thus form an antiparallel pleated sheet structure, while the part *f e d b c* is parallel, and with its covering helices builds an α/β type (Levitt and Chothia, 1976) supersecondary structure: strand *b*, helix 6, strand *c*, helix 7, strand *d*, helix 8, strand *e*, helix 9, strand *f* (see also Fig. 5). Although the parallel part

Table 2. Assignment of Secondary Structure

	1	2	3	4	5	6	7	8	9	0	1	2	3	4	5	6	7	8	9	0
0	■	■	●	t	t	t	t	t	t	●	●	●	●	●	t	H	H	H	H	H
20	H	H	H	H	H	t/H	t	●	t	t	t	t	B	B	B	t/●	t/●	●	●	●
40	●	t	t	t	t	●	●	●	●	t	H	H	H	H	H	H	H	H	H	H
60	H	H	t/H	t⁺/t	■	●	●	●	●	●	●	t	t	t	t	t	H	H	H	H
80	H	H	H	H	H	H	H	H	t	t	t	t	H	H	H	H	H	t	●	B
100	B	B	B	B	B	t	t	H	H	H	H	H	H	H	H	H	H	H	H	H
120	H	H	t	t	t	t	■	■	●	B	■	■	B	B	B	t	t	t	t	t
140	H	H	H	H	H	H	H	H	H	H	t	●	■	B	B	B	B	B	●	●
160	●	t	t	t	t	t	●	●	t	H	H	H	H	H	H	H	H	H	H	t
180	t	t	B	B	B	B	B	B	B	●	●	●	t	t	t	t	t	●	●	●
200	t	H	H	H	H	H	H	H	H	H	H	H	H	H	H	●	B	B	B	B
220	B	B	B	t	t	t	t	t	t	t	t	t	H	H	H	H	H	H	H	H
240	H	H	H	H	H	H	t	t*/●	●	B	B	B	B	B	B	H	H	H	H	H
260	t	t	t	t	t	t	B	B	B	B	B	B	B	B	●	t	H	H	H	H
280	H	H	H	H	H	H	H	H	H	H	H	H	H	H	t	t	●	●	t⁺/●	H
300	H	H	H	H	H	H	H	H	H	H	H	t	H	H	H	H	H	H	H	H
320	H	H	H	H	H	H	H	H	H	H	H	H	H	H	H	H	H	H	H	H
340	H	H	H	H	t	●	●	●	●	t	H	H	H	H	H	t	●	●	B	B
360	B	●	●	●	●	t	H	H	H	H	H	H	H	H	H	H	H	B	B	B
380	●	t	t	t	t	●	B	B	B	t	t	t	t	H	H	H	H	H	H	H
400	H	H	H	H	H	H	■	H	H	t										

Structural elements (underlines spanning the residues): helices 1, 2, 3, 4, 5, 6, 7, 8, 9, 10, 11, 12, 13, 14, 15, 16; pleated-sheet strands a, a', b, b', c, c', d, d', e, f, g.

Abbreviations: B, residue in pleated sheet; H, residue in helix; t, residue in turn (bend); ■, deletion in chicken mitochondrial AAT with respect to amino acid sequence of pig cytosolic AAT; ●, residue not in element of secondary structure; / different assignment in P1 (left) and C222₁ (right); + only in subunit S; * only in subunit S*; In case of double assignments, the priority rule is H > B > t.

of the sheet reminds one of the topology of that observed in the dehydrogenases (Rossmann et al., 1974) and is, in fact, identical to that of flavodoxin (Andersen et al., 1972), the total 7-stranded sheet has, to our knowledge, a unique topology. The pleated sheet strands *a'*, *b'*, *c'*, *d'* are interconnected

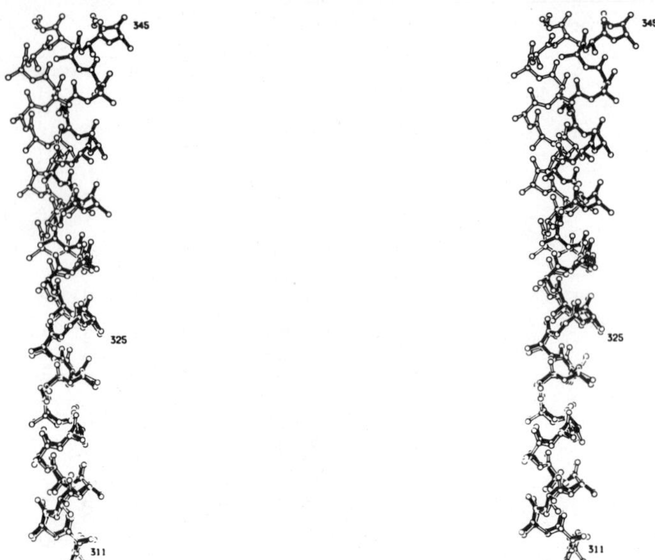

Figure 7. Stereo view of the polypeptide chain, up to and including β-carbon atoms, of residues 311−345, which encompass the long helix 13 (residues 313−344). The open and closed structure (dark bonds) are superimposed. The view direction, approximately at right angles to the helix axis and in the XZ plane, emphasizes the hinge region residues 323−325 of reorientation of the small domain. Atoms in residue 345 shift by 5.6 Å as a result of the conformational change.

through some additional hydrogen bonds as shown in Figure 8*b*. It undoubtedly contributes considerably to the integrity of the small domain, the N-terminal and C-terminal portions of which it cross-links (see also Fig. 6). With 52% helix, 20% bends, and 10% pleated sheet (leaving 18% of the residues unclassified), the secondary structure distribution of the small domain is not very different from that of the subunit overall.

3.3.3. *Hydrophobic Cores and Domain Interface*

The subunit contains three major hydrophobic cores. On both sides of the large central pleated sheet in the PLP-binding domain, hydrophobic side chains are packed between the sheet and its covering helices. The "inner core" is on the concave side of the sheet, facing the coenzyme. The "outer core" on the convex side of the sheet is larger. A third hydrophobic core is

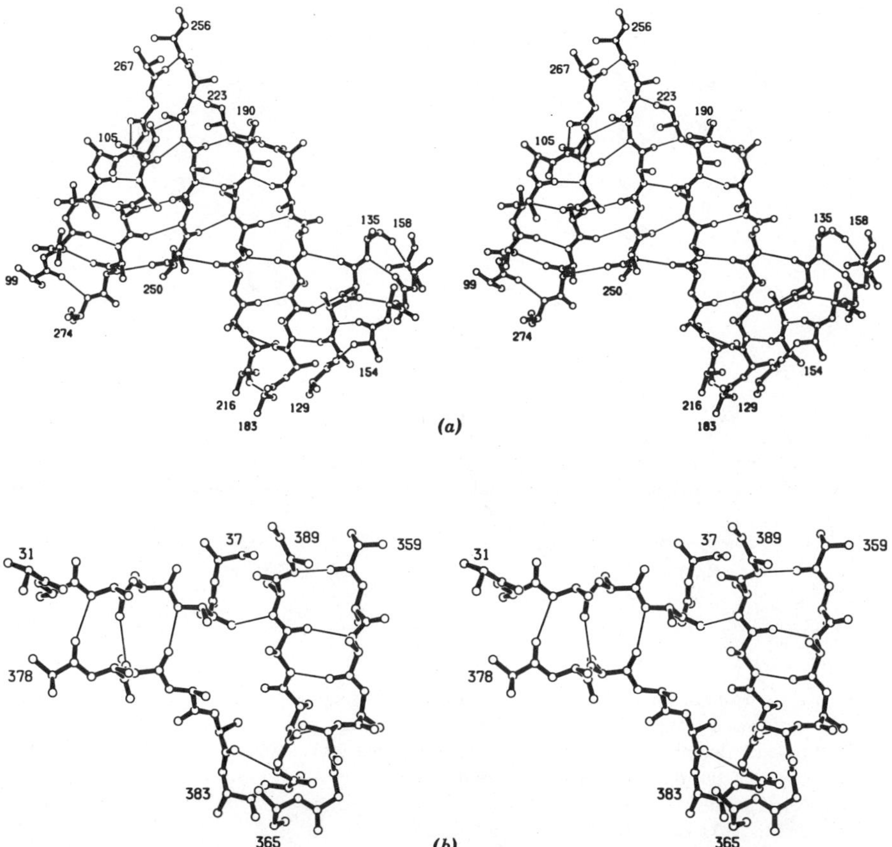

Figure 8. The pleated sheet structure of mAAT. Only main chain atoms and β-carbons are shown. Thin lines indicate hydrogen bonds. The first and last residues of each strand are labeled. (*a*) 7-stranded pleated sheet in the PLP-binding domain. Strand order from left to right: *a g f e d b c*. (*b*) The two 2-stranded pleated sheets in the small domain. Strand order from the bottom and left to right: *c'* a' d' b'. Compare with Table 2.

found in the small domain. Side chains of residues from helices 13−16 and from the pleated sheet strands *a'-d'* contribute to it. Jansonius et al. (1985) have enumerated the residues involved in the three hydrophobic cores.

The domain interface is of considerable interest because of the small domain movement occurring during catalysis (see Section 3.4.1). Figures 9*a* and 9*b* display this interface in the P1 and the C222$_1$ structures, respectively.

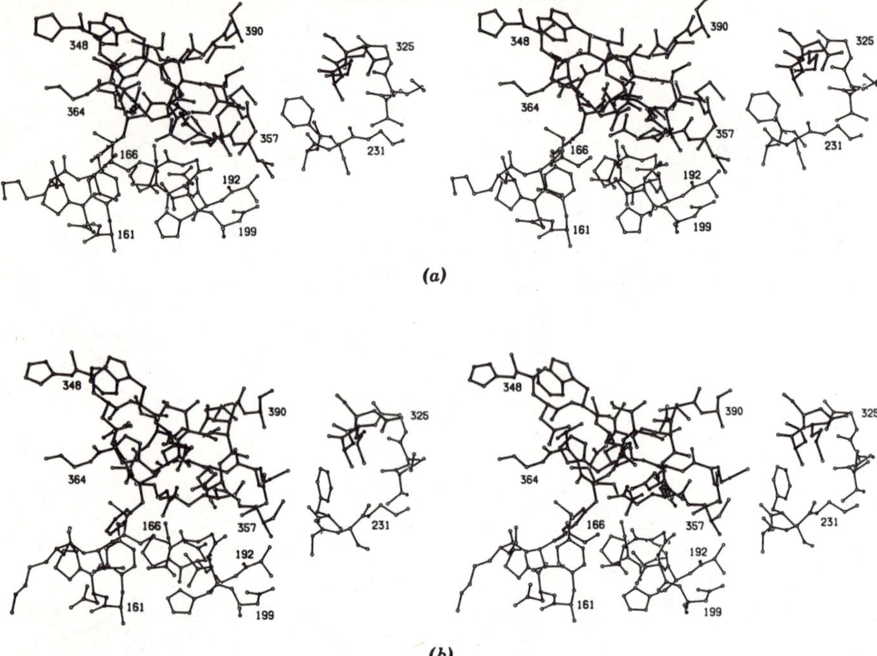

Figure 9. Domain interface in mAAT. (*a*) Open structure, space group P1. (*b*) Closed structure, space group C222$_1$. Heavy bonds: small domain residues. Light bonds: residues in PLP-binding domain. The hinge of the small domain movement is near Lys 324, and the movement in the direction open to closed is away from the viewer. Note that Cys 166 is uncovered by this conformational change. Active site residues at or near the domain interface include His 193, Asn 194, Phe 360, and Arg 386.

The interface is essentially flat, and the movement can be described as a gliding motion of the small domain surface (dark bonds) over the surface of the PLP-binding domain. We will discuss this conformational change in detail in Section 3.4.1.

3.3.4. Active Site

Figure 10*a* is a view of the active site of the PLP enzyme along the molecular dyad in the direction +Z to −Z (from the "top"). Pyridoxal phosphate is linked covalently through an aldimine double bond to the ε-amino group of Lys 258. Its A face (Ford et al., 1980) is oriented towards the protein. The pyridine moiety is surrounded by (1) Ala 224 at the "back" (Y direction);

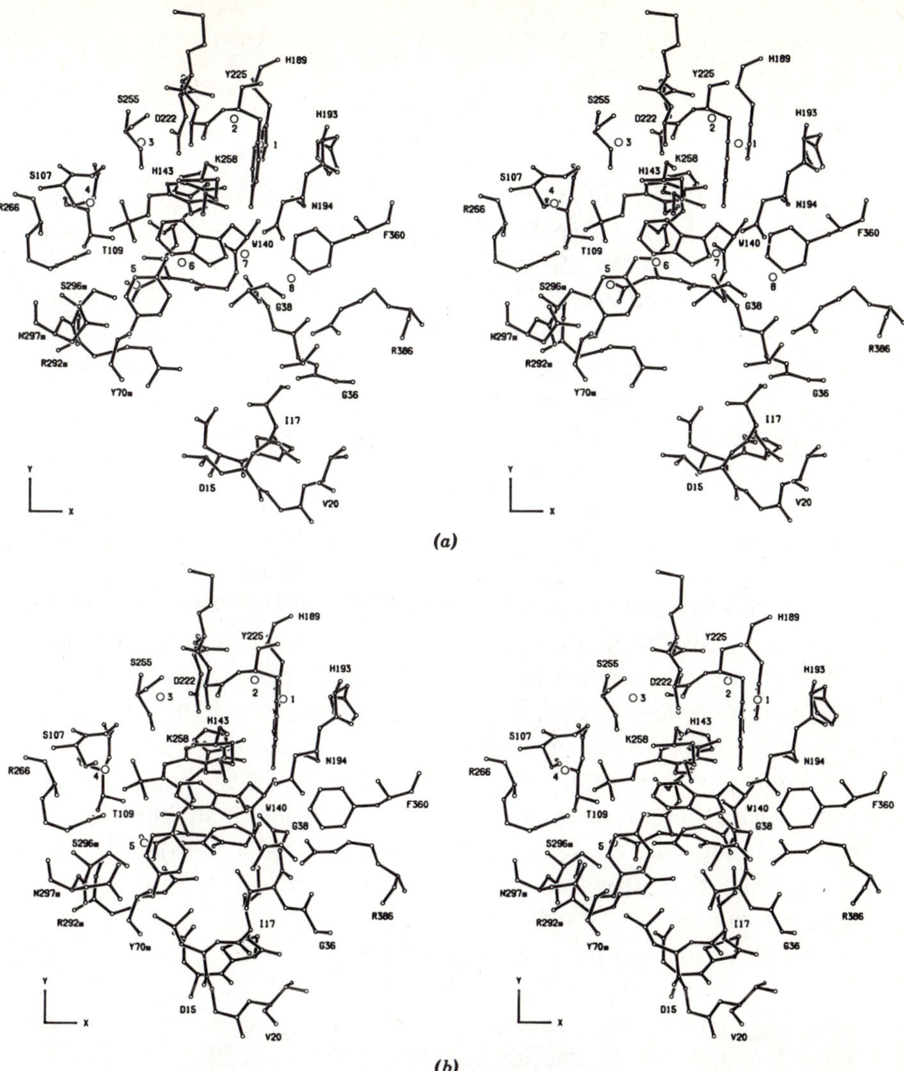

Figure 10. Active site of PLP form of chicken mAAT, including all residues in direct contact with either coenzyme or substrate, as well as some further, probably important residues in the active site area. Bound water molecules are also shown. View direction: $-Z$. (*a*) Open structure, space group P1. Eight bound water molecules have been found in this region. (*b*) Closed structure with maleate bound, space group C222$_1$. Note the large shift and reorientation of residues Ile 17, which acts as a hinge, and (especially) Leu 18. The other small domain residues including 36–39, of which Val 37 reorients completely, and Arg 386 (but not Phe 360) shift as well resulting in complete enclosure of maleate by the protein. The inhibitor is involved in hydrogen bonds with H$_2$O(5), Gly 38 N, Trp 140 Nε1, Asn 194 Nδ2, and the guanidinium groups of Arg 292* and Arg 386; the latter two also compensate the carboxylate charges. Three bound water molecules have been expelled. The coenzyme has rotated around N1–C4 upon protonation of the aldimine nitrogen, resulting in near coplanarity of both the aldimine N and OP1 with the pyridine ring (also compare Figs. 11 and 13). See text for further information.

211

by (2) Tyr 225, which makes a short H bond to the ionized (Braunstein, 1973), 3′-hydroxyl group, Phe 360 and Asn 194 at the right (X direction);and by (3) Asp 222 and Trp 140 below ($-$Z direction). Asn 194 is also hydrogen bonding to O3′, while Asp 222 makes a hydrogen bond/salt bridge to the protonated (Braunstein, 1973) pyridine N1 nitrogen. Above the coenzyme on the right, the loop Val 37-Gly 38-Ala 39 passes the Lys 258 side chain. On the other (left) side, the coenzyme phosphate is strongly bound by nine hydrogen bond donating groups. These are: (1) the γ-OH groups of Ser 107 and Ser 255 and the main chain NH of Gly 108 to oxygen OP3 (for the nomenclature, see Fig. 1); (2) the main chain NH and the γ-OH of Thr 109 together with the guanidinium Nη 1 of Arg 266 to oxygen OP4; and (3) the Nη2 of Arg 266 together with the phenolic OH of Tyr 70* (second subunit) and a water molecule to oxygen OP2. Thus, the three nonester oxygens of the double negatively charged (Braunstein, 1973) phosphate have saturated their hydrogen-bonding capacity. The negative charge is largely compensated for by Arg 266 and the positive end of the α helix dipole (Hol et al., 1978) of helix 5 (residues 108$-$122), to which the phosphate group is attached. In the apoenzyme, inorganic phosphate binds at this site with (in pig cAAT, in which the ligand Ser 107 Oγ is replaced by Ser 257 Oγ) a dissociation constant of $1.5 \times 10^{-6}M$ (Vergé et al., 1979). Replacement of this inorganic phosphate by arsenate revealed the phosphate-binding site in low-resolution work on chicken mAAT (Eichele et al., 1979a). We will see in Section 5 that strong binding of the phosphate group is essential for catalysis.

Further residues in the active site include Ser 296*, Asn 297*, and Asn 142. The indole group of Trp 140 lies in front of the bottom half of the coenzyme pyridine ring, and covers the salt bridge between N1 and Asp 222. Tyr 70*, Val 37, Gly 38, Phe 360, Asn 194, Trp 140, Ser 296*, and Asn 297* form the side walls of a pocket with dimensions $5 \times 4 \times 4$ Å in front of the coenzyme. A rather wide cleft provides access to the pocket. This cleft is partly blocked by the N-terminal residues 12$-$15 and the N-terminal part of helix 1 (residues 16$-$25). Below the entrance of the pocket we find to the left Arg 292*, which makes a salt bridge to Asp 15, and to the right Arg 386, which is rigidly built into the wall of the small domain with only its guanidinium group in contact with solvent. This arrangement strongly suggests that in the Michaelis complex, the substrate lies in the pocket with both its carboxylate groups in a "gauche" orientation closest to the entrance and interacting with the two arginines. The negative charge at the coenzyme O3′

orients the substrate by attracting the α-NH_3^+ group, causing the α carboxylate to interact with Arg 386 and the side chain carboxylate to interact with Arg 292*. The arrangement of charges in the active site guarantees the stereospecificity for L-amino acids. D-amino acids cannot be accommodated. The position of the substrate pocket with respect to the coenzyme-Lys 258 aldimine suggests that the first part of the catalytic process, transaldimination, can only take place through a rotation of the PLP pyridine ring that brings C4' further forward towards the substrate. Such a movement was suggested long ago (Karpeisky and Ivanov, 1966; Ivanov and Karpeisky, 1969).

Underneath the coenzyme is a series of three histidines: 143, 189, and 193. His 143 makes a short hydrogen bond to Asp 222. Its imidazole is in van der Waals contact with that of His 189, and the latter, in turn, with His 193. Their role is not yet quite clear, but it might be related to charge dissipation (Arnone et al., 1985a).

An interesting and long debated point is the conformation of the phosphate group with respect to the pyridoxal moiety, especially the torsion angle C4-C5-C5'-OP1 (see Fig. 1). Low-energy conformations supposedly correspond to values around 180° or ± 60° for this angle. The avidity of binding of 6-methyl-PLP to apo-cAAT suggested one of the last two possibilities (Tumanyan et al., 1974). The electron density map of mAAT at 2.8 Å resolution indicated, however, a small value for this angle (Kirsch et al., 1984). The most likely, and most reasonable, position for OP1 seemed to be behind, rather than in front of, the pyridine ring. Refinement at 2.3 Å resolution, however, has shown that OP1 is in front of the ring. The torsion angle C4-C5-C5'-OP1 is probably $\sim -37°$, which is still smaller than expected (Tumanyan et al., 1974). The *cisoid* (Dunathan, 1971; Braunstein, 1973) aldimine double bond is rotated by a considerable angle out of the pyridine ring plane towards the back (the A face). The orientation of the coenzyme ring has been described (Kirsch et al., 1984) by a "tilt" angle with respect to the XZ plane (Fig. 10a), since in all coenzyme orientations (see below) N1 was found to be approximately stationary; the line C2$-$C6, however, remained nearly parallel to X, with N1$-$C4 being roughly confined to a plane parallel to YZ. In the PLP-enzyme, the coenzyme tilt is $\sim 27°$. Figure 11 shows the coenzyme and Lys 258 (subunit S) in their density in a ($3F_{obs}$-$2F_{calc}$) electron density map at 2.3 Å resolution after partial refinement. The orientation of the pyridine ring is well defined by the density.

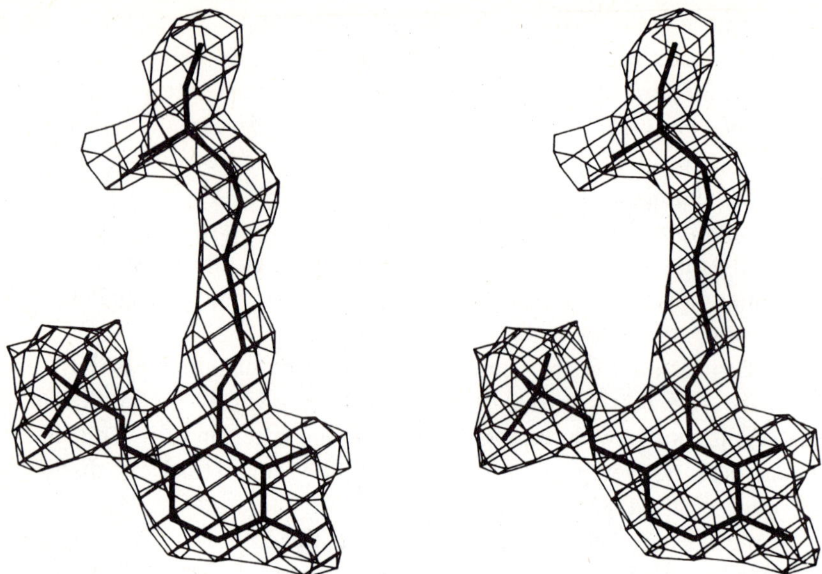

Figure 11. Electron density corresponding to Lys 258 and coenzyme (subunit S) as seen in a $(3 F_{obs}\text{-}2F_{calc})$, α_{calc} map after partial crystallographic refinement at 2.3 Å resolution of PLP enzyme in space group P1, with the model superimposed. Torsion angles around C5-C5′ (see Fig. 1) and C4-C4′ are −37° and +56°, respectively. The latter value is still open to improvement upon further refinement. View direction: +Y.

3.3.5. Pyridoxamine Phosphate Enzyme

The PMP enzyme was prepared by transamination of the PLP enzyme with cysteine sulfinate (Kirsch et al., 1984). Differences between the two structures are limited to the coenzyme and its immediate environment. Figure 12 displays the active site of the PMP enzyme, as built into a $(3F_{obs}\text{-}2F_{calc})$ map at 2.8 Å resolution in which the coenzyme and surrounding residues were omitted from the structure factor calculation. The phosphate group is essentially identically positioned and oriented as in the PLP enzyme, but the coenzyme ring is tilted about 15° further forward. The hydrogen bond to Asp 222 is retained. Small rotations around the three bonds between phosphorus and C5 can convert one structure into the other. Such a motion, however, implies that the torsion angle C4-C5-C5′-OP1 passes through 0°, since in the PMP enzyme, the "bridging" oxygen OP1 lies slightly behind (on the A side of) the pyridine ring (torsion angle ~24°). Since PMP is not covalently bound to the enzyme, the pyridine ring would be able to tilt further forward, as it

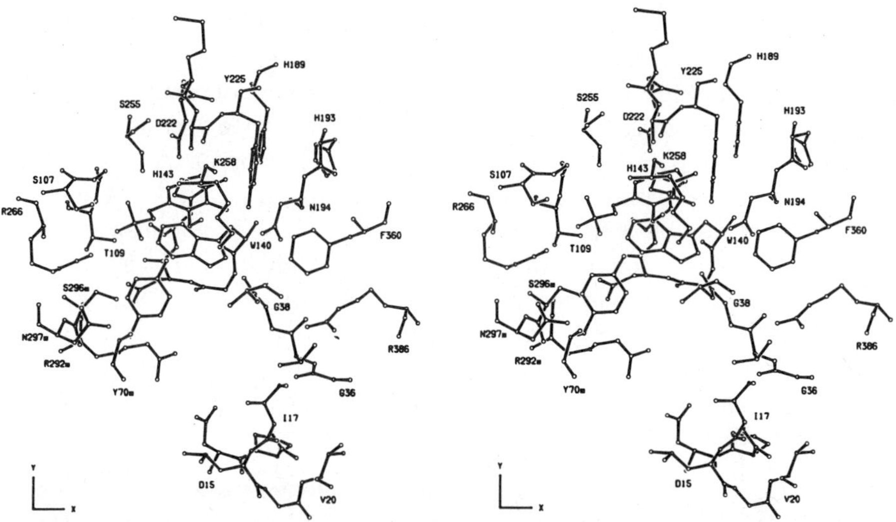

Figure 12. Active site of the PMP form of chicken mAAT, open structure, space group P1. View direction: $-Z$. Bound water molecules are not shown. Differences with Figure 10*a* include the increased coenzyme tilt and the release of Lys 258, which is H bonded to Gly 38 O, and the PMP aminogroup (which shares a third proton with it). The torsion angle around C5-C5' is $+24°$. See text for further information.

does in the external aldimine derivative with 2-methylaspartate (see Section 3.4.2). We must conclude that the short distance between OP1 and C4' does not increase the energy by much. Some preliminary quantum chemical calculations have been carried out on the conformational stability of various conformations of the PLP enzyme (J. Gerhards and E.L. Mehler, unpublished results). The results suggest that, in this case at least, a torsion angle around C5-C5' near $0°$ is not energetically unfavorable. Obviously, these studies have to be extended to other coenzyme forms.

The increase in coenzyme ring tilt is accompanied by small movements of the side chains of Tyr 225 and Trp 140, through χ rotations, in the same direction. The H bond between Tyr 225 and O3' is thus retained.

The coenzyme amino group is rotated behind the ring plane towards a *cisoid* orientation, with a torsion angle C3-C4-C4'-N4' of $-117°$ in the current model. The side chain of Lys 258 lies further forward than in the internal aldimine. Its ε-amino group makes hydrogen bonds both to the coenzyme amino group (confirming that they share a positive charge; Fasella et al., 1966), and to the carbonyl oxygen of Gly 38 above the coenzyme.

3.3.6. Apoenzyme

This enzyme form again crystallizes in space group P1, isomorphously with the PLP and PMP enzymes. Although, in itself, it is not of prime interest for the study of catalysis, the apoenzyme structure has been very useful in the search for the pyridoxal groups in the early studies on mAAT through a difference map with coefficients $(F_{holo}-F_{apo})$, $\alpha_{MIR, holo}$ at 4.4 Å resolution, and in establishing the coenzyme orientation in a similar map at 2.8 Å resolution. Furthermore, various coenzyme derivatives could be introduced directly by diffusion into apoenzyme crystals (Eichele et al., 1979a; Eichele, 1980; Kirsch et al., 1984). As mentioned earlier, inorganic phosphate from the buffer solution occupies the coenzyme phosphate position. Its essentially identical position and orientation with respect to the PLP phosphate suggests an absence of strain in the latter.

3.4. The "Closed" Structure of the Inhibited PLP Enzyme

Cocrystallization at 4°C of mAAT with various dicarboxylic acid inhibitors results in orthorhombic crystals of space group $C222_1$, with one subunit per asymmetric unit (Table 1). It is the only crystal form thus far produced of any AAT with exact molecular twofold symmetry. This has several advantages. An obvious one is that the structure determination and refinement are easier. The structure of the isomorphous 2-methylaspartate derivative has been built through a $(3F_{obs}-2F_{calc})$ Fourier map at 2.3 Å resolution, in which the coenzyme and its direct environment were excluded from the structure factor calculation.

3.4.1. Maleate Derivative

Maleate has long been known as one of the strongest noncovalent competitive inhibitors of both cAAT and mAAT (Jenkins et al., 1959; Michuda and Martinez-Carrion, 1970). Since the ingenious inhibition studies of Khomutov, Karpeisky, and their colleagues in Moscow (Braunstein, 1973), it is known that the carboxyl groups of successful inhibitors must be *cis* (or "gauche") and about 3–4 Å apart. The advantage of maleate over, for example, succinate is that its carboxylate groups are fixed in a *cis* conformation. Its stereoisomer fumarate is not an inhibitor (Jenkins et al., 1959).

The maleate crystals, buffered by phosphate at pH 7.5, are deeply yellow, in contrast to those of the native enzyme in P1 at the same pH. This is due to a considerable shift, to higher pH, of the pK of protonation of the "internal" aldimine in the presence of the inhibitor (Jenkins et al., 1959; Jenkins and D'Ari, 1966).

The crystals show an absorption maximum at 440 nm (Vincent et al., 1984) at a somewhat higher wavelength than the low pH form of the native PLP enzyme and the "external" aldimine with 2-methylaspartate (430 nm; Eichele et al., 1978; Vincent et al., 1984).

3.4.1.1.   Overall Comparison with the P1 Structure.   We will describe here the differences between the maleate derivative and the PLP enzyme structures. The comparison is based on the results of the superposition of the most rigid parts of the dimer, the PLP binding domains, of each structure in the molecular axis system XYZ first introduced for the P1 structure (Eichele et al., 1979a).

Direct comparison of the electron density maps or of atomic coordinates immediately reveals a large conformational change, consisting of a reorientation of the small domain with respect to the coenzyme-binding domain. The overall movement is visible in Figure 3. The following numbers illustrate its extent. The r.m.s. differences in Cα positions between the P1 structure (subunit S) and the C222$_1$ structure are 1.6, 0.4, and 0.6 Å for the whole subunit, the PLP-binding domain, and the small domain, respectively. The corresponding values for the S and S* subunits in P1 are 0.3, 0.2, and 0.3 Å. It is clear that the conformational change is, to a large extent, a rigid body movement. However, some larger changes occur locally, especially in side chains. Such rearrangements are to be expected (Lesk and Chothia, 1984). However, they are much smaller in AAT than in citrate synthase (Remington et al., 1982). Domain flexibility and its relation to function in proteins has been reviewed by Bennett and Huber (1983).

The domain movement in mAAT can be qualitatively described as a 13° rotation of the small domain towards the active site around an axis parallel to the molecular dyad near Gly 325. This residue lies at the kink in the long helix 13 (residues 313−344, see Table 2), and the effect of the movement is to straighten the helix (Fig. 7).

Of much interest is the domain interface, which must allow sliding of the small domain surface over that of the coenzyme-binding domain during the conformational change. Both surfaces are roughly flat and oriented more or less at right angles to the rotation axis. The motion of the small domain surface is demonstrated in Figure 9, which shows the interface area in both structures. The movement of the small domain in the direction open to closed is away from the viewer.

The domain interface is formed by residues 161−166, 192−199, 228−231, and 323−325 of the PLP-binding domain and by 326−328, 348−364, and 386−390 of the small domain. Figure 4 shows for the closed structure where these regions are in the molecule. The active site residues His 193, Phe 360, and Arg 386 form part of this interface. The residues from the PLP-binding domain remain stationary, as a comparison of Figures 9a and 9b indicates. An exception is the side chain of Phe 228. The single hydrogen bond between the domains, from Ile 357 N to Gly 197 O, remains intact. The van der Waals interactions change, but do not significantly increase or decrease. As an example, in the open structure, Phe 362 interacts mainly with Tyr 161 and Thr 196, and less with Pro 195; in the closed structure, however, its interaction is weaker with Tyr 161 and stronger with Pro 195. In the small domain, side chain reorientations occur at Met 326 and Met 359, optimizing the van der Waals interactions with Phe 228, which reorients its side chain also, as mentioned above. The imidazole of His 352 rotates ~90°, changing its stacking interaction with Cys 166 to a two-atom interaction between His 352 Cδ2 and Cys 166 Sγ. The uncovering of Cys 166 upon domain closure is quite obvious. This explains the "syncatalytic" increase in reactivity of this residue (Section 4.5.1). One may conclude that domain closure does not cost significant energy.

That the secondary structure of mAAT does not change during the conformational change is demonstrated in Table 2. The few small differences observed are by no means concentrated in the small domain and interface regions, and might for the greater part be the result of incomplete refinement. The only significant difference occurs at residues 36 and 37. The loop 37−39 rearranges considerably in addition to its movement with the small domain towards the coenzyme (see below).

The observation that a large conformational change can occur essentially without altering the secondary and supersecondary structure shows that the potential of circular dichroism to signal such structural changes has its limitations.

The most dramatic effect of the domain movement is in the active site area, where helix 1 (residues 16−25) moves towards the coenzyme and locks in the inhibitor maleate. Residues of this helix make van der Waals contact both with the opposite wall of the active site, formed by helix 11 (residues 277−294) of the neighboring subunit, and with the inhibitor. Side chain conformational changes amplify the already large shifts of some of these residues, notably Ile 17 (of which Cδ shifts by 6.5 Å) and Leu 18 (of which Cδ1 and Cδ2 shift by 6.7 and 7.9 Å respectively; compare Figs. 10a and 10b).

3.4.1.2. Hinge Regions. Since the small domain is defined as residues 15−47 and 326−410, hinges can be expected at or near residues 15, 47, and 326. The N-terminal hinge, connecting the N-terminus, which interacts with the second subunit, with the small domain is formed by the sequence Glu 10-Met 11-Gly 12-Pro 13-Pro 14-Asp 15-Pro 16. Changes in dihedral angles of more than 15° occur at Glu 10 (both φ and ψ) and in the φ angles of residues Met 11, Gly 12, Asp 15, and Ile 17. All changes are smaller than 50°. What occurs, therefore, is a gradual change in conformation over a large length of chain. A comparison of Figures 10*a* and 10*b* clearly illustrates, however, that the actual hinge is Pro 16-Ile 17.

The changes near residue 47 are more localized. Comparison of dihedral angles in the two structures revealed significant differences only for residues 37−39 and 43−44. The hinge, therefore, seems to be Asp 43-Asn 44. The change in the stretch Val 37-Gly 38-Ala 39 is an isolated reorientation directly related to its interaction with the inhibitor maleate. Val 37 is part of the lid on the active site pocket (see Figs. 10*a* and 10*b*, and the next section).

The hinge in the long helix 13 has already been mentioned and is illustrated in Figure 7. The main chain dihedral angles indicate that, in fact, the hinge involves the residues Val 323 and Lys 324, rather than Gly 325. The ψ angle of Val 323 and the φ angle of Lys 324 both change by ~20°, while the dihedral angles of Gly 325 are not changed.

In summary then, hinges occur in the positions 16−17, 43−44, and 323−324, while additional structural rearrangements take place in the regions 10−15 and 37−39.

3.4.1.3. Active Site. Figure 10b is a stereo view along −Z of the active site with maleate bound. It can be compared with the corresponding view of the open PLP enzyme active site in Figure 10*a*. The tightening of the protein around the maleate ion is remarkable. The important differences in the closed active site with respect to the open form may be described as follows:

1. The coenzyme pyridine ring has rotated roughly around N1−C4, shifting the atoms C5, C5′, and C6 forward (towards the inhibitor) and the atoms C2, C2′, C3, and O3′ backward. This reorientation of the pyridine ring makes the aldimine double bond more coplanar with it. The ketoenamine tautomer (Metzler, 1979; Christen & Metzler, 1985) of the protonated aldimine probably enforces this rotation towards (near) coplanarity of the aldimine double bond. As a result, the

torsion angle around C5-C5′ is reduced to $\sim -16°$. Figure 13 shows coenzyme and maleate in their electron density.

2. The phenolic group of Tyr 225 is pushed 0.5 Å in the $+X$ direction and 0.3 Å in the $+Z$ direction.

3. Residues 37−39 have changed their conformation and have moved to the left ($-X$ direction). Val 37, the Cγ1 and Cγ2 atoms of which have shifted by 4.5 and 7.3 Å, respectively, is now in van der Waals contact with Tyr 70*, while the peptide bond 37−38 is reoriented so that Gly 38 O is no longer available for H bonding to the ε-NH$_2$ of Lys 258 after transaldimination. Instead, the Gly 38 NH is positioned properly for hydrogen bonding to the carboxylate group that interacts with Arg 386.

4. The indole ring of Trp 140 has rearranged itself slightly, following the movement of the coenzyme.

5. The main chain of Phe 360 has moved ~ 1.5 Å, but the ε and ζ carbon atoms of the phenyl ring, nevertheless, have similar positions as in the unliganded PLP enzyme.

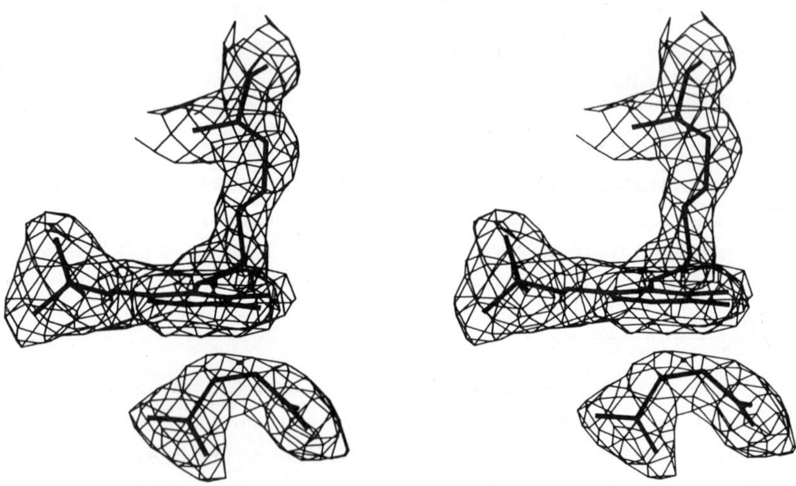

Figure 13. Same as in Figure 11, but for the closed structure with maleate, space group C222$_1$. Density and model of maleate are also shown. The dihedral angles around C5-C5′ and C4-C4′ are −16° and +11°, respectively. View direction: −Z.

6. The guanidinium group of Arg 386 has moved with the small domain and is now 2.8 Å closer to the coenzyme. It makes an H bond to Asn 194 Oδ1, while Asn 194 Nδ2, in turn, is hydrogen bonded to O3′ of the coenzyme.

7. Helix 1, as mentioned above, has shifted with the small domain towards the coenzyme. Residues 15–18 now block the active site, with the hydrophobic side chains of Ile 17 and Leu 18 forming a "front wall."

8. The side chain of Arg 292* has reoriented its guanidinium group into the active site, where it makes two H bonds to the leftmost maleate carboxylate group. In the open structure, it forms a hydrogen bond/salt bridge to Asp 15 through an η nitrogen. In the closed structure, Nε makes this hydrogen bond.

9. Last, but not least, maleate makes H bonds/salt bridges to both Arg 386 and Arg 292*, and additional H bonds to Gly 38 N, Tyr 225 Oη, Asn 194 Nδ2, and Trp 140 Nε1. It is in van der Waals contact with the coenzyme C4′ and aldimine N, and with the side chains of Tyr 70*, Val 37, Ile 17, and Leu 18.

This list clearly shows the large consequences of the small domain movement for tight binding of maleate and, presumably, for productive binding of substrates.

3.4.1.4. Trigger of Domain Movement.   An obvious question is: What triggers the open-to-closed conformational change? Inspection of the domain interface does not indicate an energetic preference for either conformation. Thus, the trigger must be sought in the active site itself.

Upon binding of the inhibitor, three things happen. First, the negative maleate charges compensate the positive charges of Arg 292* and Arg 386, which originally repelled each other. Arg 386, and thus the small domain, must move closer to maleate for optimum interaction. Second, Asp 15 has the tendency to follow the movement of the positively charged guanidinium group of Arg 292*. Small domain closure allows this to happen, too. And third, a small local reorientation of Gly 36, Val 37, and Gly 38 and side chain conformational changes of Ile 17 and Leu 18 provide, when they approach with the small domain favorable H bond and van der Waals interactions with maleate and Tyr 70*. The third movement might occur either in concert with the first and second movements, or later. The latter possibility is perhaps the most likely in view of the observations that L-alanine does *not* trigger the conformational change, and that in cAAT, high concentrations of salt (Borisov et al., 1978) or acetate (Arnone et al., 1984) *do* cause it. Thus, charge compensation and reorientation of Arg 292* presumably are the prime causes for the domain movement to occur. The effect of the latter on catalysis will be discussed later.

3.4.2. 2-Methylaspartate Derivative

The differences between the maleate derivative and the 2-methylaspartate derivative are limited to the immediate coenzyme environment (compare Figs. 10*b* and 14). The small domain has the same position and orientation. The main difference is that the coenzyme pyridine ring has rotated forward by 28° upon transaldimination. N and $C\alpha$ of the inhibitor lie in the coenzyme ring plane. The two carboxylate groups occupy almost exactly the same positions as those of maleate. The phosphorus atoms in the two structures also superimpose, while the atoms N1, C2' and C6 are nearly coincident. The phosphate group has rotated by 15°−20° around the bond P-OP4. The torsion angle OP1-C5'-C5-C4 is 85°. The 2-CH_3 group of the inhibitor makes van der Waals interactions with Gly 38 and the side chains of Val 37, Tyr 70*, Tyr 225, Lys 258 and Phe 360. Tyr 225 has followed O3' by rotation around $C\alpha$-$C\beta$ and remains hydrogen bonded to it, while Asn 194 seems to make H bonds to both O3' and Tyr 225 $O\eta$. The coenzyme ring is stacked at ~3.6 Å distance with the indole of Trp 140, which has moved somewhat down with it and has slightly changed its orientation to optimize the H bond between $N\varepsilon 1$ and an oxygen of the β carboxylate of the inhibitor. The pyridine N1 makes a relatively weak, bifurcated H bond to both $O\delta 1$ and $O\delta 2$ of Asp 222.

The side chain of Lys 258 "hangs" behind the 2-CH_3 of the inhibitor. No density is observed for its ε-amino group, which we interpret as meaning that it has no fixed position. This would be compatible with a proton shuttling equilibrium between Lys 258 and the aldimine N, as proposed by Fasella et al. (1966), to explain the pH-independent spectrum of this derivative, with peaks at 360 and 430 nm in the pH-range 5.0−8.0. A comparison of Figures 10*b* and 14 highlights the elegant transition between internal and external aldimine in a completely insulated environment, and would not seem to support the elaborate "oscillatory rotor" mechanism proposed for the transaldimination process by Arnone et al. (1984), in order to avoid an eclipsed conformation around C4-C4'. The described recent findings for

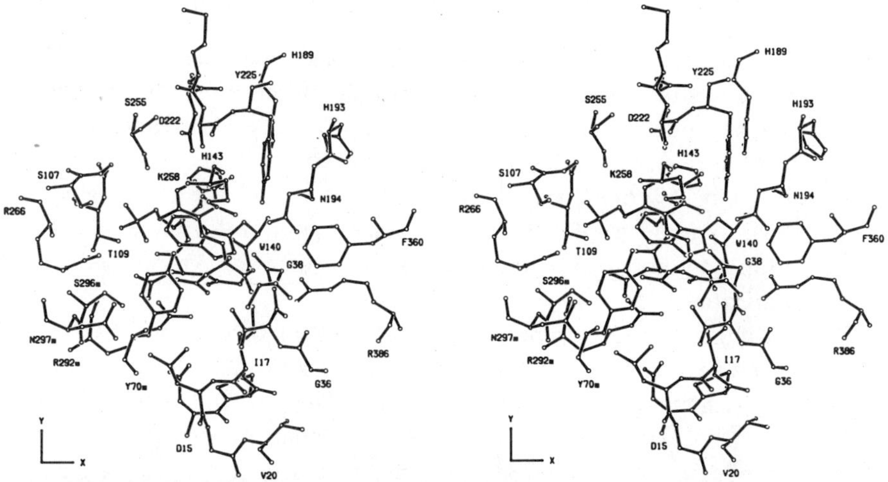

Figure 14. Active site of 2-methylaspartate (2-MeAsp) derivative of chicken mAAT, space group $C222_1$. View direction: $-Z$. Differences with Figure 10*b* include: The 2-MeAsp moiety is bound to PLP as an "external" aldimine. The coenzyme tilt has increased by 28°, with the torsion angle around C5-C5′ being +85°. The pyridine ring is stacked with Trp 140 ~ 3.6 Å distance. The phosphate has rotated ~ 20°, roughly around OP4-P, in a direction that moves OP1 away from the active site pocket. As a result, the carboxylate groups of 2-MeAsp superimpose nearly exactly on the positions of the maleate carboxylate groups. The 2-methyl group makes close van der Waals interactions with Val 37, Gly 38, Tyr 70*, and Lys 258, with distances to Tyr 225 and Phe 360 being somewhat longer. The Lys 258 ε-NH_2 is boxed in between the phosphate, Ser 255, Ala 224, Tyr 225, Gly 38, 2-MeAsp, and Tyr 70*.

mAAT clearly indicate that the coenzyme does not try to avoid small dihedral angles around C5-C5′.

3.5. The Structures of Pig and Chicken Cytosolic Aspartate Aminotransferase

The amino acid sequences of pig and chicken cAAT are both 46% homologous to that of chicken mAAT (Ovchinnikov et al., 1973; Doonan et al., 1975; Shlyapnikov et al., 1979; Graf-Hausner et al., 1983). Therefore, strong structural homology was to be expected. Indeed, this turned out to be the case, and it justifies, because of the possibility of extrapolating to cAAT, the strong emphasis that has been placed here on the description of chicken mAAT. An additional, practical reason is that the current structural information on the cytosolic enzymes is limited.

3.5.1. Pig cAAT

This enzyme crystallizes from PEG in sodium acetate buffer at pH 5.4. The yellow crystals belong to space group $P2_12_12_1$ and have one dimeric molecule per asymmetric unit (Arnone et al., 1977; see also Table 1). The structure has been solved at 2.7 Å resolution (Arnone et al., 1982), but until recently, only 75% of the amino acids could be built into the model (Hyde et al., 1984; Arnone et al., 1985b). Crystallographic refinement at 1.6 Å resolution on a now completed model is underway (A. Arnone, personal communication).

Although a detailed comparison has not yet been carried out, the folding of pig cAAT is essentially identical to that of chicken mAAT (Arnone et al., 1984; A. Arnone, personal communication). The molecule contains a twofold symmetry axis. However, there are local departures from it, for example, in the regions 15−18 and 38−39. Blurred density in these regions in one of the subunits can probably be ascribed to a static or dynamic conformational disorder. An acetate molecule is found in the substrate pocket of only one of the active sites (Arnone et al., 1984). The "bridging" oxygen (OP1) of the coenzyme phosphate lies in front of the pyridine ring, with the torsion angle around C5-C5′ being −36°, while that around C4-C4′ is +27° (Arnone et al., 1985b). These observations are now confirmed in the unliganded PLP form of mAAT at pH 7.5 at the higher resolution of 2.3 Å. Differences between the two enzymes include the replacement of Ser 107 (Gly in cAAT) by Ser 257 (Ala in mAAT) as a phosphate ligand, while the "lid" of the substrate pocket is now formed by Val 15-Leu 16-Val 17-Phe 18. This modification must have consequences for substrate binding and for the open/closed equilibrium. That these residues are critical is also indicated by their invariance within each class, with the only known exception being Ala 16 in chicken cAAT (Christen et al., 1985). Interestingly, *E. coli* AAT has Gly 107 and Ser 257 as in cAAT, but the sequence Asp 15-Pro 16-Ile 17-Leu 18 is as in the mAAT's (Kondo et al., 1984).

Pig cAAT has crystallized in the open conformation, although some distortion towards the closed structure may have occurred locally.

3.5.2. Chicken cAAT

This enzyme was crystallized from cesium chloride and ammonium sulfate in the presence of D,L-2-methylaspartate at pH 7.5 (Table 1) with the aim of stabilizing the external aldimine with 2-methylaspartate. The original crystallographic work was carried out by Borisov et al. (1977, 1978, 1980a,

1980b, 1985). The last published result is a 2.8 Å electron density map computed after application of molecular averaging and phase extension procedures (Borisov et al., 1985). Again, striking similarities with mAAT were observed, including kinks in two helices (234−247 and 312−342) in the same positions. On the other hand, locally, consistent shifts of 1.0−1.5 Å were reported between corresponding Cα positions of superimposed mAAT (Eichele, 1980) and cAAT molecules. This is not surprising since an open structure and a closed structure were compared (see below). The authors signaled small local differences around deletions in mAAT, with the largest occurring at 125−130, where cAAT forms a loop; but mAAT, with deletions at 127, 128, 131, and 132, forms a short straight connection.

The details mentioned of the active site structure also correspond to the 2-methylaspartate derivative of mAAT. Asp 222 interacts with the coenzyme N1. The coenzyme ring is stacked with the Trp 140 indole. The α carboxylate of the inhibitor interacts with Asn 194 and Arg 386 and its β carboxylate with Arg 292*, while the 2-CH_3 group is close to Tyr 70*. Ser 257 is confirmed as a "cytosolic" phosphate ligand (Borisov et al., 1985). Also in this work, a certain amount of asymmetry between the two crystallographically independent subunits was observed. This seems to be a property of cAATs, which are seemingly "softer" molecules than chicken mAAT.

Harutyunyan et al. (1982, 1984, 1985) and Torchinsky et al. (1983) have repeated the structure determination of chicken cAAT. They used the same crystals, but prior to data collection replaced the covalent inhibitor 2-methylaspartate with the noncovalent inhibitor (for the PLP enzyme) 2-oxoglutarate. Their study was also at 2.8 Å resolution. A comparison of their structure with mAAT led to conclusions similar to Borisov et al. (1985). However, some additional details were given on the active site structure. The inhibitor 2-oxoglutarate was found to bind with its α carboxylate to Arg 386 and Asn 194, and with its distal carboxylate to Arg 292* and Trp 140, just like maleate in mAAT. The carbonyl oxygen points "up" (towards +Z) and lies in front of the aldimine double bond. Thus, its mode of binding is "productive" (Kirsch et al., 1984). Val 17, Phe 18, and Val 37 take part in inhibitor binding. This corresponds to Ile 17, Leu 18, and Val 37 in the closed form of mAAT.

We conclude from the data on cAAT that only small differences occur in structure and mode of action of the mitochondrial and cytosolic isoenzymes that are sufficient to explain their difference in substrate specificity, but are not large enough to justify a proposal for a difference in catalytic mechanism. In view of this, we will, in Section 5, discuss a single catalytic mechanism for both isoenzymes of aspartate aminotransferase.

3.6. Inhibition Experiments with Crystalline AAT

Many such experiments have been carried out with the two cytosolic AATs and with chicken mAAT. The results may be summarized as follows.

Asymmetric behavior, with respect to ligand binding and conformational changes, is observed in all cases where the two subunits are crystallographically distinct; the only exception is the $C222_1$ form of mAAT (see Table 1). Chicken mAAT in P1 and pig cAAT crystallize in the open form; chicken cAAT (with 2-methylaspartate) crystallizes in the closed form. In all three cases, the small domain reorientation can only take place in one subunit without shattering the crystals. Kirsten et al. (1983) determined kinetic constants for the two subunits in triclinic microcrystals of chicken mAAT (see Section 3.7).

3.6.1. Unliganded Enzyme Forms

The PLP enzyme, the PMP enzyme, and the apoenzyme of chicken mAAT prefer, in the absence of ligands, the open conformation (Kirsch et al., 1984). This is not surprising, since it enables substrates in the holoenzyme forms and the coenzyme in the apoenzyme to enter the active sites. The pyridine ring of PMP is tilted forward by an additional $15°-20°$, as compared to PLP. In the apoenzyme, a phosphate ion replaces the ester phosphate of the coenzyme.

The PLP enzyme of pig cAAT (Arnone et al., 1982) crystallizes in the open structure, while a shift from the closed to the open form is observed in the one subunit of chicken cAAT with an unconstrained small domain upon formation of the unliganded PLP enzyme or PMP enzyme (Borisov et al., 1978; Harutyunyan et al., 1984). Surprisingly, this does not occur in the apoenzyme (Borisov et al., 1978). This latter observation may be related to the screening effect of the sulfate ions from the solvent.

3.6.2. Noncovalent Dicarboxylate Inhibitors

Maleate, succinate, glutarate, and 2-oxoglutarate have been studied with the PLP enzyme. The mode of binding of maleate to mAAT in the $C222_1$

crystal form has been described in detail in Section 3.4.1. However, 2-oxo-glutarate binds to the PLP enzyme in the open conformation (Eichele, 1980; Kirsch et al., 1984). This distinguishes the mitochondrial enzyme from its cytosolic counterpart, since all four compounds mentioned above stabi-lize the closed structure of cAAT (Borisov et al., 1980a, 1980b, 1985; Harutyunyan et al., 1984, 1985; Arnone et al., 1982, 1984, 1985b). The discrepancy may be explained tentatively by a somewhat narrower active site of mAAT and a more flexible small domain of cAAT. This last sugges-tion would also explain the locally "blurred" electron density as seen in this domain for pig cAAT, and the local departures from twofold symmetry observed in the native crystals (Arnone et al., 1984). In all three enzymes, 2-oxoglutarate was found to bind in a similar way, with the carbonyl carbon in front of the coenzyme C4' atom and the carbonyl oxygen near the aldimine nitrogen. A similar mode of binding can be expected for the Michaelis complex of 2-oxoglutarate with the PMP enzyme (Kirsch et al., 1984).

Substrate-like dicarboxylate inhibitors tend to increase the pK_a of pro-tonation of the coenzyme-lysine aldimine, while those with a different structure do not show this effect (Bonsib et al., 1975). Furthermore, 2-oxoglutarate alters the ratio 430:360 nm absorption much more in chicken cAAT (Makarov et al., 1981) than in chicken mAAT (Eichele et al., 1978). This correlates with the tendency of the structure of the 2-oxoglutarate complex of the PLP enzyme form of cAAT towards assuming the closed conformation, and that with mAAT, the open conformation (see above). It seems likely that dicarboxylate inhibitor complexes in the closed, but not in the open, conformation influence the pK_a of aldimine protonation. This could be explained by unobstructed access of water molecules and/or in-complete compensation of the charges of Arg 386 and Arg 292* by the inhibitor carboxylate groups in the open conformation. The carboxylate groups of the inhibitors are oriented *cis* (or "gauche"), as predicted on the basis of inhibition studies in solution (Jenkins et al., 1959; Khomutov et al., 1968).

3.6.3. 2-Methylaspartate

The binding to mAAT of this inhibitor, which can, in principle, produce either a Michaelis complex or an "external" aldimine (Fasella et al., 1966), has been described in detail above (see also Fig. 14). 2-Methylaspartate complexes have also been described for chicken and pig cAAT (summarized by Borisov et al., 1985; Harutyunyan et al., 1985; Arnone et al., 1985b). In all cases, similar observations were made: (1) the inhibitor binds covalently. (2) Binding involves a 20°−30° rotation of the coenzyme pyridine ring. (3) The closed conformation of the enzyme is stabilized. (4) Only the L-isomer of the racemic mixture binds. The 2-methyl group points "up" (+Z direction, Fig. 14), being positioned between Tyr 70* and Gly 38, with the α carboxylate and the β carboxylate interacting with Arg 386 and Arg 292*, respectively. (5) A salt bridge and, probably, a hydrogen bond are made between the protonated coenzyme N1 and Asp 222. (6) For both mAAT and pig cAAT, the torsion angle C4-C5-C5'-OP1 is found to be ∼ +90°.

The absorbance ratio 430:360 nm of the 2-methylaspartate derivative in solution remains constant between pH 5.5−11 (Hammes and Tancredi, 1967). The 360 nm absorption is usually ascribed to the unprotonated internal aldimine in the presence or absence of noncovalently bound 2-methylaspartate (Makarov et al., 1980, 1981; Hyde et al., 1984). However, Fasella et al. (1966) have proposed an equilibrium between a protonated external aldimine with an uncharged Lys 258 ϵ-NH$_2$ and an unprotonated external aldimine (absorbing at 360 nm) with a charged Lys 258 ϵ amino group. We believe this to be the correct interpretation. L-2-Methylaspartate binds stoichiometrically and covalently in the (orthorhombic) chicken mAAT and the chicken cAAT crystals. In pig cAAT, this inhibitor binds only (and covalently) to the subunit that is capable of making a transition to the closed conformation (Arnone et al., 1982). Furthermore, in the triclinic chicken mAAT crystals, weak binding has been observed due to conformational constraints (Eichele, 1980). The unique behavior of the 2-methylaspartate complex must be attributed to the insulation effect on Lys 258 produced by the 2-methyl group. In the external aldimine with the substrate aspartate, Lys 258 attracts the Cα proton instead. Thus, a consistent picture is obtained, and the tendency of Lys 258 to act as a proton shuttle is nicely demonstrated. The electronic environment of Lys 258 in the external aldimine is evidently such that the proton is not stable on Lys 258 either, which is an ideal situation for catalysis.

The position and orientation of 2-methylaspartate in the active site of pig cAAT has been confirmed by a 3.8 Å resolution study of a 2-hydroxy-methylaspartate derivative. A difference electron density map with the 2-methylaspartate derivative showed a single maximum for the additional oxygen atom at a position corresponding to the proper geometry of a

2-hydroxymethyl group with respect to the β-carbon position determined for the 2-methylaspartate derivative (Hyde et al., 1984).

3.6.4. *Erythro-3-Hydroxy-L-Aspartate*

This compound has very interesting properties. It quickly establishes an equilibrium between species absorbing at 330, 360, and (very strongly) at 490 nm in both cytosolic and mitochondrial AAT. Transamination to the PMP enzyme takes place very slowly. Jenkins (1964) has attributed the 490 nm absorption band to a quinonoid intermediate. Crystals of pig and chicken cAAT (Metzler et al., 1978; Makarov et al., 1980) and of chicken mAAT (Eichele et al., 1978) show the same absorption maxima with this compound, which persist even longer in crystals that stabilize the closed structure. Low-resolution X-ray studies with this derivative have been carried out, but could not be interpreted in terms of a model for binding of the inhibitor (Borisov et al., 1980b; D. Picot, M. G. Vincent, and J. N. Jansonius, unpublished results). Kirsch et al. (1984) have proposed a model for the quinonoid intermediate with *erythro*-3-hydroxyaspartate that would explain its stability. Linear dichroism studies are consistent with a coenzyme tilt for the quinonoid intermediate, which is slightly larger than in the external aldimine (Vincent et al., 1984).

3.6.5. *Monocarboxylic Acids*

Arnone et al. (1984) have collected 4.4 Å resolution X-ray data of pig cAAT at pH 5.4 in the presence of 300 mM acetate or formate. A difference map comparing these two derivatives showed a similar domain movement in the least constrained subunit, as produced by 2-methylaspartate. Thus, high concentrations of acetate stabilize the closed structure in the same way as dicarboxylic acids. Since, at lower concentrations (40 mM), acetate binds to one site near Arg 386 without causing domain rearrangement, the authors interpreted their difference map as resulting from binding of a second acetate molecule near Arg 292*, producing a maleate or glutarate-like arrangement that compensated the two arginine charges, and thus triggered the conformational change towards the closed structure. On the other hand, formate, even at 300 mM concentration, only bound to the first site.

In chicken cAAT, a buffer with Tris and 0.3 M acetate at pH 8.0 caused acetate to bind to the subunit that is constrained in the closed conformation. However, this concentration of acetate could not prevent the attainment of the open conformation by the less constrained subunit. Thus, subtle differences may exist in the behavior of these two enzymes.

3.6.6. Reaction of Crystalline Pig cAAT with Substrates

The only crystallographic data of AAT in the presence of real substrates have been reported by Arnone et al. (1982, 1985b) and Hyde et al. (1984). Cocrystallization and crystal soaking experiments of pig cAAT with glutamate (300 mM) both resulted in a hybrid structure with, as usual, one constrained open subunit. The latter did not appear to bind substrate or product, while the less constrained subunit did; it was accordingly in the closed form. Crystal spectra under the same conditions (Metzler et al., 1978) revealed three absorption bands at 330, 360, and 430 nm; all were quite intense. The combined results were interpreted as an unliganded PLP enzyme in the constrained subunit and a ketimine intermediate (closed structure) in the other. However, in view of the rather low resolution of this work (3.5 Å), a "dead end" complex of the PMP enzyme with the large excess of glutamate present can probably not yet be excluded. The experiment does again confirm the general low affinity of the open forms of AAT for substrates and inhibitors.

A similar experiment with, in addition, 2 mM 2-oxoglutarate (Arnone et al., 1982) was interpreted as having produced a Michaelis complex of PMP enzyme with 2-oxoglutarate (in the flexible subunit). This result also needs to be confirmed.

3.6.7. N- (5'-Phosphopyridoxyl)-L-Amino Acids

These compounds, PPL amino acids for short, result from NaBH$_4$ reduction of the PLP aldimines of L amino acids. Upon binding to apo-AAT for which they have a high affinity (Dempsey and Snell, 1963; Severin et al., 1969; Relimpio et al., 1975), they produce analogs of the covalent intermediates of catalysis. The enforced stoichiometric binding of the amino acid moiety is a major advantage of these compounds. PPL-Ala, PPL-3-iodo-Tyr, PPL-Asp, and PPL-Glu have been studied with chicken mAAT (Eichele et al., 1979a; Ford et al., 1980; Kirsch et al., 1984; Jansonius et al., 1984a, 1984b). The compounds were introduced into apoenzyme crystals by soaking. With PPL-3-iodo-Tyr and PPL-Ala, full occupancy in both subunits of the P1 crystal form was easily attained without deterioration of the crystals at 0.6–1.0 mM concentration and at room temperature. However, cracking of the crystals could barely be avoided for PPL-Asp, even at 0.2 mM concentration and 4°C. Thus, a conformational change could be expected for the PPL-Asp derivative. Indeed, a difference electron density map at 3.0 Å resolution revealed full occupancy of PPL-Asp in subunit S*, coupled with a large conformational change in this subunit that could be interpreted as movement of the small domain towards the coenzyme. This was the first instance in which such a domain reorientation was observed crystallographically in chicken mAAT. In the constrained subunit S, the occupancy of PPL-Asp was only ~60% (Eichele, 1980). Model building demonstrated that the PPL-Asp moiety was bound with the phosphate group in its normal position. The pyridine ring was tilted ~30° forward as compared to the situation in the PLP enzyme, with the protonated N1 retaining its hydrogen bond to Asp 222. The amine nitrogen and the $C\alpha$ atoms were coplanar with the pyridine ring, and the α and β carboxylate groups were hydrogen bonded to Arg 386 and Arg 292*, respectively. Thus, this analog provided an excellent model for the external aldimine intermediate of catalysis (Kirsch et al., 1984). The structure of the 2-methylaspartate derivative (Section 3.4.2) subsequently confirmed this.

Difference maps at 2.8 Å resolution, comparing the PPL-Ala and PPL-3-iodo Tyr derivatives with the apoenzyme and models built into these maps, had the following features in common: (1) No domain movement had occurred in either subunit, (2) the coenzyme moieties had the same orientation, very similar to that in the PMP enzyme, (3) the α carboxylate groups were positioned between the two arginines with one carboxylate oxygen H bonded to Trp 140 Nϵ1, (4) the amine N and $C\alpha$ were not coplanar with the pyridine ring, and (5) the torsion angle C4-C5-C5'-OP1 was small (~15°) and positive (OP1 behind the pyridine ring). In the PPL-Ala complex, the alanine $C\beta$ was in contact with Tyr 70*, while the iodophenol moiety of PPL-3-iodo-Tyr was in contact with the "right" wall of the active site formed by the small domain (Eichele et al., 1979a; Eichele, 1980; Kirsch et al., 1984). These models, in contrast to that of PPL-Asp, did not mimic the corresponding external aldimines. This might be significant and indicate that the mAAT active site is not complementary to these intermediates or, in other words, is less specific for them. Nevertheless, mAAT is a reasonable tyrosine as well as 3-iodo-tyrosine aminotransferase (Miller and Litwack, 1971; Shrawder and Martinez-Carrion, 1972; Mavrides and Christen, 1978). Further work towards this specificity is needed even though the environment of the side chain in the PPL-3-iodo-Tyr derivative is compatible with the observation of enhancement of (3-iodo)-Tyr transamination by high concentrations of formate (Mavrides and Christen, 1978), as observed with alanine (Saier and Jenkins, 1967; Morino et al., 1974). Another interesting observation is that triclinic crystals of pig mAAT catalyze alanine transamination equally (in)efficiently as the soluble enzyme (Mozzarelli et al., 1979), confirming that no domain reorientation is involved in this process.

Cocrystallization of apo-mAAT with PPL-Asp at room temperature gave rise to crystals in space group $P2_1$ (Table 1), with both subunits in the closed conformation, as revealed in a 4.4 Å electron density map (Jansonius et al., 1984a; C. Thaller, M. G. Vincent and J. N. Jansonius, unpublished results). Similarly, cocrystallization of PPL-Glu with apo-mAAT at 4°C resulted in crystals of space group $C222_1$ (Table 1), in which both subunits have the closed conformation. Again, a 4.4 Å (difference) map has been calculated (D. Picot, M. G. Vincent, and J. N. Jansonius, unpublished results). These PPL-amino acid derivative studies confirmed that the closed conformation of mAAT is only attained in the presence of dicarboxylate (quasi) substrates and inhibitors (Kirsch et al., 1984).

Borisov et al. (1985) soaked an apoenzyme crystal of chicken cAAT in a solution of the arsonate analog of PPL-Glu. A difference map at 5 Å resolution showed that both subunits still had the closed conformation. A maximum near Arg 292* in the least constrained subunit revealed binding of the coenzyme-substrate analog, whereas in the constrained subunit, the situation was less clear. Nonetheless, this experiment proved that the arsonate analog of PPL-Glu binds in a "productive" way and stabilizes the closed conformation.

3.6.8. *Miscellaneous X-ray Experiments*

An analog of glutamate with the γ carboxylate replaced by arsenate $(OAsO_3^{2-})$ is transaminated very slowly (2 hours) by chicken cAAT crystals. A similar arsenate derivative of 2-hydroxyglutarate could be used as an inhibitor of both the PLP and PMP form of this enzyme. However, binding occurred only to the constrained closed subunit and with low occupancy (Borisov et al., 1985). The results can perhaps be explained by the size of these molecules, which have a longer (by ~1.5 Å) as well as a bulkier side chain than glutamate and will thus not easily be accommodated.

Isonicotinic acid hydrazide caused conversion of the least constrained subunit of chicken cAAT to the open conformation (Borisov et al., 1978), again in-line with expectation. Ammonium ions in high concentration tend to produce an external aldimine analog of chicken cAAT in the constrained closed subunit, with a coenzyme orientation similar to that in the PMP enzyme. Dicarboxylate inhibitors and high concentrations of acetate effectively protect the coenzyme from reaction with ammonia (Harutyunyan et al., 1984). The compound D, L-2-amino-3-phosphonopropionate at 0.3 M concentration bound to both active sites of pig cAAT, but did not transaminate and failed to produce the closed conformation for reasons that are not yet understood (Hyde et al., 1984). D. Picot (unpublished) cocrystallized chicken mAAT with a number of ligands at 4°C. Either P1 or C222$_1$ crystals appeared. The difference in crystal habit was sufficient to establish whether a particular ligand could stabilize the closed conformation (space group C222$_1$) or not (space group P1). An interesting result was that 100 mM maleate or 2-oxoglutarate, but not 80 mM 2-methylaspartate or succinate, convert the apoenzyme to the closed form. This confirms that charge compensation is the major trigger for the small domain reorientation. Succinate probably has too much conformational freedom in the large coenzyme-free active site, and 2-methylaspartate does not have enough affinity under these conditions. 100 mM 2-oxoglutarate does *not* produce the closed form of the PLP holoenzyme, in agreement with an earlier experiment of Eichele (1980). One would be tempted to explain this by a too narrow active site in the closed form to accommodate a C$_5$ substrate. The preference of mAAT for C$_4$ substrates over C$_5$ substrates in contrast to cAAT is well known (Michuda and Martinez-Carrion, 1970). 80 mM succinate had no influence on the crystallization behavior of the PMP enzyme, while ambiguous results were obtained with the PLP enzyme.

3.6.9. Conclusions

The crystallographic inhibition studies of three different AAT isoenzymes provide a reasonably consistent picture, which we can summarize as follows: (1) Only dicarboxylate inhibitors or analogs of catalytic intermediates induce the closed structure of these enzymes, with cAAT accepting both C$_4$ and C$_5$ inhibitors, while mAAT seems to prefer C$_4$ inhibitors. (2) The binding modes of 2-oxoglutarate, 2-methylaspartate, and PPL-Asp demonstrate *cis* (or rather "gauche") orientation of the two carboxylate groups. The α carboxylate always makes a salt bridge with Arg 386, and the distal carboxylate with Arg 292*. The carbonyl oxygen of 2-oxoglutarate and the 2-methyl group of 2-methylaspartate occupy similar positions in the active site, between Tyr 70* and Gly 38. (3) Enforced open active sites, as occurring in one subunit of pig cAAT crystals and triclinic chicken (and pig) mAAT crystals, inherently have little affinity for dicarboxylate inhibitors. (4) Different coenzyme orientations were found for the PLP enzyme, the PMP enzyme, and the external aldimine with 2-methylaspartate, with the "tilt angle" increasing in that order. While there is now agreement on the values of the torsion angles around C5-C5' of the coenzyme in the unliganded PLP enzyme and the external aldimine with 2-methylaspartate in pig cAAT and chicken mAAT, the PMP enzyme and the maleate derivative of mAAT have low values for this angle, which indicates that a short distance between OP1 and C4' is not energetically unfavorable.

Although a great deal has been achieved in the crystallographic studies, much still remains to be done in this area. Michaelis complexes with the PMP enzymes have not yet been studied; a crystallographically determined model of the quinonoid intermediate with *erythro*-3-hydroxyaspartate is lacking, and the ketimine structure proposed for pig cAAT needs to be further characterized. Accurate structural information on C_5 inhibitors is not yet available for mAAT. In spite of these limitations, the results so far have provided much insight into the stereochemistry of enzymic transamination.

3.7. Optical and Kinetic Studies of AAT Crystals

Single crystal microspectrophotometry went hand in hand with crystallographic studies of AAT isoenzymes, and it has proven the catalytic competence in the crystalline form of pig cAAT (Metzler et al., 1978), chicken cAAT (Makarov et al., 1980, 1981), pig mAAT (Mozzarelli et al., 1979), and chicken mAAT (Eichele et al., 1978) through demonstration of reversible transition of the 360 nm and/or 430 nm absorbance towards 330 nm upon addition of the appropriate substrates. Moreover, with a mixture of substrates (e.g., aspartate/oxaloacetate) and with the inhibitor 2-methylaspartate and the pseudosubstrate *erythro*-3-hydroxyaspartate, qualitatively similar spectra were observed as in solution. However, the relative heights of absorption maxima often differed considerably. It is now clear that most of the discrepancies can be explained as being due to changes in affinity of substrates and inhibitors for the active site of a dynamically frozen subunit, the existence of two differently behaving subunits in the crystals of pig and chicken cAAT and in the P1 crystals of mAAT, and differences in solvents (ammonium sulfate solution was used for chicken cAAT). The existence of a "fast" and a "slow" subunit in triclinic microcrystals of chicken mAAT in catalytic conversions of specific substrates has been elegantly demonstrated by Kirsten et al. (1983). These authors showed that the "fast" subunit has a V_{max} of 15% of the solution value, but a 10-fold increase in K_m, while the "slow" subunit has normal K_m values, but a V_{max} of only 3% of its value in solution. On the other hand, alanine is transaminated equally efficiently by these crystals as in an enzyme solution. This latter finding corresponds to the already mentioned results by Mozzarelli et al. (1979) with similar triclinic crystals of pig mAAT. We may conclude from these observations that triclinic mAAT crystals contain statically intact but dynamically partially impaired enzyme. The extent of inhibition is different for the two subunits.

Linear dichroism experiments allow, in favorable cases, the determination of the directions of transition dipole moments (TDM) for the electronic transitions corresponding to the observed absorption bands, with respect to the crystallographic axes. Since the coenzyme absorption bands of AAT in the region 300−500 nm correspond to π-π^* transitions for which the TDMs lie in the coenzyme ring plane (Hofrichter and Eaton, 1976), TDM directions can be correlated with coenzyme orientations. Orthorhombic crystals are best suited for such studies. Ideally, polarized light spectra should be taken with the electric vector parallel to all three crystallographic axes. For quantitative interpretation of the data, it is necessary that there are no differently behaving chromophores in the unit cell, with a single chromophore in the asymmetric unit being the most favorable case. Although linear dichroism experiments have been carried out with orthorhombic crystals of pig and chicken cAAT and with triclinic crystals of chicken mAAT, the above requirements are only completely fulfilled for the $C222_1$ crystals of chicken mAAT. With these, it has been possible to confirm independently the crystallographically determined models of PLP (360 nm), PLP + maleate (440 nm), PMP (330 nm), and the external aldimine with 2-methyl-aspartate (430 nm), since the TDM directions were found to lie within ~5° from the coenzyme ring plane in each case. Furthermore, the comparisons established, to within ~10°, the directions of the various TDMs in the pyridine ring (Vincent et al., 1984). The angles were listed with respect to the direction N1−C4 as reference. A tentative direction in the pyridine ring was also given for the TDM of the 490 nm absorption band of the quinonoid intermediate with *erythro*-3-hydroxyaspartate; the model used is an extrapolation towards a slightly increased "tilt" angle of the external aldimine intermediate.

The only other experimental determination of a TDM for a PLP absorption band was carried out by Hyde et al. (1984) for the 430 nm absorption band of native pig cAAT at pH 5.4 and in the external aldimine with 2-methylaspartate. The determination depends on the assumption of a correct crystallographic model for the coenzyme orientations in the two independent subunits, as well as on a priori knowledge of their behavior with respect to inhibitor binding. The result agrees within $10-15°$ with that of Vincent et al. (1984). Both results will have to be updated when refined structures of the various derivatives are available.

The TDMs determined for orthorhombic mAAT crystals were transformed into the two subunits of the P1 crystal form. Assuming negligible birefringence, directions of maximum absorbance of plane-polarized light were computed with respect to the main faces (001) and 01$\bar{1}$) of two crystal habits, and compared with earlier experimental data (G. Eichele and H. Kirsten, unpublished results; Kirsten, 1982). Very good agreement was obtained (Vincent et al., 1984). Interestingly, for good agreement with the 490 nm band of the quinonoid intermediate, it had to be assumed that this intermediate is only present in the (least constrained) subunit S^*, which can assume the closed structure. This is exactly what would be expected on the basis of crystallographic experience with both chicken mAAT and pig cAAT.

Correlation of spectral data with crystallographic results has, in several instances, assisted considerably in the interpretation of either or both. Examples are the "external aldimine" that chicken cAAT forms with ammonium ions in the "closed" subunit (Makarov et al., 1980, 1981) and the studies of derivatives of the β subform of pig cAAT and of pig cAAT crystals grown in the presence of glutamate (Hyde et al., 1984).

Quantum chemical calculations to determine TDM directions of absorption bands observed in AAT have been carried out by Savin (1984) and Song (unpublished results cited by Metzler et al., 1978). The two sets of values disagree seriously. The experimental values of Vincent and colleagues agree very well with Savin for the 360 nm and 330 nm bands. Discrepancies of $15°-25°$ are found for the 430 nm and 490 nm bands. The discrepancies with Song's values range from 16° to 51°.

Savin finds only 3° difference in direction for the TDMs of the 360 nm and 430 nm bands. Makarov et al. (1980, 1981) observed, however, a significant difference in polarization ratio (ratio of absorbance parallel to a crystallographic axis and normal to it, with a diagonal direction in their case) for these two bands in unliganded and liganded PLP enzymes. The authors interpreted this difference as indicative of an increased coenzyme "tilt" upon internal aldimine protonation caused by binding of substrate (or inhibitor) in a Michaelis complex. They argued that this would be the first step towards transaldimination. Such an increase in tilt is not observed for mAAT. However, the reorientation of the PLP ring upon maleate binding that was described in Section 3.4.1 might explain the observation of Makarov et al. (1980, 1981).

Once the directions of the various TDMs have been established beyond a doubt, this knowledge will be of great help in the interpretation of polarized absorption spectra of, for example, mixtures of catalytic intermediates, as would occur when a crystal is soaked in a solution of a substrate or a mixture of substrates. The information can also be carried over to other PLP enzymes that display similar absorption bands.

4. CORRELATION OF CRYSTALLOGRAPHIC RESULTS WITH CHEMICAL AND PHYSICOCHEMICAL DATA

4.1. Amino Acid Sequences

Extensive sequence information on AATs is now available. Six sequences are complete: chicken mAAT (Graf-Hausner et al., 1983), pig mAAT (Barra et al., 1980; Kagamiyama et al., 1980), rat mAAT (Huynh et al., 1980), chicken cAAT (Shlyapnikov et al., 1979), pig cAAT (Ovchinnikov et al., 1973; Doonan et al., 1974), and *E. coli* AAT (Kondo et al., 1984). Partial sequences have been published for turkey mAAT (Barra et al., 1979), horse and human mAAT, and horse cAAT (Barra et al., 1984). These sequences allow quite a good estimate of invariant residues in all AATs to be made and correlation of invariant residues or regions with structural or functional importance. We find that 332 residues are invariant in all cAATs (80%), 326 in all mAATs (81%), 164 in all AATs excluding *E. coli* AAT (40%), and 107 in all AATs (26%). *E. coli* AAT has 147 residues in common with all mAATs and 144 with all cAATs, and thus seems to be equally distant from both vertebrate isoenzymes.

4.1.1. *Active Site*

Of the 29 residues immediately bordering the active site and the coenzyme, all but eight are invariant, if we ignore the assignment Gly 140 (for Trp) in rat mAAT, which cannot be correct. Two modifications, Ser 107 → Gly and Ala 257 → Ser, have been discussed. The others are Val 15-Leu 16-Val 17-Phe 18 in cAATs (the mAATs and *E. coli* AAT have here Asp 15-Pro 16-Ile 17-Leu 18) and Val 37 → Ile and Ala 39 → Val in *E. coli* AAT. These replacements must be related to specificity. The activity towards aromatic amino acids increases in the direction cAAT < mAAT < *E. coli* AAT. This trend suggests that Ile 17, Leu 18, and Ile 37, and perhaps Val 39 as well, influence this activity favorably. However, it cannot be excluded that the domain interface also plays a role in modifying specificity.

4.1.2. Domain Interface

The domain interface is one of the best conserved regions of the molecule, more so than the hydrophobic cores and the subunit interface (Jansonius et al., 1985). However, *E. coli* AAT deviates here strongly from the vertebrate AATs. Seven residues involved in interdomain contacts differ from mAAT. The consequences of this are unknown.

4.1.3. Subunit Interface

This area, consisting of residues from the ranges 47−70, 142−149, 262−266, and 290−301 and the isolated residue 106 (Jansonius et al., 1985), remains reasonably invariant in *E. coli* AAT. This leaves little doubt that the general structure is the same as in the vertebrate enzymes. However, the N-terminal subunit interaction in the *E. coli* enzyme is quite different from the vertebrate enzymes. The hydrophobic residues 5 and 6 are present (Met 5-Phe 6), but the pocket seems less hydrophobic with only Val 125, Leu 218, and Leu 272 contributing nonpolar side chains. Perhaps hydrogen bonds compensate for the loss of the entropy contribution towards binding.

4.1.4. Hydrophobic Core of Small Domain

The hydrophobic core of the small domain, which includes in chicken mAAT the side chains of residues 33, 35, 337, 338, 341, 350, 353, 361, 363, 365, 370, 373, 377, 379, 381, 387, 389, 392, 397, 399, 400, 401, 403, 404, and 408, is highly preserved in all AATs, including *E. coli* AAT. Thus, the small domain is probably a quite rigid structure in all these enzymes. The generally observed, rather weak electron density in this region in crystallographic maps must probably be attributed to domain reorientations.

4.1.5. Hydrophobic Cores of PLP-Binding Domain

Jansonius et al. (1985) have defined an "inner" hydrophobic core in this domain on the coenzyme side of the pleated sheet "spine" and an "outer" hydrophobic core on the opposite side. We will not list here all the residues contributing to these cores, but just mention that their hydrophobicity is conserved also in the case of *E. coli* AAT. This provides further proof that the folding of the *E. coli* enzyme must correspond closely to that of the vertebrate enzymes.

4.1.6. Proline Residues

These residues are indicators of structural *in*homology. They should not appear in pleated sheets or in helices, other than in the first turn. *E. coli* AAT has eight prolines at positions different from those in chicken mAAT. Six of these are at "allowed" positions; one, Pro 219, is in pleated sheet strand *e*, but also corresponds to a proline in the cAATs. Only one (Pro 399) is in a dubious position in the C-terminal helix 16, and will cause a local distortion, but this might not have any long-range consequences.

We have paid much attention to *E. coli* AAT in this section because it is the most distant relative to chicken mAAT. Structural correspondence between these two enzymes automatically includes the others of the family. Functional studies on this enzyme by site-directed mutagenesis are underway (Kuramitsu et al., 1985; Malcolm et al., 1985). Such studies obviously have relevance to the vertebrate enzymes as well.

4.2. The Stereochemistry of Enzymic Transamination

Enzymic conversion of oxaloacetate or 2-oxoglutarate by AAT is a chiral process, leading to only L-aspartate and L-glutamate, respectively. The questions about the stereochemistry of the action of PLP enzymes and how this could be achieved have been discussed in detail by Dunathan (1971). Since PLP enzymes catalyze reactions that involve breakage of any of three bonds at the Cα position of the amino acid in a PLP-amino acid aldimine, some mechanism must prevent the wrong bonds from being broken. Dunathan proposed that a bond would be preferentially broken if oriented about normal to the pyridine-aldimine π-bonding system, due to (1) maximum σ-π overlap, and (2) the least necessary bond reorientations towards an sp^2-hybridized Cα. The enzyme would bind the α carboxylate and possibly the side group of the amino acid substrate in such a way that the proper bond would be labilized. For AAT, this means that the Cα-H bond could have two possible orientations normal to the pyridine ring plane. Further unknowns would be: (1) The aldimine double bond could be *cisoid* (near the 3-OH) or *transoid*. (2) After deprotonation of Cα, reprotonation could be *cis* (on the same side of the plane) or *trans*. (3) In principle, the arrangement of coenzyme and amino acid relative to the aldimine double bond could also be *cis* or *trans*, but only *trans* is sterically possible. (4) Finally, the choice of L-over D-amino acids is decided by the system.

Space prevents a discussion of the methods used to answer the questions posed by this problem and the results obtained. The reader is referred to the recent review by Martinez-Carrion et al. (1985) in *Transaminases*. The results indicated that the aldimine double bond is *cisoid*, with its *re* side exposed in the internal aldimine and its *si* side in the external aldimine. The α carboxylate is on the same side of the N1-C4 axis as the 3-OH group. Proton transfer is *cis*, with the proton being attached at the pro-S position of C4′.

The X-ray results are in excellent agreement with these proposals, as inspection of Figures 10*a* and 10*b* will confirm. Reorientation of the coenzyme pyridine ring upon transaldimination exposes its formerly protected A face. The position of the 2-methyl group of 2-methylaspartate with respect to Cα and the pyridine ring is exactly as predicted by Dunathan for α deprotonation of the real substrate (see Fig. 14). Specific binding of both α and distal carboxylates enforces this orientation. Note that Lys 258 is the only reasonable choice of catalyst for both deprotonation and protonation, albeit with the possible assistance of a water molecule, in agreement with an early proposal of Snell (1962). The involvement of histidine, an alternative proposal (Peterson and Martinez-Carrion, 1970), is sterically impossible.

4.3. Binding of Substrates and Inhibitors

With the knowledge of the active site structure(s), we can now explain the results of many inhibition experiments in solution. However, since most experiments have been carried out on pig cAAT, we must extrapolate the structural information on chicken mAAT towards that enzyme.

The first, and already revealing, inhibition experiments were carried out by Jenkins et al. (1959) on pig cAAT. They observed that the following compounds (in increasing order) inhibited the enzyme: succinate < 2-methylaspartate < adipate < glutarate < maleate. Fumarate, pimelate, and 2-methylglutamate were not inhibitors. The authors concluded that the distance and orientation of the carboxylate groups define the inhibitory capacity. The carboxylate groups must be *cis*, with 2−4 carbons in between. These inhibitors increase the pK_a of the 430:360 nm absorption equilibrium of the enzyme. The 430 nm absorbance of the low pH form of the enzyme becomes stronger and is shifted to a slightly higher wavelength by the inhibitors. These observations, at the time, led to the proposal of aldimine binding of PLP to the enzyme, which was subsequently confirmed. The ionic groups on the enzyme, interacting with the inhibitor carboxylate groups, were believed to be the same that bind the substrate carboxylate groups. The reaction of the dicarboxylate inhibitors with the enzyme active site required an additional proton as compared to the amino acid substrates.

We can now explain these data on the basis of the crystallographic results. The 360 nm absorbing form of the PLP enzyme is protonated at N1 and deprotonated at O3′. The charges are stabilized by Asp 222 (N1) and, presumably, Arg 386 and Arg 292*. The charges on O3′ and the arginines are compensated by the protonated α amino group and the carboxylate groups of the substrate. Dicarboxylate inhibitors need an additional proton. Substrate and inhibitor binding is assisted entropically by the release of at least two bound water molecules (Figs. 10*a* and 10*b*). The exceptionally strong binding of maleate is due to its enforced *cis* conformation. Fumarate is not complementary to the active side pocket and pimelate is too large. We would believe, at least for mAAT, that adipate would also be too large, certainly for the closed structure. This could easily be checked by model building.

The inhibition power of dicarboxylate inhibitors parallels an increase in 430 nm absorption, which, we believe (see Section 3.6.2), indicates transition to the closed structure. The orthorhombic crystals of the maleate complex of chicken mAAT exhibit at pH 7.5 only a 440 nm absorption band (Vincent et al., 1984). The open structure has hardly any affinity for 2-methylaspartate, while its binding to the closed structure is stoichiometric (Arnone et al., 1982, 1984; our own unpublished results). The strong entropy contribution to the binding of 2-methylaspartate (Hammes and Tancredi, 1967) and glutarate (Jenkins and D'Ari, 1966) has been established. The increase in thermostability of pig cAAT in the presence of dicarboxylate inhibitors (Jenkins et al., 1959; Relimpio et al., 1981) can also be explained by transition towards the more compact closed structure.

Further studies by Jenkins and coworkers revealed that two classes of dicarboxylate inhibitors exist: one that forms a 435 nm absorbing complex at pH 8.0, and another that does not. The latter class consists of molecules that have little structural resemblance to the substrates (Bonsib et al., 1975). This fits in nicely with the correlation between 435 nm absorption and closed structure. The strong binding of some aromatic acids has been attributed to hydrophobic interaction, which is also compatible with the active site structure. For most molecules belonging to the second class defined by Bonsib et al. (1975), it is immediately obvious that they are incompatible with the closed structure. For some, for example methylmaleate and methylsuccinate, careful model building would be required to test this incompatibility.

The anomalous behavior of L-2-aminoadipate as a substrate (Jenkins, 1980a) can at least qualitatively be explained by its too long side chain. This requires conformational changes, presumably involving Arg 292*, before productive binding can occur. No 430 nm absorbance is observed. In our interpretation the reason is that domain closure is not possible.

The exact mode of inhibition of the PMP-form of pig cAAT by succinate (Dybel et al., 1972) and very strongly ($K_i \sim 9\ \mu M$) by difluoro-oxaloacetate (Briley et al., 1977) has yet to be defined by X-ray crystallography.

Harruff and Jenkins (1976, 1978) and Jenkins (1980b) have carried out detailed studies of ion binding (halides and buffer ions) to pig cAAT. Halides shift the transamination equilibrium between alanine and pyruvate in the oxoacid direction due to an accelerating effect on the forward reaction, as well as an inhibitory effect on the back reaction. This can perhaps be explained by ion binding in the PLP enzyme to Arg 292*, thereby forcing alanine towards a productive way of binding, and the binding of two ions to the more positively charged PMP enzyme with blocking of the substrate-binding site. Since specific ion binding is involved, it is no surprise that size and polarizability of the halide ions play a role, and that, due to charge neutralization of (presumably) Arg 292*, such binding leads to an increase in the pK_a of aldimine protonation. However, a structural explanation of the various relaxation times associated with ion buffer binding as observed in temperature jump experiments (Giannini et al., 1975, 1976) cannot yet be given.

Buffer ion replacement upon binding of glutarate by pig cAAT was observed by Jenkins and D'Ari (1966), who suggest that ion replacement may be the normal event in substrate binding by this and other enzymes. The proposed mode of binding involving only one cationic group and the protonated PLP aldimine has now been shown to be incorrect (Harutyunyan et al., 1984).

4.3.1. D, L-2-Methylaspartate

We have described the binding of L-2-methylaspartate (the enzyme ignores the D-isomer) in detail (see Fig. 14), and have correlated the structural data with the binding studies of Fasella et al. (1966). Hammes and Haslam (1968) carried out fast kinetic experiments on the binding reaction and observed three relaxation times. The simple model that they propose for the reaction involves three reversible steps. All six rate constants were determined. Step 1 was characterized by an increase in 430 nm and decrease in 360 nm absorbance, and was interpreted as Michaelis-type binding to the unprotonated internal aldimine. The second and third steps involve a conformational change and transaldimination in an unknown order. Since the second relaxation time involved a decrease of 430 nm and increase of 360 nm absorption, it seems logical to attribute it to transaldimination, since afterwards the external aldimine proton equilibrates with Lys 258, while (presumably) in the Michaelis complex, this proton tends to remain with the aldimine rather than the α amino group of the inhibitor. Step 3 would then be the domain closure around the inhibitor. This also correlates with the stabilization of the inhibitor complex in the closed, but not the open, structure, since the equilibrium for step 3, but not step 2, lies towards the right. The mode of binding of 2-methylaspartate in the Michaelis complex presumably is not compatible with the closed structure. Only upon fixing the inhibitor covalently does the domain reorientation take place, resulting in a more stable structure. All steps involved are much slower than measured for natural substrates (Fasella and Hammes, 1967).

Boettcher and Martinez-Carrion (1978) attached a spin label to Cys 390 of pig cAAT and subsequently studied 2-methylaspartate binding to this enzyme derivative by electron spin resonance (ESR) and nuclear magnetic resonance (NMR). Distances of inhibitor protons to the spin label could be determined. They varied between 8.7$-$9.1 Å and were identical at pH 6.0 and 8.0. No attempt has yet been made to check this result by model building.

4.3.2. Inhibition by L-Erythro-3-Hydroxyaspartate

The *threo* isomer of this aspartate derivative is a rather normal substrate, but the *erythro* form produces a covalent complex that strongly absorbs at about 490 nm, and only slowly disappears. Although the K_m values for the latter are comparable with those of aspartate, turnover is (for pig cAAT) 5000 times slower in solution (Jenkins, 1961). The species absorbing at 490 nm was assigned a quinonoid structure (Jenkins, 1961). A thorough study of the interaction of this interesting compound with pig cAAT has been made (Czerlinski and Malkewitz, 1965; Hammes and Haslam, 1969; Jenkins and Harruff, 1979).

Hammes and Haslam studied the kinetics of the reaction with temperature jump, stopped flow, and spectrophotometric techniques. They observed 11 relaxation times, of which eight probably are indirectly related to the transamination reaction, and one corresponds to a very fast equilibrium between species absorbing at 490 nm and 460 nm. A mechanism was proposed that includes a sequence of eight reversible steps on the main pathway, with the 460 nm absorbing species being the result of a side reaction. The first three steps were assigned to formation of a Michaelis complex, a conformational change, and transaldimination. The equilibrium of the first step lies very much on the side of the complex, that of the second step slightly to the left, and of the third step slightly to the right, with the rate constants of these latter two steps being very similar. All three steps are fast, of the order of milliseconds, and much faster than for 2-methylaspartate (Hammes and Haslam, 1968). It is not easy, and perhaps not relevant, to suggest whether transaldimination or transition to the closed structure occurs first in this case. The fourth step produces the quinonoid intermediate of which (at least) two forms, absorbing at 490 nm and 460 nm, are in rapid equilibrium. These intermediates involve ~20% of the enzyme molecules. Step 5, ~10 times slower, leads to a species absorbing at 330 nm, presumably the ketimine intermediate. The back reaction for this step is three times faster. Since the subsequent reaction, step 6, is again 10 times slower, the quinonoid intermediate is accumulated. The back reaction of step 6 is even five times slower than the forward reaction. Step 6 is, in fact, rate limiting. It must represent either hydrolysis of the ketimine intermediate or a conformational change. The products of both step 6 and step 7, which are equally fast as step 5, absorb at 330 nm. Their equilibrium fractions are ~30% and 20%, respectively. Step 8 would be product release, and its equilibrium constant must lie far to the left in order to explain the high fraction of intermediates beyond the rate-limiting step. The turnover number in the presence of 2-oxoglutarate is ~1 sec^{-1}.

In crystals that stabilize the closed structure, the equilibrium is completely frozen and X-ray data have been collected, as mentioned earlier (Borisov et al., 1980a, 1980b; D. Picot, unpublished results). The P1 crystals of chicken mAAT, on the other hand, transaminate *erythro*-3-hydroxyaspartate, albeit very slowly (within 1 day; Eichele et al., 1978). Exchange of ligands in the orthorhombic crystals of mAAT, in which the closed structure is stabilized, nevertheless takes place within minutes (unpublished results). Therefore, the very high affinity of substrate and product for the closed structure, rather than their complete trapping, must be the explanation for the long-term stability of 3-hydroxyaspartate complexes in these crystals. Since the spectra of AAT complexed with this compound in crystals and in solution are very similar (Eichele et al., 1978; Rossi et al., 1978; Metzler et al., 1978; Makarov et al., 1980), we may conclude that the results of Hammes and Haslam (1969) are also relevant for the crystalline state. This, in turn, allows us to suggest that step 6 represents ketimine hydrolysis and step 7 is the transition to the open structure, which is suppressed in the crystal. The alternative order presumably would reduce the fraction of species absorbing at 330 nm in the crystal. Also, one would not expect the equilibrium closed $\rightleftarrows$ open to favor the open conformation in the ketimine intermediate, as would be the case if it corresponded to step 6. Furthermore, it would not be easy to explain why the closed-to-open transition would have to be 50–100 times slower than the opposite transition (step 2 or 3). Rather, one would expect the rate-limiting step to be related to obstruction of ketimine hydrolysis, due to sterical and/or electronic interference of the 3-hydroxy group. Production of the ketimine intermediate from the oxoacid complex would be very slow for similar reasons.

The fraction of molecules representing the ketimine intermediate would be low (6%) according to Hammes and Haslam (1969) and the above interpretation. Thus, the main species in the crystal would be the quinonoid intermediate(s) and the product complex. This view is supported by two further observations. Martinez-Carrion et al. (1970b) have observed positive dichroism for the 330 nm absorption band with *erythro*-3-hydroxy-aspartate. For the β subform of pig cAAT, which is believed to be a ketimine species (Arnone et al., 1985b), only a very weak Cotton effect was observed by Martinez-Carrion et al. (1970b). Furthermore, in chicken mAAT crystals, the dichroic ratio at 330 nm of the polarized spectra with the same compound corresponds to that for the PMP enzyme (Kirsten, 1982; Vincent et al., 1984).

4.3.3. *N-(5′-Phosphopyridoxyl)-L-Amino Acids*

Dempsey and Snell (1963) introduced these compounds as inhibitors of PLP enzymes. Severin et al. (1969) made a series of such compounds and tested them with apo-AAT. Not unexpectedly, PPL-Glu and PPL-Asp were the most potent inhibitors of this type, followed by PPL-Ala. Khomutov et al. (1969) observed that PPL-Glu inhibition results in phosphorylation of a threonine residue (now known to be Thr 109), implying stress upon the phosphate ester linkage. Further work indicated that this also holds for several catalytic intermediates, suggesting that the phosphate group actively participates in catalysis (Khurs et al., 1976).

PPL-Ala (and PPL-3-iodo-Tyr) leave mAAT in its open conformation when introduced into apoenzyme crystals, while PPL-Asp causes the crystals to crack due to a transition of the enzyme towards its closed conformation (Eichele et al., 1979a; Ford et al., 1980; Kirsch et al., 1984). However, PPL-Glu cannot easily be introduced into apoenzyme crystals (G. Eichele, unpublished results). It has to be added to the apoenzyme prior to crystallization (D. Picot, unpublished results).

Strong stoichiometric binding of the L-isomers of PPL-trifluoro-Met, PPL-α-trifluoromethyl-Ala, PPL-p-fluoro-Phe, and PPL-o-fluoro-Phe to apo-cAAT was observed by Relimpio et al. (1975). The corresponding amino acids are either inhibitors (α-trifluoromethyl-Ala) or substrates (the other three) of pig cAAT.

4.3.4. "Conformational" Irreversible Inhibitors

Khomutov, Karpeisky, Severin, and colleagues in Moscow (Karpeisky et al., 1963; Khomutov et al., 1968; Braunstein, 1973) synthesized and tested several compounds that undergo partial transamination with AAT to produce a ketimine-like complex, which subsequently binds covalently to an active site group. L-aminoisoxalidone-3 ("L-cycloserine"), which is a very potent inhibitor of L-AlaAT, was the prototype of these compounds. Further examples included 4-aminohydroxyisoxazolidone-3 and the so-called cycloglutamates. These latter compounds are cycloserine derivatives, substituted with $-CH_2COOH$ at position 5 (the α cycloglutamates) or stereoisomers of these with the amino group in the side chain (the γ cycloglutamates). The inhibition constants of these glutamate analogs permitted conclusions to be drawn on the "active" conformation of L-glutamate in the active site of the pig cAAT. The carboxylate groups have to be *cis* in analogy to the situation with maleate, which is a conclusion confirmed by the recent crystallographic studies on pig and chicken cAAT (Arnone et al., 1982, 1984; Harutyunyan et al., 1984).

4.3.5. Other Irreversible Inhibitors

Many other compounds have been synthesized that are structurally related to substrates, but react covalently with the enzyme after binding to the active site. One class consists of halogenated derivatives of amino acids, and another consists of amino acids with a double or triple carbon-carbon bond in their side chain. A third class includes carbonyl reagents, such as hydrazine derivatives. Since no high-resolution X-ray data with any of these compounds has been reported, we will not discuss them further, but will refer the reader to reviews by Morino and Tanase (1985) and by Soper and Manning (1985).

4.4. Properties of the Coenzyme in AAT

4.4.1. Cotton Effect

Torchinsky and Koreneva (1963, 1964) first observed an extrinsic Cotton effect in optical rotatory dispersion, coinciding with the optical absorption bands of pyridoxal phosphate bound to pig cAAT. The signs of these Cotton effects, with maxima centered at 430 nm (pH 4.8) or at 362 nm (pH 8.2), were positive. Reduction of the coenzyme aldimine double bond, which shifts the absorption maximum to $\sim$ 330 nm resulted in a somewhat smaller positive Cotton effect with λ_{max} = 335 nm. The PMP enzyme displayed a distinct positive Cotton effect at 350 nm. Its amplitude was, however, only one-third of that observed for the PLP enzyme. These effects are caused by asymmetric interaction of the coenzyme with the protein. Quite dramatic effects were seen after reaction of the coenzyme with carbonyl reagents. For instance, the Cotton effect disappeared upon reaction with hydroxylamine or hydrazine, while larger hydrazine derivatives and hydrazides consistently gave negative Cotton effects, with a striking example being phenylhydrazine. Similar effects were observed upon binding of D,L-2-methylaspartate, which at pH 8.2 caused disappearance of the 362 nm Cotton effect but did not cause an effect at 430 nm, and also D, L-*erythro*-3-hydroxyaspartate. The latter compound gave rise to a positive Cotton effect at 333 nm, and a negative Cotton effect in the absorption band centered at 492 nm. The authors correlated the disappearance of the positive Cotton effects at 360 nm or 430 nm with displacement of the coenzyme-binding lysine (Lys 258). Later, these effects were confirmed by circular dichroism experiments (Ivanov and Karpeisky, 1969; Martinez-Carrion et al., 1970b) and played a major role in the proposal, by the former authors, of a dynamic mechanism of action of AAT, in which the coenzyme rotates around an axis C2-C5 towards its substrate. This mechanism will be discussed in Section 5. We can now say, on the basis of the X-ray data, that a decrease of positive coenzyme Cotton effect in AAT usually correlates with increased coenzyme tilt. Small decreases also result upon noncovalent inhibitor binding.

4.4.2. pK's of Coenzyme Protonation

Pyridoxal phosphate in its aldimine-bound state (to Lys 258) contains five groups with pH-dependent protonation equilibrium. These are: (1) the pyridine nitrogen, (2) the 3-hydroxyl group, (3 and 4) two phosphate hydroxyl groups, and (5) the aldimine nitrogen.

The N1 nitrogen is protonated and the 3-hydroxyl group ionized in AAT, leading to a high-energy, very reactive, dipolar ion (Heinert and Martell, 1963a, 1963b; Ivanov and Karpeisky, 1969; Metzler, 1979), which is stabilized by the hydrogen bond/salt bridge between N1 and Asp 222 and the (short) hydrogen bond of Tyr 225 to O3′, and also probably the H bond of Asn 194 to O3′ (see Fig. 10).

^{31}P NMR has been applied to the problem of the ionization state of the coenzyme phosphate group. The original work was carried out on pig cAAT by Martinez-Carrion (1975). Subsequent studies in the same laboratory on pig mAAT (Mattingly et al., 1982) revealed unexpectedly large discrepancies with the cytosolic isoenzyme. However, a reinvestigation of cAAT with more sensitive equipment (Schnackerz, 1984) produced results that were largely in agreement with those for mAAT. In comparison to free phosphate, the ^{31}P chemical shifts in the enzyme are shifted downfield, while pK_a effects are ~3 times smaller. On the other hand, bond angle distortion is known to cause large changes in chemical shift (Gorenstein, 1975). Therefore, interpretation of the pK_a values is not straightforward. If interpreted as signaling dissociations, however, they would indicate that the PLP phosphate group is deprotonated above a pK of ~6.2, while the PMP enzyme would have a monoprotonated phosphate over the whole range of pH 5.0−9.0. Inhibitor binding and aldimine reduction would, at least for mAAT, increase the pK value drastically. The strong correlation of the pK's sensed by ^{31}P NMR and those observed for the coenzyme + Lys 258 system point toward a third possible interpretation (Arnone et al., 1985a), while, of course, combinations of these effects cannot be excluded either. Thus, the issue remains open at present. The X-ray results support a double negative charge on the phosphate group, since (1) nine hydrogen bond donors are ligands to it (Section 3.3), and (2) the second negative charge would ensure stronger binding to the positively charged ($\sim +1.5e$) binding site. The low pH values, needed for resolution of PMP, would also indicate a double negative charge at pH ~ 7.0 and above for its phosphate group (Mattingly et al., 1982).

The aldimine protonation is strongly dependent on its environment. In pig cAAT, values between 5.25 in deionized water (Bergami et al., 1968) and 10.2 in 30% dimethylsulfoxide in the presence of glutarate (Haddad et al., 1977) have been observed for its pK_a, while 6.4 is the value observed under "normal" buffering conditions (Jenkins and Fonda, 1985). Protonation of the pyridine N1 lowers this pK drastically (Arnone et al., 1985a), with Arg 386 and Arg 292* also having a strong influence. Compensation of the latter two by negative charges and a hydrophobic environment (Haddad et al., 1977) brings about the rise in pK_a to 10.2. Figure 10b, which shows the binding of maleate to the active site of chicken mAAT, illustrates the above observations in structural terms. The protonation of the aldimine N is required to compensate the negative charge on O3′ when the positive electric field of the arginines is cancelled by inhibitor binding and water molecules in front of the coenzyme are displaced (compare with Fig. 10a). However, water molecules behind the coenzyme remain, and charge dissipation via the histidines 143, 189, and 193 (Arnone et al., 1985a; Metzler, 1979) is probably possible. Figure 10b very clearly demonstrates why the orthorhombic crystals of maleate-inhibited mAAT show a strong 440 nm band and no 360 nm absorption at all. The charge on the aldimine N is a necessity in the enforced closed structure. The functional role of aldimine protonation in catalysis will be discussed in Section 5.

4.4.3. *Coenzyme Derivatives*

A wide range of PLP derivatives has been tested for their binding and catalytic properties as coenzymes in cAAT. The subject has been reviewed by Braunstein (1973) and by Metzler and Fonda (1985). Here, we will correlate some of the observations with the three-dimensional structure.

N-methyl-PLP has an enforced charge on N1; therefore, the strong interaction with Asp 222 is no longer needed. However, the *N*-methyl group must cause steric hindrance, resulting in improper binding of the analog to the active site with reduction in catalytic activity, which is, in fact, observed. PPL-*N*-oxide, however, binds well and gives a 25% active enzyme. Here, the steric obstruction is much less, and the possible shift of ≤ 1.5 Å in coenzyme position is clearly tolerable.

Modifications, by nonpolar groups, at positions 2 and 6 of the pyridine ring have no electronic consequences, and therefore are well suited to probe the conformation and dynamics of motion of the coenzyme. Analogs with 2-side chain as large as isopropyl and *n*-butyl are accommodated by apo-cAAT. In the series methyl, ethyl, *n*-butyl as 2-side chain, the association constant decreases linearly with the number of bonds in this side chain, with the increments being compatible with an entropic explanation. In other words, no enthalpy changes seem to occur, indicating a hydrophobic environment (Bocharov et al., 1968). The X-ray models (Figs. 10*a* and 10*b*) indicate some space below the 2-CH_3 group, but it is surprising that three further carbons can be accommodated, without disturbing Asp 222 and His 189. They probably lie between Trp 140 and Asn 194. Even more surprising is the nearly full enzymatic activity of the reconstituted enzymes (Bocharov et al., 1968), since the coenzyme rotational movement of $30°-40°$ (Kirsch et al., 1984) should be inhibited to some extent. The 120% relative activity of 2-demethyl-PLP is indicative of enhanced rotational flexibility. Ivanov and Karpeisky (1969) concluded, from these observations, that dynamic movement of the coenzyme in catalysis should take place by rotation around C2-C5, leaving the 2-side chain and the phosphate group immobile (see Section 5).

The $6-7$ times faster combination of 6-methyl-PLP with apo-cAAT, in comparison with PLP itself (Bocharov et al., 1968), is very interesting. It can probably be explained by restricted rotation around C5-C5', which enforces 5'-side chain conformations closer to that taken up in the enzyme (Tumanyan et al., 1974).

For further substitutions at positions 2 and 6, which include halogenated compounds (e.g., 6-fluoro-PLP), the reader is referred to Metzler and Fonda (1985). Modifications in the 3-position include 3-*O*-methyl-PLP and the 3-amino analogs of PLP. Both can be accommodated; but not unexpectedly, they impair catalysis. the *O*-methyl group must cause reorientations of the side chains of Asn 194 and Tyr 225.

Modifications in the 4 position interfere with the aldehyde group, and therefore with catalysis. They include pyridoxine-phosphate (PNP), which binds to apo-mAAT in a nearly identical fashion as PMP, and the PPL amino acids. These latter compounds, already discussed in Sections 3.6.7.

and 4.3.3, gave very interesting information on substrate binding, since they are analogs of covalent catalytic intermediates.

Other acid side chains replacing the 5′-phosphate ester have also provided important information. The 5-*trans* $CH=CHPO_3^{2-}$ analog gave 14–20% activity (Metzler and Fonda, 1985). That this compound binds strongly is no surprise. One of its two preferred conformations, a 0° torsion angle around C5-C5′, should make it bind very comfortably, and major steric hindrance in coenzyme rotation is not expected either. The $5\text{-}CH_2CH_2PO_3^{2-}$ analog is 10% active (Metzler and Fonda, 1985). In this case, one would expect interference with transaldimination, especially if the final carbon is placed in front of the pyridine ring in the enzyme (nearer to the substrate, as in PLP itself).

R, S-5′-Methyl-PLP and 5′-deoxy-5′-carboxymethylene-pyridoxal were tested by Tumanyan et al. (1974). Only one isomer of the first compound binds to apo-cAAT, giving rise to a 60% active enzyme. Inspection of Figures 10 and 14 strongly suggests that it is the *S*-isomer, with the 5′-methyl group on the back side of the coenzyme. However, its distance to residues 107 and 108 would seem to be somewhat short.

The carboxymethylene derivative has little activity and its PMP form binds only weakly. The carboxylate group "rattles" in the phosphate pocket, and the single negative charge further reduces the binding constant. Nevertheless, it served the purpose, together with 6-methyl-PLP and 5′-methyl-PLP, of reducing the number of possible conformations of PLP in the enzyme to two (Tumanyan et al., 1974). One of these is very nearly correct in the unliganded PLP enzyme. However, the steric hindrance between O5′ (OP1, Fig. 1) and C4′ seems to have been overestimated (J. Gerhards and E. L. Mehler, unpublished results; see also Section 5).

4.5. Chemical Modifications

4.5.1. *"Syncatalytic" Sulfhydryl Reactivity*

Birchmeier et al. (1973) observed in pig cAAT an increase in reactivity (by two orders of magnitude) of Cys 390 towards several thiol reagents, if an amino acid/oxo acid substrate pair was present. Since a single substrate or inhibitor did not seem to have a significant effect, the phenomenon was interpreted as indicating a conformational change in the protein. Similar effects were subsequently discovered for Cys 166 in pig and chicken mAAT

(Gehring and Christen, 1975, 1978), where a 10-fold increase in reactivity resulted upon addition of the substrate pair. The validity of the interpretation was strongly supported by parallel effects in demodification of 5-thio(2-nitrobenzoate)-labeled Cys 166 by 2-mercaptoethanol in both pig and chicken mAAT (Gehring and Christen, 1978), and by retardation of peptide hydrogen-deuterium exchange caused by the presence of the substrate pair glutamate/2-oxoglutarate in pig cAAT (Pfister et al., 1978).

The above effects have been correlated with the small domain reorientation upon substrate or inhibitor binding. Cys 166, in the domain interface in mAAT at 25 Å from the coenzyme, is uncovered by the open-to-closed transition (see Figs. 9*a* and 9*b*). Cys 390 lies in the small domain behind the loop 36−39, and no doubt becomes more available for reaction by the reorientation of Val 37 (see Figs. 10*a* and 10*b*), provided this occurs in cAAT exactly as in mAAT, where residue 390 is an alanine. Indicative, in this respect, could be the observation that in chicken cAAT, which crystallizes in the closed structure, Cys 390 binds mercurials in both subunits (Harutyunyan et al., 1985). However, in pig cAAT, only one mercurial (*p*-mercury(II)benzoate) binds to Cys 390, paradoxically only in the inherently open subunit 2 (Hyde et al., 1984). Soaking times of heavy atom derivatives are long enough to allow reaction of internal side chains.

The X-ray studies on chicken mAAT have triggered further reactivity experiments with Cys 166 under conditions that promote the closed structure. Indeed, increased reactivity was found in the PLP enzyme with maleate and in the complexes with PPL-Asp and PPL-Glu, but not with PPL-Ala (Kirsch et al., 1984). Thus, we probably deal with an "induced fit" (Borisov et al., 1980b; Arnone et al., 1982; Jansonius et al., 1985), although we cannot directly prove it, since Michaelis complexes with real substrates are not stable.

Independent evidence for intramolecular flexibility in pig and chicken cAAT was obtained by measurement of rotational correlation times of two different spin labels attached to SH groups (Timofeev et al., 1980).

4.5.2. *Modification of Arg 292*

Phenylglyoxal modifies 2.7 mol Arg per mol subunit of chicken mAAT, with nearly complete loss of activity. However, in the presence of the substrate pair aspartate/oxaloacetate, only 1.3 mol Arg is lost, while the modified enzyme retains 75% activity (Sandmeier and Christen, 1982). The authors established that Arg 292 is completely protected by the substrates, while it is modified in the first experiment. The derivative with modified Arg 292 has full activity towards alanine, but the activating effect of formate (Saier and Jenkins, 1967) is lost. The activity towards the natural substrates is reduced by a factor of 10^5, but that towards Phe by only 30-fold, resulting in an activity similar to that towards Ala (Sandmeier and Christen, 1982). These results confirm: (1) Arg 292 binds the distal carboxylates of Asp and Glu, (2) the effect of formate is to guide Ala towards Arg 386 by covering Arg 292, and (3) the side chain of aromatic amino acid substrates binds to a site with little or no overlap with that of carboxylic acid substrates (compare PPL-3-iodo-Tyr, Section 3.6.7).

The dramatic reduction in Asp and Glu activity of this modified enzyme, which still has an active site large enough to accommodate Phe, demonstrates the subtleties of the structural requirements that make enzymes the enormously efficient catalysts they are. Similar arginine modification studies have been reported for pig and chicken cAAT (reviewed by Gehring, 1985).

4.5.3. mAAT 27/32−410

Native chicken mAAT is very resistant towards trypsin. However, upon prolonged treatment, nicks are made selectively after Arg 26 and Lys 31, with subsequent release of the N-terminal peptides (Sandmeier and Christen, 1980). The resulting mixture, which was termed mAAT 27/32−410, had 3% residual activity. Although its thermostability is decreased and the dissociation constant of the dimer is increased, the K_m'-values for aspartate and 2-oxoglutarate are unchanged, suggesting that the active sites are fully intact. However, the "syncatalytic" Cys 166 reactivity is reduced by a factor of three (Sandmeier and Christen, 1980). The amino terminal arm clearly has an important function in fixation of the small domain, constraining its possible conformations with respect to the active site. Without it, the small domain probably becomes more "floppy," and the conformational changes are less productive. Furthermore, the loss of helix 1 disables the enzyme from enclosing the bound substrate in the active site pocket. Therefore, one might have expected this truncated enzyme to be even less active than it is. Perhaps the wider entrance to the active site compensates, in part, a larger a priori effect of the impaired dynamic action. Comparison of the effect on catalysis of this modified enzyme with that of labeling Arg 292 indicates a predominance of the latter in imparting substrate specificity as well as catalytic efficiency. The domain movement might be a "fine tuning" that delivers the final 1−2 orders of magnitude in turnover number. This agrees quite well with the observation that in triclinic microcrystals of chicken mAAT, in which one subunit is locked into the open conformation and the other is presumably hindered in its dynamic motion, the subunit activities are still 3% and 15% (Kirsten et al., 1983). Thus, the prevention of dynamic motion does not seem to reduce the catalytic efficiency by more than a factor of, at most, 35 according to these data as well.

4.5.4. Modification of Lys 258

Lys 258 is very reactive and, therefore, is the target for many "suicide" inhibitors. It is also the nucleophile that reacts with substrate derivatives that carry a leaving group in the β position (Gehring, 1985). The only result we wish to discuss here is that carbamylated Lys 258 does not prevent PLP binding to the apoenzyme. With amino acid substrates, slow conversion into PMP is observed (Slebe and Martinez-Carrion, 1976). In D_2O, PMP is deuterated stereospecifically at C4′ in the pro-S position, exactly as in the native enzyme (Martinez-Carrion et al., 1979). These findings were used by the authors as an argument against Lys 258 catalyzing proton transfer. However, the very slow conversion to PMP does not necessarily have to be catalyzed by the same group as in the intact enzyme. In the quinonoid intermediate, the pro-S position is more accessible to water molecules than the pro-R position, even though Lys 258 is carbamylated (Kirsch et al., 1984). The reduction of an enzyme-aspartate catalytic intermediate from the *re*, rather than the *si* face in this enzyme derivative (Zito and Martinez-Carrion, 1980) is unexpected, but again one deals here with a very slow process.

4.5.5. Reactions of Histidines

Turano et al. (1966) established the presence of a histidine in or close to the active site in the following way. The PLP form of pig cAAT was treated with an acetylating agent for lysines, in order to eliminate all reactive lysines. The coenzyme was then removed, and the freed active site Lys (258) reacted with 3-bromopropionylchloride. Analysis of the (inactive) reaction product showed that one histidine per subunit was lost, while 1- and 3-carboxy-ethylhistidines could be detected. Inspection of the active site model suggests that His 143 and/or His 189 are candidates for the labeled residue.

At the time, this work introduced the belief that a histidine played a direct role in AAT catalysis. Subsequently, Martinez-Carrion et al. (1970a) proved by photo-oxidation the presence of a histidine near the coenzyme.

In the photo-oxidation product, the aldimine pK is increased by 0.4 units. After removal of the coenzyme and reconstitution with PMP, the latter binds anomalously. Transamination of photo-oxidized PLP enzyme, however, results in normally bound PMP (Martinez-Carrion et al., 1970a). These results can be understood in terms of a conformational change of the photo product in the apoenzyme, which interferes sterically or electrostatically with H bonding of the pyridine ring to Asp 222.

The photo-oxidation product binds substrates and inhibitors with somewhat smaller dissociation constants than the native enzyme. Proton transfer, however, does not take place. *Erythro*-3-hydroxyaspartate, for example, produces an external aldimine, absorbing at 360 nm, and no quinonoid intermediate is formed (Peterson and Martinez-Carrion, 1970). Not surprisingly, the authors proposed that the photo-oxidizable histidine is responsible for α deprotonation. What probably happened, though, is either an increase in the pK of Lys 258 (due to an additional negative charge nearby?) or a drastic decrease of the pK_a of N1 (Asp 222 rotated away?) with stabilization of the Cα-H bond in the external aldimine.

Yamasaki et al. (1975) studied the effect of methylene-blue-sensitized photo-oxidation of pig cAAT on alanine transamination (with and without formate) and on the α, β elimination reaction with β-chloro-L-alanine. Whereas the oxidation product was inactive against dicarboxylic acid substrates and the enhancement due to formate in Ala transamination was lost, both the transaminase activity towards Ala and the α, β elimination reactivity were fully retained. The authors interpreted these findings as a side chain-binding role of the oxidizable histidine for dicarboxylic acid substrates. These experiments point indeed towards an obstruction of the side chain-binding site. Arg 292 may have been oriented improperly as a result of a modified charge distribution near the active site. Such an effect also provides a third possible explanation for the observations of Peterson and Martinez-Carrion (1970); namely, by causing a nonproductive orientation of the substrate in the external aldimine intermediate.

We refer to Gehring (1985) for further enzyme modification studies.

4.6. Differences in Behavior of Cytosolic and Mitochondrial Isoenzymes

Thus far, we have emphasized the similarities of the cytosolic and mitochondrial isoenzymes, and often taken experimental data from one (mostly cAAT) as being valid also for the other. Since both fold identically, have comparable dynamic behavior, and have nearly identical active sites, the assumption is usually justified. Nevertheless, there are differences (there have to be) due to the different compartments and the different functions of the isoenzymes. In this section, we will summarize a number of studies in which the cytosolic and mitochondrial pig enzymes have been compared.

4.6.1. Subforms

At least four different subforms of pig cAAT have been isolated (Martinez-Carrion et al., 1965, 1967). The three main fractions were designated α, β, γ in order of increasing electrophoretic mobility towards the anode. The α form is fully active and absorbs at 360 and/or 430 nm, depending on pH. The β and γ subforms have increasingly strong 340 nm absorption bands with a corresponding reduction of the aldimine bands and of their catalytic activity.

The 340 nm absorption corresponds to PLP bound in an inactive mode. The β subform has been studied crystallographically by Arnone et al. (1985b, and references therein). It has one active and one inactive subunit. In the latter, an inhibitor is bound that obviously has two negatively charged groups and is similar to the ketimine form of the enzyme. It locks the subunit in the closed structure. These findings correlate with lack of circular dichroism in the 340 nm absorption band and with increased stability against urea denaturation (Martinez-Carrion et al., 1967).

Michuda and Martinez-Carrion (1969b) isolated three subforms of pig mAAT, A, B, and C, with different electrophoretic mobilities. Apart from the usual absorption bands associated with the internal aldimine (360 and 430 nm), these subforms had a weak 340 nm absorption, which was highest in A and lowest in C. For the aldimine bands, the order of intensity was C > B > A.

It was not possible to isolate fractions without 340 nm absorption. After removal of the coenzyme, the fractions could still be distinguished. No difference in number or location of tryptic peptides was found on peptide maps.

Gehring et al. (1977), however, isolated mAAT preparations from chicken that were devoid of 340 nm absorption in the PLP form. The X-ray work did not reveal any anomalous PLP binding in chicken mAAT either, as demonstrated by Figure 11, for example; crystal spectra show "clean" 360 nm and/or 430 nm bands (Vincent et al., 1984).

4.6.2. Amino Acid Composition

The mitochondrial isoenzyme has a higher fraction of (Arg + Lys) and a lower fraction of (Asp + Glu) residues than its cytosolic counterpart, with a correspondingly higher isoelectric point (Martinez-Carrion and Tiemeier, 1967). The differences might be related to the organelle specificity (Graf-Hausner et al., 1983).

4.6.3. Kinetic Constants

The binding constants for the four specific substrates and for *erythro*-3-hydroxyaspartate of pig cAAT and pig mAAT were compared by Michuda and Martinez-Carrion (1969a). The PLP form of the mitochondrial enzyme has a 30-fold higher affinity for aspartate than for glutamate and a > 10 times higher affinity towards oxaloacetate than towards 2-oxoglutarate, both of which act as inhibitors. Such large differences were not found for the cytosolic enzyme. Both enzymes have, in their PMP form, a very strong affinity towards oxaloacetate, a less strong affinity towards 2-oxoglutarate, and only a weak affinity for the amino acids. *Erythro*-3-hydroxyaspartate is bound more strongly. This compound has, for the PLP form of mAAT, affinity equal to aspartate; but for cAAT in its PLP-form, it has 10 and 30 times stronger affinities than Asp and Glu, respectively.

A similar trend is found in the inhibitors of the isoenzymes. Of the series maleate, succinate, glutarate, adipate, only the first two are good inhibitors of mAAT, while all four inhibit cAAT about equally well (Michuda and Martinez-Carrion, 1970; but see Jenkins et al., 1959). The cytosolic isoenzyme must be more adaptable than its mitochondrial counterpart, since in the unliganded enzymes, no obvious differences in folding—and thus in the size of the highly conserved active site—have been found. High-resolution X-ray studies of inhibitor complexes of cAAT with succinate or maleate, and with adipate, might shed more light on this matter. Indications of a greater structural flexibility in cAAT than observed with mAAT are (1) inherent asymmetry in native crystals (Arnone et al., 1982); (2) the blurred electron density in the small domain region (Borisov et al., 1978; Arnone et al., 1982); and (3) the possibility to trigger the domain closure by high ionic strength (Borisov et al., 1980b) or high acetate concentrations (Arnone et al., 1984). In contrast, unlike maleate, 2-oxoglutarate does not induce the closed structure in one subunit of triclinic mAAT; upon cocrystallization, it does not cause the production of orthorhombic crystals of the closed form of the enzyme (Eichele, 1980; D. Picot, unpublished results).

4.6.4. Aldimine pK and Anion Inhibition

The pK_a of the internal aldimine is dependent on buffer (an)ions, which are, to a varying degree, bound to the active site of AAT. With cacodylate buffer, which has the lowest affinity to the AAT active sites, both isoenzymes have a pK_a of 5.4. This pK_a increases in correspondence with the extent of anion binding. The mitochondrial enzyme has more affinity for anions than the cytosolic enzyme (Cheng et al., 1971), as has been confirmed by crystallographic observations (Eichele et al., 1979a; Eichele, 1980). IrCl$_6^{2-}$ anions bind near Arg 386 and Arg 292. Binding of heavy ions in the active site of cAAT has not been reported. The importance of anion binding in catalysis of this and other enzymes has been stressed many times (reviewed by Jenkins and Fonda, 1985).

4.6.5. Differential Active Site Reactivity

3-Bromopropionate alkylates the active site lysine (Lys 258) of the mitochondrial AAT, but not the cytosolic AAT, from pig. The kinetics of the inhibition reaction and the effective inhibition by maleate and succinate suggest that the reagent is first bound in a substrate-like manner. The reaction has a maximum speed at pH 7.0 and above. At lower pH values it decreases, with a p$K \sim 6.2$. The reaction product is inactive, absorbs at 395 nm with a negative circular dichroic band, and is not affected by NaBH$_4$. A mechanism was proposed for this reaction in which the unprotonated aldimine N carries out a nucleophilic attack on the 3-carbon of the reagent (which is anchored with its carboxylate to a positive active site group) and displaces bromide (Okamoto and Morino, 1972; Morino and Okamoto, 1972). A possible alternative to the substituted aldimine, proposed by the authors as a reaction product, is an alkylated lysine reforming the free PLP aldehyde, which is known to absorb near 390 nm. The steric interaction of the coenzyme with this modified lysine could explain the negative dichroicity. The binding site for the carboxylate group is most likely near Arg 386 in the region of H$_2$O(8) (Fig. 10a), which is more like a pocket than the alternative position near Arg 292*. The lack of reactivity of the cytosolic enzyme towards this reagent probably has to do with its lower affinity for the corresponding site, similar to what has been observed for buffer ions (see above). However, a structural explanation for this proposal cannot yet be given. Interestingly, *both* isoenzymes react with L-3-chloroalanine in the presence of formate (Morino et al., 1974; Morino and Tanase, 1978). The difference with the previous case is that first an external aldimine is formed, with an enforced proximity of the β-carbon to Lys 258. At this point, the situation in both enzymes should be essentially identical. The authors proposed a role for formate in α deprotonation. Only in 1 of 100–300 cases does alkylation of Lys 258 occur, with the normal pathway being transamination to pyruvate and ammonia. These results led to a proposal by Morino and Tanase (1978) for a catalytic role of Lys 258 in the protonation and deprotonation steps, which is a suggestion first made by Snell (1962). Iodoacetate is another reagent that is active towards both isoenzymes. It is an inhibitor of the PLP as well as the PMP enzyme. The latter is inactivated through nucleophilic displacement of the iodine by the coenzyme amino group (Morino et al., 1978). pK's of 8.3. (cAAT) and 9.1 (mAAT) were observed for this reaction and attributed to the PMP amino group. However, the structure of the PMP enzyme shows that PMP and Lys 258 share a proton, which probably resides mainly on Lys 258 (see Fig. 12). The observed pK must therefore correspond to this shared proton, the loss of which allows the PMP amino group to move in front of the coenzyme ring, where it can "reach out" towards the α-carbon of iodoacetate. Presumably the latter molecule is bound again to Arg 386. It is unclear why cAAT binds iodoacetate (much) more strongly than 3-bromopropionate. Comparison of Figures 10a and 14 reveals why iodoacetate does not inhibit the PLP enzyme: When bound to Arg 386, it is too far from the internal aldimine.

The coenzyme derivative 4′-fluorodinitrophenylpyridoxamine-5′-phosphate (FP) reacts differently with the apoforms of the two isoenzymes. The compound binds strongly in the coenzyme-binding site of both isoenzymes. Subsequently, it partly inactivates the cytosolic enzyme by labeling the ε-amino group of Lys 258 with elimination of fluoride. This reaction, however, does not take place in the mitochondrial enzyme. Slow decomposition of the coenzyme analog with restoration of the PLP enzyme occurs in both enzymes, as it does with an FP derivative lacking fluorine (Riva et al., 1980). The complexes were studied further with circular dichroism (CD). Immediately after binding, a small positive band at 320 nm (pyridoxamine moiety) and a negative band at 355 nm (dinitrofluorophenyl [DFP] moiety) were observed. In mAAT, this spectrum slowly changed into that of PLP (negative band at 300 nm; positive band at 360 nm) concomitant with the reappearance of catalytic activity. The recovery was not complete, however. The final spectrum in cAAT had a broad negative band near 300 nm, a positive band near 340 nm, and a weak negative band near 360 nm. The 360 nm band was ascribed to the dinitrophenyl moiety, as bound to Lys 258, and the 340 nm band to the coenzyme moiety in this adduct. In the absorption spectrum, a 430 nm band occurred as well, which was also attributed to the covalently bound dinitrophenyl group. The absence of circular dichroism in this band was explained by flexible binding. The defluorinated compound reacted identically with both enzymes. Both apoenzymes, carbamylated at Lys 258, still bind FP with negative CD in the 350nm band. However, decomposition of the compound no longer occurs, suggesting that Lys 258 is responsible for this observation (Carotti et al., 1982).

The inability of mAAT to react covalently with FP is undoubtedly a consequence of the way in which the DFP moiety of FP binds. For Lys 258 to be able to reach the 5 position of the ring, that side must be oriented towards its ε-amino group. This is probably only possible if the 2-nitro group interacts with Arg 386 and the 4-nitro group with Arg 292*, which corresponds sterically with the binding of a 5-carbon substrate. The mitochondrial enzyme, however, will probably orient the DFP side chain differently (compare with PPL-3-iodo-Tyr binding, Eichele et al., 1979a; Kirsch et al., 1984), which brings the 5 position of the DFP moiety out of reach of Lys 258.

The coenzyme analog, α-*N*-fluorodinitrophenyl-β-*N*-phosphopyridoxyl-diaminopropionate (FDP), a more flexible molecule than FP, affinity labels Lys 258 in both isoenzymes (Carotti et al., 1985). Our tentative guess would be that in this case the carboxylate group interacts with Arg 386 and the 2-nitro group with Arg 292* in both enzymes, bringing the 4-nitro group near Asn 297* and the fluorine close to Lys 258. However, model building would have to confirm this proposal. Ottonello et al. (1983) reacted crystals of pig cAAT (made according to Arnone et al., 1977) and pig mAAT (made according to Gehring et al., 1977) with both FP and FDP and observed that in the crystal, the same reactions occur as in solution, confirming that the active sites are not distorted in these crystals. This is especially important, because in both types of crystals, the two crystallographically independent subunits behave differently towards dicarboxylic acid substrates and inhibitors. Presumably, the open conformations are retained in both subunits.

4.6.6. *Substrate Specificity*

Shrawder and Martinez-Carrion (1972) compared transamination rates of pig cAAT and mAAT for a series of amino acids, with 2-oxoglutarate as the ketoacid substrate. Significant activity for the aromatic amino acids was observed with both enzymes, especially with mAAT, which, for example, transaminated phenylalanine 17 times faster than cAAT. The differences were smaller for tyrosine (five times) and tryptophan (twice). The inhibitory activity of these amino acids lies in the $1-10$ mM range, being about three times stronger in cAAT than in mAAT for Phe and Trp. In transamination, equilibrium between the phenylalanine and phenylpyruvate 360 and 330 nm absorption bands was observed, with the spectra being devoid of any 430 nm band. These results are consistent with the crystallographic observation that only dicarboxylate inhibitors stabilize the closed structure, and with our suggestion that 430 nm absorption indicates the closed structure. The tyrosine transaminase activity of mAAT had been reported before by Miller and Litwack (1971).

5.   CATALYTIC MECHANISM

5.1.   Early Proposals

Jenkins and Sizer (1960) were the first to confirm, for pig cAAT, the general ping-pong mechanism proposed by Snell and by Braunstein (Braunstein, 1973), in which the coenzyme shuttles between a PLP and a PMP form. Fasella et al. (1966) studied the reaction of the enzyme with 2-methylaspartate. They interpreted their results on the basis of a mechanism that included earlier proposals by Snell (1962) and Jenkins (1961, 1964), and that corresponds to the general scheme of Snell and Braunstein with external aldimine, quinonoid, and ketimine intermediates. The suggestions by these authors—that the 360 nm absorption in the 2-methylaspartate derivative is due to deprotonation of the external aldimine by the aldimine-forming Lys (258), and that in the PMP enzyme, the coenzyme and Lys 258 share a proton—are in perfect agreement with our crystallographic results. Snell's (1962) proposal that the lysine could act as an acid/base catalyst was not explicitly assumed. Peterson and Martinez-Carrion (1970) attributed this role to a histidine residue, the photo-oxidation of which is concomitant with loss of catalytic activity (see Section 4.5.5). Yamasaki et al. (1975) gave this histidine the role of a cationic binding site for the distal carboxylate group and preferred "the" lysine as the acid/base catalyst.

By far, the most detailed proposal prior to the crystallographic studies was advanced by Ivanov and Karpeisky (1969). These authors combined a number of experimental results from ultraviolet (UV) and CD spectra and from coenzyme modification studies in a very elegant dynamic model for catalysis by AAT. The coenzyme was proposed to rotate towards the substrate around an axis through C2 and C5, keeping the phosphate group and the 2-CH$_3$ group as fixed points. The arguments for this proposal, that the substrate is held by two cationic groups interacting with its carboxylates and has no reason to move, were shown to be valid by the crystallographic studies that demonstrated a coenzyme reorientation quite similar to that predicted by these authors. The process of transaldimination was explained in great detail, and the aldimine-forming lysine (258) was taken as responsible for α deprotonation. The important concept was advanced that each step in catalysis automatically changes the environment of the substrate in such a way as to promote the next step.

Furbish et al. (1969) studied the binding and catalytic properties of 10 enzyme (pig cAAT) derivatives with modified coenzymes and proposed, on the basis of their findings, a mechanism in which the phosphate group plays a role as acid/ base catalyst in the transaldimination step by accepting a proton from the substrate α amino group and offering it subsequently to the lysine ε-nitrogen prior to release of the lysine amino group. The authors emphasized the rotational motion around C4-C4′ upon formation of the geminal diamine. The active site lysine (258) was considered responsible for the 1,3 prototropic shift from the α-carbon to the C4′-carbon.

Metzler (1979) reviewed the earlier mechanistic proposals and added some new ideas, including a charge relay system to avoid charge separation.

5.2. Kinetic Studies

Equilibrium kinetic studies with aspartate by Jenkins and Taylor (1965) confirmed the observation by Velick and Vavra (1962) that V_{max} is pH-invariant, which was interpreted to mean that no protolytic dissociations are involved in the observed equilibria. It has now become apparent that this is a consequence of the active site being insulated from the outside world. Michuda and Martinez-Carrion (1969a) determined equilibrium kinetic constants for the four substrates for both isoenzymes, from which the distinct preference of the mitochondrial enzyme for the 4-carbon substrates appeared.

The fast kinetic studies by Hammes and Haslam (1968, 1969) on the interaction of cAAT with 2-methylaspartate and *erythro*-3-hydroxyaspartate, respectively, have been discussed at length in Section 4.3. We have tentatively assigned the various kinetic constants to particular steps in the reaction with 2-methylaspartate. The three steps would be formation of a Michaelis complex, transaldimination, and domain closure. The equilibrium for the second step favors the Michaelis complex. The third step stabilizes the external aldimine in the closed structure. The kinetic constants for domain closure and opening would be $\sim$70 sec^{-1} and 2.5 sec^{-1}, respectively. The slow opening rate might be related to the perfect fit with extensive van der Waals interactions and strong entropic contributions to substrate binding in this intermediate (compare Section 4.6.7). All rates in the turnover of *erythro*-3-hydroxyaspartate are considerably faster.

Fasella and Hammes (1967) carried out a temperature jump study of the reaction of the real substrates with pig cAAT. The rate-limiting step was found between intermediates absorbing at 490 nm and 330 nm and was so fast that no other rate constants could be determined. The forward and backward rates of this step were 530 and 1330 sec^{-1} for aspartate/oxaloacetate and 2940 and 1170 sec^{-1} for glutamate/2-oxoglutarate. Turnover numbers were 370 sec^{-1} for aspartate/2-oxoglutarate and 920 sec^{-1} for glutamate/oxaloacetate, testifying that AAT is a highly efficient enzyme.

What is the rate-limiting step? It would seem that it has to be the quinonoid $\rightleftharpoons$ ketimine conversion, since this step produces the first species absorbing near 330 nm. Alternatives are ketimine hydrolysis or a conformational change (domain opening), in which case the concentration of ketimine intermediate at any one time would have to be significantly less than the already very low concentration of quinonoid intermediate. This seems rather unlikely, since a ketimine structure would be expected to be more stable than a quinonoid structure. Domain opening is not expected at this point in catalysis either, since there does not seem to be a driving force to achieve it (Kirsch et al., 1984). Ketimine hydrolysis would correspond to our interpretation of the slow step for 3-hydroxyaspartate, but here we are dealing with an anomalous situation in which, presumably, hydrogen bonds made by the 3-hydroxyl group create an unproductive geometry. The most attractive hypothesis seems to be that the quinonoid $\rightleftharpoons$ ketimine equilibrium is rate limiting. However, a good structural explanation for this cannot yet be given.

5.3. pK Values of Active Site Groups

The most recent and complete data on pK values of active site groups (with pig cAAT) are from Kiick and Cook (1983). For earlier work, we refer to references therein and to Braunstein (1973) and Christen and Metzler (1985). The V_{max} of the reaction was found to be constant in the pH range 5.5−10.0. In V/K, a low pK of 6.4 and high pK's of 9.2 and 10.1 were observed with aspartate as substrate; for the opposite direction with 2-oxoglutarate as substrate, values of 5.8 and 9.2 were found. 2-Oxoglutarate acted as an inhibitor in the reaction with aspartate, with its role decreasing above a pK of 6.5. With maleate, the same value was found. Maleate inhibition was also observed with the PMP enzyme, where it decreased above a pK of 9.3. Aspartate also inhibited the PMP enzyme, increasingly so above a pK of 10.2. The pK's for inhibition by 2-methylaspartate were determined as 6.5 and 9.3. The pK of 6.3−6.5 observed with the PLP enzyme has always been ascribed to the internal aldimine protonation, undoubtedly rightfully so. It was proposed that the pK of 5.8 for 2-oxoglutarate with the PMP enzyme was due to the second protonation of the hydrogen-bonded pair of amino groups of Lys 258 and PMP. Since we now know that this pair exists, the interpretation seems reasonable, because the charged PMP amino group is not reactive. The pK of 9.3, also observed with 2-oxoglutarate as substrate, had to come from the protein and was ascribed to deprotonation of the pyridine N (Kiick and Cook, 1983; Kirsch et al., 1984). However, while it would have the observed effect on enzyme activity, this deprotonation cannot easily explain why maleate binding should be affected by this pK, since a major reorientation of the coenzyme is not to be expected. The second pK of the Lys 258-PMP amino group pair might well cause the PMP amino group to rotate forward into the active site (Fig. 12), and thus to interfere with substrate and inhibitor binding. The high pK on catalysis and inhibition of the PLP enzyme could in principle also be attributed to Lys 258; namely as the pK of aldimine protonation in the presence of substrate or inhibitor. An external aldimine would then no longer be formed. However, according to Kiick and Cook (1983), these last two pK's are above 10.0 and 8.2, respectively. The issue has not yet been settled.

5.4. Crystallographic Proposals

Kirsch et al. (1984) have made a detailed proposal for the stereochemistry of AAT catalysis. We will discuss below an updated version of that mechanism, based on improved structural knowledge.

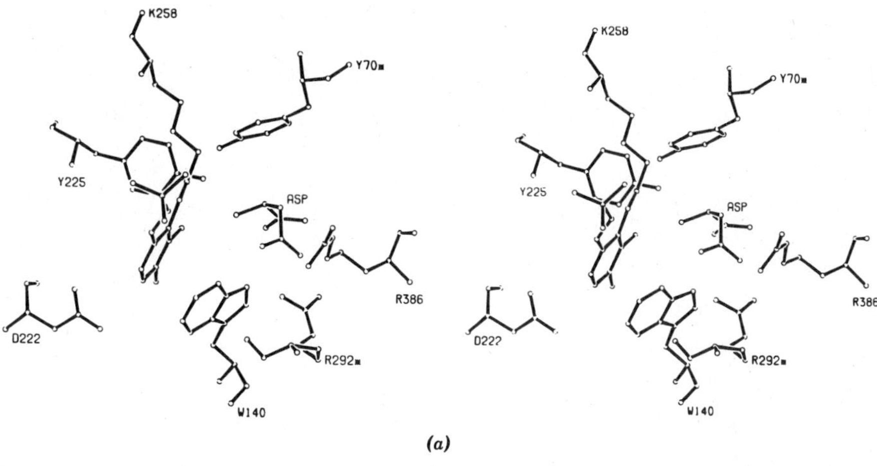

(a)

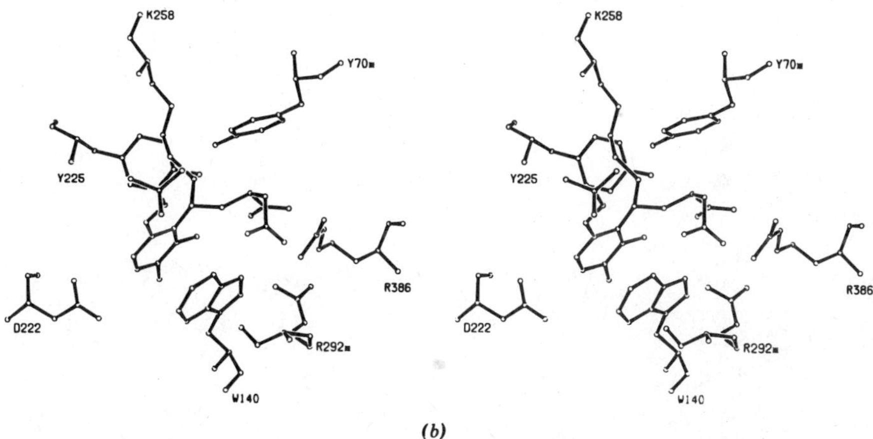

(b)

Figure 15. Models of catalytic intermediates in a half-transamination reaction (see Fig. 2) from aspartate to oxaloacetate. For clarity, only a selection of the active site groups are shown, corresponding to Figure 5 of Kirsch et al. (1984). The view direction is also the same. (*a*) Michaelis complex of PLP enzyme with aspartate. (*b*) Geminal diamine. (*c*) External aldimine. (*d*) Quinonoid intermediate. (*e*) Ketimine intermediate. The circle indicates a bound water molecule. (*f*) Carbinolamine intermediate in ketimine hydrolysis. (*g*) Michaelis complex of PMP enzyme with oxaloacetate. See text for a discussion of these models.

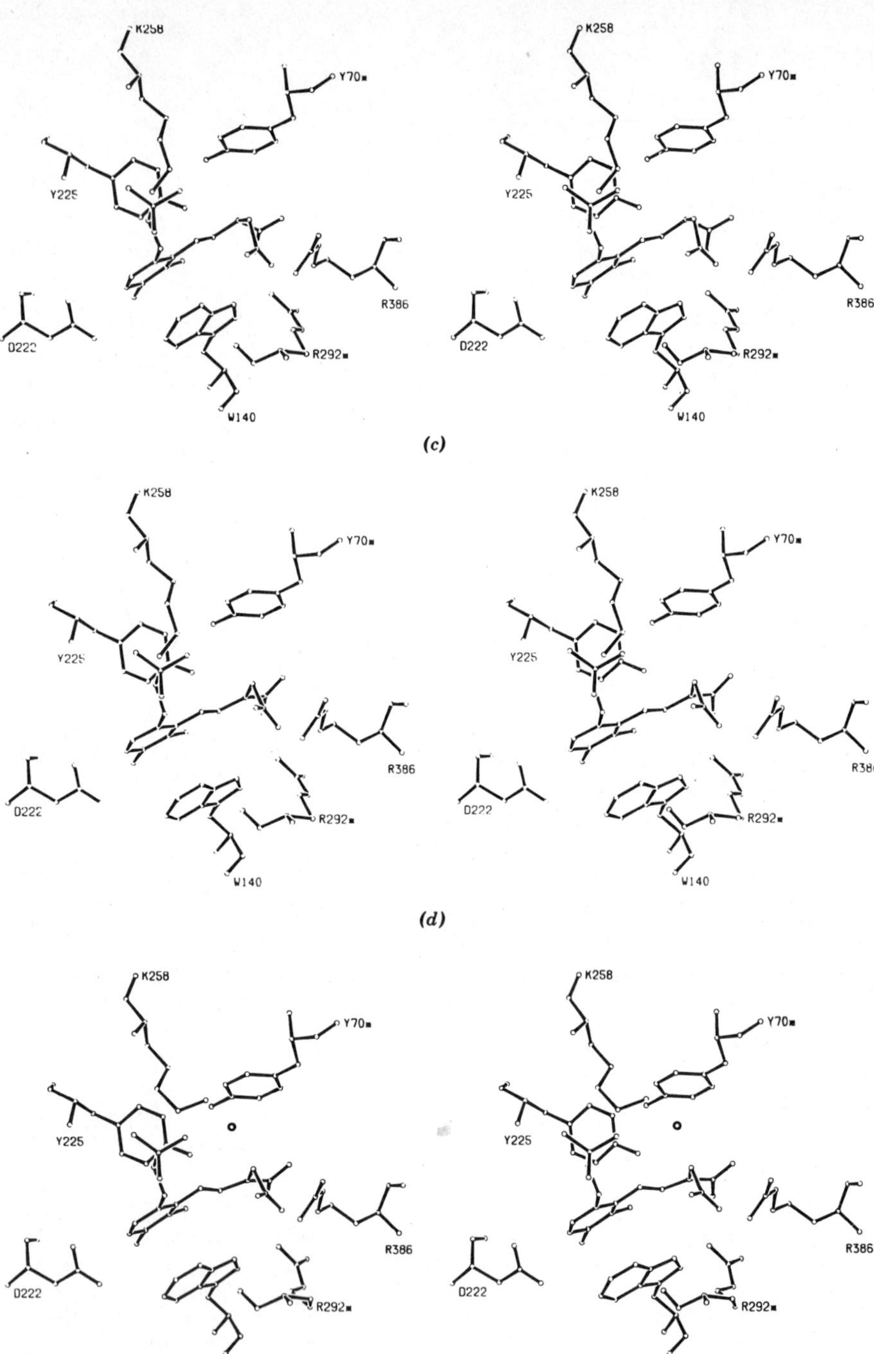

(c)

(d)

(e)

Figure 15. (c, d, e) (continued)

266

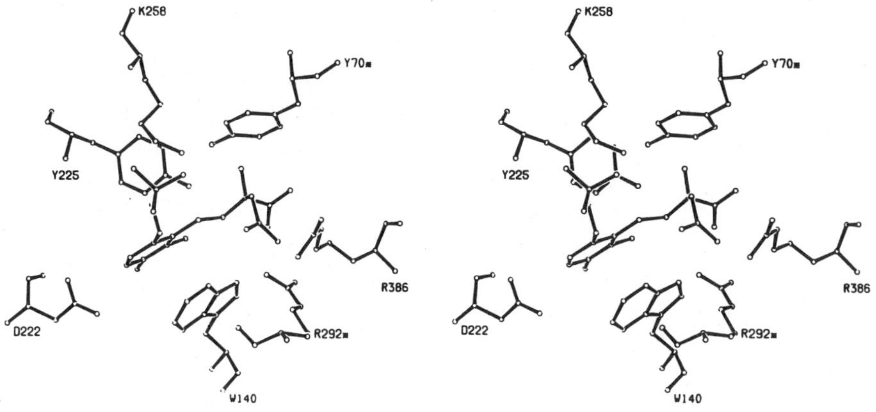

(f)

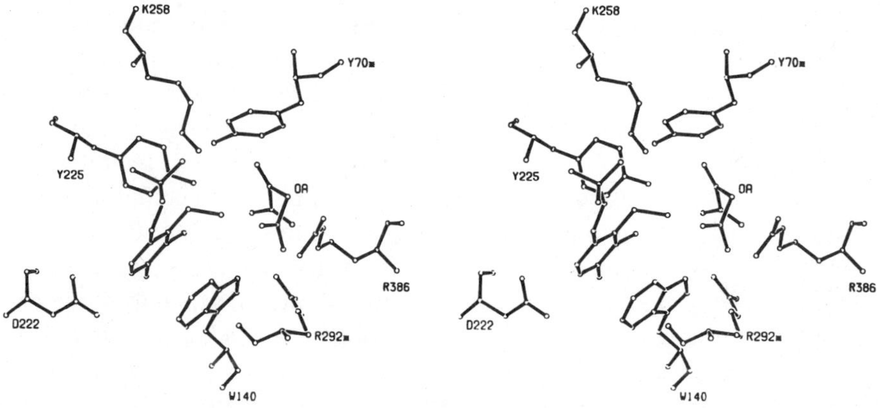

(g)

Figure 15. (f, g)

5.4.1. Model Building

The recently acquired structural information from the complexes of mAAT with maleate and 2-methylaspartate made it possible to build improved models of putative catalytic intermediates. The closed structure observed in the above derivatives was taken as a basis for these models. In the Michaelis complex with aspartate, the coenzyme was positioned as in the maleate derivative. For the Michaelis complex with oxaloacetate, the position of PMP was taken as that of its counterpart in the unliganded structure in space group P1, and was transformed into the closed structure. Tyr 225 and Trp 140 were adapted by rotations around Cα-Cβ and Cβ-Cγ bonds. The external aldimine with aspartate was based on the 2-methylaspartate structure, but the distance between the pyridine N1 and Asp 222 was reduced from 3.2. to 3.0 Å by rotation around the bonds between C5 and the phosphorus atom. This was possible because of the absence of a 2-methyl group on the substrate, and it corresponds more to the situation in the PPL-Asp derivative (Kirsch et al., 1984). The quinonoid and ketimine intermediates were subsequently built by adapting the geometry to the changes in hybridization and by rotations around single bonds in the coenzyme-substrate molecule. Structural adaptations of the active site to these catalytic intermediates were carried out by (usually small) side chain rotations of Ile 17, Leu 18, Val 37, Tyr 70*, Trp 140, Asn 194, Tyr 225, Lys 258, and Arg 292*. The position of Arg 386 was not altered. In the external aldimine and quinonoid models, Lys 258 was built as in the 2-methylaspartate derivative structure. In this structure, no density was seen for the ε-amino group (Section 3.4.2). Therefore, this group should not be taken as fixed in any one position, but rather its considerable flexibility to "rattle" in a box should be realized. Only in the models of the ketimine intermediate and the Michaelis complex with oxaloacetate has this side chain been placed in the most "productive" position. The resulting models are shown in Figure 15. For clarity and consistency with Kirsch et al. (1984), only the most important residues are illustrated. In describing the mechanism, we will go step by step through a half-cycle of catalysis, from the substrate complex with aspartate to the product complex with oxaloacetate.

5.4.2. Michaelis Complex with Aspartate

The substrate enters the active site cavity in the open PLP enzyme (Fig. 10*a*) from underneath the ion pair Asp 15-Arg 292*. It is oriented by attraction of its α-NH$_3^+$ group to the coenzyme O3$'^-$, and of its α and distal carboxylate groups to Arg 386 and Arg 292*, respectively. These charges are properly positioned to accommodate L amino acids, but not D amino acids. A number of water molecules are displaced, including the structural waters 6, 7 and 8 (Figs. 10*a* and 10*b*). H$_2$O(5) remains and probably helps in guiding Asp towards its "productive" position. The β carboxylate rotates to a gauche orientation with respect to the α carboxylate in order to optimize the charge interaction with the two arginines.

Now several things happen, probably in concert. The carboxylate negative charges allow the two arginines to approach each other more closely since, by doing so, they can make more favorable hydrogen bonding and charge interactions. The transition to the closed structure makes this possible. Arg 292* reorients its guanidinium group to make two hydrogen bonds with the β carboxylate. This is probably the main trigger of the domain closure, since it allows not only Asp 15, but also Arg 386 to make favorable hydrogen bonding and charge interactions. The release of $H_2O(7)$ probably also plays a role, as it allows the approach and reorientation of residues 36–38. The two arginines optimally orient the substrate while van der Waals contacts with the reoriented Val 37 and with Ile 17 and Leu 18 that have all approached with the small domain push the substrate towards the coenzyme (compare Figs. 10*a* and 10*b*).

The nonpolar environment now created around the substrate intensifies the charge interactions and leads to a proton shift from the α amino group to the aldimine nitrogen. pK shifts, resulting from reciprocal charge compensations of the carboxylates and the arginines (Ivanov and Karpeisky, 1969; Kirsch et al., 1984) and the need for compensation of the now somewhat isolated charges on the α-NH_3^+ group, and probably the pyridine $O3'^-$ cause this to happen. At this point we have the situation depicted in Figure 15*a*. The protonation of the aldimine nitrogen forces a rotation of the pyridine ring towards (near) coplanarity with the aldimine double bond. This causes a decrease of the torsion angle around C5-C5′ to about $-16°$, closer to the transition state for transaldimination—the geminal diamine intermediate—in which this torsion angle is $\sim 0°$ (compare Figs. 15*a* and 15*b*). The α carboxylate of the substrate is involved in one H bond to Arg 386 and in one to Asn 194 Nδ2 (not shown; compare with Fig. 10). The remaining α amino protons orient themselves towards the carboxylate groups to make internal hydrogen bonds. As a result, the lone electron pair is oriented exactly towards the coenzyme C4′, and allows an optimal close approach for overlapping with the π-orbital at C4′. A nucleophilic attack on C4′ can now take place (Kirsch et al., 1984).

5.4.3. Geminal Diamine

The first step in the transaldimination process is the production of the geminal diamine. This intermediate is characterized by single bonds between C4′ and both the nitrogens of Lys 258 and of the substrate. Its formation is expected to result in a forward tilt of the pyridine ring (Ivanov and Karpeisky, 1969), achieved by small rotations around the bonds between C5 and phosphorus, leaving N1 essentially in place (Kirsch et al., 1984). The process involves passage through an eclipsed configuration around C5-C5', with a close contact between OP1 (see Fig. 1) and C4', which is considered energetically unfavorable (Tumanyan et al., 1974).

This problem triggered the alternative proposal by Arnone et al. (1984, 1985b) of an "oscillatory rotor mechanism." However, we feel that in light of the recent findings of small torsion angles around C5-C5'—even in structures such as the PMP enzyme (+24°) where an "allowed" stronger tilt would avoid this situation, and the PLP enzyme with maleate (−16°) as compared with the unliganded PLP enzyme (−37°)—there is less strain in the eclipsed conformation than has so far been believed. Recent quantum chemical calculations (J. Gerhards and E. L. Mehler, unpublished results) confirm this hypothesis.

The slightly different orientation of the phosphate group (15°−20° rotation) in the 2-methylaspartate derivative in comparison to the maleate derivative, which can be seen as a rotation around P-OP4 (see Fig. 1) in a sense that moves OP3 to the right and OP1 to the left in Figure 15*a*, also points in this direction. The geminal diamine is a very short-lived intermediate that kinetically and spectroscopically has not been identified. A rotation of the two C-N bonds around C4-C4', to bring the α nitrogen in the pyridine ring plane and the C4'-Nζ bond almost normal to it, is necessary to break the latter bond (Furbish et al., 1969; Federiuk and Shafer, 1983; Kirsch et al., 1984). The rotation also requires a shift of ~ 1 Å (away from the viewer in Fig. 15*a*) of the α-N, with some reorientation of the amino acid moiety. Thus, there is a gradual structural change rather than a defined intermediate at this point. If the pyridine ring continues to rotate forward towards its position in the external aldimine (Fig. 15*c*), the above-described reorientations of the C−N bonds will automatically occur. Before Lys 258 can be released, a proton has to shift from the amino N probably *via* O3' to Nζ of Lys 258. The critical moment for this proton shift might be the symmetrical situation of equal distance between both nitrogens and O3', as is shown in Fig. 15*b*.

5.4.4. *External Aldimine*

Figure 15*c* shows the situation upon release of Lys 258. The pyridine ring has an increased tilt angle (always measured with respect to the XZ plane; e.g., see Fig. 4) of ~ 30° compared to the Michaelis complex. It stacks with the indole of Trp 140 at ~ 3.5 Å distance, while remaining H bonded to Asp

222. Tyr 225 has followed the coenzyme and remains H bonded to O3′. N and Cα of the substrate now lie in the pyridine ring plane. Free rotation around N-Cα is possible, and the carboxylates orient themselves to make two H bonds each to Arg 386 and Arg 292*. One β carboxylate oxygen is also H bonded to Trp 140 Nε1, and the other presumably to $H_2O(5)$ (not shown). The carboxylates are oriented such that the Cα-H bond is optimally oriented, about normal to the pyridine ring plane, for being broken. The electronic arrangement, with positive charges on N1 and the aldimine N and a negative charge on O3′, is optimal for release of the bonding electrons of the Cα-H bond into the mesomeric π system of the coenzyme aldimine. Lys 258, which has been released without charge from its aldimine bond, "hangs" above the coenzyme plane. It has the choice of making H bonds to OP1, to Tyr 70* Oη, to Tyr 225 Oη, and to Gly 38 O (not shown; see Fig. 14). H bonds to the first two oxygens bring it close to C4′, while H-bonds to the last two oxygens place it near the substrate Cα. It probably shuttles between these positions. The second possibility causes the lone electron pair on Lys 258 to orient itself towards the α proton and should lead to α deprotonation, producing the quinonoid intermediate.

5.4.5.   Quinonoid Intermediate

The resonating π-bond system in this intermediate extends to the α carboxylate. At least Cβ and C′, but probably also the α carboxylate oxygens, lie in the ring plane. In model building, starting from the external aldimine with 2-methylaspartate, no rotation around Cα-C′ was carried out. Lifting the Cα-C′ bond into the ring plane automatically brought the oxygens near to this plane. The transition from external aldimine to quinonoid is brought about by a slight (≤ 10°) rotation around C5-C5′. Trp 140 follows this movement by rotating slightly away, concomitant with the release of its H bond to the β carboxylate. Furthermore, slight reorientations around C5-C5′ and P-OP1 occur. The movement minimizes nonbonding interactions in the substrate moiety and allows even better H bonds to Arg 292* to be formed. The shifts of the oxygens of the α carboxylate and of the β carboxylate from their positions in the external aldimine are 0.2−0.3 Å and 0.5−0.7 Å, respectively. The N1 nitrogen is now sp^3 hybridized, is uncharged, and carries a lone electron pair. A strong H bond (2.7−2.9 Å) to any of the two Asp 222 oxygens is possible. The model is shown in Fig. 15d. Insofar as the very short living quinonoid intermediate can be compared with a "transition" state, the view that enzymes are complementary to transition states seems to hold for AAT. The hydrogen bonds that the carboxylates make are close to ideal. Excellent stacking interaction with Trp 140 and a strong H bond to Asp 222 complete the picture. Thus, everything seems to have been done to stabilize the quinonoid intermediate. A water molecule again fits in the position of $H_2O(7)$, as seen in the PLP enzyme (Fig. 10a; in the external aldimine, there is no room for it). It must enter from the top of the active site and would push Lys 258 backwards towards its position above C4′, where it can protonate this atom, thus producing the ketimine intermediate. The presence of the water molecule, however, is not essential at this point in catalysis.

5.4.6. *Ketimine Intermediate*

The protonation of C4′ (in the pro-S position, as determined by radioactive labeling; see Section 4.2) uncouples the double bond between N and Cα from the pyridine π system. A slight back rotation "relaxes" the pyridine ring and improves the hydrogen bond of the again charged aromatic N1 to the lower oxygen of Asp 222. The substrate moiety, coplanar with C4′, optimizes itself for making an internal H bond between the protonated ketimine N and the unprotonated O3′, and an H bond between a β carboxylate oxygen and Trp 140 Nε1. The internal H bond resembles that in the protonated PLP enzyme, where the aldimine N is also somewhat out of the pyridine ring plane. Good hydrogen bonds are still made with the two arginines. A water molecule is placed near the position of $H_2O(7)$ in the open structure, and Lys 258, pushed away from its previous position by some back rotation of the coenzyme ring, accepts a hydrogen bond from this water molecule while donating H bonds to Tyr 70* and Gly 38 O. Figure 15e illustrates the situation that sets the scene for the next step, the formation of the carbinolamine intermediate. Lys 258 takes over a water proton (Metzler, 1979) that produces a highly nucleophilic hydroxyl ion at the proper distance and in the proper orientation above Cα to add to it. The emerging sp^3 hybridization around Cα pushes the nitrogen further down, while the coenzyme rotates backwards in reaction to its approach towards Trp 140.

Figure 15f shows the structure of the carbinolamine intermediate (Metzler, 1979). The nitrogen atom must acquire an additional proton, probably again from Lys 258, before it can be released. Its pK increases when it approaches the two carboxylates. Having acquired a second proton, the nitrogen orients itself between the two carboxylates to optimize the charge interactions and to make internal H bonds to two carboxylate oxygens. Lys 258, again in need of a proton, takes it from the α-hydroxyl group. The Cα-N bond is then broken, producing PMP and oxaloacetate as the product.

5.4.7. Michaelis Complex with Oxaloacetate

Figure 15g shows the probable situation immediately after formation of the oxaloacetate product. The carbonyl oxygen is positioned between Tyr 70* and Gly 38 (not shown). Two H bonds are still made with Arg 292*, but only one with Arg 386. The two hydrogens of the PMP amino group make weak H bonds to the carboxylates of the oxoacid. Its lone electron pair is oriented towards the carbonyl carbon. The charged ε-amino group of Lys 258 makes an H bond to the carbonyl oxygen. The resulting structure is not very stable; unless a bond is rapidly made between nitrogen and carbonyl carbon (first step in the back reaction), the oxoacid would tend to move away and be released after the transition to the open structure.

5.4.8. Further Considerations

Model building showed that good fits can be obtained of the covalent catalytic intermediates in the closed structure. However, it was much less easy to make convincing models of the Michaelis complexes. The PLP enzyme/aspartate complex seemed to allow several, slightly different binding modes, of which only one is productive. On the other hand, steric hindrance made it difficult to find the probable productive complex with oxaloacetate. These findings are probably significant. The substrates move about in the active site, until they more or less accidentally reach the productive position. Then the attack by the enzyme is fast and catalysis proceeds. The first step in each direction involves aldimine or ketimine production, which is a fast process. Therefore, exact fitting of the substrate is not required as long as the orientation is more or less correct. This is ensured by the charged interactions. The question of whether the closed structure is attained before the first covalent bond is made cannot be answered with certainty, although it still seems probable. That the equilibrium open $\rightleftharpoons$ closed lies more to the left for substrates than for the most potent noncovalent inhibitors, however, is very likely. It is also desirable for effective product release.

5.5. Kinetic Isotope Effects

The crystallographic and model-building studies have shown that in the closed structure, which we believe to be assumed at least by all covalent intermediates of catalysis, Lys 258 "sees" only a single water molecule (in a position near $H_2O(7)$ in the open structure; Fig. 10a). There is room for this molecule in the structures of the ketimine and quinonoid intermediates, but probably not in the structure of the external aldimine intermediate. It has no contact with other water molecules. Thus, a deuterium or tritium label on Lys 258 can exchange with a total of four other protons on the time scale of a single catalytic half-reaction. A solvent isotope effect applies to all forward and backward steps after α deprotonation of the external aldimine. On the other hand, a substrate isotope effect on the Cα proton should only be observed in the α deprotonation step, if this step is at least partially rate limiting. The structural data can nicely explain the isotope effects observed in the kinetic study of the transamination, by cAAT, of *erthyro*-3-hydroxy-aspartate (Jenkins and Harruff, 1979). The kinetic steps leading to the quinonoid intermediate (k_1) and the reverse process (k_2) are fast, while those leading to the product complex (k_3) and its back reaction (k_4) are slow. Each of these constants is the resultant of several other kinetic constants. A strong substrate isotope effect was observed for k_1 and a weaker one for k_2, while the combination of α-deuterated substrate and D_2O as solvent resulted in a strong isotope effect on k_2. Equally strong solvent isotope effects were measured for k_3 and k_4. These facts are readily explained by the present mechanism. In the reaction of α-deuterated substrate in H_2O, the rate constant of enzyme-catalyzed exchange of the deuterons with solvent was identical to k_2, which is the rate that determines substrate release. This is also the rate at which the 80% of the deuterons that are not bound to the coenzyme (not counting the isotope effect) exchange in the open structure with solvent, again as expected for a reaction taking place in the secluded active site.

Gehring (1984) was the first to demonstrate transfer of tritium from α-tritiated glutamate to the pro-S position of the coenzyme in chicken mAAT. A fast mixing and quenching device was used to minimize the number of turnovers of each molecule.

In different experiments, 0.08−4.0% of the label was recovered on the coenzyme. This work proved that a *cis* mechanism applies to the transamination reaction, and in comparison with the active site model, that only Lys

258, Tyr 70*, or a bound water molecule can act as an acid/base catalyst (Gehring, 1984; Kirsch et al., 1984). The results are fully compatible with the notion of Lys 258 and one water molecule equilibrating during a catalytic half-cycle. This single water molecule, together with an isotope effect on protonation at C4', can explain the low percentage recovery of tritium on the coenzyme.

6. FUTURE STUDIES

Many questions on aspartate aminotransferase catalysis have been answered, but many others remain open. Inhibitor binding to the PMP enzyme has not been studied at all by crystallography, and high-resolution studies on the binding mode of 5-carbon inhibitors to the PLP enzyme will have to be carried out. Substrate binding to inactive analogs of the PLP enzyme (e.g., the reduced form) and of the PMP enzyme (e.g., the pyridoxine-phosphate analog) might be worth studying crystallographically if the binding constants warrant it.

Another problem that can be attacked crystallographically is the equilibrium between the open and closed structures. For the mitochondrial enzyme, the crystallization conditions of the P1 (open) and the $C222_1$ (closed) structures, are nearly identical. However, the crystals have very different habits. Therefore, upon crystallization with an inhibitor, the type of crystals that appear is an indication of whether the open or the closed structure predominates.

The question of the structural basis for the different specificities of the cytosolic and mitochondrial isoenzymes can hopefully be answered when refined structures of both are compared. The absence of Asp 15 in cAAT should have some influence on the open $\rightleftharpoons$ closed equilibrium.

The importance of domain closure for catalysis cannot be settled by crystallography. The indications for an upper limit of, at most, two orders of magnitude in catalytic rate should be confirmed by solution studies on an enzyme in which domain movement is blocked. The difference in catalytic rate of transamination between aspartate and alanine, a factor of 10^4 to 10^5 would have to be explained, as would the factor of ~500 between aspartate and tyrosine.

Two approaches may substantially augment our understanding of enzymic PLP catalysis. One is the availability, through DNA sequence studies, of additional amino acid sequences of enzymes of one family. A correlation of enzyme structure and specific activity for a given substrate should give excellent clues toward the explanation of such a specificity. The other approach is the study of enzyme derivatives that are produced by cloning and manipulation of its gene. In principle, an unlimited number of point mutations can be produced, and the corresponding products can be tested for their stability and catalytic properties. This, more than anything else, will deepen our understanding of enzyme catalysis. Fortunately, at least four laboratories are already engaged in cloning aspartate aminotransferase genes, and the first mutant enzymes are already under study (Malcolm and Kirsch, 1985). It is feasible that many of the remaining questions on structure-function relationships in this enzyme will be answered within the next 5 years.

The next higher level of understanding is that of functional specificity within the same family of PLP enzymes. Although Dunathan's hypothesis has now been confirmed for AAT, and should also hold for enzymes acting on one of the two other potentially labile bonds at the amino acid α-carbon, it will be exciting to learn how nature manipulates an enzyme to change its specificity without extensive structural modifications. Although there is a general belief that this is what actually happens, a proof in the form of a homologous sequence for a PLP enzyme with a different function has, to our knowledge, not yet been given.

Finally, PLP enzyme catalysis in general, including enzymes that are structurally unrelated to AAT, is of considerable interest. Does the coenzyme rotation mechanism—as first proposed by Ivanov and Karpeisky and later determined in a somewhat modified form for AAT by X-ray crystallography—also apply to other enzyme families? Is domain movement a general phenomenon in PLP catalysis? The answer to such questions will again have to be given by X-ray crystallography.

Crystallographic studies are underway on at least two bacterial PLP enzymes: ω-amino acid: pyruvate aminotransferase from *Pseudomonas* sp. F-126 (Morita et al., 1979) and tryptophan synthase from *Salmonella typhimurium* (Ahmed et al., 1985). In our laboratory, well-diffracting crystals of a phosphoserine aminotransferase from *E. coli* have recently been grown. In all three cases, no obvious sequence homology with AAT is apparent, so that either very distant or no evolutionary relationship with

AAT exists.Therefore, we may hope that in the next few years, answers to our last two questions will also be given.

We may conclude that nearly 50 years after the discovery of enzymic transamination and 40 years after the proposal of covalent pyridoxal catalysis, a good understanding of the physical basis of enzymatic PLP catalysis is emerging that can be expected to expand rapidly in the near future.

ACKNOWLEDGMENTS

We would like to thank Ms. U. Abrecht for her invaluable help in the preparation of this manuscript. We are indebted to our former colleagues who made important contributions to the structural studies on chicken mAAT, a joint project with P. Christen and colleagues, University of Zürich. We are grateful to the staff of the Basel University Computer Center for extensive computer and plotting facilities. The research on mAAT was supported, in part, by grants from the Swiss National Science Foundation to one of us (JNJ).

REFERENCES

Ahmed, S. A., Wilson Miles, E., and Davies, D. R. (1985). *J. Biol. Chem.* **260**, 3716–3718.

Andersen, R. D., Apgar, P. A., Burnett, R. M., Darling, G. D., LeQuesne, M. E., Mayhew, S. G., and Ludwig, M. L. (1972). *Proc. Natl. Acad. Sci. USA* **69**, 3189–3191.

Arnone, A., Rogers, P. H., Schmidt, J., Han, C.-N., Harris, C. M., and Metzler, D. E. (1977). *J. Mol. Biol.* **112**, 509–513.

Arnone, A., Briley, P. D., Rogers, P. H., Hyde, C. C., Metzler, C. M., and Metzler, D. E. (1982). In *Molecular Structure and Biological Activity*, J. F. Griffin, and W. L. Duax, Eds., Elsevier/North Holland, New York, pp. 57–77.

Arnone, A., Rogers, P. H., Hyde, C. C., Makinen, M. W., Feldhaus, R., Metzler, C. M., and Metzler, D. E. (1984). In *Chemical and Biological Aspects of Vitamin B_6 Catalysis*, A. E. Evangelopoulos, Ed., Part B. Alan R. Liss, New York, pp. 171–193.

Arnone, A., Christen, P., Jansonius, J. N., and Metzler, D. E. (1985a). In *Transaminases*, P. Christen, and D. E. Metzler, Eds. John Wiley & Sons, New York, pp. 326–357.

Arnone, A., Rogers, P. H., Hyde, C. C., Briley, P. D., Metzler, C. M., and Metzler, D. E. (1985b). In *Transaminases,* P. Christen, and D. E. Metzler, Eds. John Wiley & Sons, New York, pp. 138–155.

Arrio-Dupont, M., and Vergé, D. (1982) *J. Mol. Biol.* **157,** 383–394.

Barra, D., Bossa, F., Doonan, S., Fahmy, H. M. A., Hughes, G. J., Kakoz, K. J., Martini, F., and Petruzelli, R. (1977). *FEBS Lett.* **83,** 241–244.

Barra, D., Martini, F., Montarani, G., Doonan, S., and Bossa, F. (1979). *FEBS Lett.* **108,** 103–106.

Barra, D., Bossa, F., Doonan, S., Fahmy, H. M. A., Hughes, G. J., Martini, F., Petruzzelli, R., and Wittmann-Liebold, B. (1980). *Eur. J. Biochem.* **108,** 405–414.

Barra, D., Angelaccio, S., Martini, F., Doonan, S., and Bossa, F. (1984). In *Chemical and Biological Aspects of Vitamin B$_6$ Catalysis,* A. E. Evangelopoulos, Ed., Part B. Alan R. Liss, New York, pp. 285–291.

Bennett, W. S., and Huber, R. (1983). *CRC Crit Rev Biochem.* **15,** 291–384.

Bergami, M., Marino, G., and Scardi, V. (1968). *Biochem. J.* **110,** 471–473.

Birchmeier, W., Wilson, K. J., and Christen, P. (1973). *J. Biol. Chem.* **248,** 1751–1759.

Blundell, T. L., and Johnson, L. N. (1976). *Protein Crystallography.* Academic Press, New York.

Bocharov, A. L., Ivanov, V. I., Karpeisky, M. Ya., Mamaeva, O. K., and Florentiev, V. L. (1968). *Biochem. Biophys. Res. Commun.* **30,** 459–464.

Boettcher, B., and Martinez-Carrion, M. (1975). *Biochemistry* **14,** 4528–4531.

Boettcher, B., and Martinez-Carrion, M. (1976). *Biochemistry* **15,** 5657–5664.

Boettcher, B., and Martinez-Carrion, M. (1978). *J. Biol. Chem.* **253,** 4642–4647.

Bonsib, S. M., Harruff, R. C., and Jenkins, W. T. (1975). *J. Biol. Chem.* **250,** 8635–8641.

Borisov, V. V., Borisova, S. N., Kachalova, G. S., Sosfenov, N. I., Voronova, A. A., Vainshtein, B. K., Torchinsky, Yu.M., Volkova, G. A., and Braunstein, A. E. (1977). *Dokl. Akad. Nauk SSSR* **235,** 212–215.

Borisov, V. V., Borisova, S. N., Kachalova, G. S., Sosfenov, N. I., Vainshtein, B. K., Torchinsky, Yu.M., and Braunstein, A. E. (1978). *J. Mol. Biol.* **125,** 275–292.

Borisov, V. V., Borisova, S. N., Sosfenov, N. I., Vagin, A. A., Nekrasov, Yu.V., Vainshtein, B. K., Kochkina, V. M., and Braunstein, A. E. (1980a). *Dokl. Akad. Nauk SSSR* **250,** 988–992.

Borisov, V. V., Borisova, S. N., Sosfenov, N. I., and Vainshtein, B. K. (1980b). *Nature* **284,** 189–190.

Borisov, V. V., Borisova, S. N., Kachalova, G. S., Sosfenov, N. I., and Vainshtein, B. K. (1985). In *Transaminases,* P. Christen and D. E. Metzler, Eds. John Wiley & Sons, New York, pp. 155–164.

Boyd, J. W. (1961). *Biochem. J.* **81,** 434–441.

Braunstein, A. E. (1973). In *The Enzymes*. P. D. Boyer, Ed., 3rd ed., Vol. 9. Academic Press, New York, pp. 379−481.

Braunstein, A. E., and Kritzmann, M. G. (1937). *Enzymologia* **2**, 129−146.

Braunstein, A. E., and Shemyakin, M. M. (1952). *Dokl. Akad. Nauk SSSR* **85**, 1115−1118.

Braunstein, A. E., and Shemyakin, M. M. (1953). *Biokhimiya* **18**, 393−411.

Briley, P. A., Eisenthal, R., Harrison, R., and Smith, G. D. (1977). *Biochem. J.* **161**, 383−387.

Capasso, S., Garzillo, A. M., Marino, G., Mazzarella, L., Pucci, P., and Sannia, G. (1979). *FEBS Lett.* **101**, 351−354.

Carotti, D., Riva, F., Santucci, R., Ascoli, F., and Fasella, P. (1982). *Eur. J. Biochem.* **124**, 589−593.

Carotti, D., Andria, F., Giartosio, A., Turano, C., and Riva, F. (1985). *Eur. J. Biochem.* **146**, 619−623.

Cheng, S., Michuda-Kozak, C., and Martinez-Carrion, M. (1971). *J. Biol. Chem.* **246**, 3623−3630.

Christen, P., and Metzler, D. E., Eds. (1985). *Transaminases*. John Wiley & Sons, New York.

Christen, P., Jaussi, R., Sonderegger, P., Gehring, H., Graf-Hausner, U., Behra, R., Flückiger, J., and Skoda, R. (1984). In *Chemical and Biological Aspects of Vitamin B$_6$ Catalysis*. A. E. Evangelopoulos, Ed. Part B. Alan R. Liss, New York, pp. 17−25.

Christen, P., Graf-Hausner, U., Bossa, F., and Doonan, S. (1985). In *Transaminases*, P. Christen, and D. E. Metzler, Eds. John Wiley & Sons, New York, pp. 173−185.

Cleland, W. W. (1963). *Biochim. Biophys, Acta* **67**, 104−137.

Czerlinski, G., and Malkewitz, J. (1965). *Biochemistry* **4**, 1127−1137.

Dempsey, W. B., and Snell, E. E. (1963). *Biochemistry* **2**, 1414−1419.

Doonan, S., Doonan, H. J., Hanford, R., Vernon, C. A., Walker, J. M., Bossa, F., Barra, D., Carloni, M., Fasella, P., Riva, F., and Walton, P. L. (1974). *FEBS Lett.* **38**, 229−233.

Doonan, S., Doonan, H. J., Hanford, R., Vernon, C. A., Walker, J. M., Airoldi, L. P., Bossa, F., Barra, D., Carloni, M., Fasella, P., and Riva, F. (1975). *Biochem. J.* **149**, 497−506.

Dunathan, H. C. (1971). *Advan. Enzymol.* **35**, 79−134.

Dybel, M. W., Cheng, S., and Martinez-Carrion, M. (1972). *Arch. Biochem. Biophys.* **148**, 320−321.

Eichele, G. (1980). Ph.D. Thesis, University of Basel.

Eichele, G., Karabelnik, D., Halonbrenner, R., Jansonius, J. N., and Christen, P. (1978). *J. Biol. Chem.* **253**, 5239−5242.

Eichele, G., Ford, G. C., Glor, M., Jansonius, J. N., Mavrides, C., and Christen, P. (1979a). *J. Mol. Biol.* **133**, 161−180.

Eichele, G., Ford, G. C., and Jansonius, J. N. (1979b). *J. Mol. Biol.* **135**, 513–516.

Fasella, P., and Hammes, G. G. (1967). *Biochemistry* **6**, 1798–1804.

Fasella, P., and Turano, C. (1970). *Vitam. Horm.* **28**, 157–195.

Fasella, P., Giartosio, A., and Hammes, G. G. (1966). *Biochemistry* **5**, 197–202.

Federiuk, C. S., and Shafer, J. A. (1983). *J. Biol. Chem.* **258**, 5372–5378.

Ford, G. C., Eichele, G., and Jansonius, J. N. (1980). *Proc. Natl. Acad. Sci. USA* **77**, 2559–2563.

Furbish, F. S., Fonda, M. L., and Metzler, D. E. (1969). *Biochemistry* **12**, 5169–5180.

Gehring, H. (1984). *Biochemistry* **23**, 6335–6340.

Gehring, H. (1985). In *Transaminases*, P. Christen, and D. E. Metzler, Eds. John Wiley & Sons, New York, pp. 317–326.

Gehring, H., and Christen, P. (1975). *Biochem. Biophys. Res. Commun.* **63**, 441–447.

Gehring, H., and Christen, P. (1978). *J. Biol. Chem.* **253**, 3158–3163.

Gehring, H., Christen, P., Eichele, G., Glor, M., Jansonius, J. N., Reimer, A.-S., Smit, J. D., and Thaller, C. (1977). *J. Mol. Biol.* **115**, 97–101.

Giannini, I., Baroncelli, V., and Boccalon, G. (1975). *FEBS Lett.* **54**, 307–310.

Giannini, I., Baroncelli, V., Boccalon, G., and Fasella, P. (1976). *Eur. J. Biochem.* **71**, 475–481.

Gorenstein, D. G. (1975). *J. Am. Chem. Soc.* **97**, 898–900.

Graf-Hausner, U., Wilson, K. J., and Christen, P. (1983). *J. Biol. Chem.* **258**, 8813–8826.

Green, D. E., Leloir, L. F., and Nocito, V. (1945). *J. Biol. Chem.* **161**, 559–582.

Haddad, L. C., Thayer, W. S., and Jenkins, W. T. (1977). *Arch. Biochem. Biophys.* **181**, 66–72.

Hammes, G. G., and Tancredi, J. F. (1967). *Biochim. Biophys. Acta.* **132**, 312–313.

Hammes, G. G., and Haslam, J. L. (1968). *Biochemistry* **7**, 1519–1524.

Hammes, G. G., and Haslam, J. L. (1969). *Biochemistry* **8**, 1591–1598.

Harruff, R. C., and Jenkins, W. T. (1976). *Arch. Biochem. Biophys.* **177**, 394–401.

Harruff, R. C., and Jenkins, W. T. (1978). *Arch. Biochem. Biophys.* **188**, 37–46.

Harutyunyan, E. G., Malashkevich, V. N., Tersyan, S. S., Kochkina, V. M., Torchinsky, Yu. M., and Braunstein, A. E. (1982). *FEBS Lett.* **138**, 113–116.

Harutyunyan, E. G., Malashkevich, V. N., Kochkina, V. M., and Torchinsky, Yu. M. (1984). In *Chemical and Biological Aspects of Vitamin B_6 Catalysis*, A. E. Evangelopoulos, Ed. Part B. Alan R. Liss, New York, pp. 205–212.

Harutyunyan, E. G., Malashkevich, V. N., Kochkina, V. M., and Torchinsky, Yu. M. (1985). In *Transaminases*, P. Christen, and D. E. Metzler, Eds. John Wiley and Sons, New York, pp. 164–173.

Heinert, D., and Martell, A. E. (1963a). *J. Am. Chem. Soc.* **85**, 183–188.

Heinert, D., and Martell, A. E. (1963b). *J. Am. Chem. Soc.* **85**, 188–193.

Hendrickson, W. A., and Konnert, J. H. (1980a). In *Biomolecular Structure, Function, Conformation and Evolution*. R. Srinivasan, Ed. Vol. 1. Pergamon Press, Oxford, pp. 43–57.

Hendrickson, W. A., and Konnert, J. H. (1980b). In *Computing in Crystallography*, R. Diamond et al., Eds. Indian Academy of Sciences, Bangalore, pp. 13.01–13.23.

Hofrichter, J., and Eaton, W. A. (1976). *Ann. Rev. Biophys. Bioeng.* **5**, 511–559.

Hol, W. G. J., Van Duijnen, P. T., and Berendsen, H. J. C. (1978). *Nature* **273**, 443–446.

Hughes, R. C., Jenkins, W. T., and Fischer, E. H. (1962). *Proc. Natl. Acad. Sci. USA* **48**, 1615–1618.

Huynh, Q. K., Sasakibara, R., Watanabe, T., and Wada, H. (1980). *Biochem. Biophys. Res. Commun.* **97**, 474–479.

Hyde, C. C., Rogers, P. H., Briley, P. D., Arnone, A., Metzler, C. M., and Metzler, D. E. (1984). In *Peptide and Protein Reviews*, M. T. W. Hearn, Ed. Vol. 4. Marcel Dekker, New York, pp. 47–113.

Isogai, Y., Némethy, G., Rackovsky, S., Leach, J., and Scheraga, H. A. (1980). *Biopolymers* **19**, 1183–1210.

Ivanov, V. I., and Karpeisky, M. Ya. (1969). *Adv. Enzymol.* **32**, 21–53.

Jansonius, J. N., Eichele, G., Ford, G. C., Kirsch, J. F., Picot, D., Thaller, C., Vincent, M. G., Gehring, H., and Christen, P. (1984a). *Biochem. Soc. Trans.* **12**, 424–427.

Jansonius, J. N., Eichele, G., Ford, G. C., Kirsch, J. F., Picot, D., Thaller, C., Vincent, M. G., Gehring, H., and Christen, P. (1984b). In *Chemical and Biological Aspects of Vitamin B_6 Catalysis*, A. E. Evangelopoulos, Ed., Part B. Alan R. Liss, New York, pp. 195–203.

Jansonius, J. N., Eichele, G., Ford, G. C., Picot, D., Thaller, C., and Vincent, M. G. (1985). In *Transaminases*, P. Christen, and D. E. Metzler, Eds. J. Wiley & Sons, New York, pp. 109–138.

Jaussi, R., Sonderegger, P., Flückiger, J., and Christen, P. (1982). *J. Biol. Chem.* **257**, 13334–13340.

Jenkins, W. T. (1961). *J. Biol. Chem.* **236**, 1121–1125.

Jenkins, W. T. (1964). *J. Biol. Chem.* **239**, 1742–1747.

Jenkins, W. T. (1980a). *Arch. Biochem. Biophys.* **205**, 57–66.

Jenkins, W. T. (1980b). *Arch. Biochem. Biophys.* **205**, 579–586.

Jenkins, W. T., and Sizer, I. W. (1957). *J. Am. Chem. Soc.* **79**, 2655–2656.

Jenkins, W. T., and Sizer, I. W. (1960). *J. Biol. Chem.* **235**, 620–624.

Jenkins, W. T., and Taylor, R. T. (1965). *J. Biol. Chem.* **240**, 2907–2913.

Jenkins, W. T., and D'Ari, L. (1966). *J. Biol. Chem.* **241**, 5667–5674.

Jenkins, W. T., and Harruff, R. C. (1979). *Arch. Biochem. Biophys.* **192**, 421–429.

Jenkins, W. T., and Fonda, M. L. (1985). In *Transaminases*, P. Christen, and D. E. Metzler, Eds. John Wiley & Sons, New York, pp. 216–234.

Jenkins, W. T., Yphantis, D. A., and Sizer, I. W. (1959). *J. Biol. Chem.* **234**, 51–57.

Kagamiyama, H., Sakakibara, R., Wada, H., Tanase, S., and Morino, Y. (1977). *J. Biochem. (Tokyo)* **82**, 291–294.

Kagamiyama, H., Sakakibara, R., Tanase, S., Morino, Y., and Wada, H. (1980). *J. Biol. Chem.* **255**, 6153–6159.

Karpeisky, M. Ya., and Ivanov, V. I. (1966). *Nature* **210**, 493–496.

Karpeisky, M. Ya., Khomutov, R. M., Severin, E. S., and Breusov, Yu. N. (1963). In *Chemical and Biological Aspects of Pyridoxal Catalysis*, E. E. Snell et al., Eds. Pergamon Press, Oxford, pp. 323–331.

Khomutov, R. M., Severin, E. S., Kovaleva, G. K., Gulyaev, N. N., Gnuchev, N. V., and Sastchenko, L. P. (1968). In *Pyridoxal Catalysis: Enzymes and Model Systems*, E. E. Snell et al., Eds. Interscience Publishers, New York, pp. 631–650.

Khomutov, R. M., Severin, E. S., Khurs, E. N., and Gulyaev, N. N. (1969). *Biochim. Biophys. Acta* **171**, 201–202.

Khurs, E. N., Severin, E. S., Dixon, H. B. F., and Khomutov, R. M. (1976). *Mol. Biol. (USSR)* **10**, 740–747.

Kiick, D. M., and Cook, P. F. (1983). *Biochemistry* **22**, 375–382.

Kirsch, J. F., Eichele, G., Ford, G. C., Vincent, M. G., Jansonius, J. N., Gehring, H., and Christen, P. (1984). *J. Mol. Biol.* **174**, 497–525.

Kirsten, H. (1982). Ph.D. Thesis, University of Zürich.

Kirsten, H., Gehring, H., and Christen, P. (1983). *Proc. Natl. Acad. Sci. USA* **80**, 1807–1810.

Kondo, K., Wakabayashi, S., Yagi, T., and Kagamiyama, H. (1984). *Biochem. Biophys. Res. Commun.* **122**, 62–67.

Kuramitsu, S., Okuno, S., Ogawa, T., Ogawa, H., and Kagamiyama, H. (1985). *J. Biochem. (Tokyo)* **97**, 1259–1262.

Lesk, A. M., and Chothia, C. (1984). *J. Mol. Biol.* **174**, 175–191.

Levitt, M., and Chothia, C. (1976). *Nature* **261**, 552–557.

Lewis, P. N., Momany, F. A., and Scheraga, H. A. (1973). *Biochim. Biophys. Acta.* **303**, 211–229.

Lichstein, H. C., Gunsalus, I. C., and Umbreit, W. W. (1945). *J. Biol. Chem.* **161**, 559–582.

Makarov, V. L., Kochkina, V. M., and Torchinsky, Yu. M. (1980). *FEBS Lett.* **114**, 79–82.

Makarov, V. L., Kochkina, V. M., and Torchinsky, Yu. M. (1981). *Biochim. Biophys. Acta* **659**, 219–228.

Makarov, V. L., Kochkina, V. M., Rosenberg, M. V., and Torchinsky, Yu. M. (1984). In *Chemical and Biological Aspects of Vitamin B$_6$ Catalysis*. A. E. Evangelopoulos, Ed. Part B. Alan R. Liss, New York, pp. 213–221.

Malcolm, B. A., and Kirsch, J. F. (1985). *Biochem. Biophys. Res. Commun.* **132**, 915–921.

Malcolm, B., Brauer, M., Ames, G. F-L., and Kirsch, J. F. (1985). *Fed. Proc.* **44**, 1061.

Martinez-Carrion, M. (1975). *Eur. J. Biochem.* **54**, 39–43.

Martinez-Carrion, M., and Tiemeier, D. (1967). *Biochemistry* **6**, 1715–1722.

Martinez-Carrion, M., Riva, F., Turano, C., and Fasella, P. (1965). *Biochem. Biophys. Res. Commun.* **20**, 206–211.

Martinez-Carrion, M., Turano, C., Riva, F., and Fasella, P. (1967). *J. Biol. Chem.* **242**, 1426–1430.

Martinez-Carrion, M., Kuszenski, R., Tiemeier, D. C., and Peterson, D. L. (1970a). *J. Biol. Chem.* **245**, 799–805.

Martinez-Carrion, M., Tiemeier, D. C., and Peterson, D. L. (1970b). *Biochemistry* **9**, 2574–2582.

Martinez-Carrion, M., Slebe, J. C., and Gonzales, M. (1979). *J. Biol. Chem.* **254**, 3160–3162.

Martinez-Carrion, M., Hubert, E., Iriarte, A., Mattingly, J. R., and Zito, S. W. (1985). In *Transaminases*, P. Christen, and D. E. Metzler, Eds. J. Wiley & Sons, New York, pp. 308–317.

Martini, F., Angelaccio, S., Barra, D., Doonan, S., and Bossa, F. (1983). *Comp. Biochem. Physiol.* **76B**, 483–487.

Martini, F., Angelaccio, S., Barra, D., Doonan, S., and Bossa, F. (1984). *Biochim. Biophys. Acta* **789**, 51–56.

Matthews, B. W. (1968). *J. Mol. Biol.* **33**, 491–497.

Mattingly, M. E., Mattingly, J. R., and Martinez-Carrion, M. (1982). *J. Biol. Chem.* **257**, 8872–8878.

Mavrides, C., and Christen, P. (1978). *Biochem. Biophys. Res. Commun.* **85**, 769–773.

McPherson, A. (1976). *J. Biol. Chem.* **251**, 6300–6303.

McPherson, A. (1982). *Preparation and Analysis of Protein Crystals.* John Wiley & Sons, New York.

Metzler, D. E. (1977). *Biochemistry*, Chapter 8. Academic Press Inc., New York.

Metzler, D. E. (1979). *Advan. Enzymol. relat. areas Mol. Biol.* **50**, 1–40.

Metzler, D. E., and Snell, E. E. (1952). *J. Am. Chem. Soc.* **74**, 979–983.

Metzler, D. E., and Fonda, M. L. (1985). In *Transaminases*, P. Christen, and D. E. Metzler, Eds. John Wiley & Sons, New York, pp. 235–250.

Metzler, D. E., Ikawa, M., and Snell, E. E. (1954). *J. Am. Chem. Soc.* **76**, 648–652.

Metzler, C. M., Metzler, D. E., Martin, D. S., Newman, R., Arnone, A., and Rogers, P. (1978). *J. Biol. Chem.* **253**, 5251–5254.

Michuda, C. M., and Martinez-Carrion, M. (1969a). *J. Biol. Chem.* **244**, 5920–5927.

Michuda, C. M., and Martinez-Carrion, M. (1969b). *Biochemistry* **8**, 1095–1105.

Michuda, C. M., and Martinez-Carrion, M. (1970). *J. Biol. Chem.* **245**, 262−269.

Miller, J. E., and Litwack, G. (1971). *J. Biol. Chem.* **246**, 3234−3240.

Morino, Y., and Okamoto, M. (1972). *Biochemistry* **11**, 3196−3201.

Morino, Y., and Tanase, S. (1978). *J. Biol. Chem.* **253**, 252−256.

Morino, Y., and Tanase, S. (1985). In *Transaminases,* P. Christen, and D. E. Metzler, Eds. John Wiley & Sons, New York, pp. 251−265.

Morino, Y., Osman, A. M., and Okamoto, M. (1974). *J. Biol. Chem.* **249**, 6684−6692.

Morino, Y., Okamoto, M., and Tanase, S. (1978). *J. Biol. Chem.* **253**, 6026−6030.

Morita, Y., Aibara, S., Yonaha, K., Toyama, S., and Soda, K. (1979). *J. Mol. Biol.* **130**, 211−213.

Mozzarelli, A., Ottonello, S., Rossi, G. L., and Fasella, P. (1979). *Eur. J. Biochem.* **98**, 173−179.

Okamoto, M., and Morino, Y. (1972). *Biochemistry* **11**, 3188−3195.

Ottonello, S., Mozzarelli, A., Rossi, G. L., Carotti, D., and Riva, F. (1983). *Eur. J. Biochem.* **133**, 47−49.

Ovchinnikov, Yu. A., Egorov, C. A., Aldanova, N. A., Feigina, M. Yu., Lipkin, V. M., Abdulaev, N. G., Grishin, E. V., Kiselev, A. P., Modyanov, N. N., Braunstein, A. E., Polyanovsky, O. L., and Nosikov, V. V. (1973). *FEBS Lett.* **29**, 31−34.

Peterson, D. L., and Martinez-Carrion, M. (1970). *J. Biol. Chem.* **245**, 806−813.

Pfister, K., Kägi, J. H. R., and Christen, P. (1978). *Proc. Natl. Acad. Sci. USA* **75**, 145−148.

Relimpio, A., Slebe, J. C., and Martinez-Carrion, M. (1975). *Biochem. Biophys. Res. Commun.* **63**, 625−634.

Relimpio, A., Iriarte, A., Chlebowski, J. F., and Martinez-Carrion, M. (1981). *J. Biol. Chem.* **256**, 4478−4488.

Remington, S. J., Wiegand, G., and Huber, R. (1982). *J. Mol. Biol.* **158**, 111−152.

Richardson, J. (1981). *Advan. Prot. Chem.* **34**, 167−339.

Riva, F., Carotti, D., Barra, D., Giartosio, A., and Turano, C. (1980). *J. Biol. Chem.* **255**, 9230−9235.

Rossi, G. L., Ottonello, S., Mozzarelli, A., Tegoni, M., Martini, F., Bossa, F., and Fasella, P. (1978). In *Protein: Structure, Function and Industrial Applications. FEBS Proceedings,* E. Hoffmann et al., Eds. Vol. 52. Pergamon Press, Oxford, pp. 249−258.

Rossmann, M. G., Moras, D., and Olsen, K. W. (1974). *Nature* **250**, 194−199.

Saier, M. H., Jr., and Jenkins, W. T. (1967). *J. Biol. Chem.* **242**, 101−108.

Sandmeier, E., and Christen, P. (1980). *J. Biol. Chem.* **255**, 10284−10289.

Sandmeier, E., and Christen, P. (1982). *J. Biol. Chem.* **257**, 6745−6750.

Savin, F. A. (1984). In *Physico-Chemical Problems of Enzymatic Catalysis*, Yu. M. Torchinsky, Ed. Nauka, Moscow, pp. 69−83.

Schlegel, H., and Christen, P. (1978). *Biochim. Biophys. Acta* **532**, 6–16.

Schlegel, H., Zaoralek, P. E., and Christen, P. (1977). *J. Biol. Chem.* **252**, 5835–5838.

Schnackerz, K. D. (1984). *Biochim. Biophys. Acta* **789**, 241–244.

Severin, E. S., Gulyaev, N. N., Khurs, E. N., and Khomutov, R. M. (1969). *Biochem. Biophys. Res. Commun.* **35**, 318–323.

Shlyapnikov, S. V., Myasnikov, A. N., Severin, E. S., Myagkova, M. A., Torchinsky, Yu. M., and Braunstein, A. E. (1979). *FEBS Lett.* **106**, 385–388.

Shrawder, E., and Martinez-Carrion, M. (1972). *J. Biol. Chem.* **247**, 2486–2492.

Slebe, J. C., and Martinez-Carrion, M. (1976). *J. Biol. Chem.* **251**, 5663–5669.

Snell, E. E. (1945). *J. Am. Chem. Soc.* **67**, 194–197.

Snell, E. E. (1962). *Brookhaven Symp. Biol.* **15**, 32–51.

Soper, T. S., and Manning, J. M. (1985). In *Transaminases*, P. Christen, and D. E. Metzler, Eds. John Wiley & Sons, New York, pp. 266–285.

Steigemann, W., Deisenhofer, J., and Huber, R. (1976). In *Crystallographic Computing Techniques*, F. R. Ahmed, Ed. Munksgaard, Copenhagen, pp. 302–306.

Thaller, C., Weaver, L. H., Eichele, G., Wilson, E., Karlsson, R., and Jansonius, J. N. (1981). *J. Mol. Biol.* **147**, 465–469.

Timofeev, V. P., Dudich, I. V., and Volkenstein, M. V. (1980). *Biophys. Struct. Mech.* **7**, 41–49.

Torchinsky, Yu. M. (1985). In *Coenzymes and Cofactors,* D. Dolphin, Ed. Vol. 1. John Wiley & Sons, New York (in press).

Torchinsky, Yu. M., and Koreneva, L. G. (1963). *Biokhimiya* **28**, 1087–1096.

Torchinsky, Yu. M., and Koreneva, L. G. (1964). *Biochim. Biophys. Acta* **79**, 426–429.

Torchinsky, Yu. M., Harutyunyan, E. G., Malashkevich, V. N., Kochkina, V. M., Makarov, V. L., and Braunstein, A. E. (1983). *Cell Function and Differentiation. Part C. Proc. Special FEBS Meeting (Symp. 12, Athens, 1982)*, A. E. Evangelopoulos, Ed. Alan R. Liss, New York, p. 13.

Tumanyan, V. G., Mamaeva, O. K., Bocharov, A. L., Ivanov, V. I., Karpeisky, M. Y., and Yakovlev, G. I. (1974). *Eur. J. Biochem.* **50**, 119–127.

Turano, C., Giartosio, A., Riva, F., Baroncelli, V., and Bossa, F. (1966). *Arch. Biochem. Biophys.* **117**, 678–680.

Velick, S. F., and Vavra, J. (1962). *J. Biol. Chem.* **237**, 2109–2122.

Vergé, D., Tenu, J-P., and Arrio-Dupont, M. (1979). *FEBS Lett.* **100**, 265–268.

Vincent, M. G., Picot, D., Eichele, G., Jansonius, J. N., Kirsten, H., and Christen, P. (1984). In *Chemical and Biological Aspects of Vitamin B$_6$ Catalysis,* A. E. Evangelopoulos, Ed. Part B. Alan R. Liss, New York, pp. 233–243.

Yamasaki, M., Tanase, S., and Morino, Y. (1975). *Biochem. Biophys. Res. Commun.* **65**, 652–657.

Zito, S. W., and Martinez-Carrion, M. (1980). *J. Biol. Chem.* **255**, 8645–8649.

5

Catalysis by Phospholipases A2

JAN DRENTH
B. W. DIJKSTRA
R. RENETSEDER
Laboratory of Chemical Physics
University of Groningen
Groningen
The Netherlands

CONTENTS

1. OVERVIEW
  1.1.   Introduction
  1.2.   Phospholipids
  1.3.   The Phospholipases A2

2. STRUCTURAL RESULTS
  2.1.   Three-dimensional Structure
  2.2.   X-Ray Diffraction Results
  2.3.   The Structure of Phospholipase A2 From Bovine and
         Porcine Pancreas
  2.4.   The Structure of *Crotalus atrox* Phospholipase A2

3. THE CATALYTIC MECHANISM

4. CONCLUSION

REFERENCES

1. OVERVIEW

1.1. Introduction

The catalytic action of the phospholipases is strongly dependent on the nature of their substrates: the phospholipids. Therefore, phospholipid molecules and their conformation are discussed first. Then, the molecular properties and kinetics of substrate hydrolysis of the phospholipases A2 are reviewed briefly. Additional details are found in reviews on phospholipases A2 that have appeared recently (Volwerk and De Haas, 1982; Slotboom et al., 1982; Dennis, 1983). The principal text of the present review, however, focuses on a description of the structural details of the enzyme as they emerge from X-ray crystallographic studies. In the final section, the structural results are correlated with the results from a variety of chemical and physicochemical studies.

1.2. Phospholipids

The natural substrates for phospholipases are the phospholipids. They are detergent-type molecules with a lipophilic and a hydrophilic moiety (Fig. 1). The backbone of a phospholipid molecule is formed by the three carbon atoms of glycerol, which are esterified in positions 1 and 2 with a fatty acid and in position 3 with a phosphate. This phosphate group can be further esterified with an additional glycerol, a serine, or an amino alcohol. The second carbon atom in the glycerol moiety (position 2) forms a chiral center and, in the naturally occurring phospholipids, this carbon atom has the configuration as given in Figure 2. The official name of the natural phospholipids is *sn*-3-phosphoglycerides. Due to their insolubility in water and their polar/nonpolar detergent type of structure, the phospholipids form organized structures in water, such as lamellar bilayer structures, micelles, and vesicles. An important function of the phospholipids is the formation of cellular membranes that enclose all cells and surround their internal organelles.

The structure of phospholipids has been studied by X-ray and spectroscopic methods. These studies suggest that the conformation of the *sn*-3-phospholipids in membranes has the glycerol moiety oriented almost perpendicular to the membrane surface. The fatty acid chain at the C1 atom of glycerol continues in the same direction. The fatty acid chain at C2 starts in a direction parallel to the membrane surface, but bends sharply towards the membrane interior at its first CH_2 group (Fig. 1). This conformation was derived from single-crystal X-ray diffraction studies (Hitchcock et al., 1974;

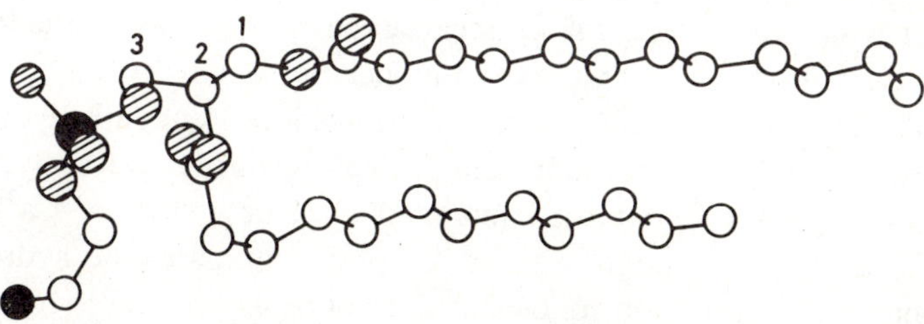

Figure 1. Conformation of a phospholipid molecule. The open circles represent carbon atoms; the shaded ones, oxygen atoms. The doubly hatched circle represents a nitrogen atom; the filled circle, a phosphorus atom. Atoms 1, 2, and 3 are from glycerol. (After Hitchcock et al., 1974.)

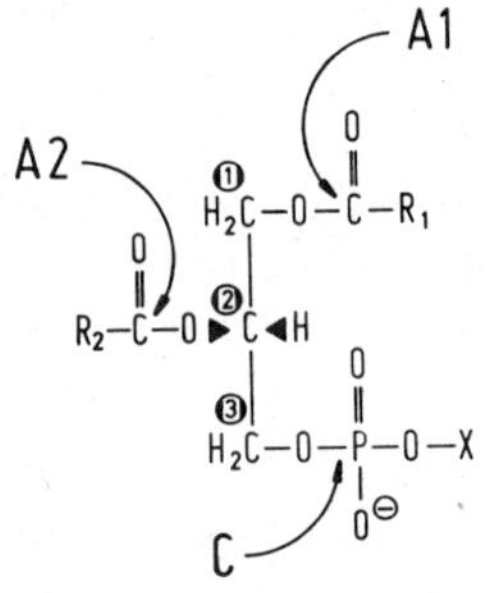

Figure 2. A phospholipid molecule in the *sn*-3 configuration. R1 and R2 are alkyl chains. X represents any of the naturally occurring substituents (such as, choline, ethanolamine, glycerol, and serine) or a proton. The arrows indicate the site of attack by the phopholipases A1, and A2, and C.

Pearson and Pascher, 1979). ^{2}H NMR and neutron diffraction data suggest that the same conformation as in the single-crystal structure is also present in the liquid crystalline state, and that it is a predominant conformation of phospholipids in fluid membranes (Seelig and Seelig, 1974, 1975; Seelig and Browning, 1978).

The hydrolysis of the various ester bonds in phospholipids is catalyzed by different enzymes (Fig. 2). Phospholipase A1 is specific in removing the fatty acid at C1, while phospholipase A2 does the same with the fatty acid at C2. The hydrolysis of the phosphate group from the glycerol is catalyzed by phospholipase C. The conformation of the phospholipid molecules as they occur in membranes and micelles is such that the targets for phospholipase A2 and C can be reached from the water phase and, consequently, most of these enzymes are water soluble. Because the ester bond with the first fatty acid chain is buried in the lipid layer, the specific cleaving enzyme, phospholipase A1 must penetrate this layer. The enzyme is consequently hydrophobic in character, insoluble in water, and difficult to purify. Recently, interest in phospholipase C has been stimulated by a new view of Ca^{2+}-linked receptor action. Activation of these receptors stimulates the hydrolysis of phosphatidylinositol-4,5-bisphosphate by a phospholipase C to form diacylglycerol and inositol trisphosphate, both of which act as second messengers that evoke the cell's responses. Structural work on phospholipase C is in progress, but has not yet led to a high-resolution structure.

1.3. The Phospholipases A2

Phospholipase A2 is the most extensively investigated of all phospholipases. It occurs in small quantities as an intracellular protein in virtually all cells and tissues, but the extracellular phospholipases A2 are far more abundant. Particularly rich sources of the secretory phospholipases are mammalian pancreatic tissue and juice, and the venoms of bees and a great variety of snakes.

The intracellular phospholipases are involved in physiologically important processes because in response to different stimuli, they release arachidonic acid from phospholipids. Arachidonic acid is the starting compound for the synthesis of prostaglandins, prostacyclin, thromboxanes, and leukotrienes. Unfortunately, the low concentration of the endocellular phospholipases A2 have hampered detailed studies, but the enzymes do seem to have the same essential properties as the more abundant pancreatic and venom enzymes (Crews et al., 1981; McGivney et al., 1981; De Winter et al., 1982; Gray and Strickland, 1982).

1.3.1. Molecular Properties

Phospholipases A2 from pancreatic and venom sources are well characterized. They are highly stable with respect to heat, variations in pH, and other denaturing conditions. Consequently, the secretory phospholipases A2 are fairly simple to isolate and purify (Slotboom et al., 1982). The enzymes are relatively small, with one polypeptide chain of 120−130 amino acid residues. The chain contains at least 14 cysteines connected in 7 disulfide bridges, which undoubtedly contributes to the high stability of the molecules. The amino acid sequences of approximately 40 phospholipases A2 have now been determined, and it appears that one-quarter of the residues are absolutely conserved; and for another one-quarter, only conservative replacements have occurred (Slotboom et al., 1982). Therefore, the overall three-dimensional structures of all these phospholipases A2 are likely to be very similar, although local differences could occur and do occur (see Sections 2.3 and 2.4). In Figure 3, the amino acid sequence of bovine pancreatic (pro)phospholipase A2 (Fleer et al., 1978) is presented as a representative for the whole family. The bee venom phospholipases A2 form an exception because although their sequence has similar features, it differs too greatly from all other phospholipases A2 to allow a homology comparison (Shipolini et al., 1974a, and 1974b).

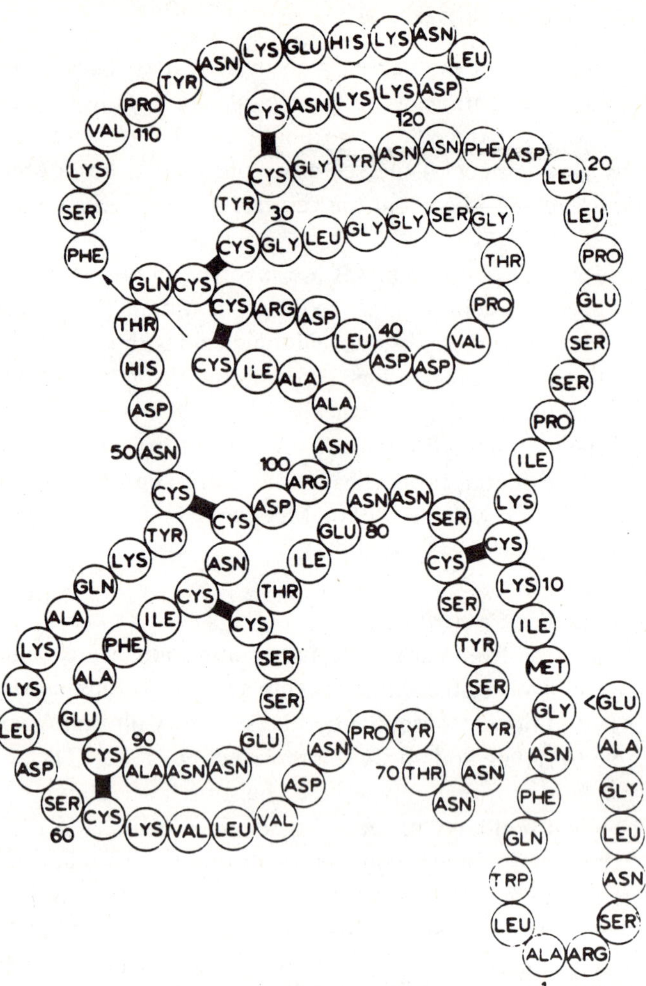

Figure 3. The amino acid sequence of bovine pancreatic prophospholipase A2 (Fleer et al., 1978). Ala 1 is the first residue in the active enzyme after removal of the extra seven residues in the precursor. (Reprinted by permission from p. 366, Hawthorne and Ansell, 1982.)

The venom enzymes tend to form dimers (Wells, 1971; Shen et al., 1975) or higher order aggregates in solution, whereas the pancreatic enzymes exist as monomers. Another difference between the two groups of enzymes is that the venom gland secretes the native enzyme and the pancreas secretes a zymogen. The prophospholipase A2 differs from the active enzyme in having an extra heptapeptide at the N-terminus. A typical purification procedure for prophospholipase A2 is given by Volwerk and De Haas (1982). The pure zymogen can be converted into the active enzyme by limited proteolysis with trypsin and by a subsequent purification step. Purification of the venom enzymes is described in Wells and Hanahan (1969) and in Joubert and Van der Walt (1975).

1.3.2.   Kinetic Studies

Phospholipase A2 is highly stereospecific in its hydrolysis of *sn*-3-phospholipids. Van Deenen and De Haas (1963) found that the C1 atom of the glycerol and the fatty acid chain attached to it is not essential in the catalytic process. The minimal substrate requirements include one fatty acid ester bond in a position adjacent to the alcohol-phosphate ester group and a well-defined stereochemical orientation in space of the adjacent acyl chain. (Fig. 2). Ca^{2+} is also an absolute requirement in the catalytic process. Ba^{2+} and Sr^{2+} were found to be fully competitive with Ca^{2+}, but do not induce enzymatic activity (Pieterson et al., 1974a). However, some transition metal ions, such as Gd^{3+}, can replace Ca^{2+} with retention of some activity (Hershberg et al., 1976).

The action of phospholipase A2 can be considered as an example of heterogeneous catalysis, because its natural substrate is insoluble in water and is present as membranes, micelles, liposomes, or vesicles, whereas the enzyme acts from the water phase. It is difficult to use the natural long chain phospholipids as substrates in kinetic studies. Therefore, Pieterson et al. (1974b) utilized artificial short chain phospholipids in their kinetic studies of pancreatic phospholipases A2. A remarkable dependence of the enzymatic activity on the physical state of the substrate was found (Fig. 4). Below the critical micelle concentration (CMC), where the substrate is in monomeric form or present as small clusters, the enzyme displays a moderate activity. Above the CMC, the enzymatic activity rises sharply. The proenzyme, on the other hand, has a low activity on either side of the CMC. Subsequent studies have demonstrated that the difference in catalysis between the active enzyme and the precursor is due to the failure of the zymogen to bind to lipid-water interfaces (Pieterson et al., 1974b; Van Dam-Mieras et al., 1975; De Haas et al., 1978).

The behavior of the enzyme towards a monomolecular surface film of phospholipid molecules on water has also been studied (Verger and De Haas, 1973; Verger et al., 1976). The packing density of the substrate

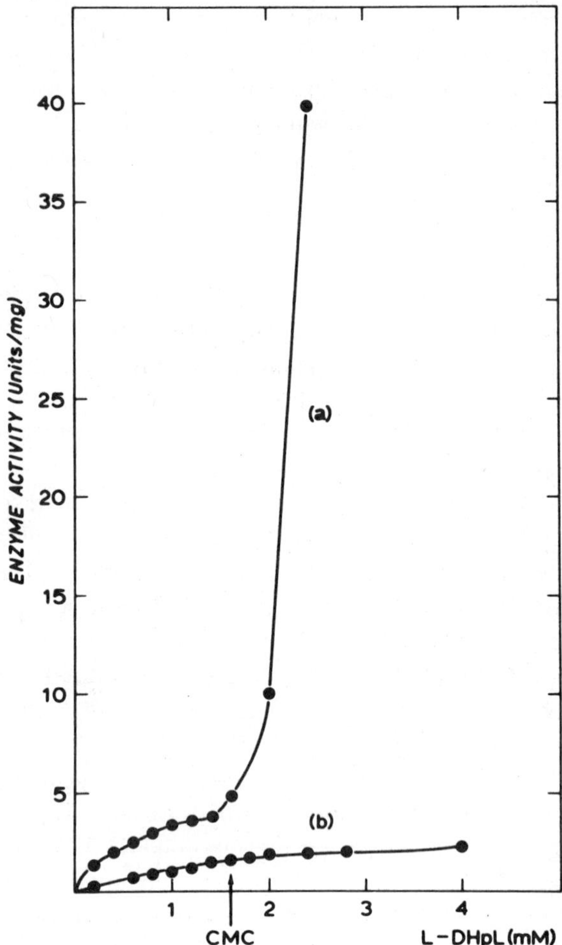

Figure 4. The activity of porcine pancreatic phospholipase A2 (*a*) and its zymogen (*b*) as a function of the substrate concentration. L-DHpL is the abbreviation for 1, 2-diheptanoyl-*sn*-3-glycerophosphocholine and CMC is for critical micelle concentration. (Reprinted with permission from p. 1456, Pieterson et al., 1974b.)

molecules in the film has a dramatic effect on the enzymatic activity. Above a well-defined surface pressure, the activity drops suddenly. Apparently, a dynamic interaction between the enzyme and its substrate is no longer possible. The venom enzymes are usually active to a much higher surface pressure than the enzymes from the pancreas. The latter ones prefer loosely packed phospholipid molecules, such as in micelles, but they are unable to attack membranes in which the phospholipids are more tightly packed.

From studies on phospholipase A2, De Haas and colleagues concluded that two topographically distinct sites play a role in the catalytic process. The two sites include the active center where the ester bond is hydrolyzed, and the "interface recognition site" (IRS) where binding to the phospholipid surface occurs (Verger et al., 1973; Pieterson et al., 1974b; Van Dam-Mieras et al., 1975). It was suggested that the pancreatic enzyme and its precursor have the same active center, but the zymogen lacks an intact IRS, and therefore cannot bind to a phospholipid surface. It was further shown that His 48 is part of the active center, because p-bromophenacyl bromide reacts specifically with this residue, resulting in an irreversibly inactivated enzyme, which still binds to micellar substrates with the same affinity as the native enzyme (Volwerk et al., 1974; Pieterson et al., 1974b). This proves the separate location of the active center and the IRS. Because Ca^{2+} protects the enzyme against inactivation by p-bromophenacyl bromide (Pieterson et al., 1974a), the Ca^{2+}-binding site and His 48 are probably in close proximity. Dennis introduced a dual phospholipid model for the cobra venom enzyme in which the enzyme binds to more than one phospholipid molecule in order to function more effectively (Dennis, 1983).

2.   STRUCTURAL RESULTS

2.1.   Three-Dimensional Structure

The catalytic action of an enzyme can only be fully understood if its three-dimensional structure is known. Electron microscopy is the most direct way to observe the structure of a protein molecule. Unfortunately, the resolution that can be obtained is usually limited to about 20 Å and only in favorable cases, can it be extended to 10 Å. This is excellent for studying the organization of very large biological structures, but it is impossible to obtain details at the atomic level. Nuclear magnetic resonance (NMR) and extended X-ray absorption fine structure (EXAFS) are two spectroscopic methods that can provide information on the structural details. The limitation in using NMR is the size of the protein molecule: the molecular weight should be less than 10,000 in order to distinguish and identify the peaks in the NMR spectrum.

Phospholipase A2, with a molecular weight of 14,000, is just too big for a full structure determination with NMR. However, local structural details, such as accessibility of aromatic rings, can be studied quite conveniently in the larger biological structures with NMR.

In EXAFS, the X-ray absorption spectrum near the absorption edge of a metal ion is observed. It provides information on the local structure around the metal ion, which is especially important if the metal ion is in the active site of an enzyme. However, extensive structural information cannot be obtained with this method.

X-ray crystallography leads to a protein structure in atomic detail, provided one major requirement is met: *the protein must be crystallized as single crystals with a size of 0.2−0.5 mm.* Because X-rays are scattered by electrons in the structure, X-ray crystallography results in an electron density map that is interpreted in terms of the polypeptide main chain, amino acid side chains, solvent molecules, and other structural details. Depending upon the quality of the crystals and the X-ray pattern, a resolution of 1.5−2.5 Å can be obtained. This does not imply that the accuracy in the atomic coordinates is limited to 1.5−2.5 Å. The accuracy is approximately 10 times greater due to the fact that in the interpretation and refinement of the structure, one can take advantage of the chemical and structural information already known, such as the amino acid sequence, atomic bond lengths, and bond angles.

2.2. X-Ray Diffraction Results

At present, structures of the following pancreatic A2 phospholipases have been determined: the active bovine enzyme at 1.7 Å resolution, the active porcine enzyme at 2.6 Å, the bovine proenzyme at 3.0 Å and a transaminated bovine enzyme at 2.1 Å in which the terminal amino group is replaced by an oxygen atom.

All of these structures were solved by Drenth and colleages in Groningen. Sigler and his coworkers in Chicago have determined the structure of a dimeric venom phospholipase A2 from western Diamond-backed rattlesnake (*Crotalus atrox*) at 2.5 Å resolution. The structure of the active bovine and porcine pancreatic enzymes will be discussed first, and then the venom enzyme. In the final section of this review, a relation between the enzyme's structure and the mechanism of the catalytic process is proposed.

2.3.   The Structure of Phospholipase A2 From The Bovine And Porcine Pancreas

The bovine and porcine pancreas enzymes were isolated by De Haas and coworkers in Utrecht according to well-established procedures (De Haas et al., 1968; Dutilh et al., 1975). Crystallization of the bovine enzyme was accomplished as follows: freeze-dried protein was dissolved in Tris buffer (100 mM Tris, 5 mM CaCl$_2$, pH 7.6) to a final concentration of 3% (w/v). Then 50 μl of the protein solution was frozen at $-18°$C and 50 μl of 2-methyl-2,4-pentanediol was layered on top of the frozen protein solution (Dijkstra et al., 1978). After 1$-$2 weeks at room temperature, crystals had grown with sizes up to 0.5 $\times$ 0.5 $\times$ 0.8 mm. They belong to the ortho-rhombic space group P2$_1$2$_1$2$_1$ with cell dimensions a = 47.07 Å, b = 64.45 Å, and c = 38.15 Å, with four molecules in the unit cell. As is usual for protein crystals, a large part of the crystal volume (39%) is occupied by solvent. The porcine enzyme was crystallized from 0.05 M Trismaleate buffer at pH 7.2 with 5 mM CaCl$_2$. Added to 50 μl of a 1% w/v enzyme solution was 10$-$15 μl of methanol. The crystals belong to space group P3$_1$21 with a = b = 69.82 Å and c = 67.66 Å (Dijkstra et al., 1983).

2.3.1.   The Molecular Structure

In the first stage, the structure of the bovine enzyme was determined to a resolution of 2.4 Å (Dijkstra et al., 1978). The peptide chain was traced easily and the side chains were identified in the electron density map. Seven disulfide connections, two of which had not been found chemically, were determined unambiguously. The high quality of the X-ray pattern allowed an extension of the data to 1.7 Å resolution. Crystallographic refinement of the structure at this resolution was carried out by Dijkstra et al. (1981a) and reduced the errors in the atomic coordinates down to a r.m.s. value of 0.12 Å. Moreover, solvent molecules were positioned more reliably and information was obtained on the disorder or motional freedom in the structure.

The improvement of the model is evident from a comparison of Ramachandran plots (Ramakrishnan and Ramachandran, 1965) before and after the refinement (Fig. 5). The main chain dihedral angles of most amino acids have shifted to or near the fully allowed regions. The five exceptions are glycines, which are residues that can occupy a larger range in the Ramachandran plot because they lack a side chain. These residues include

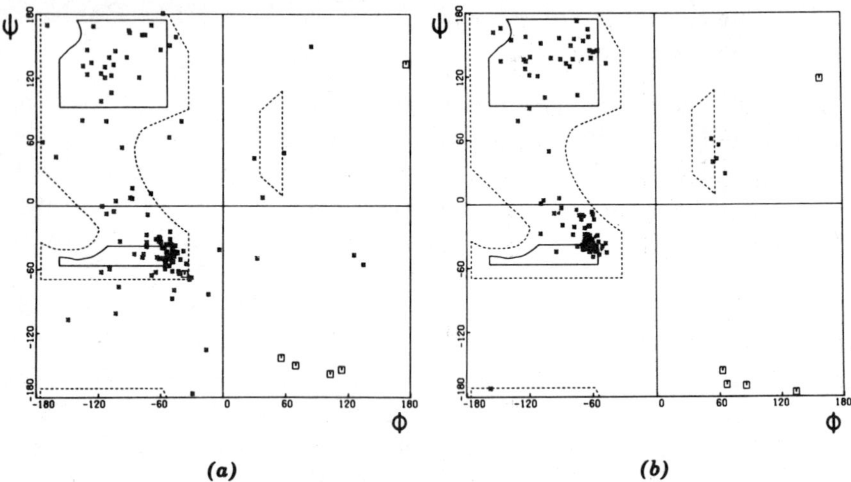

Figure 5. Ramachandran plot of the polypeptide chain conformation in bovine phospholipase A2 (*a*) before refinement and (*b*) after refinement. Glycine is denoted by open squares; all other residues, by asterisks. Continous lines enclose regions with fully allowed conformations for τ (N,Cα, C) = 110°, and dashed lines enclose the outer limit boundary for τ (N, Cα, C) = 115°. (These lines are taken from Ramakrishnan and Ramachandran, 1965. Reprinted by permission from *J. Mol. Biol.* **147**, 108; copyright 1981, Academic Press, London.)

Gly 26, 30, 32, 33, and 35. Substitution of these glycines by any other amino acid residue, while maintaining the chain-folding pattern, would be energetically highly unfavorable. The five glycines are found in all enzymatically active pancreatic and snake venom phospholipases A2 sequenced to date, giving further support to the assumption that all phospholipases A2 have a close structural similarity. From the venom of the Australian tiger snake, *Notechis scutatus scutatus*, a phospholipase A2 type of protein was isolated that shows the usual homology in its amino acid sequence except for the replacement of Gly 30 by a Ser. Very interestingly, this protein is devoid of enzymatic activity (Lind and Eaker, 1980), which may be due to a different conformation for its calcium-binding loop (Fig. 6). A sixth glycine residue in the bovine enzyme at position 7 is incorporated into an α helix and is usually replaced by another type of residue in the other pancreatic and venom phospholipases A2; for example, by a Ser in the porcine enzyme (Puijk et al., 1977).

The structure of the porcine enzyme has been refined at a resolution of

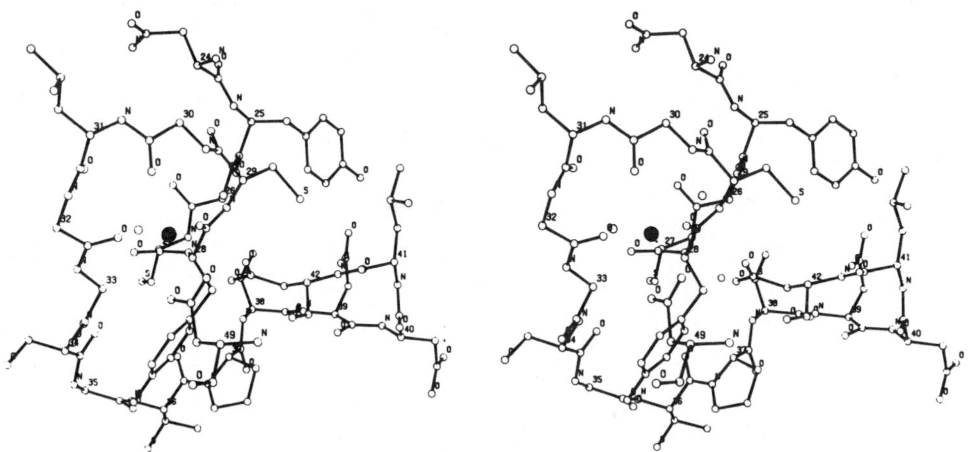

Figure 6. A stereo pair showing the calcium binding loop. The black dot represents the Ca^{2+}. Nonconnected open circles represent water molecules. (Reprinted by permission from *J. Mol. Biol.* **147**, 116; copyright 1981, Academic Press, London.)

2.6 Å, and it appears to be very similar to the structure of the bovine enzyme except for small differences near residues 31 and 33 and near the C-terminus, which is caused by an insertion. There are somewhat larger differences in the loop comprising residues 59–70 (Dijkstra et al., 1983).

The bovine enzyme molecule is kidney shaped, and has dimensions of approximately 22 × 30 × 42 Å. It has an appreciable content of secondary structure: 50% of the residues are in an α helical conformation and 10% in an antiparallel 2-stranded piece of β structure. The core of the molecule consists of two antiparallel α helices, each having five full turns (C and E in Fig. 7). The N-terminal region is also in an α helical conformation (helix A) with the N-terminal NH_3^+ of Ala 1 buried in the interior of the protein. Helices B and D are short α helices of approximately one turn only (Fig. 8). In the bovine enzyme, helix D runs from residues 59–64. Surprisingly, this helix is not present in the porcine enzyme, where instead residues 67–74 form 1.5 turns of a 3_{10} helix. This difference is probably caused by the replacement of Val at position 63 in the bovine enzyme by a Phe in the porcine enzyme. Whereas the side chain of Val 63 in the bovine enzyme is at the surface of the molecule, the Phe 63 side chain finds a position in the molecule's interior. The r.m.s. deviation between the positions of the α-

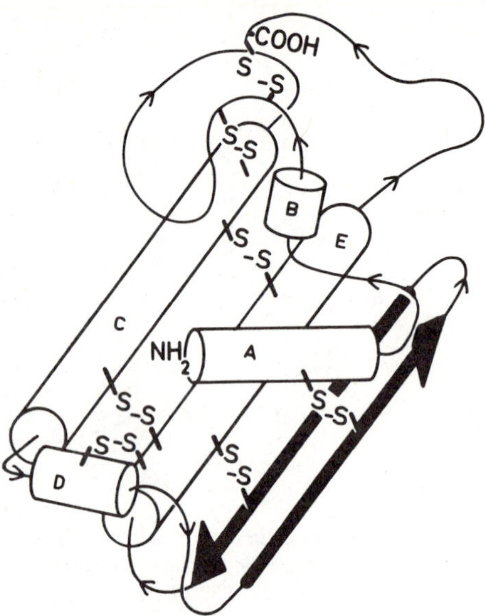

Figure 7. A schematic representation of the phopholipase A2 molecule. A−E are helical regions. The arrows indicate the double-stranded piece of β sheet. (Reprinted by permission from *J. Mol. Biol.* **124**, 57; copyright 1978, Academic Press, London.)

carbon atoms of the 12 residues from 59−70 in both enzymes is 4.7 Å, whereas for the 112 equivalent α-carbon positions, it is 0.47 Å (59−70 and 120−121 excluded).

Two antiparallel β strands are at the surface of the molecule from residues 74−78 and 81−85. As seen in Figure 7, all parts of the molecule are knitted together by seven disulfide bonds that certainly stabilize the molecule. In the high-resolution bovine structure, it was found that staggered conformations are preferred for the side chains with χ_1 angles near 60°, 180°, and 300°. This is quite common in protein structures. Neutron diffraction studies on crambin (Kossiakoff and Shteyn, 1984) have shown that even methyl rotations are quantized in 120° steps.

2.3.2.   The Active Site

The active site is located in a cavity at the molecular surface. His 48 is located at the bottom of this cavity. Volwerk et al. (1974) have shown that this residue plays a role in the catalytic mechanism because its modification

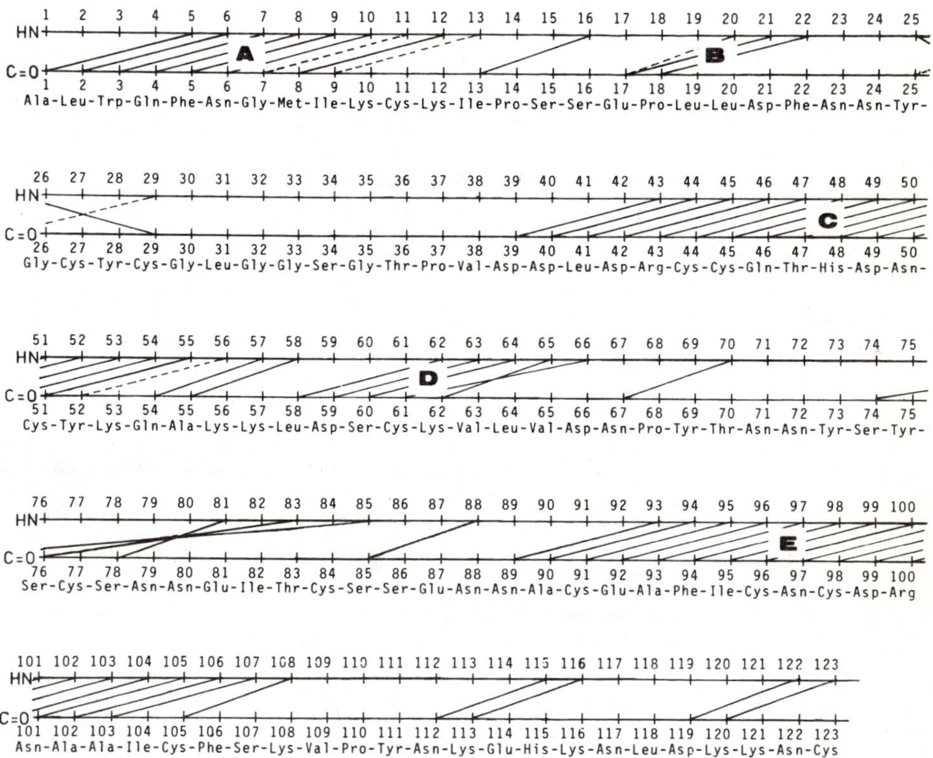

Figure 8. Secondary structure in bovine phospholipase A2. Hydrogen bonds are indicated. Dashed lines represent possible hydrogen bonds. (Reprinted by permission from *J. Mol. Biol.* **147**, 111; copyright 1981, Academic Press, London.)

destroys enzymatic activity. The imidazole ring of His 48 forms a hydrogen bond with the carboxyl group of Asp 99 that is similar to what is found in the proteolytic serine enzymes. The wall of the active site is covered by hydrophobic residues: Phe 5, Ile 9, Phe 22, Ala 102 and 103, Phe 106, and the disulfide bond between Cys 29 and 45. Also in the active site, near Asp 49, is the position for the essential Ca^{2+} (Fig. 9). It is surrounded by seven oxygen ligands (Fig. 10). Five ligands are at the corners of an octahedron at an average distance of 2.39 Å from the Ca^{2+}. The sixth octahedral position is occupied jointly by one of the Asp 49 carboxyl oxygen atoms ($O\delta_1$ at 2.69 Å) and a water molecule at 2.66 Å from the Ca^{2+}. The presence of only one full negative charge in the ligand sphere of a Ca^{2+} is not a common situa-

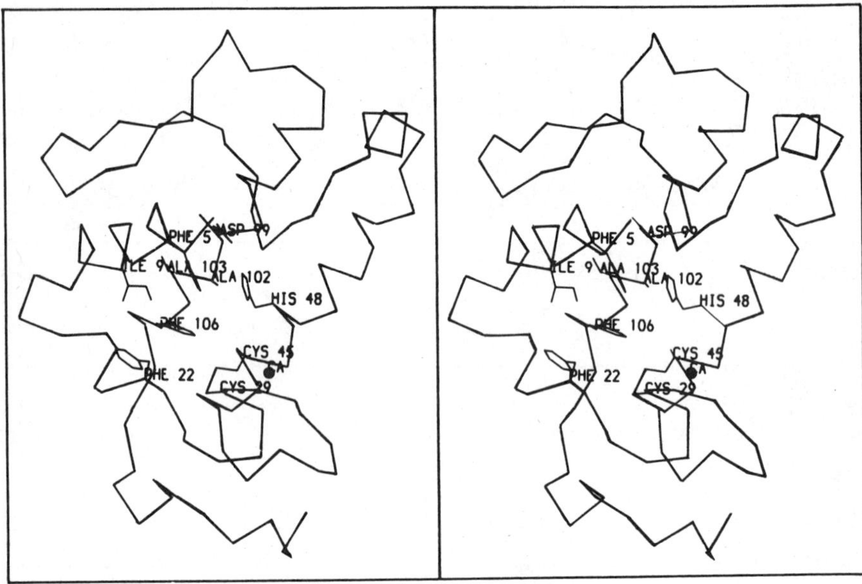

Figure 9. A stereo pair showing the essential active site residues. The black dot represents the Ca^{2+}. The view is directed towards the active site. (Reprinted by permission from *Nature*, **289**, No. 5798, 605; copyright 1981, Macmillan Journals Limited.)

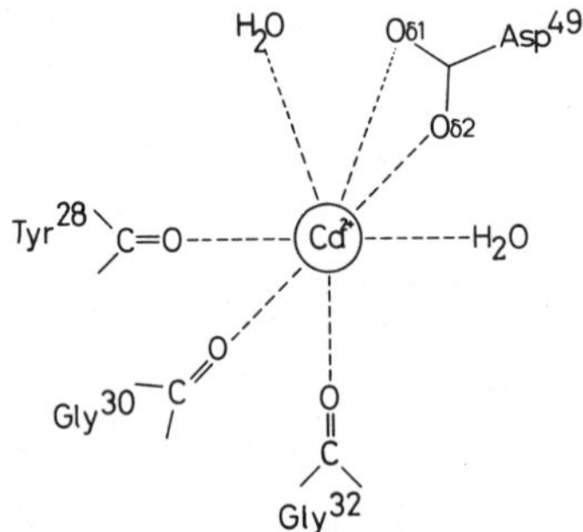

Figure 10. Schematic representation of the ligandation of the essential Ca^{2+}. (Reprinted by permission from *J. Mol. Biol.* **147**, 115; copyright 1981, Academic Press, London.)

302

tion. That it does occur in phospholipase A2 may point to a role for the Ca^{2+} in binding the substrate's phosphate group.

The main chain oxygen ligands of the Ca^{2+} belong to residues 28, 30, and 32. They are part of the calcium-binding loop that runs from residues 25–42 (Fig. 6). Many residues in this sequence are completely or nearly completely invariant in the phospholipase A2 sequences published to date (Slotboom et al., 1982). They make many hydrogen bonds with each other. In addition, the calcium-binding loop is stabilized by two disulfide bonds, Cys 27 with 123 and Cys 29 with 45. The stabilization of the calcium-binding loop guarantees a precise geometry of the Ca^{2+} with respect to the active site residues. In addition to the Ca^{2+} bound in the active site, the presence of a second calcium-binding site has been reported for porcine phospholipase A2 (Slotboom et al., 1978). This second Ca^{2+} is only weakly bound. In the crystal structure, it is located away from the active site on a crystallographic dyad, and two symmetry-related molecules contribute to its coordination sphere. Thus, there is only half of a Ca^{2+} per enzyme molecule at this site in the crystal structure. The metal ion is coordinated by at least four ligands: two carbonyl oxygen atoms of Ser 72 and two Oε1 atoms of Glu 92.

Recently, Maraganore et al. (1984) have isolated and sequenced monomeric phospholipases A2 from the venoms of *Agkistrodon piscivorus piscivorus* and *Bothrops atrox*. Although the sequences of these enzymes are, in general, homologous with the other phospholipases A2, they show interesting deviations from the usual pattern of conservatively occupied positions. The most intriguing one is the replacement of the calcium ligand Asp by a Lys. As a consequence, in the absence of substrate, these venom enzymes hardly bind Ca^{2+}. Moreover, they have a higher affinity towards aggregated substrate than their Asp 49 counterparts when calcium is not present. The simplest explanation is that the positively charged Lys side chain fulfils the same function as the Asp-calcium combination, that is, the binding of the substrate's phosphate group in the active site. However, more information is required to definitely state the function of residue 49, Asp or Lys in phospholipase A2.

The active sites of the bovine and porcine enzymes are identical within the accuracy of the molecular models. A slight difference exists for residue 31 in the calcium-binding loop. The α-carbon atoms of these residues are 1.06 Å apart in the models for unknown reasons. Both enzymes share the property of a relatively low mobility of the active site residues. Their crystallographic temperature factors are about half of the average temperature factor of all protein atoms.

Although not part of the active site proper, Ala 1 is of extreme importance for the activity of the enzyme on micellar substrates. A loss of the α-NH$_3^+$ group (Pieterson et al., 1974a, 1974b) and even the replacement of L-Ala by D-Ala (Slotboom et al., 1977) results in an enzyme inactive on these substrates. In the molecular structure, Ala 1 has its amino group buried in the interior of the enzyme, and this amino group has various hydrogen bond interactions (Fig. 11):

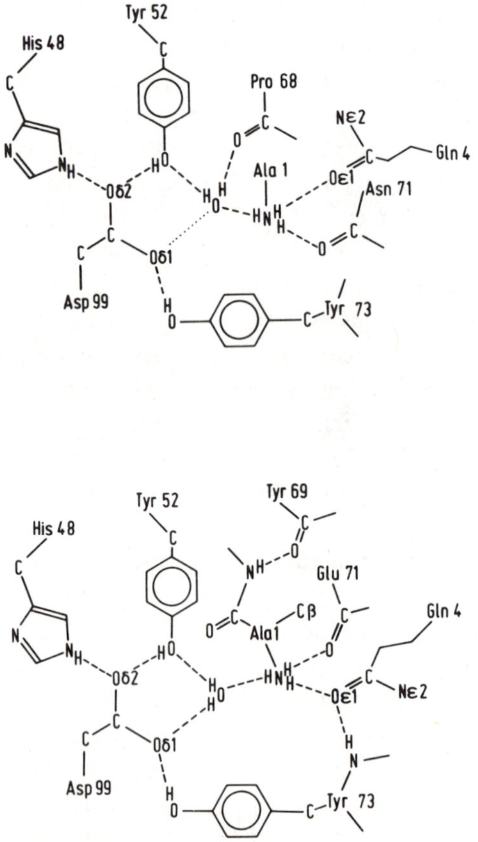

Figure 11. Hydrogen-bonding networks around the N-terminus of bovine (upper) and porcine (lower) phospholipase A2. (Reprinted by permission from *J. Mol. Biol.* **168**, 175; copyright 1983, Academic Press, London.)

1. With the side chain of Gln 4;

2. With the carbonyl oxygen of residue 71; and

3. With an internal water molecule that provides a link with several active site residues.

In the bovine and porcine enzymes, a similar hydrogen-bonding network is present around the internal water molecule, but they differ in the finer details. In the porcine enzyme, the N-terminus is somewhat more open to the solvent than in the bovine enzyme.

When the terminal amino group of Ala 1 is absent (as in the precursor) or replaced by an oxygen atom (as in the transaminated enzyme), or when it is not in its internal position (L-Ala → D-Ala), the hydrogen-bonding network is disrupted. This has a profound effect on the mobility of the main chain in the 59–70 region, and will be discussed in Section 3.

2.3.3. The Interface Recognition Site

To explain the interaction of the phospholipase A2 molecule with aggregated substrates such as micelles, De Haas and coworkers proposed the presence of a special binding site distinct from the active site (Verger et al., 1973; Pieterson et al., 1974a, 1974b; Van Dam-Mieras et al., 1975). From kinetic and amino acid modification experiments, it was possible to assign several residues to this interface recognition site (IRS) (Volwerk and De Haas, 1982). This was completed when the X-ray structure became available. The IRS residues form a broad ring around the active site. There are 20 of these residues: 2, 3, 6, 17, 19, 20, 23, 24, 31, 56, 65, 67, 69, 70, 72, 116, 117, 119, 121, and 123 (numbering for the bovine enzyme). In the porcine enzyme, the inserted Lys at position 121 is also part of the IRS as well as Leu 64, due to a different conformation of the 59–70 loop. None of the IRS residues is absolutely conserved in all phospholipases A2, and these differences may explain the different affinity of the various phospholipases A2 for aggregated substrates. In the porcine enzyme, the IRS has a more basic character than the bovine enzyme. This suggests a simple electrostatic explanation for its tighter binding to the phosphate groups in the surface layer of the substrate.

2.4. The Structure of *Crotalus atrox* Phospholipase A2

Sigler and coworkers have determined the three-dimensional structure of phospholipase A2 from the venom of *Crotalus atrox* (Keith et al., 1981).

They grow crystals from a dilute sodium fumarate buffer at pH 4.2 containing some polyethylene glycol (PEG) 6000. These crystals do not contain Ca^{2+}. They belong to space group $P2_12_12_1$ with a = 53.4 Å, b = 100.2 Å, c = 48.6 Å, and diffract to 1.4 Å. The interpretation of the electron density map at 2.5 Å resolution immediately showed the expected structural similarity between the venom and pancreatic enzymes. In the crystal structure, the venom enzyme exists as a dimer in which the protomers are related by a rotation of approximately 180°. The mean distance between the α-carbon positions after optimal superposition of the two protomers is 2.2 Å for the unrefined structure at 2.5 Å resolution. A striking feature in the structure is that the catalytic centers of the protomers face each other in the dimer. Even more striking is the absence of an obvious pathway for the substrate to enter the active sites.

In the pancreatic enzymes, the Ca^{2+} is liganded, in addition to two water molecules, to Asp 49 and three residues in the calcium-binding loop: 28, 30, and 32. The venom enzyme crystals do not contain Ca^{2+} and Asp 49 is now interacting with Lys 64 of the dyad-related protomer, thus stabilizing the dimer. Further stabilization of the dimer interaction is provided first by the calcium-binding loop because residues 27–34 penetrate deeply into the opposite protomer, and second, by polar interactions between Glu 6 of one monomer with Tyr 31 and His 34 of the related monomer; and finally, by hydrophobic interactions between Tyr 31 and Leu 2.

It is not known whether this intimate pair of protomers is equal to the dimer that is enzymatically active in solution. Neither is it known what effect Ca^{2+}, essential for activity, would have on the structure as it is found in the crystals. Further work is required to solve these problems and to explain fully the relation between structure and catalytic action of this venom phospholipase A2.

3.   THE CATALYTIC MECHANISM

Phospholipase A2 acts as a catalyst for the hydrolysis of the second ester bond in phospholipids. If the phospholipids are present in an aggregated form, as they usually are, the chemical reaction in the catalytic process is preceded by binding of the enzyme molecule to the surface of the phospholipid aggregate. Because the active center and binding site (IRS) are two distinct areas on the enzyme's surface, it is convenient to discuss them separately.

In organic chemistry, the mechanisms for ester and for peptide hydrolysis are closely related. Therefore, esterases and proteases may have common features in their active sites. Moreover, the presence of the His 48/Asp 99 couple in the active site of phospholipase A2 suggests a relationship with the proteolytic serine enzymes, although phospholipase A2 lacks the serine residue. Instead, phospholipase A2 has a water molecule in the position where the proteolytic enzymes have their serine OH. It lies in the plane of the imidazole ring at a distance of 3.1 Å from His 48 δ-N1. This water molecule is probably the nucleophile that attacks the ester bond in the substrate. If this is true, then ester hydrolysis by phospholipase A2 is chemically identical to the deacylation step in the serine protease mechanism, which also starts with nucleophilic attack of a water molecule on an ester bond. The difference is that in the serine proteases, the substrate is covalently linked to the enzyme. Substrate-binding studies with phospholipase A2 are required to find the exact orientation of the substrate in the active site. They should also provide us the location of the carbonyl group of the ester bond and what the role of the Ca^{2+} is. This ion has only one full negative charge in its ligand sphere, and therefore is available for binding the substrate's phosphate group. Figure 12, in a schematic form,

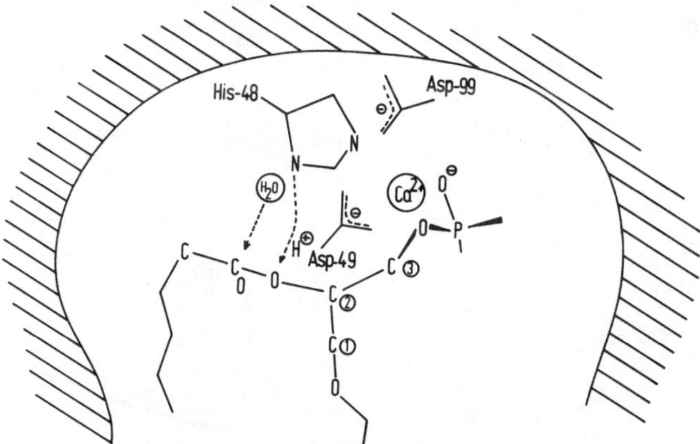

Figure 12. Schematic drawing of the proposed fitting of a substrate molecule in the active site of phospholipase A2.

gives the proposed position of a phospholipid molecule in the active site. The fatty acid chain at C2 of glycerol has the sharp bend near its first CH_2 group, which was found in structural studies on phospholipids in crystalline form and in membranes. The hydrophobic residues in the wall of the active site serve to hold the first 4−6 CH_2 groups of the fatty acid chains, but the exact orientation of these chains is not yet known.

One of the most striking features of phospholipase A2 is the enormous difference in activity towards phospholipid substrates below and above the critical micelle concentration (Fig. 4). This is true for the active enzyme, but only when it has a protonated α amino group at its first amino acid residue. If this α amino group has disappeared or is modified, as in the proenzyme and in the transaminated enzyme, the high catalytic activity towards micelles is lost together with the ability of the enzyme to bind to micelles. Therefore, binding, N-terminus, and high activity are closely related.

One can assume that the enzyme molecule binds to a micelle with its active site depression oriented towards the micelle surface. Figure 13 gives a

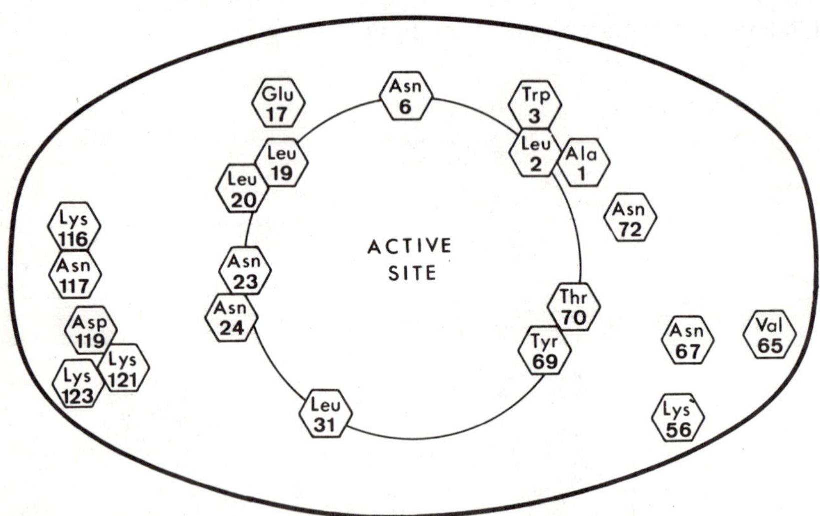

Figure 13. Schematic representation of the face of the molecule, which is directed towards the phospholipid layer. The residues in boxes, in the bovine enzyme, have their side chain pointing to the substrate layer. Residue 1 is incorporated because its free NH_3^+ is essential for binding. (Reprinted by permission from *Nature*, **289**, No. 5798, 605; copyright 1981, Macmillan Journals Limited.)

schematic representation of the face of the molecule, which comes in contact with the phospholipid layer. These should be the residues that together form the IRS. They are located in a circular region surrounding the active site. Two other regions are at the extreme left and right. The residues in the IRS are not conserved in all phospholipases A2, but their hydrophobic and polar character is usually maintained. It is also true that there is always an excess of basic over acidic residues in the IRS, suggesting salt bridge interactions with the phosphate groups on the micelle surface. This would orient the enzyme molecule when it approaches a micelle surface, with its active site directed towards the phospholipid surface, ready to pick up a substrate molecule. This preferred orientation could explain, at least partly, the high activity towards aggregated substrates (Dijkstra et al., 1981b).

An undisturbed N-terminus is required for the binding of phospholipase A2 to micelles. The role of the N-terminus became evident from the structures of the bovine pancreatic proenzyme and the transaminated enzyme. The proenzyme differs from the native enzyme by the presence of seven extra residues at the N-terminus (Fig. 3). The catalytic properties of precursor and native enzymes are quite similar with respect to monomeric substrates, but completely different towards aggregated substrates such as micelles. Contrary to the native enzyme, the precursor does not show the dramatic increase in activity above the critical micelle concentration.

The structure of the bovine proenzyme at 3 Å resolution (Dijkstra et al., 1982) was found to be very similar to the native enzyme structure. The active site residues are positioned in a similar fashion, and the Ca^{2+} is bound in the same way. This is in agreement with the observed similarities of precursor and native enzyme in their catalytic activity with respect to monomeric substrates. However, a striking difference between the two structures was the disorder in the proenzyme in two parts of the polypeptide chain. Whereas in the native enzyme the entire chain is quite rigid, in the proenzyme the N-terminus from residues -7 to $+4$ is so disordered that it does not show up in the electron density map. The same is true for the loop from $62-73$. A similar result was obtained for the 2.1 Å resolution structure of transaminated bovine phospholipase A2 (Dijkstra et al., 1984). In this derivative, the N-terminal residue Ala is converted into a pyruvoyl group, thus replacing the terminal NH_3^+ group of the native enzyme by an oxygen atom. It appears again that the derivative structure differs from the native enzyme structure in one respect only: disorder at the N-terminus (Ala 1, Leu 2, and the side chain of Trp 3) and in the loop from residue $63-72$. These residues form part of the IRS (Fig. 13).

The conclusion is that removal of the terminal NH_3^+ in the native enzyme causes this part of the IRS to become disordered or mobile, impairing the binding to micelles. The function of the terminal NH_3^+ group, therefore, seems to be to lock part of the IRS into a proper conformation required to bind to phospholipid aggregates. This is effected by the hydrogen-bonding network (Fig. 11), in which the terminal NH_3^+ is interacting directly with residue 4 and residue 71 and indirectly with residue 68 in the bovine enzyme and residue 73 in the porcine enzyme.

4. CONCLUSION

A combination of extensive biochemical and biophysical studies on phospholipase A2 has revealed the complete structure of the bovine and porcine pancreatic enzymes and of a venom enzyme. From these studies, a proposal for the catalytic mechanism is derived. Further studies are required to investigate the catalytic process in detail. The work should also be extended to the highly interesting, but difficult to isolate, tissue phospholipases A2, which catalyze the release of arachidonic acid as the starting compound for the synthesis of prostaglandins, thromboxanes, and leukotrienes.

REFERENCES

Crews, F. T., Morita, Y., McGivney, A., Hirata, F., Siraganian, R. P., and Axelrod, J. (1981). *Arch. Biochim. Biophys.* **212**, 561−571.

De Haas, G. H., Postema, N. M., Nieuwenhuizen, W. N., and Van Deenen, L. L. M. (1968). *Biochim. Biophys. Acta* **159**, 103−107.

De Haas, G. H., Slotboom, A. J., Verhey, H. M., Jansen, E. H. J. M., De Araujo, P. S., and Vidal, J. C. (1978). In *Advances in Prostaglandin and Thromboxane Research*, C. Galli, G. Galli, and G. Porcellati, Eds., Vol. 3. Raven Press, New York, pp. 11−21.

Dennis, E. A. (1983). In *The Enzymes*, P. D. Boyer, Ed., Vol. 16. Academic Press, New York, pp. 307−353.

De Winter, J. M., Vianen, G. M., and Van den Bosch, H. (1982). *Biochim. Biophys. Acta* **712**, 332−341.

Dijkstra, B. W., Drenth, J., Kalk, K. H., and Vandermaelen, P. J. (1978). *J. Mol. Biol.* **124**, 53−60.

Dijkstra, B. W., Kalk, K. H., Hol, W. G. J., and Drenth, J. (1981a). *J. Mol. Biol.* **147**, 97−123.

Dijkstra, B. W., Drenth, J., and Kalk, K. H. (1981b). *Nature* **289**, 604–606.

Dijkstra, B. W., Van Nes, G. J. H., Kalk, K. H., Brandenburg, N. P., Hol, W. G. J., and Drenth, J. (1982). *Acta Crystallogr.* **B38**, 793–799.

Dijkstra, B. W., Renetseder, R., Kalk, K. H., Hol, W. G. J., and Drenth, J. (1983). *J. Mol. Biol.* **168**, 163–179.

Dijkstra, B. W., Kalk, K. H., Drenth, J., De Haas, G. H., Egmond, M. R., and Slotboom, A. J. (1984). *Biochemistry* **23**, 2759–2766.

Dutilh, C. E., Van Doren, P. J., Verheul, F. E. A. M., and De Haas, G. H. (1975). *Eur. J. Biochem.* **53**, 91–97.

Fleer, E. A. M., Verheij, H. M., and De Haas, G. H. (1978). *Eur. J. Biochem.* **82**, 261–270.

Gray, N. C. C., and Strickland, K. P. (1982). *Can. J. Biochem.* **60**, 108–117.

Hawthorne, J. N., and Ansell, G. B., Eds. *Phospholipids.* Elsevier Biomedical Press, New York, p. 366.

Hershberg, R. D., Reed, G. H., Slotboom, A. J., and De Haas, G. H. (1976). *Biochemistry* **15**, 2268–2274.

Hitchcock, P. B., Mason, R., Thomas, K. M., and Shipley, G. G. (1974). *Proc. Natl. Acad. Sci. USA* **71**, 3036–3040.

Joubert, F. J., and Van der Walt, S. J. (1975). *Biochim. Biophys. Acta* **379**, 317–328.

Keith, C., Feldman, D. S., Deganello, S., Glick, J., Ward, K. B., Jones, E. A., and Sigler, P. B. (1981). *J. Biol. Chem.* **256**, 8602–8607.

Kossiakoff, A. A., and Shteyn, S. (1984). *Nature* **311**, 582–583.

Lind, P., and Eaker, D. (1980). *Eur. J. Biochem.* **111**, 403–409.

Maraganore, J. M., Merutka, G., Cho, W., Welches, W., Kézdy, F. J., and Heinrikson, R. L. (1984). *J. Biol. Chem.* **259**, 13839–13843.

McGivney, A., Morita, Y., Crews, F. T., Hirata, F., Axelrod, J., and Siraganian, R. P. (1981). *Arch. Biochem. Biophys.* **212**, 572–580.

Pearson, R. H., and Pascher, I. (1979). *Nature* **281**, 499–501.

Pieterson, W. A., Volwerk, J. J., and De Haas, G. H. (1974a). *Biochemistry* **13**, 1439–1445.

Pieterson, W. A., Vidal, J. C., Volwerk, J. J., and De Haas, G. H. (1974b). *Biochemistry* **13**, 1455–1460.

Puijk, W. C., Verheij, H. M., and De Haas, G. H. (1977). *Biochim. Biophys. Acta* **492**, 254–259.

Ramakrishnan, C., and Ramachandran, G. N. (1965) *Biophys. J.* **5**, 909–933.

Seelig, A., and Seelig, J. (1974) *Biochemistry* **13**, 4839–4845.

Seelig, A., and Seelig, J. (1975). *Biochim. Biophys. Acta* **406**, 1–5.

Seelig, J., and Browning, J. L. (1978). *FEBS Lett.* **92**, 41–44.

Shen, B. W., Tsao, F. H. C., Law, J. H., and Kézdy, F. J. (1975). *J. Am. Chem. Soc.* **97**, 1205–1208.

Shipolini, R. A., Callewaert, G. L., Cottrell, R. C., and Vernon, C. A. (1974a). *Eur. J. Biochem.* **48**, 465–476.

Shipolini, R. A., Doonan, S., and Vernon, C. A. (1974b). *Eur. J. Biochem.* **48**, 477–483.

Slotboom, A. J., Van Dam-Mieras, M. C. E., and De Haas, G. H. (1977). *J. Biol. Chem.* **252**, 2948–2951.

Slotboom, A. J., Jansen, E. H. J. M., Vlijm, H., Pattus, F., Soares de Araujo, P., and De Haas, G. H. (1978). *Biochemistry* **17**, 4593–4600.

Slotboom, A. J., Verheij, H. M., and De Haas, G. H. (1982). In *Phospholipids*, Hawthorne, J. N., and Ansell, G. B., Eds. Elsevier Biomedical Press, pp. 359–434.

Van Dam-Mieras, M. C. E., Slotboom, A. J., Pieterson, W. A., and De Haas, G. H. (1975). *Biochemistry* **14**, 5387–5394.

Van Deenen, L. L. M., and De Haas, G. H. (1963). *Biochim. Biophys. Acta* **70**, 538–553.

Verger, R., and De Haas, G. H. (1973). *Chem. Phys. Lipids* **10**, 127–136.

Verger, R., Mieras, M. C. E., and De Haas, G. H. (1973). *J. Biol. Chem.* **248**, 4023–4034.

Verger, R., Rietsch, J., Van Dam-Mieras, M. C. E., and De Haas, G. H. (1976). *J. Biol. Chem.* **251**, 3128–3133.

Volwerk, J. J., and De Haas, G. H. (1982). In *Lipid-Protein Interactions*, P. C. Jost, and O. H. Griffith, Eds., Vol. 1. John Wiley & Sons, New York, pp. 69–149.

Volwerk, J. J., Pieterson, W. A., and De Haas, G. H. (1974). *Biochemistry* **13**, 1446–1454.

Wells, M. A. (1971). *Biochemistry* **10**, 4074–4078.

Wells, M. A., and Hanahan, D. J. (1969). *Biochemistry* **8**, 414–424.

The Thiol Proteases: Structure and Mechanism

6

EDWARD N. BAKER
Department of Chemistry and Biochemistry
Massey University
Palmerston North
New Zealand

JAN DRENTH
Laboratory of Chemical Physics
University of Groningen
Groningen
The Netherlands

CONTENTS

1. INTRODUCTION
   1.1. Occurrence
   1.2. Molecular Properties
   1.3. Biological Roles

2. STRUCTURAL STUDIES
   2.1. Crystallographic Analyses of Papain and Actinidin
   2.2. Structural Organization
   2.3. Amino Acid Sequence Comparison, Papain and Actinidin
   2.4. Detailed Comparison of Three-Dimensional Structures
   2.5. The Active Site
   2.6. Flexibility of the Structure
   2.7. Structural Studies on Other Thiol Proteases

3. CATALYSIS AND MECHANISM
   3.1. The General Catalytic Mechanism
   3.2. The Catalytically Important Groups
   3.3. Electrostatic Effects
   3.4. Role of Asp 158
   3.5. Crystallographic Binding Studies
   3.6. The Catalytic Mechanism in Structural Terms
   3.7. Specificity

4. CONCLUSIONS

5. FUTURE DIRECTIONS

REFERENCES

1. INTRODUCTION

The thiol proteases comprise a group of enzymes whose proteolytic activity is dependent on the thiol group of a cysteine residue. They are widely distributed throughout living systems, being found in animals, plants, and bacteria, in which they serve a variety of roles. There is good evidence that most of them belong to a family of homologous enzymes similar in structure as well as in function. The purpose of this chapter is to describe their main structural features (and any specific differences), and then to use the latter, together with chemical, kinetic, and spectroscopic results, to discuss aspects of their biological activities.

1.1. Occurrence

The best known of the thiol proteases are those derived from plant sources, with the enzyme papain being regarded as the archetypal example. Thus, previous reviews on the thiol proteases (e.g., Glazer and Smith, 1971; Lowe, 1976; Brocklehurst et al., 1981a) have concentrated on the plant enzymes, particularly on papain. One reason for this emphasis is that the plant enzymes are fairly readily obtained in quantity, as certain fruits yield very large amounts. For example, papain is derived from the dried latex of the tropical papaya fruit (*Carica papaya*), with 180 g of latex giving more than 1 g of crystalline papain (Smith and Kimmel, 1960). Similarly, in the preparation of actinidin from kiwifruit (*Actinidia chinensis*), more than 300 mg of pure crystalline enzyme are obtained from 1 kg of fruit (McDowall, 1970). Again, because of these large yields, thiol proteases from plants are widely used industrially in processes as diverse as meat tenderizing, "chill haze" prevention in beer-making, and carpet manufacturing.

Other sources of plant thiol proteases include the pineapple, *Ananus comosus* (the enzyme bromelain); the latex of species of fig, *Ficus* spp. (ficin); roots or latex of milkweed, *Asclepias syriaca* (asclepain); and the latex in the Indian madar plant, *Calotropis gigantea* (calotropin and calotropain). In some cases, multiple enzymes can be isolated from the same source. At least three distinct enzymes, papain, chymopapain, and papaya peptidase A, have been characterized from papaya latex (Brocklehurst et al., 1981a; Brocklehurst and Salih, 1983), and the fraction known as chymopapain appears to contain a number of imperfectly differentiated enzymes (e.g., Brocklehurst, 1983; Khan and Polgar, 1983; Brocklehurst et al., 1984a; Polgar, 1984). Similarly, two groups of enzymes (asclepains A and B), each with a number of members, have been characterized from milkweed latex (Brockbank and Lynn, 1979); two groups (calotropins and calotropains) from madar plant latex (Abraham and Joshi, 1979; Pal and

Sinha, 1980); and separate enzymes from the juice and the stem of pineapples (Ota et al., 1964). These multiple enzymes from a single source are not simply degradation products since, in most cases, there are clearly established differences in amino acid sequences and/or physical properties.

The other main group of characterized thiol proteases comprises those derived from the lysosomes of animals, chiefly from organs such as liver, spleen, and kidneys. These are generally given the names cathepsins (note that not all of the lysosomal proteases are thiol proteases, though many of them are). Some have alternative names, denoting a specific biological function, for example, cathepsin N has also been known as collagenolytic cathepsin because it attacks the N-terminal peptides of collagen. Many of these thiol-dependent cathepsins, the best-characterized ones of which are listed in Table 1, have multiple forms, with different electrophoretic properties or different functional properties. Classification of the cathepsins from different sources is generally based on their activities and inhibition properties. Thus, the enzyme from bovine spleen, originally named cathepsin S (Locnikar et al., 1981), has been reclassified as a cathepsin L on this basis (Turk et al., 1983).

Finally, a number of bacterial thiol proteases have been identified, with the two best known being streptococcal proteinase (Liu and Elliott, 1971) and clostripain (Mitchell and Harrington, 1971), both of which are released into the culture filtrate during the growth of the appropriate bacterial strains (hemolytic streptococci and *Clostridium histolyticum*, respectively).

1.2. Molecular Properties

The best-characterized thiol proteases are listed in Table 1, together with published values for molecular weights and isoelectric points. Nearly all of the enzymes have estimated molecular weights in the range 20,000 to 28,000, which is close to the value for papain (23,350). Some of the enzymes are glycoproteins, notably both stem and fruit bromelain, chymopapain, some ficin variants, and calotropain FI among the plant enzymes, and most of the cathepsins. The isoelectric points are quite variable, probably reflecting the different biological environments in which the various enzymes operate and indicating that when their three-dimensional structures are compared, there are likely to be distinct differences on the molecular surface. All of the enzymes from papaya are very basic, as are all the ficins, calotropins, and stem bromelain, whereas actinidin, asclepains, and fruit bromelain are very acidic. On the other hand, the cathepsins all have isoelectric points much closer to neutrality, in the pH range 5.0–7.0. This is consistent with their lysosomal origin, with the pH in lysosomes being about 6.3 (Kirschke et al., 1977b).

Only a small number of the enzymes have molecular properties that distinguish them from papain. Among the plant enzymes, chymopapain differs in that it apparently has two free thiol groups, only one of which is enzymatically active (Baines and Brocklehurst, 1982a; Brocklehurst and Salih, 1983). It also has some differences in kinetic properties (Brocklehurst et al., 1980), but it probably does belong to the same structural family as the other plant enzymes. Of the mammalian thiol proteases, cathepsin B2 has a molecular weight of 52,000; but it consists of two approximately equal-sized subunits and has two thiol groups, so that it may, in fact, consist of two papain-like subunits. Calpain is the name given to a group of enzymes that are also referred to as calcium-dependent neutral proteases because of their requirement for Ca^{2+} for activation. Of the bacterial enzymes, clostripain, with an overall molecular weight of 55,000, has recently been found (Gilles et al. 1979) to exist as two chains (M_r 43,000 and 12,500) and to have no sequence homology with papain (Gilles et al., 1983). It is therefore probable that clostripain bears the same relationship to the other thiol proteases as does subtilisin to the mammalian serine proteases; that is, completely different molecular structure with any similarity of the active site probably due to convergent, rather than divergent, evolution. The other bacterial enzyme, streptococcal proteinase, appears to be more like the plant enzymes, with a similar molecular weight and some similarities in kinetic properties (Liu and Elliott, 1971). It does, however, differ in that (1) the active enzyme is derived from an inactive precursor of considerably higher molecular weight, 44,000 (Liu and Elliott, 1971), and (2) its amino acid sequence shows no detectable homology except in the immediate vicinity of the catalytically important Cys and His residues (Tai et al., 1976).

All of the plant enzymes listed in Table 1 are endopeptidases. Most of the cathepsins are also endopeptidases, although some (e.g., cathepsins B and L) have been reported to have bifunctional activities, being able to act as exopeptidases as well (Kalnitsky et al., 1983), while cathepsin B2 is primarily a carboxypeptidase. All share the property that their optimal activity is in the mildly acidic pH range 5.0–7.0. One exception is cathepsin N

Table 1. Thiol Proteases From Various Sources

Enzyme	Source	Molecular Weight	Isoelectric Point	References
		(a) Plant Enzymes (Glazer and Smith, 1971)		
Papain	*Carica papaya*	23,350	8.75	Glazer & Smith, 1971
Chymopapain(g)	*C. papaya*	35,000 26−28,000	10.1−10.6	Brocklehurst et al., 1981a Robinson, 1975
Papaya peptidase A	*C. papaya*	24,000	11.0	Brocklehurst et al., 1981a Robinson, 1975
Actinidin	*Actinidia chinensis*	23,500	3.1	McDowall, 1970 Carne & Moore, 1978
Ficin(g)	*Ficus* spp.	24−26,000	9.1	Kortt et al., 1974 Sugiura & Sasaki, 1974
Bromelain(stem)(g)	*Ananas comosus*	33,500 26−28,000	9.55	Glazer & Smith, 1971 Takahashi et al., 1983
Bromelain(fruit)(g)	*A. comosus*	31,000	4.6	Yamada et al., 1976
Asclepain A	*Asclepias syriaca*	23,000	3.1	Brockbank & Lynn, 1979
Asclepain B	*A. syriaca*	21,000	3.1	Lynn et al.,1980a
Calotropin DI	*Calotropis gigantea*	23,400	9.6	Pal & Sinha, 1980
Calotropain FI(g)	*C. gigantea*			Abraham & Joshi, 1979

(b) Mammalian Enzymes (Barrett and McDonald, 1980)

Cathepsin B(g)	Rat liver	23–26,000	4.9–5.3	Towatari et al., 1979
	Human liver	24–27,500	4.5–5.5	Barrett, 1973
Cathepsin B2(g)	Rabbit lung	52,000	5.0	Lones et al., 1983
Cathepsin H(g)	Rat liver	28,000	7.1	Kirschke et al., 1977a
	Human liver	28,000	6.0, 6.4	Schwartz & Barrett, 1980
Cathepsin L(g)	Rat liver	23–24,000	5.8–6.1	Kirschke et al., 1977b
	Bovine spleen[a]	23–25,000	6.3–6.9	Locnikar et al., 1981
Cathepsin I	Rabbit lung	26,000	5.9–6.5	Kalnitsky et al., 1983
Cathepsin N	Bovine spleen	18–20,000	6.55	Ducastaing & Etherington, 1978
	Human placenta	34,600	5.1	Evans & Etherington, 1978
Cathepsin T	Rat liver	33–35,000		Gohda & Pitot, 1981
Cathepsin P	Rat islets	31,500		Docherty et al., 1982
Calpain	Rat liver	90,000	4.5	Murachi et al., 1981

(c) Bacterial Enzymes

Streptococcal proteinase	Hemolytic streptococci	32,000	8.4	Liu & Elliott, 1971
Clostripain	*Clostridium histolyticum*	55,000	4.8–4.9	Mitchell & Harrington, 1971 Gilles et al., 1979

Enzymes known to be glycoproteins are indicated (g).
[a]Cathepsin L from bovine spleen was originally named cathepsin S.

(collagenolytic cathepsin), whose specific substrate can apparently be attacked only at low pH, thus leading to a low pH optimum of 3.5 (Evans and Etherington, 1978).

Early kinetic studies on papain showed a dependence on ionizable groups on the enzyme with apparent pK_a's of about 4.2 and 8.2 (Lowe, 1970), and similar results were obtained for others of the plant enzymes, that is, 4.3 and 8.5 for ficin (Holloway et al., 1971) and 3.8 and 8.1 for actinidin (Boland and Hardman, 1973). The precise interpretation of these values has been controversial (e.g., Lowe, 1976), but they are generally attributed to the presence of a cysteine-histidine interactive pair in the active site of each enzyme. More recent studies using reactivity probes have shown that such a Cys-His system appears to be a common feature of both plant and animal enzymes (e.g., Brocklehurst, 1982; Baines and Brocklehurst, 1982b; Willenbrock and Brocklehurst, 1985a). When taken together with the similar specificities shown by many of the enzymes (e.g., Glazer and Smith, 1971; Barrett and Kirschke, 1981; Lynn, 1983) and similar binding modes, as detected by Resonance Raman spectroscopy (Carey et al., 1983; Storer et al., 1984; Brocklehurst et al., 1984b), these observations indicate that they all share similar active site structures and similar mechanisms of action.

1.3. Biological Roles

The precise biological roles of many of the thiol proteases remain unknown. The plant enzymes tend to have quite broad specificities, attacking a variety of substrates. Their function may be to protect the ripening fruit against attack by insects, fungi, etc., but why such large quantities are produced and why multiple enzymes with very similar properties are produced by a given plant remains a mystery. The thiol proteases secreted by certain bacteria probably act as external digestive enzymes, breaking down protein to small peptides and amino acids that can then be assimilated by the bacteria. The cathepsins tend to have rather narrower activities than the plant enzymes, but collectively, they cover a very wide range of substrates. These thiol-dependent cathepsins account for a large part of the proteolytic activity found in lysosomes of mammalian tissues, and are considered to play an important role in intracellular protein degradation. Some are recognized in terms of specific functions, for example, collagenolytic cathepsin (cathepsin N) for its action on collagen (Evans and Etherington, 1978). Others are apparently involved in the post-translational processing of some biologically important molecules, for example, cathepsin P is thought to be the enzyme that converts proinsulin to insulin (Docherty et al., 1982). In general, the cathepsins remain confined within lysosomes; and, in fact in all external tissues, specific thiol protease inhibitors are found (Lenney, 1983), presumably to complex with any enzymes that leak out.

2. STRUCTURAL STUDIES

While the physical data and kinetic properties outlined in Section 1 suggest a broad similarity between most of the thiol proteases, very specific similarities have been demonstrated by structural studies on the two plant enzymes papain and actinidin. For each protein, the complete amino acid sequence has been determined (Mitchel et al., 1970; Carne and Moore, 1978), and the three-dimensional structure has been elucidated by X-ray crystallography and refined at high resolution (Baker, 1980; Kamphuis et al., 1985a). More recently, amino acid sequence data on other thiol proteases has added to this information, so that common structural features and structural variations are now more clearly delineated still. In this section, we shall compare papain and actinidin with respect to both sequence and three-dimensional structure, and then show how these results relate to other thiol proteases.

2.1. Crystallographic Analyses of Papain and Actinidin

Relevant details of the structural analyses of papain and actinidin are given in Table 2. The crystals used in the two analyses were grown at different pH values in totally different media (and, incidentally, with completely different crystal packing)—observations that are very important for subsequent comparison of the structures. In both cases, the initial structure analyses were by multiple isomorphous replacement methods at a resolution of 2.8 Å (Drenth et al., 1968, 1971; Baker, 1977), but the structures have since been refined at high resolution: 1.65 Å in the case of papain (Kamphuis et al., 1985a) and 1.7 Å in the case of actinidin (Baker, 1980; Baker and Dodson, 1980). Again, it is noteworthy that data collection methods and refinement methods were quite different for the two proteins.

For the papain structure refinement, data were collected by oscillation

Table 2. Crystallographic Data—Papain and Actinidin

	Papain	Actinidin
Crystallization	62% w/w methanol/water, pH 9.3 (Drenth et al., 1971, 1976)	20% saturated $(NH_4)_2\,SO_4$, pH 6.0 (Baker, 1973)
Unit cell	Orthorhombic, $P2_12_12_1$ $a = 45.20$, $b = 104.64$, $c = 50.88$ Å	Orthorhombic, $p2_12_12_1$ $a = 78.2$, $b = 81.9$, $c = 33.03$ Å
Data collection	Oscillation photography	Diffractometer
Refinement method	Restrained least-squares	Free-atom least-squares with regularization
Resolution	1.65 Å	1.7 Å
No. of reflections	24,350	23,990
R factor	0.161	0.165
Protein atoms	1654	1657
Solvent molecules	195 water, 29 methanol	272 water
Mean B value (protein atoms)	15.8 Å	11. 4 Å^2
r.m.s. deviation in bond lengths	0.022 Å	0.015 Å

photography (24,350 unique reflections to 1.65 Å resolution), and the structure was refined using the restrained least-squares procedures of Hendrickson and Konnert (1980). In this method, stereochemical data, which in the papain refinement comprised bond lengths and angles, the planarity of groups, and the chirality of chiral centers (but not torsion angles, non bonded contacts, or B values), are included as observations in the refinement to supplement the X-ray data. The result, after 38 cycles of refinement, is a model that includes 1654 protein atoms, 195 water molecules, and 29 methanol molecules, and for which there is little deviation from expected geometry (the root-mean-square deviation of bond lengths from standard values is 0.022 Å). The final R factor* of 0.161 indicates extremely good agreement with the X-ray data. The estimated average error in the atomic coordinates is about 0.1 Å (0.08−0.12 Å); but this must be qualified by the fact that for any given atom, the uncertainty in its position depends on its B value and those of its bonded neighbors, since the B values express their mobility and/or static disorder. Thus, atoms in the protein

---

*$R = \Sigma||Fo|-|Fc||/\Sigma|Fo|$, where $|Fo|$ and $|Fc|$ are the observed and calculated structure amplitudes.

interior or in the active site, or atoms participating in multiple hydrogen bond interactions, all tend to have low B values and their positional accuracy is likely to be 0.1 Å or better. On the other hand, some external side chains are disordered or highly mobile (including the side chains of Arg 59, 98, and 145, Lys 10, 139, and 156, Tyr 94 and 197, Gln 9 and 114, and Asn 84), and the uncertainty in their positions is much greater.

In contrast with papain, the structure refinement of actinidin was based on diffractometer data (23,990 reflections to 1.7 Å resolution) and a different refinement method. The latter was essentially a free-atom refinement, using fast Fourier least-squares methods, with the geometry being regularized (i.e., brought back close to standard geometry) at intervals (Baker and Dodson, 1980). The results are very similar to those for papain: (1) the final model includes 1657 protein atoms (24 of partial occupancy, where side chains are disordered) and 272 water molecules, (2) the r.m.s. deviations of bond lengths and angles from standard values are 0.015 Å and 3°, respectively, and (3) the final crystallographic R factor is 0.165. The estimated error in the positions of most atoms is 0.06−0.10 Å, but as discussed above, there is a much greater uncertainty associated with some residues. In actinidin, no density at all was seen for the two C-terminal residues, 219 and 220, nor for the side chain of Glu 87 beyond C_β, while the side chains of Glu 21, 172, and 191, Asp 97 and 142, Tyr 4 and 130, Ser 3, Lys 145, and Gln 146 have weak density, high B values, and much lower accuracy. One change to the amino acid sequence was clearly indicated (Asp 86 → Glu 86).

A clear conclusion from these structure refinements is the importance of using refined models in drawing structural conclusions. For papain, the average shift in atomic positions as a result of refinement (starting from an already partly-refined model) was 0.42 Å, while for actinidin, it was 0.50 Å for main chain atoms and 0.84 Å for side chain atoms (starting from an unrefined structure). Furthermore, superposition of the papain coordinates on those of actinidin before and after refinement showed far closer agreement between the structures after refinement (Baker, 1981; Kamphuis et al., 1985b).

Finally, comment should be made on the state of the catalytically essential Cys 25 in the above crystallographic studies, since the thiol groups of these enzymes are very easily oxidized. The preparations of papain contained a mixture of inactive enzyme (in which Cys 25 was oxidized) and activatable enzyme. For the high-resolution refinement of the structure, crystals of the purified, inactive oxidized enzyme were used. Preparations of actinidin also contained some inactive oxidized material, but the use of tetrathionate as a reversible inhibitor during purifications meant that most of the enzyme was obtained in active form. The tetrathionate was removed prior to crystallization, however, and the X-ray analysis showed that in the crystals used for data collection, Cys 25 was mostly oxidized. However, two points must be made. First, earlier work by Drenth et al. (1976) has carefully examined the (very small) changes in the active site structure accompanying changes in the state of Cys 25; these are discussed in Section 2.5. Second, the lack of appreciable disorder in the positions of the active site residues of actinidin (which contained a mixture of active and inactive species) suggests that very little structural change accompanies the oxidation of Cys 25.

2.2. Structural Organization

Papain consists of a single polypeptide chain of 212 residues, while actinidin is slightly larger, with 220 residues. The crystallographic studies (described above) show that both proteins share the same overall polypeptide chain fold. This is shown schematically in Figure 1 (see also Fig. 4 for a plot of α-carbon positions). The main features can be summarized as follows (the description is in terms of the papain numbering).

The two proteins are classic examples of two-domain structures. Although the exact definition of crossover points is necessarily indistinct, the L-domain (following the terminology used in most of the published work on papain) can be described as comprising residues 12−112 and 208−212, and the R-domain comprises residues 1−11 and 113−207. Thus, the L-domain consists principally of residues from the first half of the chain, but with the N-terminal end crossing over into the R-domain, while the latter consists principally of residues from the second half of the chain, but with the C-terminal end crossing back over into the L-domain. It is easy to imagine the molecule being folded as two distinct globular units, with the N-terminal and C-terminal sections bound into the R- and L-domains, respectively, as "straps" holding the structure together.

The L-domain is mostly α helical, with three α helices, A (24−42, a long α helix through the center of the molecule at the interface between the two domains), B (50−57), and C (67−78), and several discrete α helical turns. The R-domain, on the other hand, is based upon a twisted antiparallel β sheet, which is folded over to form a barrel with its interior filled with

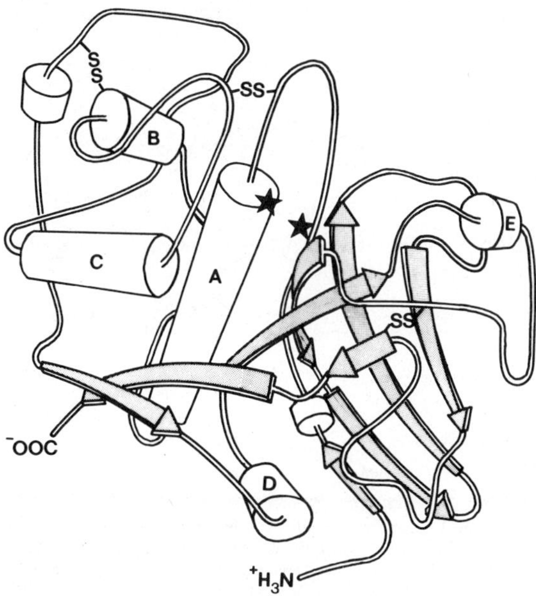

Figure 1. Schematic diagram showing the polypeptide chain folding in papain and actinidin. α helices are represented by cylinders, and strands of β sheet by arrows. One domain (L) contains three principal α helices, A, B, and C, while the other domain is based on a double β-sheet structure with helices D and E at opposite ends. The positions of the catalytic Cys and His residues are shown by ★. The three disulfide bridges are also shown.

hydrophobic side chains. Two α helices, D (117−127) and E (138−143), on the molecular surface seal opposite ends of the barrel. A schematic representation of the main chain hydrogen bonding is given in Figure 2.

Each domain has a central core of hydrophobic residues (14 in the L-domain and 15 in the R-domain), but the interface between the two is predominantly polar in nature (see Fig. 3). It includes four buried, charged residues (Lys 17, Glu 35, Glu 50, and Lys 174) that interact across the interface, a number of neutral polar side chains, and a network of buried water molecules (9 in papain, 8 in actinidin). These complete an extensive domain-domain interface that is notable for a large number of internal hydrogen bonds, almost all with highly favorable bond lengths (2.7−3.0 Å) and angles and few hydrophobic interactions, except at the outer edges. Here the side chains of Trp 7 and Ile 41 at one end, and Phe 28 at the other,

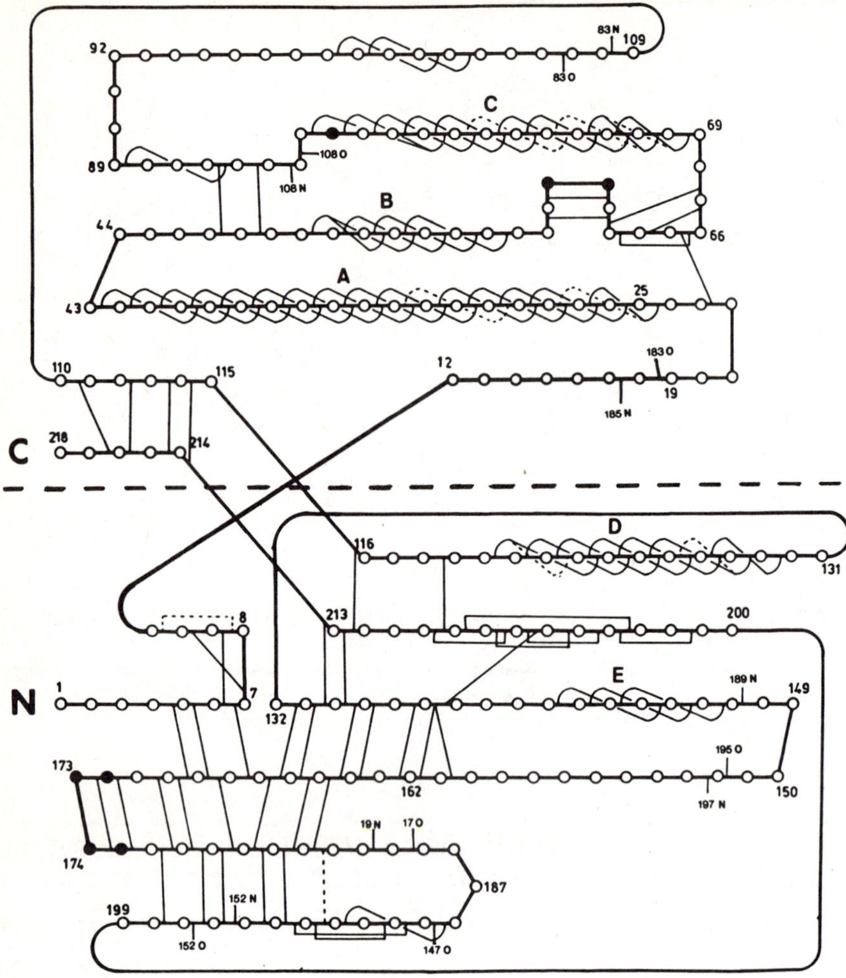

Figure 2. Schematic diagram showing the main chain hydrogen bonding in actinidin. The hydrogen bonding in papain is essentially the same, except at positions of insertions or deletions (shown by filled circles). The separation of the two domains is shown by a dashed line. Numbering is that of actinidin.

appear to "seal off" the interface. In general, however, the large number of polar groups within the interface reinforce the impression of two globular units, with polar surfaces, that have come together to create the overall structure.

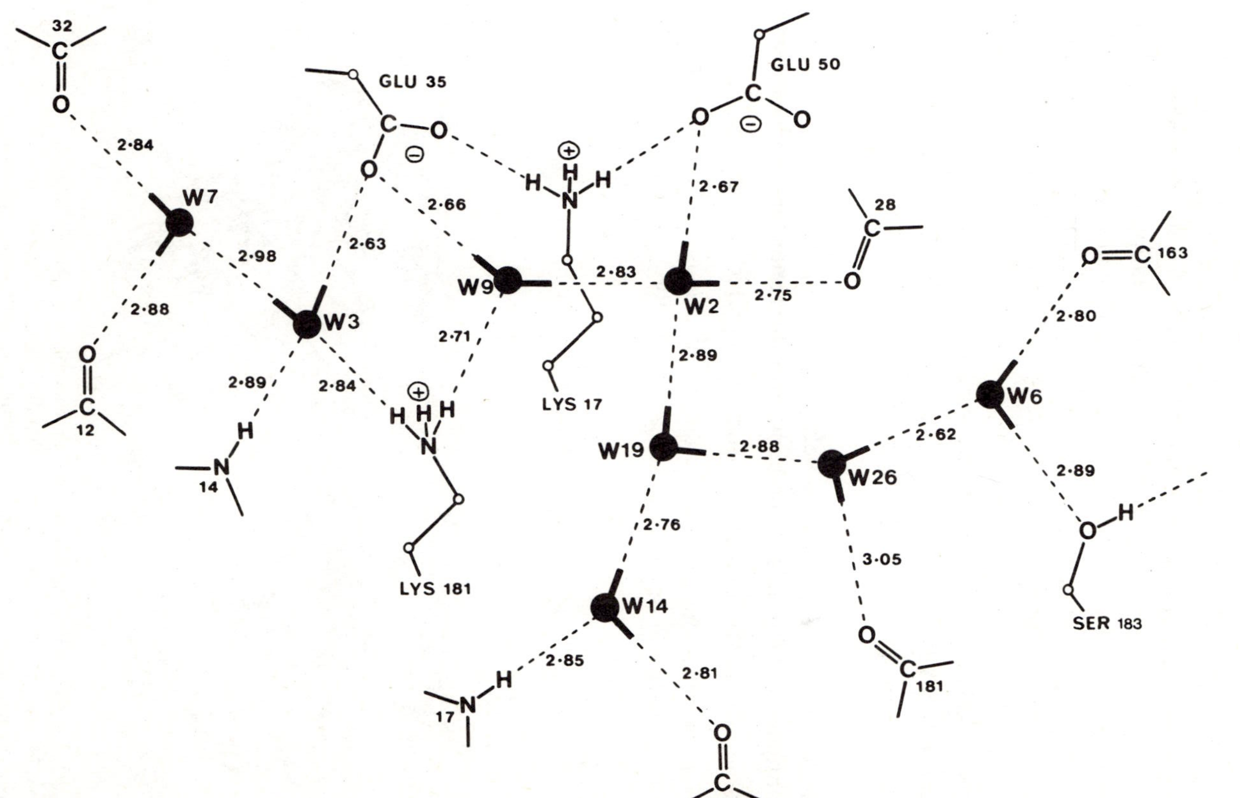

Figure 3. The cluster of water molecules associated with the four buried charged residues, Lys 17, Glu 35, Glu 50, and Lys 181, in the interior of the domain-domain interface of actinidin. An almost identical pattern is found in papain (see also Fig. 6). Other polar side chains within the interface include Ser 29, Thr 33 and Ser 134, Gln 131, and Ser 183. The nonpolar side chains of Ala 32 and Val 165, and Ala 12 and Pro 132, are in contact within the interface, while the periphery is sealed off by Trp 7, Ile 40, Leu 46, Val 14, Phe 28, Ala 163, and Ile 70.

The active site is located in a cleft between the two domains, with one of the catalytically important residues, Cys 25, contributed by the L-domain and the other, His 159 (*162* in actinidin), contributed by the R-domain. Adjacent to these residues is a pocket lined with hydrophobic side chains.

Within this overall folding pattern, additional residues in one protein or the other are all accommodated in external loops. Thus, comparing actinidin with papain, insertions and deletions are as follows (see also Fig. 4):

1. Two residues are inserted between residues 59 and 60, just after the end of helix B (50−57), extending a surface loop.

2. One residue is added at the C-terminus of helix C (67−78), extending it slightly.

3. Four residues are inserted between 168 and 169, extending two antiparallel β strands in the β sheet of the R-domain.

4. One residue (194) is deleted from an irregular surface loop.

5. Two extra residues are added at the C-terminus (but are disordered and not seen in the actinidin electron density).

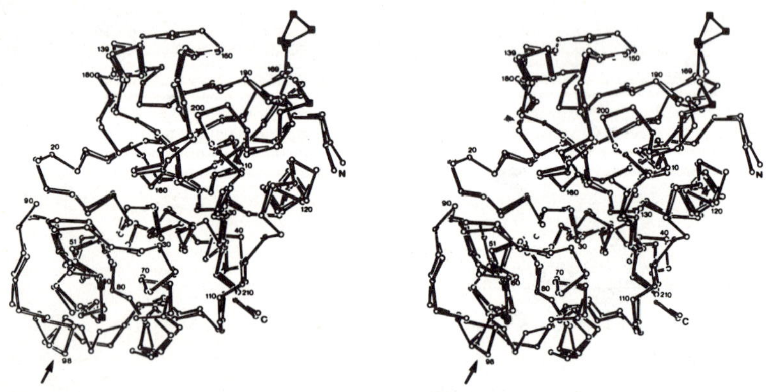

Figure 4. Superposition of α-carbon atoms of papain (open bonds) and actinidin (filled bonds). Insertions in one molecule or the other are identified with filled squares, and the one major difference in conformation in an external loop with virtually no sequence identity is arrowed. Numbering follows that of papain.

2.3.   Amino Acid Sequence Comparison, Papain and Actinidin

Alignment of the amino acid sequences of papain and actinidin is straightforward, given the three-dimensional structures as a guide, and given the small number of insertions and deletions. More detailed comparisons have been made elsewhere (Carne and Moore, 1978; Baker, 1981; Kamphuis et al., 1985b), and only the main features will be summarized here.

The overall level of sequence identity in the two proteins is 46% (98 of the 212 residues in papain are identical in actinidin), but the sequence differences are not evenly distributed. For example, there is a high degree of sequence identity in helices A and B of the L-domain (18 of 26 identical, i.e., 69%), in the β strands of the R-domain (31 of 45, i.e., 69%), and in the residues making up the domain-domain interface (17 of 21, i.e., 81%). On the other hand, the residues of the hydrophobic core are not so strongly conserved (12 of 29 identical, i.e., 41%), probably because they do not make specific interactions and there are holes in the centers of the cores that allow some rearrangement of side chains without affecting the main chain conformation.

At the other extreme is the middle section of the polypeptide chain, residues 82–116. This folds loosely over the surface of the L-domain forming a shell that separates the interior residues from solvent and eventually, in a long stretch of extended chain, crosses to the R-domain. In this section, only 4 of 35 residues are identical (11%), and many of the changes responsible for the difference in isoelectric point occur (papain has many basic side chains in this region; actinidin has many acidic ones). Contiguous to this is helix C, which also shows low sequence identity (2 of 12, i.e., 17%).

2.4.   Detailed Comparison of Three-Dimensional Structures

The two molecular structures have been compared in detail elsewhere (Baker, 1981; Kamphuis et al., 1985b). We shall simply review the main features here. Superposition of the coordinates shows a remarkably high degree of similarity. Using all main chain atoms (C_α, C, O, N) the r.m.s. deviation for 638 atoms (76% of the polypeptide chain) is only 0.40 Å. Moreover, when those side chains that are identical are added in, the agreement is equally good, indicating that the side chain positions of identical residues are conserved to the same extent as the main chain. A plot of α-carbon positions (Fig. 4) shows the remarkable similarity of the two molecules, and this can also be seen from a plot of the average deviation (between papain and actinidin) for the main chain atoms of each residue as a function of residue number (Fig. 5).

With one exception, the only major disagreements between the main chain atomic positions are in the immediate vicinity of insertions and deletions. The latter make minimal disturbances in the structure, with the effects extending only one or two residues on either side. This is achieved by compensating changes in the main chain torsional angles, ϕ and ψ, that is, the polypeptide chain conformation is self-correcting at these points. The one difference that results from amino acid sequence changes is at residues 97−101. These residues are remote from the active site, in the middle section of the chain where amino acid sequence identity is extremely low, and the change in conformation arises from a series of changed interactions accompanying the sequence changes (see Baker, 1981; Kamphuis et al., 1985b). This region is also near the C-terminal end of helix C, where an extra residue is inserted in actinidin. In a series of concerted movements arising from sequence changes, the extra residue, and the different conformation at nearby 97−101, helix C is shifted bodily by 1−2 Å, with the displacements increasing towards the C-terminus.

When the two domains are superimposed separately, the agreement is somewhat better than for the molecule as a whole. The R-domain agrees better (r.m.s. deviation 0.27 Å for 75% of its main chain atoms) in the two molecules than does the L-domain (r.m.s. deviation 0.35 Å for 63% of main chain atoms), but the latter does include most of the "middle section" of chain (82−116) with its lack of secondary structure and low-sequence conservation. The closeness of the figures when the domains are superimposed individually (0.27−0.35 Å) with that for the molecule as a whole (0.40 Å) indicates that there is little difference in the relative dispositions of the two domains in papain and actinidin, if any. It might be imagined that the network of water molecules in the domain-domain interface might allow flexing or hinge-bending of the two-domain structure, but there is no evidence of that here.

Residues near the active site and those actually involved in the catalytic process also superimpose to an extent that approaches the atomic coordinate accuracy of both structures. Thus, for 20 residues at and around the active site (with main chain atoms, and identical side chains included), the r.m.s. deviation is only 0.29 Å for 69% of the atoms.

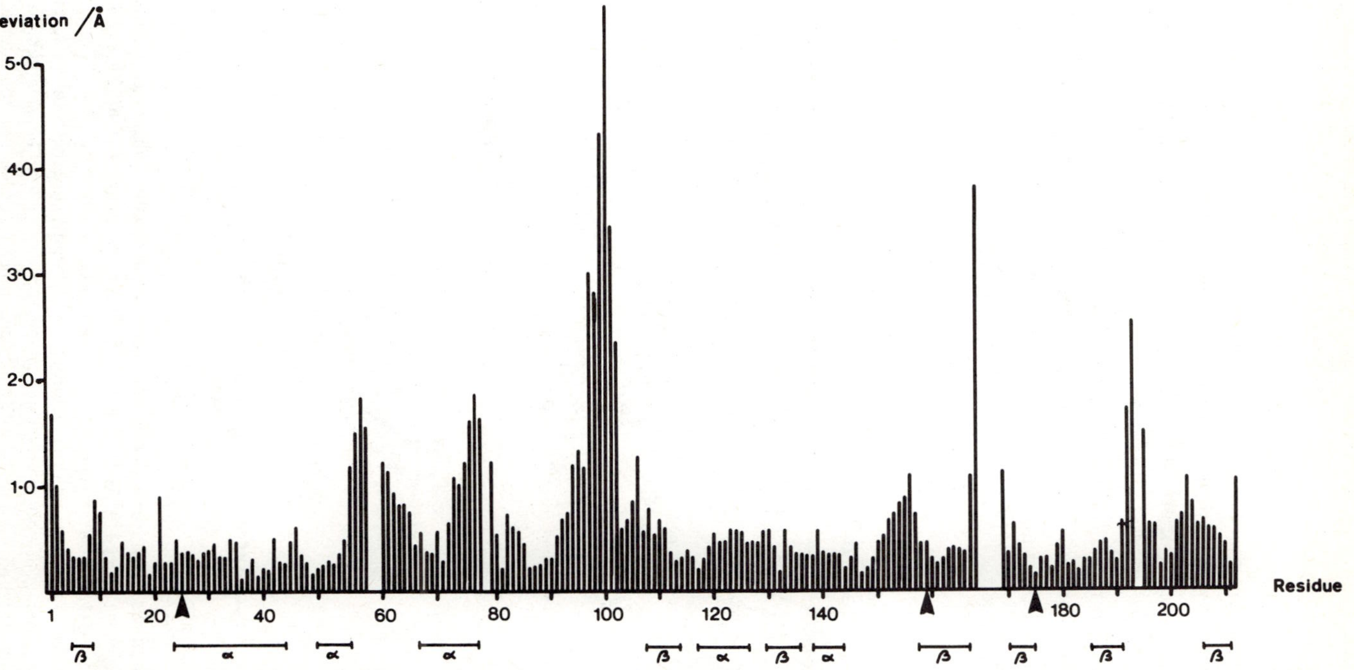

Figure 5. Plot of the deviations in the positions of main chain atoms in papain and actinidin after superposition of the two structures. For each residue, the deviation is the average for its N, C_α, C, and O atoms. Numbering refers to that in papain. Locations of the α helices and β strands are shown, and arrows indicate the positions of Cys 25, His 159, and Asn 175. Gaps are left where insertions occur in one structure or the other. The major disagreement is at residues 97−102, where sequence differences lead to a local change in main chain conformation, but for 60% of all main chain atoms, the deviation is less than 0.5 Å.

331

Hydrogen-bonding interactions in the two proteins are highly conserved. Of the 93 main chain-main chain hydrogen bonds in papain, 90 are also found in actinidin. Main chain-side chain hydrogen bonds are also generally conserved (to the extent of about 80%), but side chain-side chain hydrogen bonds are much less so. A feature of the hydrogen bonding is the way in which the spatial positions of hydrogen bonds in the protein interior are conserved even when individual residues change. Two examples are shown in Figure 6. Other observations highlight the similarity of the two molecular

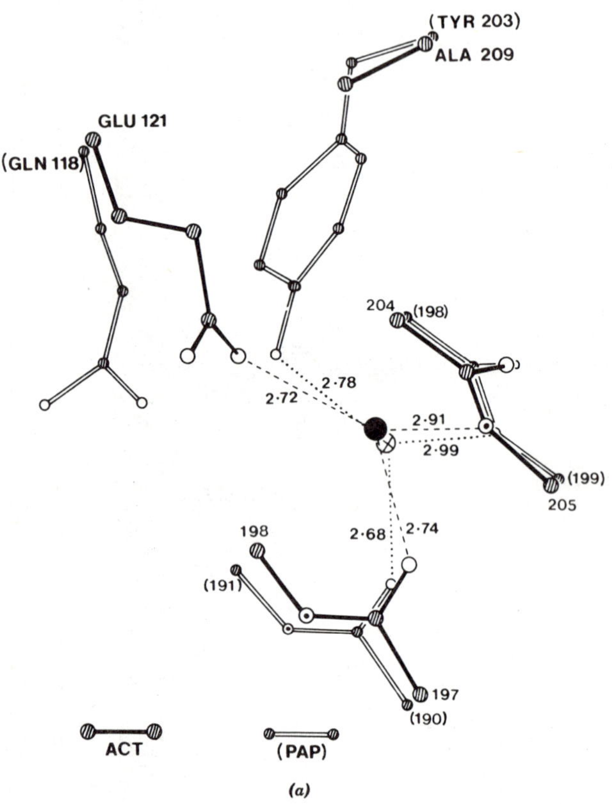

Figure 6. Two examples of the conservation of hydrogen bonding in papain and actinidin. In (*a*) Tyr 203 in papain is replaced by Ala 209 in actinidin, but the side chain of Glu 121 (Gln 118 in papain) moves in so that its carboxyl group occupies the position left by the phenolic oxygen of Tyr 203 and preserves the same pattern of hydrogen bonds with an internal water molecule. In (*b*) Thr 14 in the domain-domain interface in papain is replaced by Val 14 in actinidin. Now Lys 17 can no longer hydrogen bond with it, and instead moves further into the interface, displacing a water molecule, but maintaining a very similar network of hydrogen bonds within the interface.

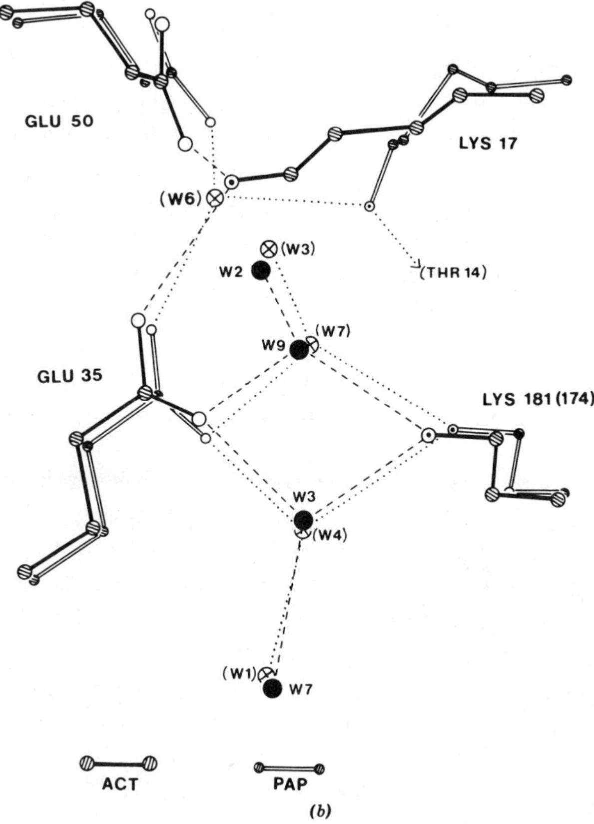

Figure 6. (*b*) (*continued*)

structures. Of the 17 internal water molecules in actinidin, 15 are also found in papain (with an r.m.s. displacement of only 0.48 Å), one is replaced by a methanol molecule, and the other is displaced by sequence changes. A bend. in helix A of papain (a break in the hydrogen bonding at 29 C=O) occurs in the identical place in actinidin, and bifurcated hydrogen bonds at helix termini are the same in both proteins.

Again, the above similarities in three-dimensional structure are rendered all the more significant when it is remembered that the two proteins crystallize from totally different mother liquors at different pH (9.3 and 6.0), with different crystal packing, and the crystallographic studies have used different modes of data collection and different refinement techniques.

2.5. The Active Site

The main features of the active site are shown in Figures 7 and 8. The catalytically essential thiol group is that of Cys 25, which is situated at the bottom of the cleft between the two domains, at the N-terminus of the long central α-helix A (residues 24—42). Adjacent to the side chain of Cys 25 is one nitrogen atom, $N_{\delta 1}$, of the imidazole ring of His 159, which is also implicated in the catalytic mechanism (see Section 3). The other nitrogen atom of His 159, $N_{\epsilon 2}$, is hydrogen bonded to the amide oxygen, $O_{\delta 1}$, of the side chain of Asn 175, creating a triad Cys...His...Asn somewhat reminiscent of the triad Ser...His...Asp found in the serine proteases. In the thiol proteases, the His...Asn hydrogen bond almost certainly fulfils an orientational role, and is shielded from the external solvent by the indole ring of Trp 177.

Drenth et al. (1976) have used difference electron density maps to compare (1) oxidized, nonactivatable papain with activatable papain (in which a cysteinyl group is attached to S_γ of Cys 25), and (2) oxidized papain with derivatives in which chloromethyl ketone inhibitors are bound to S_γ. These studies have shown that there are changes in the orientation of the imidazole ring of His 159 according to the state of the thiol group of Cys 25. These changes are possible because the His...Asn hydrogen bond is approximately colinear with the His 159 C_β-C_γ bond—thus, rotation about C_β-C_γ can occur without disrupting the His...Asn interaction. Both in the oxidized protein, in which the thiol group is oxidized to SO_2^- or SO_3^-, and in

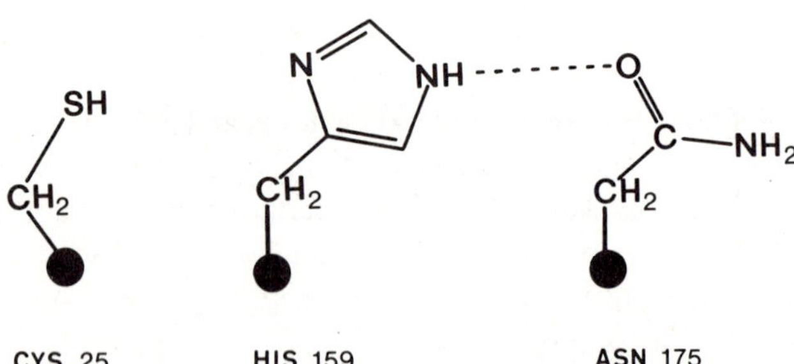

Figure 7. The triad of residues in the active site of papain and actinidin. Cys 25 and His 159 (*162*) are adjacent, with the latter hydrogen bonded to Asn 175 (*182*). The triad Cys . . . His . . . Asn is similar to that of Ser . . . His . . . Asp in the serine proteases, but the His . . . Asn interaction probably merely orients the His side chain.

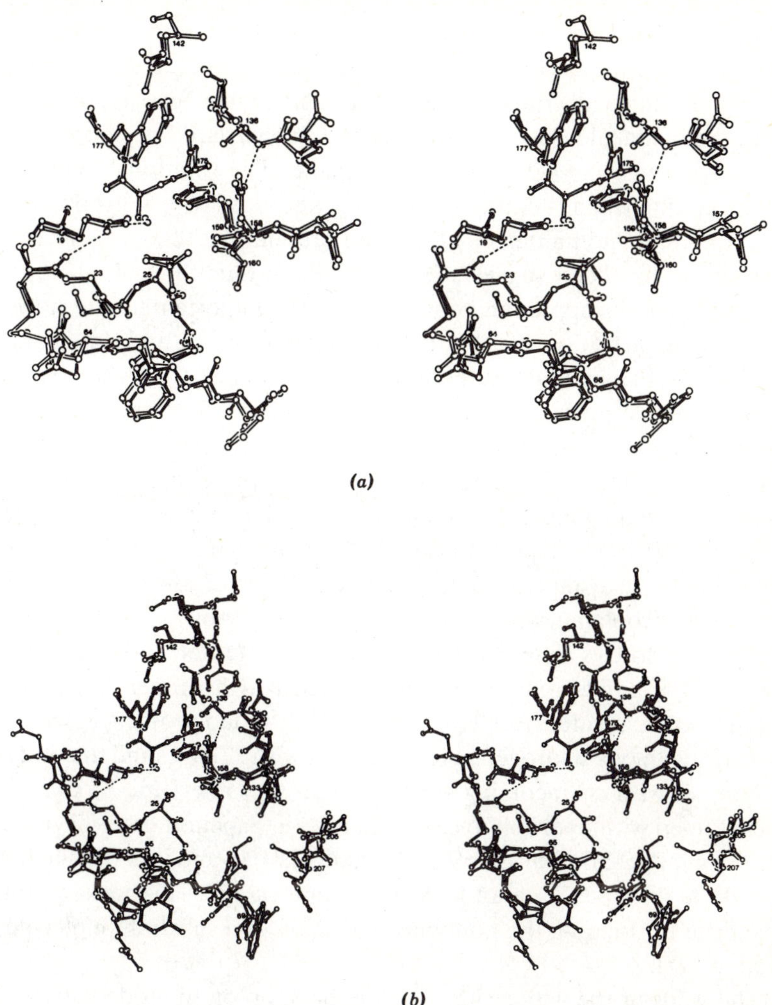

(a)

(b)

Figure 8. Two stereo views showing the active sites of papain (open bonds) and actinidin (filled bonds) superimposed. Numbering is as for papain. In (a) the oxidized sulfur of Cys 25 is shown, with three attached oxygens in the case of papain, and two in actinidin. In each case, one oxygen occupies the "oxyanion hole," hydrogen bonded to Gln 19 NH_2 and 25 NH. The imidazole ring of His 159 is rotated away from coplanarity with S_γ, that is, it is in the "upward" position (see text). In (b) the whole active site cleft is shown. The oxygens are removed from S_γ of Cys 25. The nonpolar S_2 binding pocket is to the right of Cys 25, while Asp 158 lies above His 159, turned away to hydrogen bond with 136 NH. Note the hydrophobic environment of the His . . . Asn interaction, and the location of residues 21, 64, 135, and 142, which change from uncharged in papain (Ser, Asn, Gln, Gln) to charged in actinidin (Glu, Asp, Asp, Lys) around the periphery of the active site.

335

the chloromethyl ketone derivatives, the imidazole ring is oriented in the "upward" position, in which the sulfur atom lies about 2.5 Å below a plane through the imidazole ring (see Fig. 8). In "activatable" papain, on the other hand, the imidazole ring is rotated 30° to a "downward" position, in which S_γ lies somewhat above the imidazole plane. The rotational angle of the imidazole ring when the thiol group is fully reduced to $-SH$ is not known, but if the thiol and imidazole groups are present as a thiolate-imidazolium ion pair (as has been suggested, Polgar, 1974; Drenth et al., 1975), they should perhaps be approximately coplanar. The important points, however, are the demonstration that (1) the imidazole ring can and does rotate in response to changes at the sulfur atom of Cys 25, and (2) only small changes in the positions and orientations of Cys 25 and His 159 occur in the different forms of papain.

For both papain and actinidin, the structure refinements have been carried out on the oxidized proteins, in which the imidazole rings are in the "upward" position. The distances $S_\gamma...N_{\delta 1}$ are 3.6 Å and 3.3 Å respectively, but these distances would decrease if the imidazole ring were rotated until S_γ was coplanar with it (Kamphuis et al., 1985a).

Adjacent to Cys 25 in the floor of the active site cleft is the side chain of Gln 19. The amide oxygen of this side chain is hydrogen bonded to an internal serine residue, Ser 176, but its $-NH_2$ group projects upwards into the cleft, with one hydrogen able to interact with 22 C=O and the other free to interact with an incoming substrate. On the other side of Cys 25 is a pocket lined with hydrophobic side chains—in papain those of Tyr 67, Pro 68, Trp 69, Phe 207, Val 133, Val 157, and Ala 160, and in actinidin those of Tyr 69, Ile 70, Met 211, Ala 136, Val 160, and Ala 163. This pocket forms the S_2 subsite for binding the nonpolar side chain of a substrate molecule (see Section 3.7).

The walls of the active site cleft are made up on one side with the side chains of Trp 177 and His 159 and the peptide 158−159, and on the other side with residues 22−24 and 64−67. Residues 64−67 form a piece of extended chain with 64 C=O, 66 NH, and 66 C=O all projecting into the cleft; these groups together with 158 C=O (also projecting into the cleft) are potentially able to interact with an entering substrate. Two glycine residues, Gly 23 and Gly 66, are probably required to be Gly because their orientation is such that a side chain would seriously crowd the cleft.

The only charged group within 10 Å of Cys 25 is the carboxyl group of Asp 158, about 7.5 Å away. For a variety of reasons (see Section 3.4), a direct involvement in catalysis has been suggested for this residue—note here, however, that it is turned away from the active site, so that it forms a hydrogen bond with the main chain NH group of residue 136. Other charged groups that are more remote from the active site, but could conceivably influence substrate binding are (in papain) Lys 156 ($\sim$18 Å away from Cys 25) and (in actinidin) Glu 21 ($\sim$14 Å away), Asp 66 ($\sim$12 Å), Asp 138 ($\sim$11 Å), Asp 142 ($\sim$17 Å), Lys 145 ($\sim$13 Å), and Asp 185 ($\sim$16 Å). All of these are surface residues near the active site cleft and are, at least, likely to influence its electrostatic properties (see Section 3.3).

Finally, the oxidation of the sulfur atom of Cys 25 shows a similar pattern in both papain and actinidin. In papain, positions were found for three oxygens (implying oxidation to SO_3^-); in actinidin, only two oxygen positions were identified, but a "bump" in the density could be attributed to a third oxygen (Baker, unpublished results), and this may also represent oxidation to SO_3^-. One oxygen is hydrogen bonded to Gln 19 NH_2 and 25 NH, that is, it occupies the "oxyanion hole" (Section 3.5). The existence of such a site, able to stabilize a negatively charged oxygen lying adjacent to Cys 25, undoubtedly explains the very ready oxidation of the SH group in the thiol proteases. The other oxygen positions are also stabilized by hydrogen bonding: one with 26 NH, the other with 160 NH and His $159N_{\delta1}$.

2.6. Flexibility of the Structure

The atomic B values derived during the refinement of protein structures give valuable information about protein flexibility (for a review, see Petsko and Ringe, 1984). The overall level of B values is of less significance than their relative values in different parts of the structure, indicating more or less flexible regions. For papain and actinidin, the variations in B values along the polypeptide chain are very similar, indicating that they represent a molecular property of these enzymes. The only major differences, apart from those at the positions of insertions or deletions, are at residue 21 and residues 97–101. The former is probably due to imperfect refinement (the fit of the main chain carbonyl group of this residue in papain was unsatisfactory), while the latter is at the site of the one major conformational difference between the two proteins, with associated differences in hydrogen bonding and other interactions.

The pattern of B values is generally typical of most protein molecules, with low values (and presumably low mobility) in internal regions and ele-

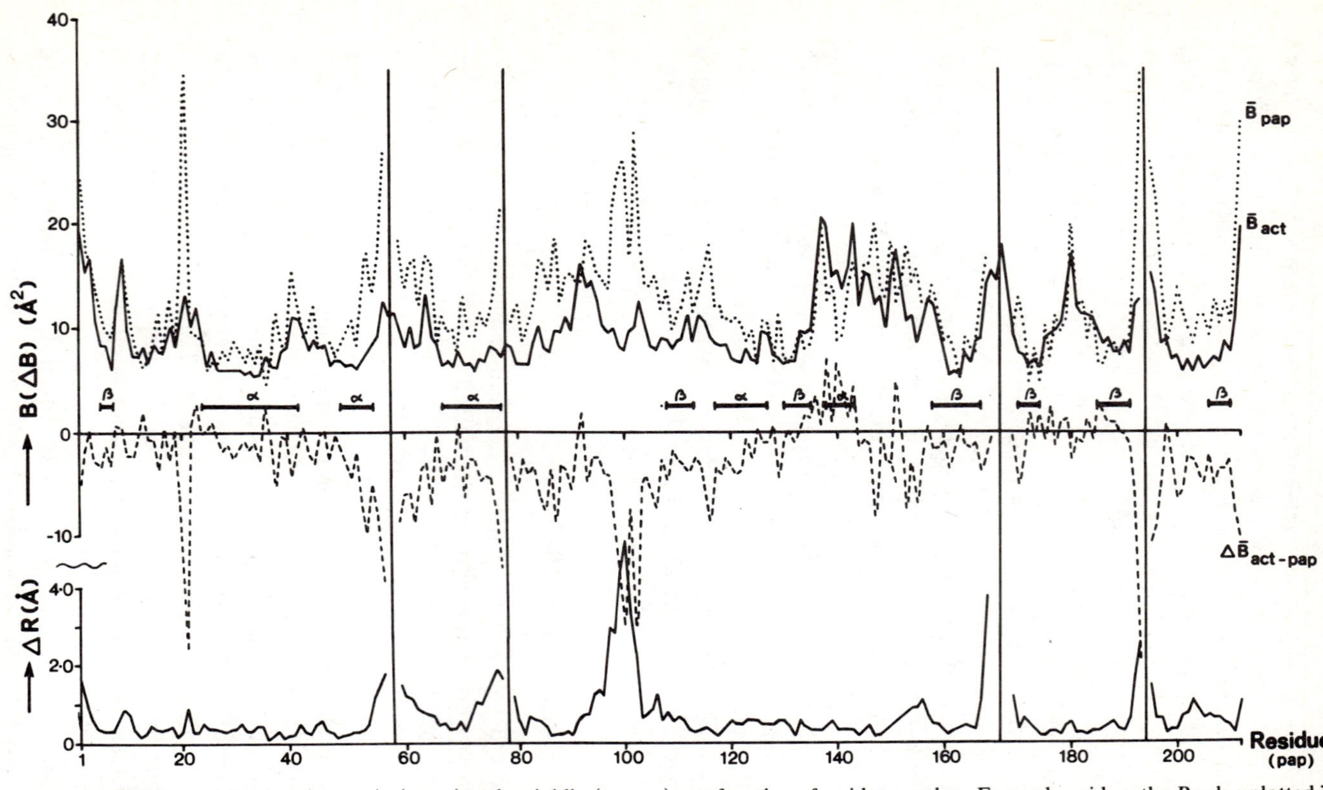

Figure 9. Variation of B values in papain ($\cdots$) and actinidin (———) as a function of residue number. For each residue, the B value plotted is the average for its N, C_a, C, and O atoms. The difference $\Delta\bar{B}$ ($= \bar{B}_{act} - \bar{B}_{pap}$) is shown by ($-\,-\,-$). The average difference in position for corresponding atoms is depicted in the lower graph—note the similarity between B values in the two proteins and the correlation between structural differences and B value differences. Positions of insertions and deletions are shown with vertical lines.

ments of secondary structure, and higher values (and presumably greater flexibility) in external loops. The variation in different parts of the structure can be seen graphically in Figure 9. Of most interest in connection with catalysis are the residues in and around the active site. For these, the B values are given (as the averages for main chain and side chain atoms) in Table 3, and can be compared with the overall average values of 12.3 and 19.5 Å^2 for papain and 9.5 and 13.6 Å^2 for actinidin.

The two catalytically essential residues, Cys 25 and His 159, have relatively low B values for their main chain atoms, because both are fixed into secondary structures: Cys 25 to helix A in the L-domain and His 159 to the β sheet of the R-domain. Their side chains have higher B values, suggesting some mobility; it is particularly interesting to note that in both papain and actinidin, the atoms $N_{\delta 1}$ and $C_{\epsilon 1}$ of His 159 have higher B values than the other ring atoms, exactly as would be expected if there is rotation about the C_β-C_γ...$N_{\epsilon 2}$ direction.

Of most interest are the residues that make up the walls of the active site cleft: Trp 177 and residues 157−159 on one side and residues 22−24 and 64−67 on the other. Inhibitor (and presumably also substrate) binding

Table 3. B Values[a] for Active Site Residues

	Gln 19	Cys 22	Gly 23	(Ser) 24	Cys 25	Trp 26
Papain	8.1, 11.8	9.7, 16.0	9.2	8.7, 10.4	6.0, 13.7	6.8, 7.2
Actinidin	8.3, 12.3	10.1, 10.8	12.0	8.6	6.2, 19.0	7.8, 6.6
	(Asn) 64	Gly 65	Gly 66	Tyr 67	(Pro) 68	(Trp) 69
Papain	16.8, 25.2	16.0	8.5	11.2, 23.2	9.6, 10.7	9.9, 26.4
Actinidin	13.9, 32.4	9.1	8.2	6.4, 16.5	7.0, 8.2	6.1, 8.2
	(Val) 133	Val 157	Asp 158	His 159	Ala 160	
Papain	9.5, 9.8	12.0, 9.6	14.0, 17.7	11.7, 15.4	8.8, 4.8	
Actinidin	9.6, 9.5	12.7, 13.7	12.2, 20.2	10.1, 12.6	8.5, 6.3	
	Asn 175	Ser 176	Trp 177	(Ser) 205	(Ser) 206	(Phe) 207
Papain	9.9, 10.7	7.0, 9.2	11.1, 11.7	11.0, 15.9	9.1, 14.8	12.4, 10.1
Actinidin	8.8, 8.0	9.5, 6.8	9.6, 13.7	5.9, 24.2	6.6, 6.4	7.1, 7.1

[a]For each residue, the average B value for main chain atoms ($C\alpha$, C, O, N) is given, followed by that for side chain atoms.
Where residues differ in actinidin, the papain residue is in parentheses.

causes the active site cleft to widen (Section 3.5) and these residues to be pushed apart, with the movement being especially noticeable for the main chain between 64–67. Despite this, the B values for these residues are lower than average in both proteins, and there is no suggestion of any extra mobility in the "lips" of the active site cleft as seen, for example, in lysozyme (Artymiuk et al., 1979). Moreover, analysis of the B values for papain gives no support for a hinge-bending model (Kamphuis et al., 1985a), and superposition of the papain and actinidin structures (Section 2.4) shows that the two domains have the same relative dispositions in both cases. Thus, in these proteins, two equal-sized domains, with an extensive interface between, seem to produce a relatively rigid active site cleft that must be pushed apart by inhibitors or substrates.

2.7.   Structural Studies on Other Thiol Proteases

2.7.1.   Crystallographic

Only one other thiol protease has been studied crystallographically: calotropin DI, another plant enzyme for which a structure analysis at 3.2 Å has been reported (Heinemann et al., 1982). Unfortunately, no amino acid sequence information is yet available for this enzyme, so that individual residues cannot be positively indentified. Nevertheless, it is apparent that calotropin DI has the same polypeptide chain-folding pattern as papain and actinidin (with some deletions on the molecular exterior, since calotropin DI has 14 fewer residues than papain). Density for side chains analogous to those of the papain active site residues Gln 19, Cys 25, His 159, Asn 175, and Trp 177 has also been found at positions 17, 21, 147, 167, and 169. Of most interest in this structure are that (1) its activity is rather different from that of some of the other plant thiol proteases, in that it does not hydrolyze synthetic amide and ester substrates hydrolyzed by papain, and it is not inhibited by some of the natural protein inhibitors that act on papain, ficin, and bromelain; and (2) in the crystal structure analysis, the active thiol is reversibly inactivated by a bound thiosulfate group—thus there is the opportunity to look in detail at the active site residues in another situation.

2.7.2.   Amino Acid Sequence Comparisons: Evidence for Structural Homology in Plant and Animal Thiol Proteases

In addition to papain and actinidin, complete amino acid sequences are available for two mammalian thiol proteases, cathepsin H and cathepsin B, and an almost complete sequence for the plant enzyme, stem bromelain. Numerous N-terminal sequences have also been determined, and these are given in Table 4 with references to the original literature. Some alignments of these sequences have been presented previously (e.g., Drenth, 1976; Takio et al., 1983; Kamphuis et al., 1985b).

The N-terminal sequences offer compelling evidence for structural homology. Residues 1−17 form the N-terminal "strap" that helps to hold the L- and R-domains together in papain, actinidin, and calotropin DI, and some of its residues contribute specific interactions to the domain-domain interface. It is immediately apparent that there is a very high degree of similarity that is strongly suggestive of a common evolutionary origin and a similar three-dimensional structure for all the enzymes, both plant and animal. Thus, 3 of 17 residues are absolutely invariant (Asp 6 and Arg 8, which form an ion pair in papain and actinidin, and Pro 2), residues 1 and 5 are always nonpolar, 13 and 16 are always Val or Ile, 7 is always Trp except in cathepsin B, 11 is almost always Gly (replaced only by Asn or Pro, like Gly often found in turns), and residue 17, which in papain and actinidin is one of the four charged residues in the domain-domain interface, is always Lys or Arg. Insertions of one and four residues, respectively, are found in cathepsins H and B at about residue 10, where papain and actinidin have a short piece of 3_{10}-helix (residues 7−11) on the surface of the molecule.

Similar homologies can be seen when the full sequences of the animal enzymes (cathepsins B and H) are aligned with those of the plant enzymes (papain, actinidin, and stem bromelain) as in Kamphuis et al. (1985b). Comparing cathepsin H first, the level of sequence identity is almost as great as between actinidin and papain. Altogether, 39% of the residues in papain are identical in cathepsin H (compared with 46% in actinidin); and when the very high degree of similarity in the three-dimensional structures of papain and actinidin is recalled (Section 2.4), there is little doubt that cathepsin H will also have the same structure. Sequence identity is less in cathepsin B, although still considerable (29% of residues in papain are identical in cathepsin B), and the level of sequence identity in the incomplete sequence of stem bromelain is about 40%.

If a common polypeptide chain fold is assumed and the sequences are aligned as in Kamphuis et al. (1985b), insertions and deletions are all on the molecular surface, and they frequently occur at the same sites already identified in the structures of papain and actinidin. Cathepsin B, with 252 residues, is considerably larger than papain or actinidin—and here most of

Table 4. N-Terminal Amino Acid Sequences

	1				5					10					15					
Papain[1]	I	P	E	Y	V	D	W	R	Q	K	—	—	—	—	G	A	V	T	P	V	K
Actinidin[2]	L	P	S	Y	V	D	W	R	S	A	—	—	—	—	G	A	V	V	D	I	K
Stem bromelain[3]	V	P	Q	S	I	D	W	R	D	Y	—	—	—	—	G	A	V	T	S	V	K
Asclepain A[4]	L	P	N	S	I	D	W	R	Q	K	—	—	—	—	G	V	V	F	P	I	K
Asclepain B[4]	L	P	N	F	V	D	W	R	K	N	—	—	—	—	G	V	V	F	P	I	R
Chymopapain[5]	Y	P	E	S	I	D	W	R	A	K	—	—	—	—	G	A	V	T	P	V	K
Papaya peptidase A[5]	L	P	E	N	V	D	W	R	K	K	—	—	—	—	G	A	V	T	P	V	R
Papaya peptidase B[5]	L	P	E	S	V	D	W	R	A	K	—	—	—	—	G	A	V	T	P	V	K
Cathepsin H (rat liver)[6]	Y	P	S	S	M	D	W	R	K	K	G	—	—	—	N	V	V	S	P	V	K
Cathepsin L (bovine spleen)[7]	L	P	D	S	V	D	W	R	E	K	—	—	—	—	G	G	V	T	P	V	K
Cathepsin B (rat liver)[6]	L	P	E	S	F	D	A	R	E	Q	W	S	N	C	P	T	I	A	Q	I	R

References: [1]Mitchel et al. (1970); [2]Carne and Moore (1978); [3]Goto et al. (1976, 1980); [4]Lynn et al. (1980b); [5]Lynn and Yaguchi (1979); [6]Takio et al. (1983); [7]Turk et al. (1983).

the extra residues, as well as the attached carbohydrate, are accommodated in the region of least homology in papain and actinidin, that is, in the L-domain, in the long external loop of residues 82−116, and at the adjacent C-terminus of helix C. The carbohydrate attachments in cathepsin H and bromelain are also in this region. The presumed common folding pattern is shown schematically in Figure 10, with positions of insertions and deletions indicated.

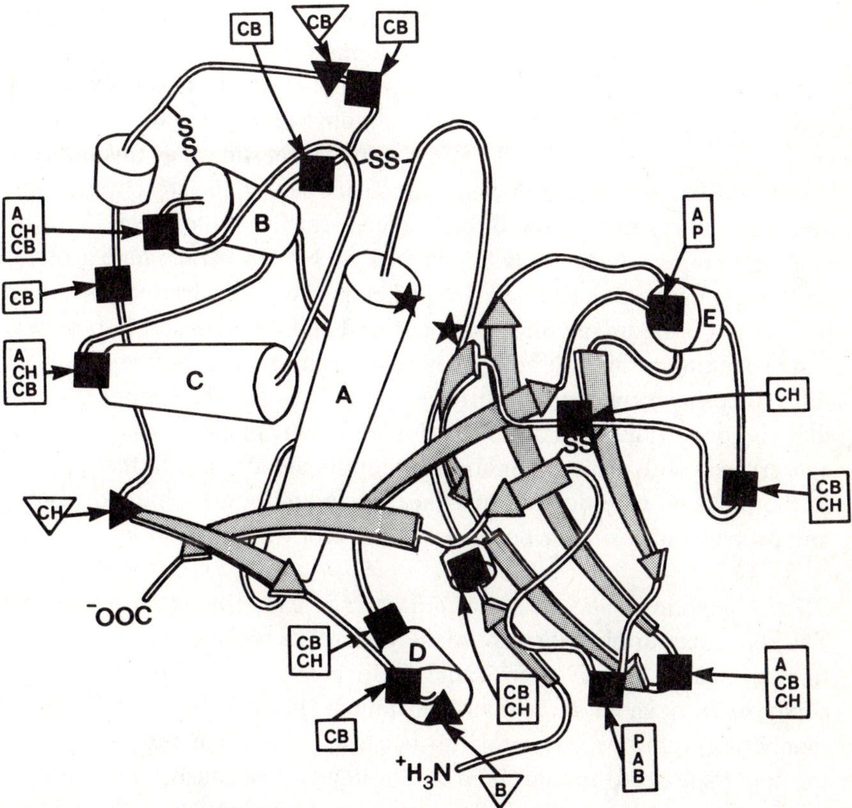

Figure 10. Schematic diagram showing the approximate positions of insertions and deletions in various thiol proteases (P, papain; A, actinidin; B, stem bromelain; CH, cathepsin H; CB, cathepsin B) deduced from crystallographic studies (papain, actinidin) and amino acid sequence alignments (cathepsins H and B, and stem bromelain). Carbohydrate sites are designated ▲.

Looking at specific structural features, almost all of the hydrophobic residues that form the cores of the two domains remain hydrophobic in all five proteins, although only three remain identical (these are all aromatic, Trp 26, Phe 141, and Trp 188). Of the four charged residues in the domain-domain interface in papain and actinidin, Lys 17, Glu 35, Glu 50, and Lys 174, all are found in cathepsin H, and two (corresponding to Lys 17 and Glu 35) are present in stem bromelain. In cathepsin B, Lys 17 becomes Arg, but the residues equivalent to 35, 50, and 174 all lose their charge (changed to Ser, Ala, and Ala). However, other residues within the interface change from uncharged to charged: Val 32 to Glu, Gly 36 to Asp, Ser 131 to Glu, and Ala 162 to Arg. Thus, five charged residues (2 Glu, 2 Arg, and 1 Asp) should then be found in the domain-domain interface, although not in sequentially equivalent positions. The nonpolar residues that seal off the ends of the interface in papain and actinidin, Trp 7, Ile 40, and Phe 28, appear to be conserved in all five proteins, suggesting that this interface remains essentially similar in character. Elsewhere in the structure, among glycine residues, nine are totally conserved in the five proteins; and a further five, all in turns, change only to Asn or Pro. No fewer than nine aromatic residues remain aromatic in all five proteins; and of the disulfide bridges in papain and actinidin, all three can be traced in bromelain and cathepsin H, and two in cathepsin B (although it is not known whether the Cys residues in the cathepsins form disulfide bridges or exist as free thiol groups). Thus, there is good evidence that not only is the overall polypeptide chain folding conserved in both plant and animal proteins, but so are specific features such as the polar domain-domain interface (including buried charged residues and probably also similar networks of internal water molecules).

2.7.3.   Amino Acid Sequence Similarities in and Around the Active Site
The sequences around the essential Cys and His residues for a number of thiol proteases are given in Table 5. In both cases, there is a marked similarity through all of the plant and animal enzymes. Residue 23 (papain numbering) is always Gly, probably because a side chain at this position in the active site near the side chain of Gln 19 would seriously block substrate binding. Residue 19, which is thought to hydrogen bond with the carbonyl oxygen atom of the scissile peptide bond of a substrate, thus stabilizing the oxyanion intermediate (see Section 3.6), appears to be always Gln. The one apparent exception is chymopapain B, but the general lack of homology of this sequence with the others and the recent sequence data of Lynn suggest

Table 5. Amino Acid Sequences Around Catalytic Residues

(1) Active Cys Sequences

			20					25					30					35				
Papain[1]	N	Q	G	S	C	G	S	C	W	A	F	S	A	V	V	T	I	E	G	I	I	K
Actinidin[2]	S	Q	G	E	C	G	G	C	W	A	F	S	A	I	A	T	V	E	G	I	N	K
Stem bromelain[3]	N	Q	N	P	C	G	A	C	W	(A F G)	A	I	A	T	V	E	S	V	A	S		
Fruit bromelain[4]	N	Q	N	P	C	G	A	C														
Ficin[5]	Q	Q	G	Q	C	G	S	C	W													
Chymopapain B[6,a]	R	V	P	A	S	G	E	C	Y													
Chymopapain[7]	N	Q	G	S	C	G	S	C	W	A	F	S	T	I	A	T	V	E	G	I	N	K
Cathepsin H[8]	N	Q	G	A	C	G	S	C	W	T	F	S	T	T	G	A	L	E	S	A	V	A
Cathepsin L[9]	Y	Q	G	A	C	G	S	C	W	A	F	S	A	V	V	L	A	Q				
Cathepsin B[8]	D	Q	G	S	C	G	S	C	W	A	F	G	A	V	E	A	M	S	D	R	I	C
Streptococcal proteinase[10]	G	Q	A	A	T	G	H	C	V	A	T	A	T	A	Q	I	M	K	Y	H	N	Y

(2) Active His Sequences

	155					160					165		
Papain[1]	N	K	V	D	H	A	V	A	A	V	G	Y	G
Actinidin[2]	T	A	V	D	H	A	I	V	I	V	G	Y	G
Stem bromelain[3]	D	K	L	N	H	A	V	T	A	I	G	Y	N
Ficin	T	S	L	D	H	A	V	A	L				
Cathepsin H	D	K	V	N	H	A	V	L	A	V	G	Y	G
Cathepsin B	V	M	G	G	H	A	I	R	I	L	G	W	G
Streptococcal proteinase	K	V	G	G	H	A	F	V	I	D	D	G	A

References: [1]Mitchel et al. (1970); [2]Carne and Moore (1978); [3]Goto et al. (1976, 1980); [4]Yamada et al. (1976); [5]Wong and Liener (1964); Husain and Lowe (1970); [6]Tsunoda and Yasunobu (1966); [7]Lynn, K.R. (personal communication); [8]Takio et al. (1983); [9]Turk et al. (1983); [10]Tai et al. (1976).
[a]May be sequence around the second, nonessential Cys of chymopapain.

that the sequence lacking Gln 19 may, in fact, correspond to that around the second, nonessential Cys in chymopapain. Note that even the bacterial enzyme streptococcal proteinase, which has virtually no sequence homology elsewhere (Tai et al., 1976), does retain Gln 19 as well as Gly 23 and Cys 25. Of the other highly conserved residues in the plant and animal enzymes, Cys 22 forms a disulfide bridge, Trp 26 is in the nonpolar core of the L-domain, and Phe 28 is at one end of the domain-domain interface.

The sequences about the essential His residue contain only three absolutely conserved residues in the six plant and animal enzymes for which data exist. Apart from His 159, these are Ala 160, whose side chain forms part of the wall of the nonpolar substrate-binding pocket (see Section 3.5), and Gly 165, which is in one of the central β strands in the R-domain. On the other hand, Asp 158, which has been ascribed a direct role in catalysis by some authors (Section 3.4) because of its charge and proximity to the active site, is not conserved. Papain, actinidin, and ficin all have Asp in this position, but bromelain and cathepsin H have Asn, and cathepsin B has Gly.

The conservation of other residues in and around the active site is less easily examined, since full sequences are available for only four plant or animal proteins, together with most of the sequence of a fifth—bromelain. Data for active site residues in these proteins are summarized in Table 6.

Table 6. Homology Among Active Site Residues in Papain, Actinidin, Cathepsin H, Cathepsin B, and Stem Bromelain

	19	22				26	64					69	133
Papain	Q	C	G	S	C	W	N	G	G	Y	P	W	V
Actinidin	Q	C	G	G	C	W	D	G	G	Y	I	T	A
Bromelain	Q	C	G	A	C	W							A
Cathepsin H	Q	C	G	S	C	W	Q	G	G	L	P	S	A
Cathepsin B	Q	C	G	S	C	W	N	G	G	Y	P	S	A
	○	Δ				○	Δ	Δ	☆	☆	☆		

	157			160	175		177	205		207
Papain	V	D	H	A	N	S	W	S	S	F
Actinidin	V	D	H	A	N	S	W	M	P	S
Bromelain	L	N	H	A	(S	G)	K	D	P	L
Cathepsin H	V	N	H	A	N	S	W	C	A	S
Cathepsin B	G	G	H	A	N	S	W	E	I	V
	☆		○	☆	○		Δ	☆		☆

○, Catalytic site; ☆, nonpolar side chain-binding pocket; and Δ, walls of active site cleft.

The piece of extended chain, 64–67, which makes up part of one wall of the active site cleft, is seen to have Gly at 65 and 66 in each case. The retention of Gly at position 66 is likely to be most important in light of inhibitor-binding experiments (Section 3.5). The residues that make up the main part of the nonpolar side chain-binding pocket, 67, 68, 133, 157, and 160, remain nonpolar and are mostly similar (although 157 changes from Val or Leu to Gly in cathepsin B). The residues at the back of the pocket, 69 and 205–207, are much more variable, and these appear to affect the detailed specificity of this site (Section 3.7). Residue 175, which has an orientational role in hydrogen bonding to His 159, is Asn in each case except bromelain, where it becomes Ser (presumably, however, the hydrogen bond can still be formed). Residue 176, which hydrogen bonds to the side chain amide oxygen of Gln 19 in papain and actinidin, is Ser in all cases except bromelain. Likewise, Trp 177, which covers the His...Asn hydrogen bond and is a prominent feature of the active site wall in papain and actinidin, is conserved except in bromelain (→ Lys). These changes suggest some structural and functional differences in the active site in bromelain.

The one bacterial enzyme for which the full amino acid sequence is known, streptococcal proteinase (Tai et al., 1976), is less easily compared. Although it is quite similar in size (253 residues, compared with 212 for papain), it has no disulfide bridges and no obvious sequence homology—with two important exceptions. The sequence leading up to the essential Cys residue, Cys 47, contains Gln and Gly residues analogous to the functionally important Gln 19 and Gly 23 in papain (See Table 5). A histidine residue at position 195 (148 residues away, compared with 134 residues away in papain) may be analogous to His 159 in papain—it also has an Ala residue adjacent (Table 5) and further on a Trp that could be analogous to Trp 177 (Tai et al., 1976). Either this enzyme has diverged so far from the plant and animal enzymes that this is the only detectable sequence homology remaining (although the structure remains of the same type) or the structure is quite different, but is an example of convergent evolution in which the same catalytically important groups have been selected.

3.   CATALYSIS AND MECHANISM

It has long been recognized (e.g., see Glazer and Smith, 1971) that the free thiol group of papain is actively involved in catalysis. Crystallographic studies (Drenth et al., 1971) and chemical studies (Husain and Lowe, 1968) also implicated the imidazole group of His 159, and subsequent studies on other thiol proteases have confirmed that an SH…imidazole pair forms the essential catalytic apparatus in this whole group of enzymes. However, there has been much discussion about the details of the catalytic mechanism, both electronic and structural. Reviews include those of Lowe (1976), Polgar (1977), Brocklehurst et al. (1981a), and Polgar and Halasz (1982), and some of the most recent results are discussed by Willenbrock and Brocklehurst (1985b). Here we shall briefly attempt to relate kinetic and mechanistic results to the structures observed crystallographically.

3.1. The General Catalytic Mechanism

The general features employed by the thiol proteases in the hydrolysis of polypeptide substrates are analogous to those of other types of proteolytic enzymes (for a comparative review, see Drenth, 1980). The basic steps are shown schematically in Figure 11.

Formation of an enzyme-substrate complex is followed by attack by a nucleophilic group of the enzyme on the carbon atom of the peptide bond to

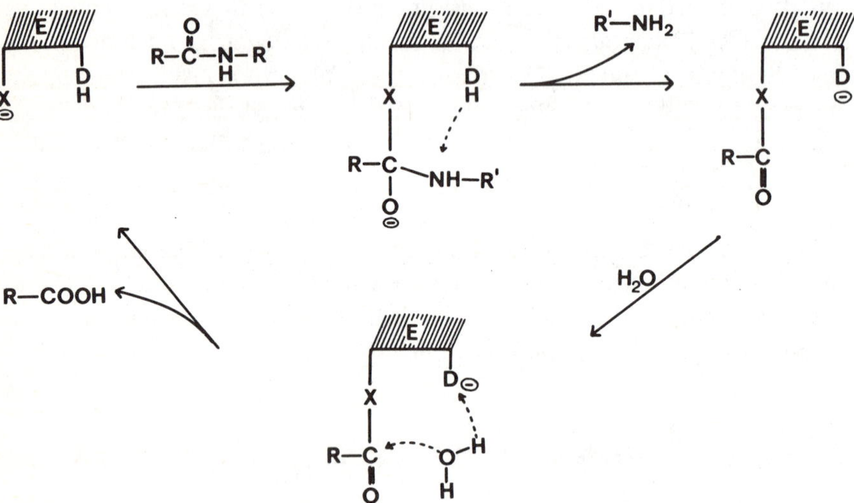

Figure 11. A generalized scheme for proteolytic cleavage. X is the essential nucleophile, while DH is a proton donor that is able to protonate the leaving group.

be cleaved. This results in the formation of a tetrahedral intermediate that may be stabilized by interaction of the oxyanion $(-C-O^-)$ with an appropriate group on the enzyme. Bond cleavage is completed by donation of a proton (by some donor group D) to the leaving group. The latter can then diffuse away, leaving the acylated enzyme (step 3), which in this case is a thioester. This can subsequently be hydrolyzed by nucleophilic attack of OH^- (from a water molecule).

3.2. The Catalytically Important Groups

In the thiol proteases, the essential nucleophile is provided by a cysteine residue. Although the SH group itself is a poor nucleophile, the deprotonated form, S^-, is a good one. The problem, however, is that the normal pK_a for a thiol group is about 9; therefore, there should be very little tendency for the formation of S^- in the pH range of maximal activity $(4-7)$. There is abundant evidence, however, that the adjacent imidazole and thiol groups in fact exist predominantly as an ion pair $S^-...^+HIm$ in this pH range. Evidence came from fluorescence-quenching measurements (Sluyterman and de Graaf, 1970; Lowe and Whitworth, 1974), which indicated an imidazolium ion below pH ~ 8.0, from direct detection of the S^- ion by ultraviolet (UV) difference spectroscopy (Polgar, 1974), from proton nuclear magnetic resonance (NMR) spectroscopy (Lewis et al., 1981), and from other measurements (see Drenth, 1975; Polgar, 1977; Polgar and Halasz, 1982 for more detailed discussion). Three different states of the thiol-imidazole pair can then be identified as the pH is changed:

$$SH.....^+HIm \qquad\qquad S^-.....^+HIm \rightleftarrows SH....Im \qquad\qquad S^-.....Im$$
$$pH < 4.0 \qquad\qquad\qquad pH\ 4.0-8.5 \qquad\qquad\qquad\qquad pH > 8.5$$
$$\text{inactive} \qquad\qquad\qquad\qquad \text{active} \qquad\qquad\qquad\qquad\qquad \text{inactive}$$

The doubly protonated, low pH, form is inactive because the nucleophilicity of SH is too low, while the nonprotonated high pH form is inactive because although S^- is a good nucleophile, the nonprotonated Im cannot act as a proton donor. In the medium pH range, the ion pair $S^-.....^+HIm$ is assumed to be the predominant and catalytically competent species. This is consistent with the fact that plots of rate of acylation as a function of pH are bell-shaped with a maximum at about pH 6.0, and the two pK values of about 4.0 and 8.5 refer to the formation and decomposition of the ion pair, that is, pK ~4 associated with the protonation of Cys 25 and pK ~8.5 with protonation of His 159.

For the deacylation step, only one ionizing group appears to be involved (the sulfur of Cys 25, of course, is acylated), and the pK of about 4 for this process is that of the imidazole group, which acts as a general base accepting a proton from water. The pK is low because (1) now there is no neighboring S^- to raise it, and (2) His 159 is in a fairly hydrophobic environment, and this is expected to reduce its pK_a significantly (Lowe, 1976). Although the low pK_a indicates that His 159 is only a weak base here, this does not matter, as the relatively unstable thioester hydrolyzes easily anyway.

3.3.   Electrostatic Effects

An important question concerns the unexpected existence of the thiolate-imidazolium ion pair in the middle pH range, 4.0−8.0. A probable explanation can be found from consideration of the protein environment. First, energy calculations on model systems (van Duijnen et al., 1979) suggest that the presence of the imidazole group adjacent to S_γ of Cys 25 is enough to increase the probability of an ion pair (reducing the pK_a for Cys 25 and increasing that for His 159). Much more significantly, however, Cys 25 is situated at the N-terminus of the long α helix, 24−42. Consideration of the electric field that results from the α helix dipole (Hol et al., 1978) suggests that its positive charge should favor the transfer of the proton from the sulfur to the imidazole ring, giving the system:

$$\text{Helix N-terminus } \delta^+ ....S^- ..... {}^+HIm$$

Molecular orbital calculations using models for Cys 25, His 159, Asn 175, and the α helix confirm this (van Diujnen et al., 1979, 1980), and predict that the ion pair is the more stable arrangement.

A more detailed study, in which the contributions of all surface groups to the overall electrostatic field were computed (Lavery et al., 1983), has confirmed the importance of the α helix field in facilitating the proton transfer (and consequent ion pair formation); but it has suggested that the α helix itself accounts for less than half the effect. Neighboring groups supply the remaining part, with the most important single residue being Asp 158. Support comes from reactivity probe kinetic studies showing that the ion pair formation in the various thiol proteases is modulated by environmental effects. These include (1) local electric field effects, as seen in comparisons of papain and actinidin (Brocklehurst et al., 1983, 1984c; Salih and Brocklehurst, 1983); and (2) ionizations of groups more remote from the active site, with the contributions of these two types of effects differing from one enzyme to another (Willenbrock and Brocklehurst, 1985b).

These observations probably explain the variations seen in the pK_a values of the ionizing groups in the various thiol proteases. Although values of ~4.0 and ~8.5 are found in every case except chymopapain A (Brocklehurst et al., 1980), differences occur. For example, for reactions of 2,2′-dipyridyl disulfide, values of 3.1 and 9.6 are found for actinidin and 2.7 and 7.7 for cathepsin H, compared with analogous values of 3.8 and 8.6 for ficin and 3.9 and 8.8 for papain (Brocklehurst et al., 1981b; Willenbrock and Brocklehurst, 1985a). A consideration of the surface groups that contribute to the electrostatic field in the environment of the thiolate-imidazolium ion pair in papain (Fig. 12) shows that many are changed in actinidin. For example, Ser 21 changes to Glu 21, Asn 64 to Asp 66, Gln 135 to Asp 138, and Gln 142 to Lys 145, and these will undoubtedly change the electrostatic field, and presumably the pK_a values.

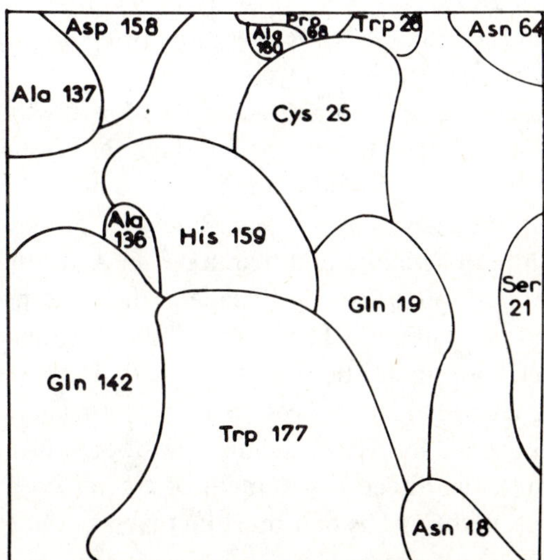

Figure 12. Surface residues contributing to the electric field that influences the formation of the S^- . . . . $^+$HIm ion pair in papain, according to Lavery et al. (1983). In actinidin, Ser 21 becomes Glu, Asn 64 becomes Asp, and Gln 142 becomes Lys, presumably with consequent changes in the electric field.

3.4. Role of Asp 158

A constant source of controversy in mechanistic studies on papain has been the role of the carboxyl group of Asp 158. In early studies, it was assumed that a carboxyl group was responsible for the lower pK_a of about 4 observed in pH activity studies (e.g., Smith, 1958). More recently, however, this pK_a value has been shown to be associated with the thiolate-imidazolium pair (see above); the crystal structures of both papain and actinidin show that the carboxyl group of Asp 158 (Asp *161* in actinidin) is turned away from the active site to hydrogen bond with the main chain NH of residue 136 (*139* in actinidin) (see Fig. 8) (Baker, 1981). This makes it far too distant from the catalytic site (7.5 Å from Cys 25; 5.5 Å from His 159) to be involved directly in catalysis without major conformational changes.

Nevertheless, some kinetic studies continue to implicate a carboxyl group (e.g., Zannis and Kirsch, 1978; Angelides and Fink, 1978; Allen et al., 1978). The most discussed mechanism is that of Angelides and Fink (1978, 1979a, 1979b), based principally on cryoenzymological studies. This proposes a pH-dependent conformational change involving substantial rotations of His 159 and Asp 158 to form a His...Asp ion pair after substrate binding. Further discussions of the question of involvement of Asp 158 can be found elsewhere (e.g., Lowe, 1976; Baines and Brocklehurst, 1982a, 1982b; Polgar and Halasz, 1982). Here we note two points that argue against any direct role by Asp 158 in catalysis.

(1) The structures of papain and actinidin show precisely the same arrangement of Asp 158 and His 159, even though the crystals were grown at pH 9.3 and 6.0, respectively (and in the two-state mechanism of Angelides and Fink [1978], the Asp...His ion pair should be present at pH 6.0).

(2) Although both actinidin and ficin have an Asp residue in a sequentially equivalent position to Asp 158 in papain, the corresponding residue in both bromelain and cathepsin H is Asn, while in cathepsin B, it is Gly. There are no neighboring substitutions that could supply a carboxyl group in a compensating change. Thus, unless the mechanisms in these enzymes are radically different (and this is unlikely in view of their broad structural and kinetic similarity), the direct involvement of a carboxyl group in catalysis can be ruled out, and Cys-His pair must be taken as the essential catalytic apparatus.

There remains evidence for conformational changes during catalysis (Holloway and Hardman, 1973; Angelides and Fink, 1978, 1979a, 1979b;

Brocklehurst et al., 1979), but these are more likely to be small-scale movements of the type observed crystallographically by Drenth et al. (1976). Likewise, observations of other ionizations that control the reactivity of the essential thiol group (e.g., Baines and Brocklehurst, 1982a, 1982b; Willenbrock and Brocklehurst, 1984, 1985a, 1985b) are likely to derive from the more subtle environmental effects of residues such as Asp 158.

3.5. Crystallographic Binding Studies

Berger and Schechter (1970) have shown, in a series of classic experiments, that the active site of papain includes seven "subsites," each able to accommodate one amino acid residue of the substrate. Four subsites (S_1 to S_4) are located on the acyl side of the peptide bond to be cleaved, and three (S_1' to S_3') are on the amino side—and the corresponding substrate residues are labeled P_1 to P_4 and P_1' to P_3' (Fig. 13). They further showed that the primary specificity of papain is for binding a hydrophobic residue (Phe in particular) in the S_2 subsite. It is clear that other thiol proteases show a similar specificity for the S_2 subsite, and the structural similarities (Sections 2.4 and 2.6) argue for a similar extended active site. Crystallographic binding studies have naturally aimed at exploiting this specificity.

Early crystallographic binding studies on papain, using inhibitors that resembled the acylating portion of good substrates were handicapped by low occupancies and lack of isomorphism and crystalline order (Wolthers et al., 1970). Similar problems have handicapped inhibitor-binding experiments

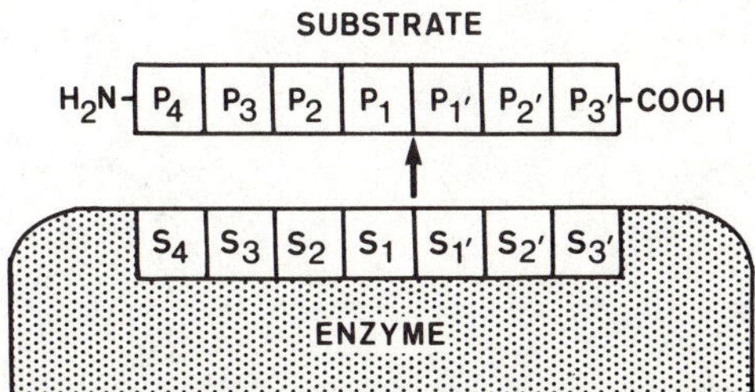

Figure 13. The extended binding site in papain in the terminology of Berger and Schechter (1970). The seven subsites in the extended active site cleft are designated S_1 to S_4 and S_1' to S_3', with corresponding substrate residues P_1 to P_4 and P_1' to P_3'. Cleavage is at the P_1-P_1' peptide bond. (Adapted from Berger and Schechter 1970.)

on actinidin—in this case, low occupancies due to weak binding (binding of phenylalanine derivatives to the S_2 subsite is about 100 times weaker than for papain; Baker et al., 1980). A series of X-ray studies using chloromethyl ketone inhibitors of papain, however, have given a clear picture of the way these molecules bind (Drenth et al., 1976). The results for three different inhibitors (di-, tri-, and tetrapeptides) are consistent and agree well with the earlier work of Wolthers. Moreover, there is good reason to believe that their binding mode is similar to that of real substrates.

The chloromethyl ketone inhibitors used—Z-Phe-Ala-CK, Z-Gly-Phe-Gly-CK, and Ac-Ala-Ala-Phe-Ala-CK (where Z = benzyloxycarbonyl, Ac = acetyl, and CK = chloromethyl ketone)—all bond covalently to the sulfur atom of the essential Cys 25 as follows:

$$\underset{\displaystyle O}{R\text{-}\overset{\|}{C}\text{-}CH_2\text{-}Cl} + HS\text{-papain} \longrightarrow \underset{\displaystyle O}{R\text{-}\overset{\|}{C}\text{-}CH_2\text{-}S\text{-papain}} + HCl$$

All had L-phenylalanine at the P_2 position (Fig. 13). The binding mode of the Z-Phe-Ala-CK inhibitor is shown in a stereo view in Figure 14, and the

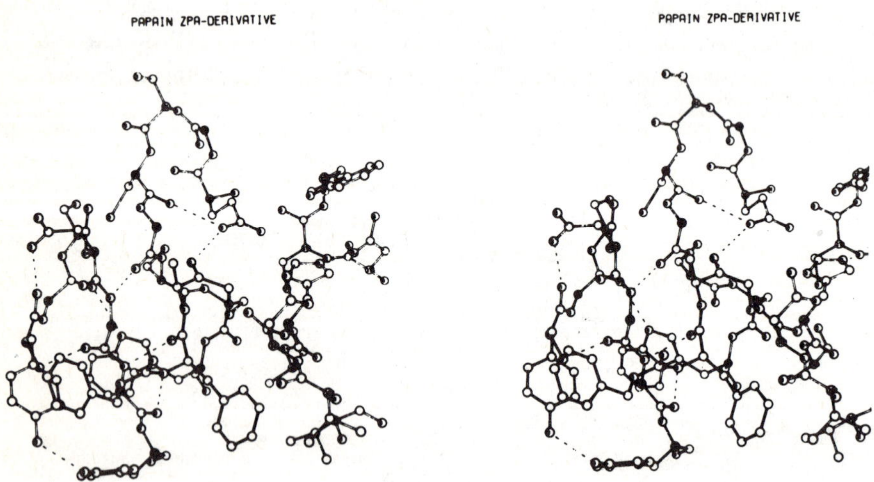

Figure 14. Stereo diagram showing the binding of the chloromethyl ketone inhibitor Z-Phe-Ala-CK in the active site of papain. (Inhibitor shown with filled bonds; papain with open bonds.) (From Drenth et al., 1976.)

various interactions made are represented schematically in Figure 15. For each inhibitor, the fit to the positive density in the active site was essentially the same, although in several places density was weak because of the displacement of solvent molecules present in the parent enzyme. The terminal methylene group is attached to S_γ of Cys 25, and the carbonyl oxygen adjacent to it (i.e., the P_1 C=O, corresponding to the C=O of the scissile bond of a substrate) is directed towards two NH groups: the main chain NH of residue 25, at the N-terminus of α helix 24–42, and an NH group of the amide side chain of Gln 19. The side chain of the Ala residue in position P_1 projects outward from the center of the cleft, from where it would be easy to build a Lys side chain to interact with the carboxyl group of Asp 158, thus

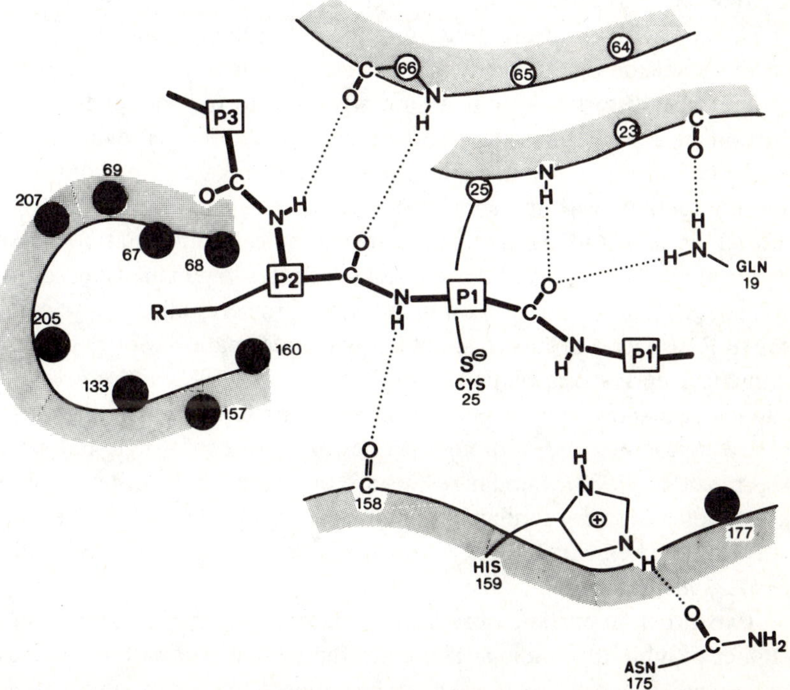

Figure 15. Schematic diagram showing the probable enzyme-substrate interactions that occur in papain (deduced from inhibitor binding studies of Drenth et al., 1976). Only interactions on the acyl side of the scissile bond are shown. Residues most intimately involved in the nonpolar binding pocket (S_2 subsite) are Val 133 and Val 157, and to a slightly lesser extent, Tyr 67, Pro 68, and Ala 160.

nicely explaining an apparent preference for basic side chains at P_1 (Glazer and Smith, 1971). The peptide between P_1 and P_2 runs past the extended piece of chain comprising residues 65−67 in the wall of the active site cleft in such a way that the P_2 NH and C=O groups can hydrogen bond with the C=O and NH groups of Gly 66, while the P_1 NH is directed towards the C=O of residue 158 on the opposite wall of the cleft. The side chain of residue P_2 (phenylalanine) is arranged so that C_β lies between the side chains of Pro 68 and Ala 160, and beyond it the phenyl ring is adjacent to the side chains of Val 133 and Val 157; these residues, together with Tyr 67, Trp 69, and Phe 207 form a hydrophobic-binding pocket. Beyond P_2, the inhibitor binding is less clear-cut. The P_3 residue is situated near the side chains of Tyr 67 and Tyr 61, but there are small differences in the binding of the different inhibitors, while beyond P_3 no convincing interpretation could be made as the density (and presumably binding) became very weak.

The above observations give a nice picture of the binding of the residues on the acyl side of the point of cleavage. The inhibitor structure has no geometrical distortions, and a similarly undistorted model for the acyl portion of a true substrate could be constructed by removing the extra methylene group and joining up the thioester linkage. The conformation and interactions with the surrounding protein remained essentially undisturbed. It is gratifying that an almost identical conformation has been deduced for the binding of thionoester substrates to papain using resonance Raman (RR) spectroscopy combined with X-ray crystallography (comparison of RR spectra of substrates of known conformation with RR spectra of dithioacyl derivatives of papain) (Huber et al., 1982). Moreover, these studies also show that the substrate conformation is *not* significantly distorted by the enzyme, although it does exercise conformational selection (Storer et al., 1983). Similar results are obtained for six other thiol proteases, indicating very similar binding of the acyl portion of a substrate and very similar chemistry of deacylation (Carey et al., 1983, 1984; Brocklehurst et al., 1984b).

Two other important observations come from the difference Fourier studies of inhibitor binding. These are that (1) distinct patterns of positive and negative density were seen in the walls of the active site cleft, in the region of residues 59−68, 21−23, and the 22−63 disulfide bridge in one wall of the cleft, and (less pronounced) in residues 158−159 on the other wall; and (2) the S_γ atom in Cys 25 is required to move up into the cleft by about 0.9 Å in order to bind to the inhibitor, with a consequently very slight adjustment of the residues of the first turn of the α helix 24–42. The movements in the walls of the cleft are consistent with a widening of about 1 Å to accommodate the inhibitors, thus supporting the suggestion of Lowe and Yuthavong (1971) that binding of residues such as phenylalanine in the S_2 site (the hydrophobic pocket) should force the cleft to open somewhat. The movement of the sulfur of Cys 25, together with the widening of the cleft, are probably required because the essential catalytic residues have almost zero solvent-accessible area in the native protein (Kamphuis et al., 1985a).

3.6.   The Catalytic Mechanism in Structural Terms

The above inhibitor-binding studies, together with the investigations of the state of the Cys-His couple (Sections 3.2 and 3.3), have led to detailed proposals for the mechanisms of action of the thiol proteases (Drenth et al., 1975, 1976; Drenth, 1980). The main features are as follows, and are shown schematically in Figure 16.

(1) In the resting enzyme the Cys-His couple is predominantly in the ion pair form RS^-.....^+HIm (with the sulfur and imidazole ring probably coplanar, Section 2.4). The substrate can bind, in an unstrained conformation, so that the strongly nucleophilic thiolate sulfur attacks the carbonyl carbon of the peptide bond to be cleaved. At the same time, the carbonyl oxygen is directed towards two proton donors that can hydrogen bond to it; the side chain NH_2 group of Gln 19 and the main chain NH of residue 25.

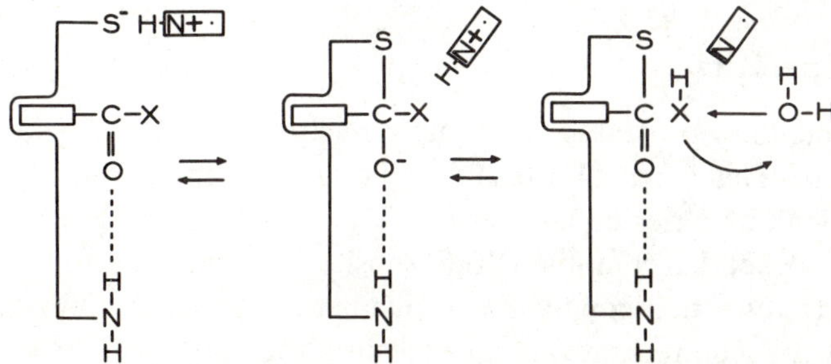

Figure 16. Proposed mechanism of action of papain and other thiol proteases, with attack by the nucleophilic S^- ion, stabilization of the tetrahedral intermediate by Gln 19-NH_2, and rotation of the protonated imidazole ring to act as proton donor to the leaving group.

These interactions should increase the polarization of the C=O bond, facilitating the attack of the sulfur, and should then help to stabilize the tetrahedral intermediate, which carries a formal negative charge on the oxygen. The two groups, Gln 19 NH_2 and 25 NH, thus provide in the thiol proteases the equivalent of the "oxyanion hole" found in the serine proteases (Henderson et al., 1971; Robertus et al., 1972). Note that although the kinetic importance of this "oxyanion hole" in thiol proteases has recently been questioned (Asboth and Polgar, 1983), Gln 19 is invariant in all thiol proteases thus far sequenced, even in the otherwise nonhomologous streptococcal proteinase (Table 5).

(2) In the tetrahedral intermediate, the NH group of the scissile peptide bond is in the position to hydrogen bond to the main chain C=O of residue 158 (thus further stabilizing this intermediate). The rotational freedom of the imidazole ring (demonstrated crystallographically, Section 2.5) allows it to rotate about 30° about C_β-C_γ, so that its protonated $N_{\delta 1}$ atom is now pointing at the lone pair on the nitrogen atom of the leaving group, that is, it is ideally placed to donate a proton to the leaving group (His 159 is now a strong proton donor, with pK_a ~4, because the sulfur is acylated).

(3) After departure of the leaving group, the acyl group is removed by attack by OH^- from a water molecule, with the imidazole group acting as a base (albeit weak, because of its low pK_a) in accepting a proton.

(4) Binding of the substrate beyond the scissile peptide bond is assumed to follow that observed for the chloromethyl ketone inhibitors (Section 3.5). That is, the side chain of the P_2 residue (Fig. 13) fits into the hydrophobic-binding pocket (S_2 site), and the P_2 NH and C=O groups can hydrogen bond with the C=O and NH groups of Gly 66.

3.7. Specificity

The plant thiol proteases are usually regarded as having a fairly broad specificity. This is partly a result of the fairly deep, extended active site groove with its series of subsites (see Fig. 13). Although the hydrophobic binding pocket that comprises the S_2 subsite does confer a strong preference for substrates with a nonpolar side chain at this position, this can be overruled by the requirements at other subsites, thus giving rise to a wide range of proteolytic cleavages.

The reactivity and inhibition studies of Berger and Schechter (1970) showed that papain has a strong affinity for Phe, Tyr, Val, or Leu at the P_2

position of a substrate. A similar preference is shown by all the plant thiol proteases, as judged by comparisons of their cleavage of the B chain of insulin (Lynn, 1983)—the most common sites, shown by almost all of the eight enzymes examined were:

Val-Asn-Gln Leu-Cys-Gly Val-Glu-Ala Val-Cys-Gly Phe-Phe-Tyr
  ↑   ↑   ↑   ↑   ↑

(The residues that would occupy S_2 are underlined). The animal proteins, cathepsins B, H, and L, also show a preference for a nonpolar side chain on the P_2 residue to bind in the S_2 subsite (Barrett and Kirschke, 1981).

The residues in the S_2 subsite that make most intimate contact with such a side chain are those of Val 133, Val 157, Ala 160, Pro 68, and Tyr 67. In the four enzymes for which full amino acid sequences are available (papain, actinidin, and cathepsins B and H; Table 6), these all remain hydrophobic. The nonpolar character, and general specificity, of the S_2 subsite would therefore be conserved. There are, however, differences of detail. In actinidin, the bulky side chain of Met 211 (corresponding to Ser 205 in papain) lies across the back of the binding pocket, making it significantly shorter (Fig. 17), and kinetic experiments show that substrates with aromatic groups at P_2 bind $10-100$ times less well than to papain (Baker et al., 1980). Valine or leucine residues at P_2 should still bind well, but tyrosine would be expected to bind poorly because the hydroxyl group would clash with Met 211. In cathepsin B, the back of the pocket is occupied by Glu in place of Ser 205. Now substrates with aromatic groups at P_2, for example, Z-Phe-Arg derivatives, still bind well because the hydrophobic part of the pocket remains—but so do substrates with basic groups at P_2, for example, Z-Arg-Arg-naphthylamide (Barrett and Kirschke, 1981), presumably because the new acidic group at the back of the pocket can interact with the basic guanidinium group, while the nonpolar region binds the nonpolar part of the side chain. In cathepsin H, on the other hand, Ser 205 is replaced by Cys, and only the nonpolar preference remains.

The S_1 site has no binding pocket as such, and has a much less clearly defined specificity. Low molecular weight substrates indicate some preference for Lys or Arg (over Ala) at P_1 in papain (Smith and Kimmel, 1960) and actinidin (Boland and Hardman, 1972). This and the slow hydrolysis of substrates with Glu at P_1 suggest that interaction with the carboxyl group of Asp 158 could be involved. However, cathepsins B and H, which lack an

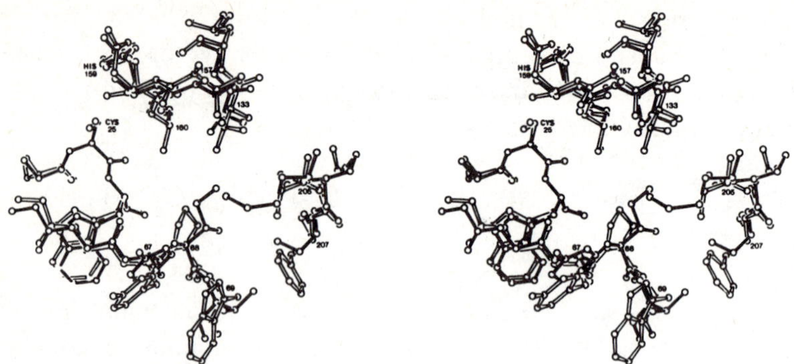

Figure 17. Stereo picture showing the nonpolar binding pockets (S_2 subsites) in papain (open bonds) and actinidin (filled bonds). The change of Ser 205 in papain to Met in actinidin shortens the pocket and reduces the binding of Phe side chains of substrates. In cathepsin B, Ser 205 becomes Glu, probably explaining the good binding of Arg side chains in S_2.

equivalent to Asp 158, appear to show a similar specificity. Berger and Schechter (1970) showed that Leu and Phe bind equally well at S_1, so the interaction may be primarily hydrophobic. The one striking feature of the S_1 subsite is that β-branched side chains such as Val are not accepted. Thus, the peptide Phe-Val-Ala is not hydrolyzed, that is, the spatial restriction preventing the Val side chain from occupying S_1 abolishes the Phe-X ↓ -Y specificity. Similarly, only bromelain of eight plant enzymes hydrolyzes the insulin B chain at Phe-Val ↓ -Asn (Lynn, 1983).

Other specificities are also understandable in terms of the three-dimensional structure. There is a preference for Trp or Ile at S_1' (Alecio et al., 1974), and this presumably involves contact with the side chain of Trp 177 in the active site wall. Bromelain may show a different specificity here, as Trp 177 is replaced by Lys. The absolute requirement for L-amino acids at P_1 arises because when a peptide binds to the enzyme, as shown by Drenth et al. (1976), the hydrogen on the α-carbon points towards the enzyme (in the vicinity of Gly 23). This is presumably why D-amino acids cannot bind properly (their side chain would clash with Gly 23), and also why residue 23 is always Gly. Finally, it has been suggested (Lowe and Yuthavong, 1971) that it is the presence of Asp 158 that prevents carboxypeptidase activity by papain (due to repulsion of the terminal carboxy group by that of Asp 158). This could explain the dipeptidylcarboxypeptidase activity of cathepsin B (Kalnitsky et al., 1983), in which Asp 158 is replaced by Gly. (On the other hand, the same residue is replaced by Asn in bromelain and cathepsin H, but carboxypeptidase activity has not been reported.)

4. CONCLUSIONS

The plant thiol proteases papain and actinidin are small proteins (M_r ~23,500) with a characteristic two-domain structure. Crystallographic refinement at high resolution has shown that these two proteins have a remarkably close similarity in three-dimensional structure despite less than 50% sequence identity. Insertions and deletions cause minimal perturbation of the structures, and outside these positions, there is only one significant difference in polypeptide chain conformation between the two proteins. This results from sequence changes in a region of virtually no sequence identity that is remote from secondary structures or the active site.

The catalytic apparatus in both proteins is the same, with the side chains of Cys 25 and His 159 (*162* in actinidin) situated together in a deep, extended cleft between the two domains. The imidazole ring of His 159 (*162*) is oriented by hydrogen bonding to the buried side chain of Asn 175 (*182*), but crystallographic studies have shown that it can rotate ~30° about C_β-C_γ in response to changes in the state of S_γ of Cys 25. This rotation is probably important in catalysis.

Although inhibitor-binding studies show that the active site cleft is widened by binding, the cleft itself appears to be a relatively rigid structure, as judged by the crystallographic B values of the residues comprising its walls. This is true in both papain and actinidin. Conformational flexibility within the active site is mainly confined to the side chains of Cys 25 and His 159.

Amino acid sequence comparisons show a high level of identity in all of the plant thiol proteases, as well as with the animal proteins: cathepsins B, H, and L. The sequence identity between papain and cathepsin H is almost as great as between papain and actinidin. This leaves no doubt that the plant and animal proteins form a family of enzymes with the same three-dimensional folding and very similar active sites. The sequence comparisons also show that Asp 158 cannot be directly involved in catalysis in papain and actinidin unless a different kinetic mechanism is used in some of the other thiol proteases. This is very unlikely in our view. The sequence differences also account for some differences in specificity (e.g., in cathepsin B) through changes at the back of the S_2 binding pocket.

Various spectroscopic studies (fluorescence, UV, NMR) have shown that the Cys-His couple exists predominantly as an ion pair, $RS^-.....^+HIm$, in the pH range of maximal activity. This state is facilitated by the local electric field, primarily arising from the situation of Cys 25 at the N-terminus of a 5-turn α helix, but also contributed to by neighboring side chains. It also illustrates the fact that the pK values of particular groups in proteins can deviate appreciably from normal values, with the value for Cys 25 being much lower and that for His 159 much higher than normally accepted values.

Crystallographic binding studies with chloromethyl ketones give a nice picture of the interactions that contribute to binding on the acyl side of the point of cleavage. Features of the proposed catalytic mechanism include (1) the "oxyanion hole," comprising 25 NH and Gln 19 NH_2, which may stabilize the oxyanion tetrahedral intermediate; and (2) the rotational flexibility of the imidazole ring of His 159, which allows it to form the crucial ion pair with Cys 25, and also to act as proton donor to the leaving group.

5.   FUTURE DIRECTIONS

There are a number of questions left unanswered by the above structural studies. Although there are good indications (Sections 2.1 and 2.5) that the active site is essentially unchanged in two different forms of papain and in actinidin, it remains true that no structural work has yet been done on a thiol protease with its active Cys fully reduced. Comparison of the refined structure of a third form of papain, the 2-hydroxyethylthio derivative of a different crystal form, D (Priestle et al., 1984) will be of considerable interest. We do not have any direct evidence on the details of substrate-binding modes—low temperature X-ray studies offer one approach. There is also no crystallographic information about binding on the amino side of the point of cleavage. Numerous natural inhibitors of the thiol proteases exist—both small molecule and protein inhibitors—and complexes of these would be extremely interesting. There are also significant variations in the properties of some of the enzymes, for example, the chymopapains and some of the cathepsins. There remains the intriguing question of whether the bacterial enzyme, streptococcal proteinase, is related to the plant and animal proteins, but has diverged to the point where virtually no amino acid identity remains, or whether it provides another example of convergent evolution to a similar active site. Finally, studies with reactivity probes have suggested differences in the electrostatic properties of the active sites of different enzymes (Brocklehurst et al., 1983, 1984a, 1984b). Two things are needed here: X-ray studies to define how these probes bind, and a better understanding of the contribution that neighboring or more distant groups make to the electrical properties of the protein surface.

ACKNOWLEDGMENTS

We wish to acknowledge the considerable contribution made by Rene Kamphuis in refinement of the papain structure and in the structure comparisons. We are grateful to Drs. Wolfram Saenger, Keith Brocklehurst, Paul Carey, Andrew Storer, and K.R. Lynn for their help in sending both published and unpublished results. E.N.B. wishes to thank colleagues at Massey University for their help and encouragement over a number of years, and Dr. Brian Matthews for his hospitality and support throughout a most enjoyable year at the University of Oregon during which this chapter was written.

REFERENCES

Abraham, K. I., and Joshi, P. N. (1979). *Biochim. Biophys. Acta* **568**, 111–119, 120–126.

Alecio, M. R., Dann, M. L., and Lowe, G. (1974). *Biochem. J.* **141**, 495–501.

Allen, K. G. D., Stewart, J. A., Johnson, P. E., and Wettlaufer, D. G. (1978). *Eur. J. Biochem.* **87**, 575–582.

Angelides, K. J., and Fink, A. L. (1978). *Biochemistry* **17**, 2659–2668.

Angelides, K. J., and Fink, A. L. (1979a). *Biochemistry* **18**, 2355–2363.

Angelides, K. J., and Fink, A. L. (1979b). *Biochemistry* **18**, 2363–2369.

Artymiuk, P. J., Blake, C. C. F., Grace, D. E. P., Oatley, S. J., Phillips, D. C., and Sternberg, M. J. E. (1979). *Nature* **280**, 563–568.

Asboth, B., and Polgar, L. (1983). *Biochemistry* **22**, 117–122.

Baines, B. S., and Brocklehurst, K. (1982a). *J. Protein Chem.* **1**, 120–139.

Baines, B. S., and Brocklehurst, K. (1982b). *Biochem. J.* **205**, 205–211.

Baker, E. N. (1973). *J. Mol. Biol.* **74**, 411–412.

Baker, E. N. (1977). *J. Mol. Biol.* **115**, 263–277.

Baker, E. N. (1980). *J. Mol. Biol.* **141**, 441−484.

Baker, E. N. (1981). In *Structural Studies on Molecules of Biological Interest*, G. G. Dodson, J. P. Glusker, and D. Sayre, Eds. Clarendon Press, Oxford, pp. 339−349.

Baker, E. N., and Dodson, E. J. (1980). *Acta Crystallogr.* **A36**, 559−572.

Baker, E. N., Boland, M. J., Calder, P. C., and Hardman, M. J. (1980). *Biochim. Biophys. Acta* **616**, 30−34.

Barrett, A. J. (1973). *Biochem. J.* **131**, 809−822.

Barrett, A. J., and McDonald, J. K. (1980). *Mammalian Proteases. A Glossary and Bibliography*, Vol. 1. Academic Press, New York, 416 pp.

Barrett, A. J., and Kirschke, H. (1981). *Methods Enzymol.* **80**, 535−561.

Berger, A., and Schechter, I. (1970). *Phil. Trans. Roy. Soc. Lond.* **B257**, 249−264.

Boland, M. J., and Hardman, M. J. (1972). *FEBS Lett.* **27**, 282−284.

Boland, M. J., and Hardman, M. J. (1973). *Eur. J. Biochem.* **36**, 575−582.

Brockbank, W. J., and Lynn, K. R. (1979). *Biochim. Biophys. Acta* **578**, 13−22.

Brocklehurst, K. (1982). *Methods Enzymol.* **87C**, 427−469.

Brocklehurst, K. (1983). *Biochem. J.* **213**, 559−560.

Brocklehurst, K., and Salih, E. (1983). *Biochem. J.* **213**, 559−560.

Brocklehurst, K., Malthouse, J. P. G., and Shipton, M. (1979). *Biochem. J.* **183**, 223−231.

Brocklehurst, K., Baines, B. S., and Mushiri, M. S. (1980). *Biochem. J.* **189**, 189−192.

Brockelhurst, K., Baines, B. S., and Kierstan, M. P. J. (1981a). *Top. Enzyme Ferment. Biotechnol.* **5**, 262−335.

Brocklehurst, K., Baines, B. S., and Malthouse, J. P. G. (1981b). *Biochem. J.* **197**, 739−746.

Brocklehurst, K., Mushiri, S. M., Patel, G., and Willenbrock, F. (1983). *Biochem. J.* **209**, 873−879.

Brocklehurst, K., Baines, B. S., Salih, E., and Hatzoulis, C. (1984a). *Biochem. J.* **221**, 553−554.

Brocklehurst, K., Carey, P. R., Lee, H-H., Salih, E., and Storer, A. C. (1984b). *Biochem. J.* **223**, 649−657.

Brocklehurst, K., Salih, E., and Lodwig, T. S. (1984c). *Biochem. J.* **220**, 609−612.

Carey, P. R., Ozaki, Y., and Storer, A. C. (1983). *Biochem. Biophys. Res. Comm.* **117**, 725−731.

Carey, P. R., Angus, R. H., Lee, H-H., and Storer, A. C. (1984). *J. Biol. Chem.* **259**, 14357−14360.

Carne, A., and Moore, C. H. (1978). *Biochem. J.* **173**, 73−83.

Docherty, K., Carroll, R. J., and Steiner, D. F. (1982). *Proc. Natl. Acad. Sci. USA* **79**, 4613−4617.

Drenth, J. (1976). In *CRC Handbook of Biochemistry and Molecular Biology*, Vol. 3, *Proteins*, G. D. Fasman, Ed., 3rd ed. CRC Press, Cleveland, pp. 356–357.

Drenth, J. (1980). *Recl. Trav. Chim. Pays-Bas* **99**, 185–190.

Drenth, J., Jansonius, J. N., Koekoek, R., Swen, H. M., and Wolthers, B. G. (1968). *Nature* **218**, 929–932.

Drenth, J., Jansonius, J. N., Koekoek, R., and Wolthers, B. G. (1971). *Adv. Protein Chem.* **25**, 79–115.

Drenth, J., Swen, H. M., Hoogenstraaten, W., and Sluyterman, L. A. Ae. (1975). *Proc. Koninkl. Akad. Wetensch. C* **78**, 104–110.

Drenth, J., Kalk, K. H., and Swen, H. M. (1976). *Biochemistry* **15**, 3731–3738.

Ducastaing, A., and Etherington, D. J. (1978). *Biochem. Soc. Trans.* **6**, 938–940.

Evans, P., and Etherington, D. J. (1978). *Eur. J. Biochem.* **83**, 87–97.

Gilles, A-M., Imhoff, J-M., and Keil, B. (1979). *J. Biol. Chem.* **254**, 1462–1468.

Gilles, A-M., DeWolf, A., and Keil, B. (1983). *Eur. J. Biochem.* **130**, 473–479.

Glazer, A. N., and Smith E. L. (1971). In *The Enzymes*, P. D. Boyer, Ed., 3rd ed., Vol. 3. Academic Press, New York, pp. 501–546.

Gohda, E., and Pitot, H. C. (1981). *J. Biol. Chem.* **256**, 2567–2572.

Goto, K., Murachi, T., and Takahashi, N. (1976). *FEBS Lett.* **62**, 93–95.

Goto, K,. Takahashi, N., and Murachi, T. (1980). *Int. J. Pept. Prot. Res.* **15**, 335–341.

Heinemann, U., Pal, G. P., Hilgenfeld, R., and Saenger, W. (1982). *J. Mol. Biol.* **161**, 591–606.

Henderson, R., Wright, C. S., Hess, G. P., and Blow, D. M. (1971). *Cold Spring Harbor Symp. Quant. Biol.* **36**, 63–70.

Hendrickson, W. A., and Konnert, J. H. (1980). In *Biomolecular Structure, Function and Evolution*, Srinivasan, R., Ed. Vol. 1. Pergamon Press, Oxford, pp. 43–57.

Hol, W. G. J., van Duijnen, P. T., and Berendsen, H. J. C. (1978). *Nature* **273**, 443–446.

Holloway, M. R., and Hardman, M. J. (1973). *Eur. J. Biochem.* **32**, 537–546.

Holloway, M. R., Antonini, E., and Brunori, M. (1971). *Eur. J. Biochem.* **24**, 332–341.

Huber, C. P., Ozaki, Y., Pliura, D. H., Storer, A. C., and Carey, P. R. (1982). *Biochemistry* **21**, 3109–3115.

Husain, S. S., and Lowe, G. (1968). *Biochem. J.* **108**, 855–859.

Husain, S. S., and Lowe, G. (1970). *Biochem. J.* **117**, 333–340.

Johnson, F. A., Lewis, S. D,. and Shafer, J. A. (1981). *Biochemistry* **20**, 44–48.

Kalnitsky, G., Chatterjee, R., Singh, H., Lones, M., and Paszkowski, A. (1983). In *Proteinase Inhibitors: Medical and Biological Aspects*, N. Katunuma, H. Umezawa, and H. Holzer, Eds. Springer-Verlag, Berlin, pp. 263–273.

Kamphuis, I. G., Kalk, K. H., Swarte, M. B. A., and Drenth, J. (1985a). *J. Mol. Biol.* **179**, 233–256.

Kamphuis, I. G., Drenth, J., and Baker, E. N. (1985b). *J. Mol. Biol.* **182**, 317–329.

Khan, I. U., and Polgar, L. (1983). *Biochim. Biophys. Acta* **760**, 350–356.

Kirschke, H., Langner, J., Wiederanders, B., Ansorge, S., Bohley, P., and Hanson, H. (1977a). *Acta Biol. Med. Ger.* **36**, 185–199.

Kirschke, H., Langner, J., Wiederanders, B., Ansorge, S,. and Bohley, P. (1977b). *Eur. J. Biochem.* **74**, 293–301.

Kortt, A. A., Hinds, J. A., and Zerner, B. (1974). *Biochemistry* **13**, 2023–2028.

Lavery, R., Pullman, A., and Wen, Y. K. (1983). *Int. J. Quantum Chem.* **24**, 353–371.

Lenney, J. F. (1983). In *Proteinase Inhibitors: Medical and Biological Aspects*, N. Katunuma, H. Umezawa, and H. Holzer, Eds. Springer-Verlag, Berlin, pp. 113–123.

Lewis, S. D., Johnson, F. A., and Shafer, J. A. (1981). *Biochemistry* **20**, 48–51.

Liu, T-Y., and Elliott, S. D. (1971). In *The Enzymes*, P. D. Boyer, Ed., 3rd ed., Vol. 3. Academic Press, New York, pp. 609–647.

Locnikar, P., Popovic, T., Lah, T., Kregar, I., Babnik, J., Kopitar, M., and Turk, V. (1981). In *Proteinases and Their Inhibitors: Structure, Function and Applied Aspects*, V. Turk, and L. J. Vitale, Eds. Pergamon Press, Oxford, pp. 109–116.

Lones, M., Chatterjee, R., Singh, H., and Kalnitsky, G. (1983). *Arch. Biochem. Biophys.* **221**, 64–75.

Lowe, G. (1970). *Phil. Trans. Roy. Soc. Lond.* **B257**, 237–248.

Lowe, G. (1976). *Tetrahedron* **32**, 291–302.

Lowe, G., and Yuthavong, Y. (1971). *Biochem. J.* **124**, 107–115.

Lowe, G., and Whitworth, A. S. (1974). *Biochem. J.* **141**, 503–515.

Lynn, K. R. (1983). *Phytochemistry* **22**, 2485–2487.

Lynn, K. R., and Yaguchi, M. (1979). *Biochim. Biophys. Acta* **581**, 363–364.

Lynn, K. R., Brockbank, W. J., and Clevette, N. A. (1980a). *Biochim. Biophys. Acta* **612**, 119–125.

Lynn, K. R,. Yaguchi, M., and Roy, C. (1980b). *Biochim. Biophys. Acta* **624**, 579–580.

McDowall, M. A. (1970). *Eur. J. Biochem.* **14**, 214–221.

Mitchel, R. E. J., Chaiken, I. M., and Smith, E. L. (1970). *J. Biol. Chem.* **245**, 3485–3492.

Mitchell, W. M., and Harrington, W. F. (1971). In *The Enzymes*, P. D. Boyer, Ed., 3rd ed., Vol. 3. Academic Press, New York, pp. 699–719.

Murachi, T., Tanaka, K., Hatanaka, M., and Murakami, T. (1981). *Adv. Enzyme Regulation* **19**, 407–424.

Ota, S., Moore, S., and Stein, W. H. (1964). *Biochemistry* **3**, 180–185.

Pal., G. P., and Sinha, N. K. (1980). *Arch. Biochem. Biophys.* **202**, 321–329.

Petsko, G. A., and Ringe, D. (1984). *Ann. Rev. Biophys. Bioeng.* **13**, 331–371.

Polgar, L. (1974). *FEBS Lett.* **47**, 15–18.

Polgar, L. (1977). *Int. J. Biochem.* **8**, 171–176.

Polgar, L. (1981). *Biochim. Biophys. Acta* **658**, 262–269.

Polgar, L. (1984). *Biochem. J.* **221**, 555–556.

Polgar, L., and Halasz, P. (1982). *Biochem. J.* **207**, 1–10.

Priestle, J. P., Ford, G. C., Glor, M., Mehler, E. L., Smith, J. D. G., Thaller, C., and Jansonius, J. N. (1984). *Abstracts*, XIII IUCr Congress, Hamburg, C-17.

Robertus, J. D., Kraut, J., Alden, R. A., and Birktoft, J. J. (1972). *Biochemistry* **11**, 4293–4303.

Robinson, G. W. (1975). *Biochemistry* **14**, 3695–3700.

Salih, E., and Brocklehurst, K. (1983). *Biochem. J.* **213**, 713–718.

Schwartz, W. N., and Barrett, A. J. (1980). *Biochem. J.* **191**, 487–497.

Silverstein, R. M., and Kezdy, F. J. (1975). *Arch. Biochem. Biophys.* **167**, 678–686.

Sluyterman, L. A. Ae., and de Graaf, M. J. M. (1970). *Biochim. Biophys. Acta* **200**, 595–597.

Smith, E. L. (1958). *J. Biol. Chem.* **233**, 1392–1397.

Smith, E. L., and Kimmel, J. R. (1960). In *The Enzymes*, P.D. Boyer, H. Lardy, and K. Myrbäck, Eds., Vol. 4. Academic Press, New York, pp. 133–173.

Storer, A. C., Lee, H., and Carey, P. R. (1983). *Biochemistry* **22**, 4789–4796.

Sugiura, M., and Sasaki, M. (1974). *Biochim. Biophys. Acta* **350**, 38–47.

Tai, J. Y., Kortt, A. A., Liu, T-Y., and Elliott, S. D. (1976). *J. Biol. Chem.* **251**, 1955–1959.

Takahashi, N., Yasuda, Y., Goto, K., Miyake, T., and Murachi, T. (1973). *J. Biochem* **74**, 355–373.

Takio, K., Towatari, T., Katunuma, N., Teller, D. C., and Titani, K. (1983). *Proc. Natl. Acad. Sci. USA* **80**, 3666–3670.

Towatari, T., Tanaka, K., Yoshikawa, D., and Katunuma, N. (1978). *J. Biochem.* **84**, 659–671.

Towatari, T., Kawabata, Y., and Katunuma, N. (1979). *Eur. J. Biochem.* **102**, 279–289.

Tsunoda, J. N., and Yasunobu, K. T. (1966). *J. Biol Chem.* **241**, 4610–4615.

Turk, V., Brzin, J., Kopitar, M., Kregar, I., Locnikar, P., Longer, M., Popovic, T., Ritonja, A., Vitale, L., Machleidt, W., Giraldi, T., and Sava, G. (1983). In *Proteinase Inhibitors: Medical and Biological Aspects*, N. Katunuma, H. Umezawa, and H. Holzer, Eds. Springer-Verlag, Berlin, pp. 125–134.

van Duijnen, P. T., Thole, B. T., and Hol, W. G. J. (1979). *Biophys. Chem.* **9**, 273–280.

van Duijnen, P. T., Thole, B. T., Broer, R., and Nieuwpoort, W. C. (1980). *Int. J. Quantum Chem.* **17**, 651−671.

Willenbrock, F., and Brocklehurst, K. (1984). *Biochem. J.* **222**, 805−814.

Willenbrock, F., and Brocklehurst, K. (1985a). *Biochem. J.* **227**, 511−519.

Willenbrock, F., and Brocklehurst, K. (1985b). *Biochem. J.* **227**, 521−528.

Wolthers, B. G., Drenth, J., Jansonius, J. N., Koekoek, R., and Swen, H. M. (1970). In *Proc. Int. Symp. Structure-Function Relationships of Proteolytic Enzymes*, P. Desnuelle, H. Neurath, and M. Ottesen, Eds. Munksgaard, Copenhagen, pp. 272−288.

Wong, R. C., and Liener, I. E. (1964). *Biochem. Biophys. Res. Comm.* **17**, 470−474.

Yamada, F., Takahashi, N., and Murachi, T., (1976). *J. Biochem.* **79**, 1223−1234.

Zannis, V. I., and Kirsch, J. F. (1978). *Biochemistry* **17**, 2669−2674.

7

Catalytic Properties of Trypsin

ANTHONY A. KOSSIAKOFF
Department of Pharmaceutical Chemistry
School of Pharmacy
University of California, San Francisco
Genentech, Inc.
San Francisco, California

CONTENTS

1. INTRODUCTION

2. THE MECHANISM OF PEPTIDE HYDROLYSIS OF THE SERINE PROTEASES

3. THE ROLES OF ASP 102, HIS 57, AND SER 195 IN CATALYSIS
    3.1. Orientation Factors: Modeling the Reaction Pathway During Deacylation
    3.2. Chemical Properties of the Catalytic Groups: Determination of Protonation States of Asp 102 and His 57
    3.3. Implications of the Protonation States of Asp 102 and His 57 on the Reaction Mechanism

4. SUBSTRATE BINDING SITES
    4.1. The Primary Site
    4.2. Oxyanion Binding Site
    4.3. Secondary Binding Site
    4.4. Protein Inhibitors

5. STRUCTURE-FUNCTION RELATIONSHIPS IN TRYPSINOGEN
    5.1. Structure of the Binding Pocket Region
    5.2. Inhibitor Complex Studies
    5.3. Active Site Groups

6. WATER STRUCTURE
    6.1. Chemical Influences on Catalysis
    6.2. Influences on Protein Conformation

7. STRUCTURE-FUNCTION INVESTIGATIONS USING GENETIC ENGINEERING
    7.1. Residue 102 (Asp to Asn)
    7.2. Residue 189 (Asp to Glu)

REFERENCES

1. INTRODUCTION

The focus of this chapter is the structure/function relationship in trypsin, which is a primary enzyme in the digestive degradation of proteins. Because of the important role that trypsin and other members of the serine protease family play in digestion and other essential biological processes (Stroud, 1974), these enzymes are among the most thoroughly studied of all protein systems by structural as well as chemical methods. The recent appearance of important new information from X-ray and neutron diffraction investigations makes this an appropriate time to review the current knowledge of the role of structure with regard to the enzyme's function and physicochemical properties.

The serine protease family consists of two distinct classes of enzyme structures: (1) enzymes that are trypsin-like in their tertiary folding, and (2) those whose structure resembles that of subtilisin. Both classes possess three conserved residues at the catalytic site—an aspartic acid, a histidine, and a serine—arranged in a configuration shown to be essential for their enzymatic activity (Jansen et al., 1949; Fahrney and Gold, 1963; Ong et al., 1964; Shaw et al., 1965; Henderson, 1971; Martinek et al., 1971). Structural studies of the two classes indicate that their mode of substrate binding is also analogous (Kraut et al., 1971; Segal et al., 1971). Aside from these similarities, however, the two classes have no apparent sequence homology or common folding pattern, presumably because they come from entirely independent lines of evolutionary development.

The structure of trypsin was first determined by Stroud and his colleagues in the early 1970s (Stroud et al., 1971, 1974). Since then, a number of additional studies have been performed, investigating the binding of synthetic substrates (Krieger et al., 1974; Fehlhammer and Bode, 1975; Bode and Schwager (1975) and natural protein inhibitors (Ruhlmann et al., 1973; Huber et al., 1974; Sweet et al., 1974) to the molecule. Trypsin is also one of the few proteins that has been studied by neutron diffraction (Kossiakoff and Spencer, 1980, 1981), which is a method uniquely capable of locating hydrogen atoms in molecules as large as proteins (Kossiakoff, 1983).

Figure 1 shows a schematic representation of trypsin, emphasizing its hydrogen-bonding structure. The molecule consists of two domains of nearly equal size, the major constituent of each being a set of six antiparallel strands of polypeptide chains laced together into a β-sheet unit by a network of H bonds. Analysis of three-dimensional folding shows that these struc-

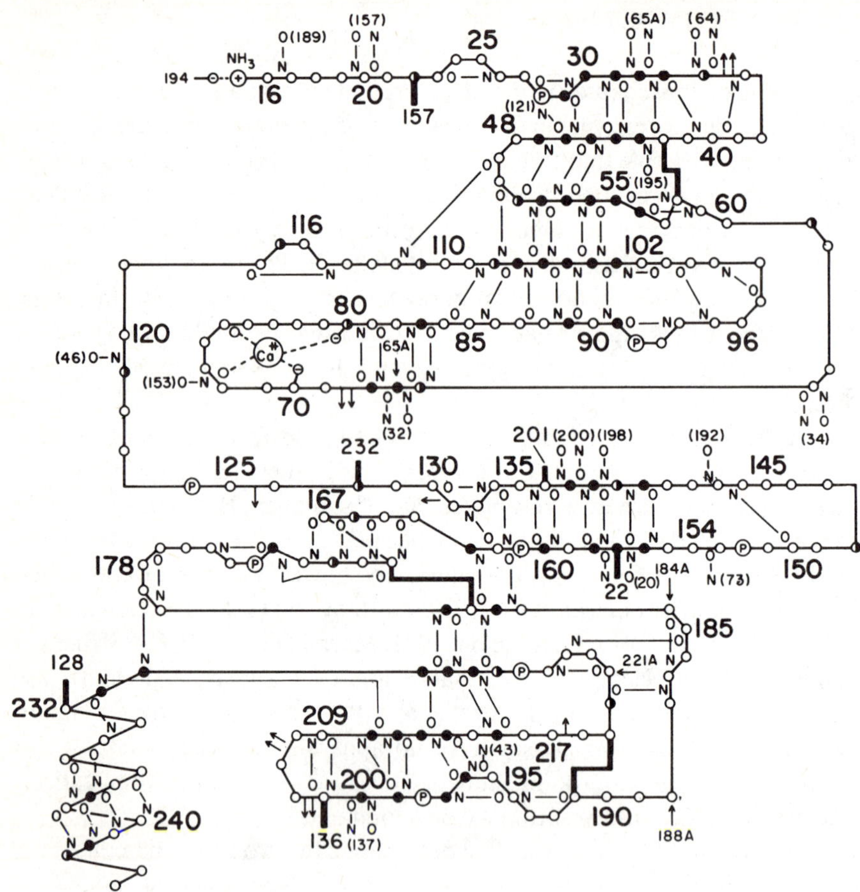

Figure 1. Schematic representation of the main chain H-bonding structure of trypsin. H bonds are denoted by lines connecting N and O atoms. The molecule is divided into two large domains (1–120, 121–245). The hydrogen exchange (H/D) characteristics of the amide peptide groups are represented as follows; full exchange, ○; partial exchange, ◑; unexchanged, ●; proline, Ⓟ. Of the 215 exchangeable amide groups, 68% were found to be fully exchanged, 8% partially exchanged, and 24% unexchanged (Kossiakoff, 1982). This corresponds to the cumulative result of the exchange process over the entire span of exchange soaking (1 year at pH 7.0 20°C). The exchangeable sites form a distinctive pattern whose most prominent feature is the clustering of unexchangeable sites in regions corresponding to the β-sheet structures of the protein.

tures are actually cylindrical, being linked between their outer most strands, and are appropriately called "β barrels." The two barrels are interlinked by several H bonds and form a quite rigid arrangement.

The bulk architecture and tertiary folding of the molecule is represented in a space-filling way by the 5 Å model pictured in Figure 2. The model is viewed looking directly toward the active site region. This model gives a level of information greater than that contained in the H-bonding schematic by showing the tertiary folding pattern; but at this resolution, it does not contain the detail necessary to answer most structure/function questions. These questions are most satisfactorily answered at resolutions of 2 Å or below.

Figure 3 shows the active site of an electron density map of monoiso-propylphosphoryl (MIP) trypsin, which is a close analog of the postulated structure of the transition state of the trypsin-catalyzed reaction (Hirono et al., 1985). The map was computed at 1.5 Å resolution using a highly refined model, (Chambers and Stroud, 1977a), and shows that at this level of resolution, it is a fairly straightforward task to derive accurate positions for the well-ordered atoms in the polypeptide main and side chains. The electron density map also serves as an indication of the high level of confidence concerning structural detail that can be placed on the trypsin model. This type of structural data for trypsin and a number of its derivatives and related family members (Cohen et al., 1969; Shotten and Watson, 1970; Alden et al., 1970; Birktoft et al., 1970; Tulinsky et al., 1973; Delbaere et al., 1975; James et al., 1980a), as well as recent neutron diffraction data (Kossiakoff

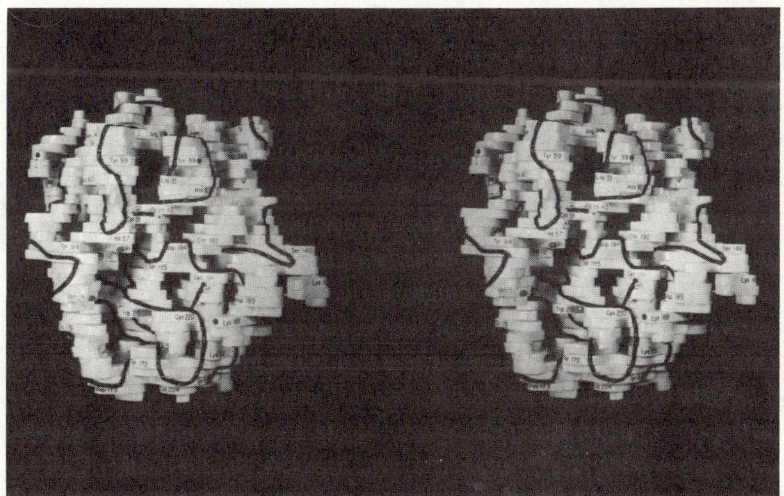

Figure 2. 5 Å model of trypsin, viewing the active site.

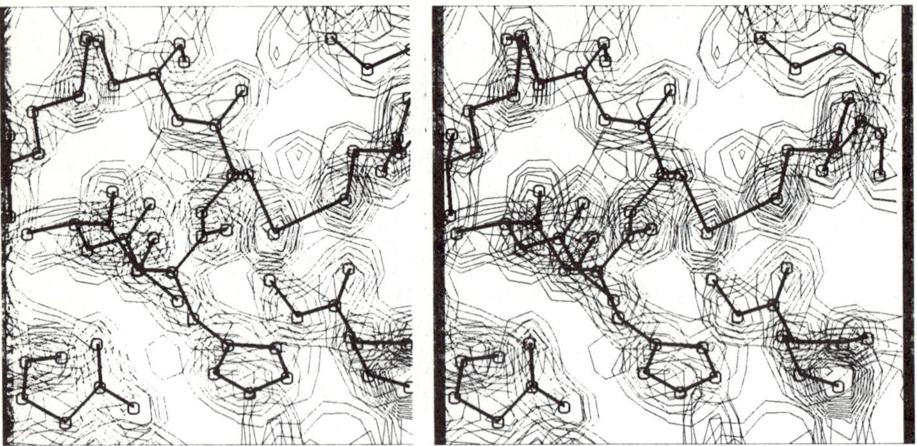

Figure 3. Electron density map in the region of the catalytic site residues of MIP-Trypsin. Density map was calculated at 1.5 Å resolution. (From Chambers et al., 1977a.)

and Spencer, 1980, 1981), provide a wealth of direct information on a number of fundamental issues regarding the relationship between structure and function.

The sections of this review are organized according to the following topics: (1) The mechanism of action of serine proteases describing orientation and chemical factors encountered in the reaction pathway. (2) The interaction of specific substrates with trypsin's binding pocket using data derived from structural studies of inhibitor complexes of trypsin and those of its close relatives, chymotrypsin and elastase. (3) The role of water in substrate binding and how it affects protein conformation. (4) Application of genetic engineering to structure/function studies of the enzyme.

There have been several previous reviews on various structural aspects of the serine proteases (Stroud, 1974; Stroud et al., 1975; Blow, 1976; Kraut, 1977; Huber and Bode, 1978). Accordingly, this chapter will deal with the relationship between structure and function, with emphasis given to the most recent findings.

2. THE MECHANISM OF PEPTIDE HYDROLYSIS OF THE SERINE PROTEASES

The mechanism of peptide hydrolysis consistent with most experimental data is depicted in Figure 4. The reaction sequence consists of two steps

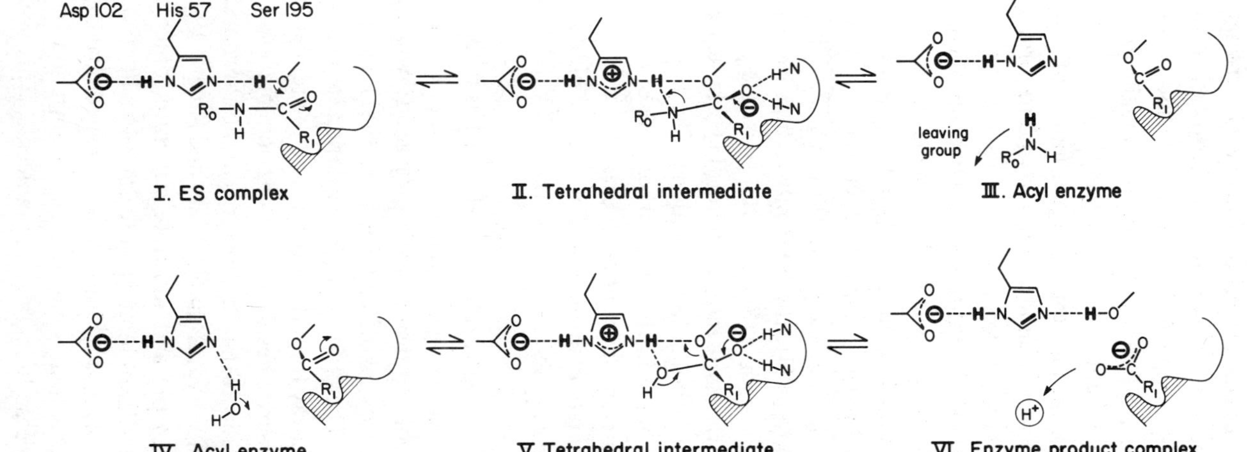

Figure 4. Schematic showing the intermediate configurations of the acylation (top) and deacylation (bottom) steps of the reaction mechanism.

(Bender and Kezdy, 1964; Inward and Jencks, 1965). The step pictured in the upper half of the figure is referred to as "acylation," and results in formation of an amine-leaving group (on the C-terminal segment of the substrate) and an acyl intermediate. The hydrolysis is completed in the second step, deacylation, in which a water molecule attacks the acyl group and displaces the Ser 195 OG, thereby decoupling the remaining portion of the cleaved substrate chain from the enzyme.

Each step proceeds through three intermediate configurations. In configuration I, the substrate is bound to the enzyme to form a Michaelis (ES) complex, in which the peptide group to be cleaved (the so-called scission peptide) is precisely oriented through stereochemical constraints imposed by its interaction with the enzyme's substrate-binding apparatus. Configuration II is the result of nucleophilic attack on the peptide carbonyl carbon by the Ser 195 hydroxyl oxygen and the concurrent transfer of the displaced hydroxyl proton to the imidazole (NE2) of His 57. (The NE2 nitrogen is sometimes referred to as Nτ, the ND1 nitrogen as Nπ.) In this stage, the nucleophilic addition product (Ser 195 OG, substrate peptide group) exists as a short-lived tetrahedral intermediate (TI) whose negative charge is stabilized by several H-bonding interactions. The imidazole is protonated and thus positively charged, while the carboxylate of Asp 102 remains in its anionic state. This arrangement of charges of the active site groups (Asp$^-$, His$^+$, TI$^-$) is postulated to be an important feature of the reaction pathway by lowering the energy of the activated complex (Warshel, 1981; Umeyama et al., 1981).

In configuration III, the proton on NE2 of His 57 is transferred to the substrate nitrogen, facilitating the breaking of the N-C bond. The timing and mechanism of this proton transfer have been at issue (Wang, 1970; Komiyama and Bender, 1979; Kossiakoff and Spencer, 1981) and are discussed in a subsequent section. The products of the acylation step are an amine-leaving group that diffuses away from the enzyme and the remaining bound part of the substrate that rearranges to an acyl enzyme intermediate. Kinetic isotope experiments indicate that the breaking of the C-N bond is normally rate determining for amide hydrolysis (O'Leary and Kluetz, 1972); however, for certain conditions of pH and substrate structure, other steps have been reported to be rate determining (Fastrez and Fersht, 1973).

The breakdown of the acyl intermediate, called deacylation (configurations IV–VI), approximates the reverse of acylation reaction, with the nucleophile in this case being a water molecule. This breakdown process involves displacement of the Ser 195 OG and occurs when the His 57 transfers the proton it extracted from the incoming water to the OG. The resulting hydolyzed product has a carboxylate group, and it has been proposed that this negatively charged group is repelled by the charge on Asp 102 to help dissociate the product from the enzyme (Johnson and Knowles, 1966). In this final step, the enzyme is regenerated and is ready to begin the process again.

3. THE ROLES OF ASP 102, HIS 57, AND SER 195 IN CATALYSIS

A summary of the possible roles of the catalytic groups in mediating the hydrolysis reaction is shown in Figure 5. The functions (deacylation is chosen here, for example) can be grouped into two main categories: the first dealing with orientation factors, and the second with the chemical properties of their side chains. For the sake of clarity, these functions will be treated separately in the section below, although many of their effects are highly correlated.

3.1. Orientation Factors: Modeling the Reaction Pathway During Deacylation

The structures of the natural inhibitor complexes show that in the acylation step, the favorable positioning of the nucleophile (Ser 195 OG) relative to a productively bound substrate is an inherent feature of the enzyme's structure (Ruhlmann et al., 1973; Huber et al., 1974; Sweet et al., 1974). In contrast to this, during deacylation, the nucleophile is a water molecule, which approaches the bound acyl group from the surrounding solvent (Fig. 4). If the deacylation process involved simply unassisted attack by water at the acyl carbon site, the resulting entropic factors would probably make this process rate determining even though the alcohol (Ser 195 OG) is a much better leaving group than the peptide amide (acylation). However, this is not the case, since acylation with concomitant breaking of the C-N bond is rate determining for peptide or amid hydrolysis (O'Leary and Kluetz, 1972). Therefore, it is reasonable to suspect that there exists some direct stereochemical features of the enzyme structure that substantially assist the deacylation step of the hydrolysis.

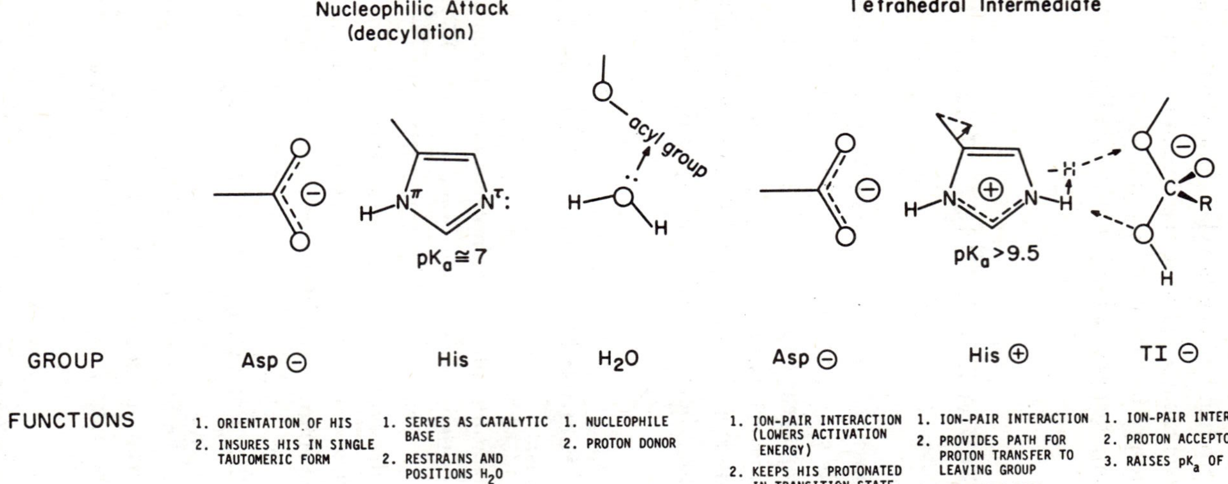

Figure 5. Functions of the catalytic groups during the deacylation step of the hydrolysis reaction. As the TI is formed, the electrostatic and dielectric environment of the His 57 imidazole is altered by influences of Asp 102 and the TI.

How the enzyme might assist nucleophilic attack by water is worth examining in some detail. A water molecule could be positioned to form a hydrogen bond to the NE2 (Nτ) of His 57, and to direct lone electron pairs of its oxygen at the acyl group; a similar arrangement of the His 57 side chain and a water molecule has been observed experimentally in the indoleacyloyl α-chymotrypsin structure by Henderson (Steitz et al., 1969; Henderson, 1970). He pointed out, however, that this indoleacyloyl derivative is not a particularly good model for an acyl enzyme, and Kraut (1977) has listed general difficulties involved in observing experimentally the binding site of the water molecule. Although the obvious choice for binding is at the equivalent site occupied by the leaving group during acylation, this is not a strict requirement, since the principle of microscopic reversibility cannot be applied rigorously to a reaction in which the entering (H_2O) and leaving (R-NH_2) groups differ (Kraut, 1977). Nevertheless, there is compelling evidence suggesting that the water-binding site and the leaving group position are essentially identical.

3.1.1. Formation of the Tetrahedral Intermediate (TI) During Deacylation

In order to model the steps leading to the formation of the TI, it is important to have a good model for the geometry relating the enzyme and substrate. The structure at the catalytic site of MIP-trypsin resembles a true TI (Stroud et al., 1971; Kossiakoff and Spencer, 1981), thus making it an ideal starting point for a detailed examination of the stereochemical parameters of the reaction and for an assessment of the role that His 57 plays in facilitating the attack on the acyl group by water. The most likely conformations at three distinct positions along the reaction coordinate were modeled to determine whether the resulting stereochemical parameters were compatible with those accepted for most nucleophilic additions to carbonyl groups (A.A. Kossiakoff, unpublished results). The principal findings of this study clearly suggest an important orienting role of the His 57-H_2O hydrogen bond during deacylation.

3.1.2. Model Configurations of the Catalytic Groups

Three stages of the reaction were modeled corresponding to successive steps along the reaction path. The individual models were comprised of the four groups pictured in Figure 6. The initial placement and orientation of the water molecule in each model was approximated by requiring that it be a specified distance from the acyl carbon (3.5 Å, 2.6 Å, and 1.53 Å, respec-

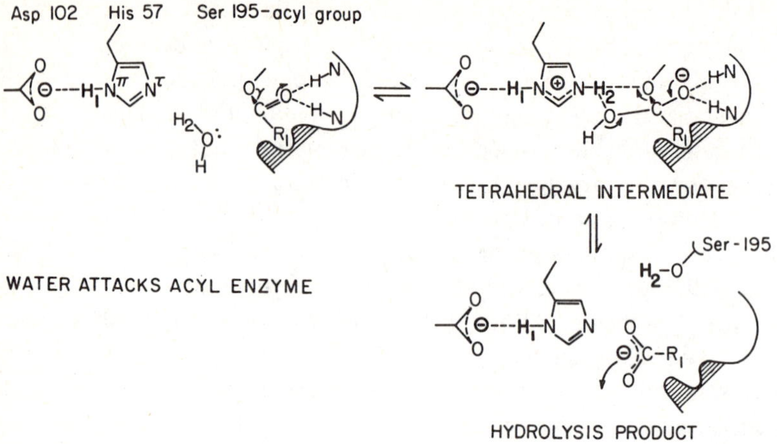

Figure 6. Schematic depicting the mechanism of the deacylation step of peptide hydrolysis by serine proteases. The groups used for the model study discussed in the text are the side chains (including CA atoms) of Asp 102 and His 57, the acyl group (Ser 195 side chain connected to RCO), and a water molecule.

tively), and that it have good hydrogen-bonding geometry with His 57, with one hydrogen directed towards NE2 and an oxygen lone pair pointing toward the acyl carbon. The orientations of the groups incorporated in each model were adjusted to the predicted stereochemistry (Burgi et al., 1974; Alagona et al., 1975; Kleier et al., 1976) at the three stages of the "reaction" by a computerized energy minimization procedure (Chambers and Stroud, 1977a). The highest weights in the parameter minimization were placed on the refinement of intragroup bond lengths and angles; lower weights were used in optimization of H bond lengths and donor-acceptor angles. H bond energies were calculated using a potential of the "6−12" form (Poland and Scheraga, 1967). Puckering of the acyl plane, which occurred at different stages of the reaction sequence, was accommodated by movement of the carbonyl carbon through bond elongation toward the nucleophile (the incoming H_2O). Table 1 gives a summary of the stereochemical parameters calculated for each model.

In stage 1, the water molecule was placed outside the van der Waals contact region, 3.5 Å from the acyl carbon. At this separation, the water molecule has little influence on the electronic character of the acyl group (Kleier et al., 1976). In this orientation, the water would be aligned through

Table 1. Summary of Model–Building Results[a]

	Model I	Model 2	Model 3
Bond lengths (A)			
O_W-C_{AC}	3.5	2.6	1.53
N^τ-O_W	2.9	2.9	2.8
O_2-N^π	2.6	2.6	2.6
Angles (degrees)			
N^τ-H_1-O_W	$149.0\ (-13.0)$[b]	$164.3\ (-15.7)$	$170.0\ (-10.0)$
N^τ-O_W-C_{AC}	$73.2\ (-35.8)$	$84.1\ (-24.9)$	$107.2\ (-1.8)$
O_W-Plane Normal[c]	$9.2\ (+9.2)$	$10.2\ (+10.2)$	
Torsion Angle (degrees)			
H_2-O_W-C_{AC}-O_γ	$14.2\ (+14.2)$	11.1	5.2

[a]Definition of symbols: O_W, water molecule; C_{AC}, acyl carbon; O_2, OD2 Asp-102; N^π and N^τ, imidazole nitrogens; H_2, proton transferred from O_W; and O_γ, Ser-195 OG.
[b]Numbers in parentheses indicate deviation from theoretical value.
[c]Plane defined by atoms O_γ-C_{AC}-O_2-R.

its H bond to His 57 to accommodate an approach to an acyl carbon nearly normal to the acyl plane ($\cong 9°$ off plane normal), which is an orientation consistent with the theoretically predicted path of lowest energy (Burgi et al., 1974; Alagona et al., 1975; Kleier et al., 1976). The 30° departure from linearity found for the H_2O-imidazole H bond would not significantly decrease the hydrogen bond energy (Kerns and Allen, 1978).

In stage 2, the water molecule was moved to a position 2.6 Å from the acyl carbon. The water is just at the edge of the van der Waals contact region, but has not begun to create changes in the acyl group geometry (Burgi et al., 1974; Alagona et al., 1975; Kleier et al., 1976). There is a good hydrogen bond between the water and the imidazole (2.9 Å, 164°), and the approach of the water remains closely perpendicular to the plane of the acyl group (10.2° from the plane normal).

Kleier et al. (1976) have studied the energy profile of the nucleophilic attack on a peptide model (formamide) as a function of angular deviation from the direction normal to the formamide plane. They found that the energy profile was defined by a rather broad reaction "funnel" that subtended about 20% of the hemispheric surface centered above the acyl carbon; and as long as the nucleophile remained within this funnel volume, it could be described as optimally oriented. With respect to the imidazole-water alignment, the approach position at 2.6 Å separation in this model is located nearly at the funnel center.

Stage 3 represents the TI geometry, where the oxygen of the water is coordinated to the acyl carbon at a distance of 1.53 Å. Essentially all the geometric and electronic changes in the intermediate occur during the transition between models 2 and 3 (Burgi et al., 1974; Alagona et al., 1975; Kleier et al., 1976). By the time the reaction coordinate reaches the formation of the TI, a proton from the water molecule has been fully transferred to the imidazole of His 57. The imidazole exists as an imidazolium cation, and the catalytic site is characterized by a three-ion configuration (Asp$^-$, His$^+$, TI$^-$).

3.1.3. *Dissociation of the Tetrahedral Intermediate*

The mechanism of breakdown of the TI involves defining the path the proton follows from the imidazolium to when it is transferred to the serine hydroxyl group, thereby regenerating the free enzyme and the hydrolyzed product. In the geometrical representation of the TI (model 3), the atoms H_1-O_W-C_{AC}-O_γ are nearly coplanar (Table 1). In terms of the stereochemistry of a proton transfer, this is the most efficient arrangement of atoms, since the plane defines the shortest path between donor (water) and acceptor (serine) sites (Kleier et al., 1976). The energy barrier for unassisted proton transfer from the donor to the acceptor would be extremely high due to the large separation distance (Schuster, 1970; Delpuech and Serratrice, 1975).

3.1.4. *Role of Imidazole in the Proton Transfer Mechanism*

Movement of the imidazole ring may play an important role in the transport of the proton between the donor and acceptor sites. The mobility of the imidazole ring was investigated by the model-building techniques described above. A rotation of the ring about the C_α-C_β bond of only $15°-20°$ (corresponding to a translation of the ring center by 0.7 Å, Figure 7) is required to move NE2 (Nτ) from a hydrogen-bonding distance from the donor to a corresponding distance from the acceptor. Such a motion was found to be stereochemically feasible. This is in contrast to the energy required to swing the imidazole in the opposite direction, into the solvent, which is directly opposed by a stretching of the hydrogen bond between Asp 102 and His 57. That the imidazole ring is free to swing into a position

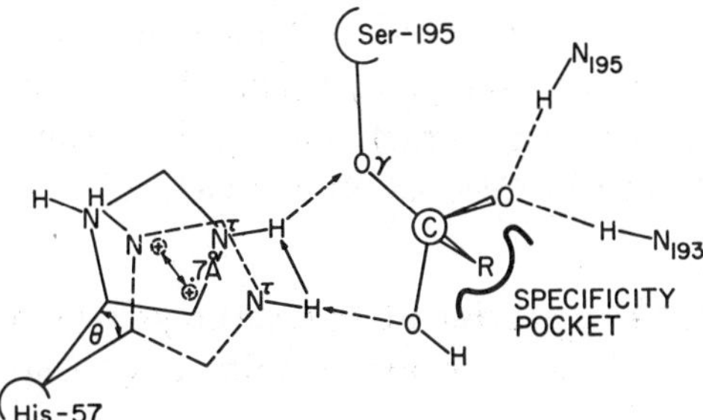

Figure 7. Schematic representation of the predicted reorientation of the imidazole ring during the breakdown of the tetrahedral intermediate. In position 1 (dashed lines), the N^τ (NE2) of the imidazole accepts the proton from the attacking water. A rotation around CA-CB of His 57 by 10°-15° repositions the ring so that it is hydrogen-bonding distance from the leaving group, Ser 195 OG. The proton will transfer at that stage of the reaction when the proton affinity of OG exceeds that of His 57.

pointing towards the Ser 195 OG is borne out by the X-ray structures of the natural inhibitor-trypsin complexes (Ruhlmann et al., 1973; Huber et al., 1974; Sweet et al., 1974) as well as trypsinogen (Kossiakoff et al., 1977; Fehlhammer et al., 1977).

Present evidence from experiments and theories does not allow a quantitative estimate of the effect of the movement of the imidazole on proton transfer. However, given the evidence for the structural mobility of the imidazole ring and the fact that breaking the Ser 195 OG-acyl carbon bond is energetically unfavorable, it would seem likely that the proton would be at least partially transported by the ring before transferring to the leaving group.

3.1.5. *Entropic Effects of His 57*

From the model-building study emerges the inference that His 57 can play a role in minimizing the unfavorable entropy of activation that would arise if the attack by water were unassisted. The H bond interaction, in effect, reduces the possible degrees of freedom of the water from 6 to 1, since only the rotation around the H bond axis is unconstrained. Bruice and Benkovic (1964) suggest that entropic effects on reaction rates can be approximated by the simple empiric relationship in which each increase in reaction order adds an increment of 4−5 kcal/mol to the system. The H_2O-His interaction effectively decreases the order from 2 to 1, so that a decrease of 4−5 kcal/mol in overall free energy of activation would be expected, translating to a rate enhancement of about 10^3. Although this is only a rough approximation, it is a reasonable assumption that the H_2O-His interaction is important in minimizing the unfavorable energy of activation resulting from restricting the degrees of freedom of the reactants in the transition state.

3.2.   Chemical Properties of the Catalytic Groups: Determination of the Protonation States of Asp 102 and His 57

It is now accepted that both the acylation and deacylation steps of the hydrolysis reaction (Fig. 4) involve general acid-base catalysis by the imidazole group of His 57. Until recently, however, there was disagreement on this point, with several different mechanisms proposed. The primary issue was whether His 57 is the actual chemical base or whether the histidine acts as an intermediary through which Asp 102 functions as the base (Fig. 8). These two possibilities are chemically quite different; the correct one depends on the relative proton affinities (or pK_a's) of the Asp and His groups within the electrostatic environment provided by the enzyme structure. Apart from its importance to understanding the mechanism of the serine proteases, the possibility that the chemical characteristics of the His and Asp side chains

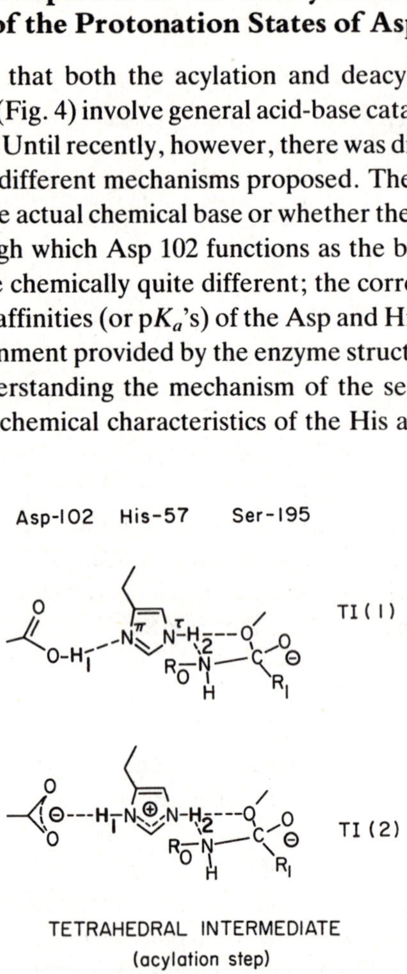

Figure 8. The two different locations proposed for proton H1 in the tetrahedral intermediate (TI) structure. The intermediate pictured here corresponds to that formed during acylation, and differs from the deacylation intermediate only by the replacement of the leaving group (R_0-NH_2) with a hydroxyl group.

could be radically altered by their local environment raised a fundamental structure/function issue of general interest.

The number of reports in the literature attempting to identify the catalytic group in this mechanism provides a good indication of the interest in answering this question. They reflect as well the experimental difficulties encountered in determining its solution. Experimental determination requires a technique capable of assigning to one of the two side chains the proton in the Asp His hydrogen bond (H1, Fig. 8) under conditions where the chemical base would be expected to be protonated. This is feasible with spectroscopic methods that take advantage of the difference in charge states (Asp $\ominus$-His $\oplus$ vs. Asp-His [neutral]) between the two possibilities. Another technique, neutron diffraction, offers a more direct approach because it has the unique capability of experimentally locating hydrogen (or deuterium) atoms in molecules as large as proteins. This section reviews the most significant of the investigations and their interpretation.

3.2.1. Spectroscopic and Kinetic Studies

Nuclear magnetic resonance (NMR) was utilized by a number of workers in the field to study the enzyme-catalyzed hydrolysis of proteins. Robillard and Shulman (1972), using ^{1}H NMR, identified a single proton in chymotrypsin that was sensitive to binding of substrates or inhibitors and titrated with a pK_a of 7.5. They interpreted their results as indicating pK_a of 7.5 for His 57, although there was no direct confirmation of the assignment. Later, ^{1}H NMR studies by these authors (Robillard and Shulman, 1974a, 1974b) and by Markley and coworkers (Markley and Porubean, 1976; Porubean et al., 1978) on a variety of serine proteases, protease-inhibitor complexes, and zymogens were interpreted to further support the conclusion that His 57 had a pKa around 7.0.

These results were called into question by the findings of Hunkapillar et al. (Hunkapiller et al., 1973), who reported that they were unable to reproduce the results of Robillard and Shulman. These authors studied α-lytic protease selectively enriched with ^{13}C at the C(2) carbon of His 57. α-Lytic protease has a single histidine (His 57), making the proton assignment more straightforward than in the case of chymotrypsin, which has three histidines. They found that the C(2) carbon exhibited significant chemical shifts at both pH 6.7 and 3.3, but the carbon-hydrogen spin-spin coupling constant characteristic of a protonated imidazole was observed only below pH 3.8. To explain this result, they proposed that environmental effects in the protein had so increased the apparent pK_a of Asp 102 that the titration at pH 6.7 reflected the loss of a proton from the Asp 102 carboxyl group in the presence of a neutral imidazole, rather than the loss of a proton from an imidazolium cation in the presence of a carboxylate anion.

An assignment of Asp 102 as the protonated species in the transition state was also made by Koeppe and Stroud from difference infrared (IR) absorption experiments (Koeppe and Stroud, 1976). By chemically modifying all the carboxyl groups except Asp 102 and Asp 194, they were able to observe that one carboxyl group titrated at pH 6.8. This was assigned as Asp 102, based on a competitive binding study using a Cu ion.

Further support for this hypothesis came from the solvent isotope experiments by Hunkapiller et al. (Hunkapiller et al., 1976). The observed second-order rate dependence was interpreted as indicating the simultaneous movement of two protons in the transition state, which is in agreement with their NMR interpretation. It was noted, however,that this result conflicted with the first-order rate dependence reported in several prior studies (Pollock et al., 1973; Elrod et al., 1975).

In a subsequent study with α-lytic protease, Bachovchin and Roberts incorporated ^{15}N histidine at the single histidine site (His 57) (Bachovchin and Roberts, 1978). Contrary to the findings of the ^{13}C work of Hunkapiller et al. (Hunkapiller et al., 1973), they found that the imidazole in the free enzyme titrated with a pK_a of approximately 7.0. It should be noted that the ^{15}N analysis gives a more readily interpretable spectra, and by monitoring changes directly at each of the nitrogen sites on the imidazole, it is a more sensitive probe of the group's protonation state. However, only the free enzyme was studied, and it is known that when substrate is bound, the local environment of the imidazole must be perturbed to some degree. Therefore, the interpretation of the imidazole having a pK_a of 7.0 cannot be generally extrapolated to the situation occurring in the transition state.

3.2.2.   Theoretical Studies

Theoretical studies reported by Scheiner et al. (1975) and Scheiner and Lipscomb (1976) were interpreted as supporting the assignment of Asp 102 as the catalytic base in the reaction. In contrast to the findings derived by the semiempirical approach of Scheiner and Lipscomb, results from higher level molecular orbital (MO) studies predicted that His 57 is the protonated species in the transition state (Scheiner and Lipscomb, 1976; Nakagawa et al., 1980; Allen, 1981; Kollman and Hayes, 1981).

The use of these higher level calculations (i.e., 4−31G ab initio calculations) seems to be important. Using an improved electrostatic model for the surrounding environment of the primary groups used in the calculation, Umeyama and his colleagues (Nakagawa et al., 1980) found that the effects of expanding the number of primary groups and including secondary charges significantly affected the stability of the ion-pair structure of the Asp-His couple and clearly favored His 57 as the catalytic base.

3.2.3. Direct Determination of the Protonation State of the Asp-His Couple by Neutron Diffraction

Neutron diffraction has been developed over the last several years to the point of being able to determine the structure of large molecules (Kossiakoff 1985a). Because of its capability to locate hydrogens (or deuteriums), the opportunity was presented to directly attack the problem of the protonation states of the Asp-His couple. With this as a primary goal, Kossiakoff and Spencer undertook a neutron diffraction study of an inhibited form of trypsin (Kossiakoff and Spencer, 1980, 1981).

The configuration of the protonation states of the catalytic residues of interest to the hydrolysis mechanism is that of the TIs (step II, V, Fig. 4). It had been determined that the expected structure and electrostatic properties of a real substrate-enzyme intermediate corresponded very closely to those of a form of bovine trypsin inhibited by an MIP group (Stroud et al., 1971; Kossiakoff and Spencer, 1981; Kossiakoff, 1983). Analysis of this complex, therefore, allowed the direct determination of the protonation state of the Asp-His couple during the most crucial stage of the reaction.

Because a major objective of the neutron study was to identify conclusively the group acting as the base in the hydrolysis, it was important that the interpretation of the proton positions in the catalytic site be definitive. Toward this end, two methods were devised to test the preferred location of proton H(1) (Fig. 8).

These methods involved the use of two types of difference maps. In method 1, proton H1 was omitted from the model to guard against phasing bias. (Proton H1, like most other exchangeable protons in the structure, was replaced by deuterium when the crystal was transferred to a D_2O solution prior to neutron analysis). Figure 9a depicts the resulting difference map that clearly shows the deuterium (H1) residing on the imidazole. In the second difference synthesis, structure factors were calculated with the deuterium placed by stereochemistry on the OD2 of Asp 102. The resulting map

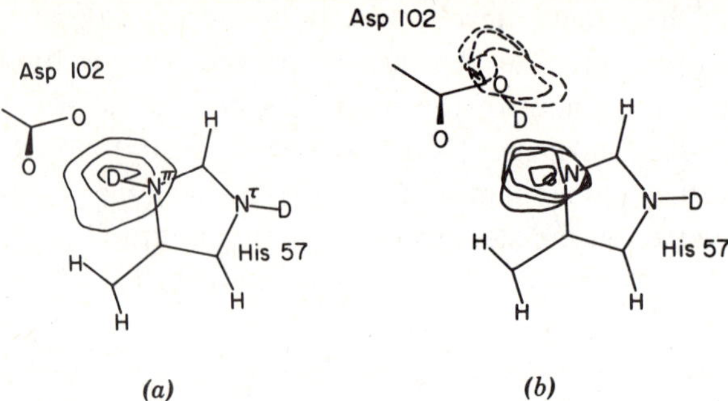

Figure 9. (*a*) A difference map [$(F_o - F_c)\exp(i\phi_c)$] calculated with only the deuterium (D) between the His 57 and Asp 102 side chains left out of the phases. The difference peak is approximately 5.5 σ above the background level and shows the deuterium to be bound to the imidazole nitrogen. (*b*) A difference map in which the deuterium was placed by stereochemistry on atom OD2 of Asp 102 (if there were one, it would be on OD2). This difference density peak clearly indicates that the preferred location of the deuterium is on the imidazole of His 57.

is pictured in Figure 9*b*; the difference peak clearly indicates that a substantial shift away from the carboxylate group towards the imidazole is warranted. This test is particularly sensitive and is unambiguous in its interpretation.

The results from each of these two independent methods definitely supported the conclusion that the mechanistically important proton is coordinated to the imidazole of His 57 in MIP-trypsin. Taken together, these tests offered compelling proof that, at physiological pH, the protonated species in the TI is the imidazole of His 57 rather than the carboxylate of Asp 102.

3.3. Implications of the Protonation States of Asp 102 and His 57 on the Reaction Mechanism

The hydrogen-bonding network between the carboxyl oxygens and other groups (OD1-NH (Robillard and Shulman, 1974a, 1974b; Markley and Porubean, 1976), OD2-OH [214], HND1 [Markley and Porubean, 1976]) acts as a very good electrostatic source to stabilize the anion (Kossiakoff and Spencer, 1981; Warshel, 1981). Thus, the environment of Asp 102 is certainly not hydrophobic, and the assumption of a large pK_a shift due to neighbor effects is not warranted.

In the often cited "charge-relay" mechanism, it was proposed that the aspartate channels some of its negative charge through the His 57 imidazole to the Ser 195 OG, effectively enhancing the nucleophilic properties of the Ser OG group (Blow et al., 1969). However, the "charge-relay" concept is not supported by either experimental or theoretical studies (Matthews et al., 1977; Nakagawa et al., 1980; Desmeules, 1980). The importance of the aspartate to the reaction mechanism, therefore, does not involve its protonation at any stage of the reaction or significant charge transfer effects.

Theoretical analyses suggest that a negative charge state for the aspartate is a key factor in the reaction sequence (Warshel and Levitt, 1976; Warshel, 1978; Umeyama et al., 1981). In the TI, there exists a charge interaction among the triad $Asp^-His^+TI^-$. It has been proposed that thermodynamic properties of a similar arrangement of ionized groups (i.e., $- + -$) in the catalytic site of lysozyme play an important role in lowering the activation energy of the reaction (Warshel and Levitt, 1976; Warshel, 1978). Nakagawa and Umeyama, using high-level ab initio calculations, similarly found supporting evidence that the Asp^--His^+ ion pair plays an important role in reducing the activation energy of the serine protease system (Umeyama et al., 1981). In their treatment, they found that other hydrogen-bonding groups also participate in stabilizing the substrate-enzyme complex.

Two other functions that Asp 102 performs are listed in Figure 5. Although only deacylation is shown in the figure, the functions apply to the acylation reaction as well. Besides its role in the charge triad, it is involved through H bonding in orienting His 57 and ensuring that during nucleophilic attack by water (deacylation) or the Ser 195 OG (acylation), the imidazole remains in the correct tautomeric form with NE2 ($N\tau$) unprotonated. These two functions increase the effectiveness of His 57 as a catalytic base by ensuring the correct stereochemistry for proton transfer from the nucleophile. The composition of these functions are sufficient to explain the importance of Asp 102 without postulating unusual chemical properties for this residue.

In both the acylation and deacylation processes, His 57 performs a bifunctional catalytic role, acting first as the base by accepting a proton from the attacking group, and then as an acid by transferring this proton to the leaving group. One factor governing the timing of the rate-determining proton transfer event is the proton affinity of the catalytic group; that is, the proper relationship must exist between the proton affinities of the nucleophile, His 57, and the leaving group in order to assure an energetically favorable pathway for proton transfer both to and from the histidine. Because of the difference in proton affinities between amides and alcohols, the formation and breakdown of the TI during acylation must be different in some respects when compared to the equivalent deacylation steps.

During the reaction, His 57 becomes hydrogen bonded to the negatively charged TI. This interaction insulates the imidazole from solvent contact altering the electrostatic environment of the group. Consequently, it would be expected that these changes would elevate the proton affinity (pK_a) of the imidazole. This proposal is supported to some extent by several NMR investigations.

The conventional definition of pK applies only to acids (or bases) in which the ionizing site is accessible to solvent. This is the case of His 57 in the native enzyme for which NMR evidence indicates that the pK_a is only slightly higher than that for a normal free histidine (Bachovchin and Roberts, 1978). Markley et al. studied chymotrypsin and trypsin inhibited by the uncharged diisopropylphosphoryl (DIP) group (Porubean et al., 1979). They observed about a 1 unit rise in the histidine pK_a by effects that presumably only insulate the group from solvent, but do not drastically alter its charge environment. Robillard and Shulman reported a series of studies with negatively charged inhibitors (derivatives of boronic acid) bound to chymotrypsin (Robillard and Shulman, 1972). They found that charged inhibitors apparently affect significantly the pK_a of an ionizing group that they identified as His 57; however, it should be pointed out that this assignment was based on indirect evidence.

In summary, these results are suggestive that the proton affinity of His 57 is raised during formation of the TI, which is an effect presumably due to changes in the local environment of the imidazole. The value of the pK_a of His 57 is indicated to be close to that of the leaving group (about $8-10$ [Satterthwait and Jencks, 1974a; Warshel, 1978], making it a more efficient catalytic group. This is consistent, from a thermodynamic standpoint, with a reaction pathway through a TI and is in agreement with kinetic investigations (Caplow, 1969; Fersht and Requena, 1971).

4. SUBSTRATE BINDING SITES

The characteristic difference in specificity between trypsin and its family members, chymotrypsin and elastase, is in their particular preference for hydrolyzing peptide bonds involving different amino acid side chain types. Although the basic overall architecture of the substrate-binding regions of these enzymes is similar, each contains a set of small individual differences that define the steric and chemical character of the side chain type it accepts.

Figure 10 shows the portion of the trypsin molecule involved in the catalysis and binding of the substrate. For purposes of this discussion, the binding site will be subdivided into three functional regions: (1) the primary site, (2) the oxyanion-binding site, and (3) secondary sites. These regions act in concert to position the carbonyl peptide bond of the substrate in the optimum orientation for nucleophilic addition by Ser 195 OG during acylation and by a water in the deacylation reaction.

4.1. The Primary Site

Of the family of pancreatic serine proteases, trypsin has the most narrow specificity and preferentially hydrolyzes those peptide bonds that follow either of the basic amino acids, arginine or lysine (Neurath and Hartleg, 1959). The specificity of trypsin is conferred primarily by an aspartic acid residue at position 189, located at the bottom of the binding pocket. The carboxylate of 189 is positioned by two hydrogen bonds (NH-221 and a H_2O), such that it can form good hydrogen bonds to the appropriate substrate side chain atoms. In the case of arginine, this bonding is made

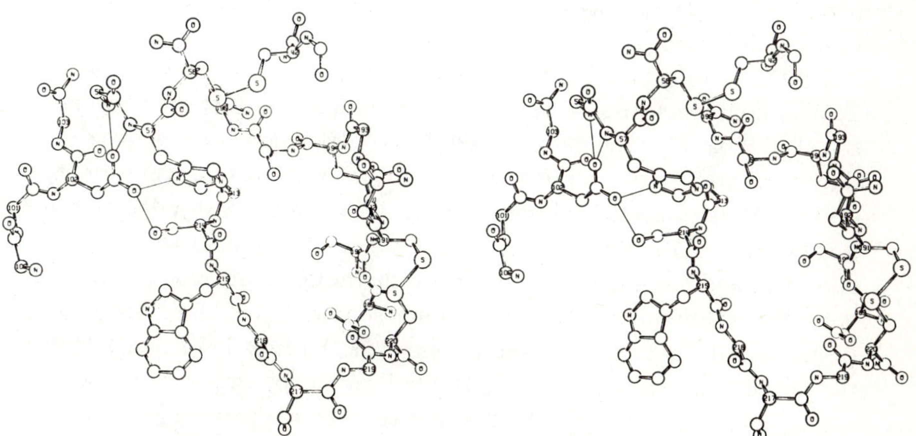

Figure 10. Structure of the active site region in trypsin.

directly to the terminal guanidinium group; however, because lysine's side chain is one atom shorter, its interaction is mediated through a water molecule (Ruhlmann et al., 1973). Additional important interactions involve hydrophobic contacts to the aliphatic portion of these side chains (Ruhlmann et al., 1973; Sweet et al., 1974).

With both natural and synthetic substrates, Arg is preferred to Lys by a ratio of about 25:1 (Fersht and Requena, 1971). This preference is probably due to both the direct H bond interaction with Asp 189 and increased van der Waals contacts due to the larger bulk of the Arg side chain. The role of the side chain binding in the specificity of trypsin has been addressed in the study of hydrolysis of a nonspecific substrate, acetylglycine ethyl ester (Keil, 1971). This substrate is hydrolyzed at a rate (K_{cat}/K_m) 10^3 lower than that measured for the equivalent specific substrate. Interestingly, it was found that the addition of ethyl ammonium ions increased that rate approximately ninefold (Inagami and Mitsuda, 1964a). This ion was thought to bind specifically to Asp 189, suggesting that activity is enhanced by interaction of a positively charge species to the carboxylate group.

Crystallographic evidence suggests that the binding of the substrate does not require a substantial change in the conformation of the backbone groups of the enzyme (Ruhlmann et al., 1973; Fehlhammer and Bode, 1975; Krieger et al., 1974; Sweet et al., 1974). The pocket is essentially conformed in the native molecule to optimize productive binding of the substrate, and only small adjustments in side chain groups need to take place. This is supported by the fact that the segments of polypeptide chain forming the pocket are extensively involved in H bonding, as illustrated by the schematic drawing in Figure 1.

All the binding pocket residues are contained within a single structural domain consisting of a large, 6-stranded antiparallel β-sheet network (a so-called β barrel). This β-barrel structure is deformed at the active site by the separation of two strands and insertion of a 3_{10}-type conformation (191−194) to form a well-defined pocket. Temperature factor data indicate a rigid conformation in this region in the time-averaged structure (Chambers, J.L. and Stroud, R. M., 1977b) furthermore, neutron H/D exchange data support this model in that the normal conformational breathing modes (small motions inherent to all protein molecules) do not significantly deform the pocket (Kossiakoff, 1982). After a 1 year exposure to D_2O, several backbone NH groups in the binding pocket region remained unexchanged (Figs. 1 and 12).

4.2. Oxyanion Binding Site

The oxyanion-binding site in trypsin is defined by the peptide NH groups of Gly 193 and Ser 195 (Fig. 11). When a substrate is productively bound at the active site, the carbonyl oxygen of the scission peptide forms a bifurcated hydrogen bond to these two proton donors. Henderson first suggested the possible importance of this interaction (Henderson, 1970), and the concept gained further support in the findings of Kraut and his group, who identified a similar hydrogen-bonding site in subtilisin (Robertus et al., 1972). In subtilisin, the oxyanion site consists of the peptide NH of Ser 221 and the side chain ND2 of Asn 155.

It has been proposed that in addition to a role in accurately positioning the scission peptide, H bonds reduce the ground state energy of the TI by helping to stabilize its charge distribution (Kraut, 1977; Warshel, 1981; Nakagawa and Umeyama, 1984). There is some theoretical basis for this hypothesis. Using a set of high-level ab initio MO calculations of trypsin, Nakagawa and Umeyama studied the effects of the 193 and 195 NH hydrogen bonds on the energy of the TI (Nakagawa and Umeyama, 1984). They interpreted the results to show that the TI state is greatly stabilized by the electrostatic effects of these bonds. On this point, Warshal has developed a general theoretical treatment indicating that the active site regions of enzymes can be considered as "supersolvents" for a particular transition state "with enzyme dipoles preoriented to create an electric field complemen-

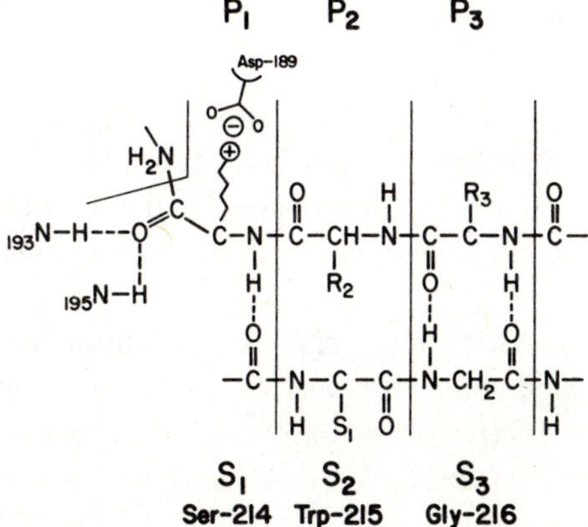

Figure 11. Schematic of the possible substrate-binding interactions with trypsin-like enzymes.

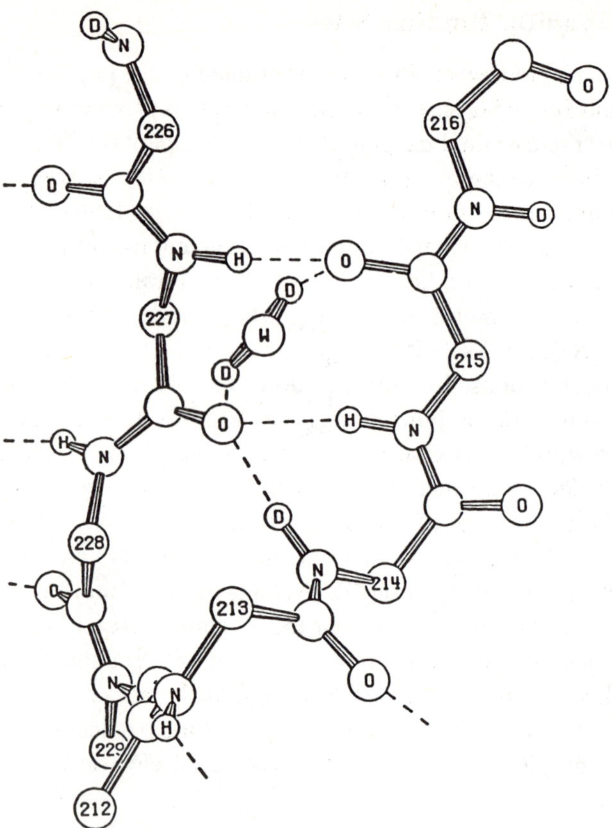

Figure 12. H bond configuration of an ordered water (D_2O) in the binding pocket. The D_2O is so highly ordered that its orientation (positions of the D's) can be deduced.

tary to the charge distribution of the transition state" (Warshel, 1981). The degree of charge stabilization predicted for the TI would be substantially greater than what could be achieved through contact with the solvent medium alone.

A recent set of genetic engineering experiments using site-directed mutagenesis has provided some direct experimental data on this point. Wells, Estell, and coworkers (personal communication) have replaced the Asn 155 side chain in subtilisin (the ND2 of this side chain forms part of the oxyanion H-bonding system) with a Thr. Modeling studies indicate that the Thr hydroxyl proton cannot arrange itself to substitute for the ND2. It was found that such a change affected the catalysis rate (k_{cat}) considerably more than binding of the substrate (K_m); K_m remained essentially unaltered, while k_{cat} was lowered by a factor close to 10^4. This suggests that the H-bonding system of the oxyanion pocket is required primarily for binding and charge stabilization of the transition state.

4.3. Secondary Binding Sites

Structural studies of a number of substrate-enzyme complexes in the trypsin family have implicated several sites on the surface of the binding pocket as being secondary binding sites for an extended substrate. The notation usually used to designate binding subsites is one developed by Schechter and Berger (1967). The scheme showing the relationship of a bound substrate to the individual subsites on trypsin is depicted in Figure 11. In this scheme, subsites Sn on the enzyme interact with substrate groups Pn. The main feature of the subsite interactions in the trypsin-like enzymes is that they form an extended β-type conformation with the three residues on the N-terminal side of the scissile bond (Segal et al., 1971; Shotton et al., 1971). The interactions of the C-terminal end of the substrate with the enzyme surface are less specific—a factor that may facilitate easy release of the first hydrolysis product.

4.4. Protein Inhibitors

A shortcoming of using covalent synthetic substrates as models for defining the details of enzyme-substrate interactions is that they can only mimic some of the possible interactions—and those only on the N-terminal side of the scissile bond. Understanding the full picture of the binding properties of a substrate requires information on how the enzyme surface interacts from both the N- and C-terminal sides of the scissile bond. Better model systems for the structural features involved in binding real substrates are complexes formed between enzymes and their natural inhibitors, which function by binding tightly to the active site region of the enzyme.

Crystallogaphic analyses have been performed on two trypsin-inhibitor complex systems. Huber and colleagues have investigated complexes of bovine trypsin with the Kunitz pancreatic inhibitor (PTI), (Ruhlmann et al., 1973; Huber et al., 1974) and have also studied the structure of this inhibitor complexed to trypsinogen (Bode et al., 1978). Sweet et al., 1974) have studied the soybean trypsin inhibitor (STI) complexed to porcine trypsin. In comparing these two systems, it is interesting to note that although these inhibitors are very different, with regard to their size and sequence, they both bind to trypsin in a similar fashion; thus, the common features probably describe the mode of binding of a true peptide substrate.

4.4.1.   Sites of Interaction

The principal interaction of these inhibitors is at the primary (P1) site. Here the carboxylate of Asp 189 of trypsin forms hydrogen bonds to either Lys 15 (through a water molecule) in PTI or Arg 63 in STI. On the basis of available data, all secondary binding interactions expected for residues in the P2 and P3 positions of a real trypsin substrate are formed in PT1 and ST1 (Ruhlmann et al., 1973; Sweet et al., 1974). Although there is no independent structure data on how real substrates bind at the C-terminal side of the scissile bond, model-building studies and kinetic data are consistent with the type of interaction observed for the P1′ group (Ala 16 [PTI]; Ile 64 [STI] in these structures.

4.4.2.   Geometrical Configuration at the Reactive Center

Whether or not the inhibitor-enzyme geometry at the active sites of these complexes corresponds to a real tetrahedral adduct has been a debated issue (Steitz and Shulman, 1982). The conclusion that they were tetrahedral and covalently attached was supported by the initial interpretations reported for the PTI and STI structures (Ruhlmann et al., 1973; Sweet et al., 1974). However, a qualification stated by both groups of investigators was that although such an interpretation fits the density maps, the degree of resolution in the analyses was not sufficient to determine definitively the geometry around the scissile bond. Inclusion of the higher resolution data and subsequent refinement suggested, in the cause of PTI, that the distortion at the carbonyl carbon of Lys 15 was not as significant as first reported, and is better described as being trigonally, rather than tetrahedrally, distorted (Huber et al., 1974). Actually, the group may not be distorted at all. A ^{13}C NMR study, in which the carbonyl carbon was selectively enriched, showed that in the PTI-trypsin complex, the carbonyl is trigonal and therefore undistorted (Richarz et al., 1980).

In an analysis of a related inhibitor system, the complex between *Streptomyces griseus* Protease B (SGPB) and the third domain of the turkey ovomucoid inhibitor, James and colleagues (1980b) interpreted the structure such that no appreciable distortion at the scission peptide occurs even though the Ser 195 OG is within van der Waals contact distance. The SGPB study is mentioned here because its structure is somewhat more refined than the PTI and STI complexes, and the finding that it is not a tetrahedral adduct may be relevant to these other structures. (James and colleagues have studied a number of inhibitor-complex systems that form an important subset of studies for trypsin like enzymes [James et al., 1980a, 1980b]). It should be noted, however, that the configuration at the scission peptide is probably not exactly the same in all inhibitor complexes because, in some cases, the peptide bond is cleaved with aging (STI) while in others it is not (PTI) (Laskowski and Sealock, 1971). This suggests that these structures have slightly different low-energy configurations.

5. STRUCTURE-FUNCTION RELATIONSHIPS IN TRYPSINOGEN

In addition to the direct studies of trypsin, another approach to understanding the structure/function relationships conferring catalytic efficiency is structural comparison of trypsin with its inactive precuser, trypsinogen (Tgen). Identifying the structural elements modified during activation may disclose features important to the enzyme's activity not evident from other studies.

5.1. Structure of the Binding Pocket Region

Prior to the existence of crystallographic data, an open question was what changes in the tertiary structure of Tgen, as compared to trypsin (T), were responsible for its inactivity. It was not known whether this involved a single structural modification at some critical site or a number of independent modifications.

The X-ray crystallographic analyses of Kossiakoff et al. (1977) and Bode et al. (1977) showed that there were several significant structural differences between Tgen and T. The most notable of these were found in the binding pocket region. In T, the binding pocket is rigidly structured by a number of hydrogen bonds; in contrast, in the same segments of polypeptide chain

(189–194, 219–221) in Tgen, a number of these hydrogen bonds are missing. As a consequence, this region of the polypeptide chain exhibits positional variation.

Two features of the Tgen-binding pocket region especially affected by the disorder are notable. In T, the side chain-binding group, the carboxylate of Asp 189, is held in position by a hydrogen bond to the NH peptide group of Ala 221; Tgen lacks this H bond, resulting in some disorder in the carboxylate group (Kossiakoff et al., 1977). This situation probably produces weaker or disoriented binding of the substrate side chain.

A second feature of the disordered binding pocket region is disruption of the oxyanion-binding site (the NH groups of Gly 193 and Ser 195). This is a feature that Tgen shares with chymotrypsinogen (CTgen) (Kossiakoff et al., 1977; Fehlhammer et al., 1977). However, it is interesting to note that the modifications of CTgen's binding pocket region, when compared to chymotrypsin (Freer et al., 1970), are significantly different than those between T and Tgen. The CTgen and chymotrypsin structures have been compared in several investigations (Freer et al., 1970; Wright, 1973a, 1973b). However, the T/Tgen comparisons are better validated because the structures have been refined at higher resolution (Kossiakoff et al., 1977; Fehlhammer et al., 1977).

5.2. Inhibitor Complex Studies

X-ray structure analyses of complexes of PTI with Tgen have been performed by Huber and colleagues (Bode et al., 1978). Comparison of T-PTI and Tgen-PTI structures shows a remarkable similarity considering the extent of the differences observed between T and Tgen. This is evidently because in forming the complex with PTI, the Tgen molecule adapts to the PTI-binding surface, thereby stabilizing the Tgen conformation. This suggests that major changes in the conformation of the polypeptide chain of Tgen need not occur for the molecule to assume its active structure. In a companion study, Bode et al. (1978) soaked the dipeptide Ile-Val into the Tgen-PTI cystals. This dipeptide is a good analog of the N-terminus of trypsin, and was found to form an H bond between its NH_2 group and the carboxylate group of Asp 194 in a manner similar to the interaction with the N-terminus in the active form of trypsin. Bode et al. (1978) report that structurally the Ile-Val derivative of Tgen-PTI is virtually indistinguishable from T-PTI.

These results suggest that given a substrate that makes a significant number of specific interactions with T, it is possible the substrate could force Tgen into an active conformation. Similarly, it is interesting to note that the related serine protease, tissue-type plasminogen activator (tPA), when bound to fibrinogen, is equally active as a zymogen as when in its activated form (Vehar et al., 1984). Although the mechanism by which it assumes an active conformation is not known, it is undoubtedly mediated by interactions with the fibrinogen molecule.

5.3. Active Site Groups

In contrast to the differences in the binding domains of Tgen and T, the relative position of the active site groups (57, 102, 195) are almost the same. The only notable difference is that in Tgen (and also in CTgen; (Freer et al., 1970); there is a good H bond between the NE2 of His 57 and the OG of Ser 195; whereas in T, the Ser 195 OG is about 1.4 Å away from an optimal H bond position (Kossiakoff et al., 1977; Fehlhammer et al., 1977). This difference plays a significant role in the activity of the enzyme.

Several studies have focused on the question of whether or not the Asp-His-Ser arrangement in Tgen (and CTgen) productively reacts with small substrates that do not directly interact with the binding pocket (Morgan et al., 1972; Robinson et al., 1973), and, therefore, whose binding would not be greatly affected by the changes in the pocket. Kinetic evidence indicates that the binding efficiency of small synthetic substrates like methane sulfonyl fluoride are not as greatly depressed in the inactive configuration (1.5 orders of magnitude vs. 5 orders or more) as those for the larger substrates (Morgan et al., 1972).

The nature of the changes in binding geometry between Tgen and T for these small substrates was determined recently by Stroud and colleagues (R. M. Stroud, unpublished results). In their analysis of the DIP-Tgen structure, they found a steric interference between the DIP group bound to Ser 195 OG and the His 57 imidazole. Their analysis showed that this contact forces the imidazole ring to rotate away from its normal H bond interaction with Asp 102 to a position in the solvent. This implies that given the His-Ser arrangement in the native Tgen molecule, there is insufficient room between them to bind a substrate productively. Thus, the arrangement of the catalytic groups in native Trypsin, with the 195 OG approximately 3.8 Å (vs. 2.8 Å for Tgen and CTgen) from the NE2 of His 57, apparently represents the most effective disposition of catalytic groups, prior to substrate binding, necessary to produce an enzyme-substrate complex.

6. WATER STRUCTURE

Water interactions are a major consideration in a protein's chemistry and conformation. Water can influence a mechanism directly, as exemplified by its role as the nucleophile in the deacylation step of the hydrolysis reaction, or indirectly by modulating group charges through solvation. From the standpoint of conformation, water-protein interactions are the thermodynamic driving force for the tertiary folding of the protein's polypeptide chain. These chemical and conformational influences are highly interrelated. For the organization of the discussion in the following subsections, however, they will be treated as separate topics.

It should be noted here that the water structure of trypsin has been investigated crystallographically in several different laboratories (Bode and Schwager, 1975; Chambers and Stroud, 1977b; Kossiakoff and Spencer, 1981), and is one of the few proteins that has been studied by both X-ray and neutron methods. These independent investigations provide a good cross-check of the accuracy of the assignments of individual water sites, and ensure that the interpretations presented below are based on reliable structure data.

6.1. Chemical Influences on Catalysis

The role of water as the nucleophile during deacylation has been discussed in Section 2. Other waters are also essential to catalysis. Crystallography has identified a number of highly ordered waters directly associated with catalytic site and binding pocket residues (Bode and Schwager, 1975; Chambers and Stroud, 1977b; Kossiakoff and Spencer, 1981). It is incorrect, however, to assign the importance of a particular water to its relative occupancy. In terms of solvation, water, so far as it exhibits properties of bulk solvent, certainly makes crucial contributions. This is evidenced by calculations showing the energy of the activated complex to be lowered by effects of neighboring bulk solvent molecules (Weiner, 1984; Weiner et al., 1985).

The binding pocket of trypsin has a number of highly ordered waters (Bode and Schwager, 1975; Kossiakoff, 1982), while no discernible water structure is observed in those parts of the pocket removed from potential H bond donor-acceptor groups. In trypsin, it is much of this unstructured water that is displaced by the substrate during formation of the Michaelis complex. The positions of several well-ordered waters in the binding pocket, however, remain unaltered by substrate binding (Bode and Schwager, 1975; Chambers and Stroud, 1977b). One of these is the water that bridges the carboxylate oxygen of Asp 189 and the hydroxyl oxygen of Try 228. This water, along with the NH of Ala 221, H bonds to the carboxylate group in the binding pocket. A second water, H bonding to the other carboxylate oxygen, interacts with the ammonium group of the bound lysine side chain. Binding of benzamidine (an analog of the guanidinium group of an arginine) displaces this water, as presumably would arginine-containing substrates.

A very interesting binding pocket water is that pictured in Figure 12, which bridges through a set of H bonds the carbonyl oxygens of 215 and 227. These carbonyls are part of the β sheet that forms a portion of the pocket. It was pointed out by Kossiakoff (1982, 1984) that the NH peptide groups of 215 and 227, which share the H bond to the carbonyls with the water, were found (from a neutron diffraction H/D exchange experiment) to remain unexchanged over a period of at least 1 year. This indicates that since the NH peptide groups are close to being solvent accessible, this region of the binding pocket is extremely rigid.

6.2.   Influences on Protein Conformation

Even though protein molecules possess a high degree of secondary structure as a result of H bonding and van der Waals interactions, much experimental evidence indicates they possess numerous segments that exhibit substantial conformational fluctuations from the native structures (Kossiakoff, 1985a). These fluctuations are a consequence of the natural equilibrium between protein-protein and protein-solvent interactions. Waters that are H bonded to protein groups in inclusions at the molecular surface or are found in interior cavities likely aid in stabilizing the native conformation.

6.2.1.   Interior Waters

Trypsin has 15–20 interior pockets that contain bound water molecules (Bode and Schwager, 1975; Kossiakoff, 1982, 1984). Some of these cavities contain single waters, while others have up to four. In most cases, these waters are highly coordinated, making from two to four hydrogen bonds to hydrophilic sites on the protein. A number of these water sites are also found in chymotrypsin and elastase, suggesting that they may be as conserved as some of the important amino acid residues in the structures (Greer, 1981; Rose et al., 1983). Their presence is apparently obligatory for a particular folding of the polypeptide chain or for the chemical environment of certain groups.

The importance of interior waters to protein folding has been pointed out by several investigators (reviewed in Finney, 1979; Edsall and McKenzie, 1983). Rose et al. (1983) surveyed a large number of refined protein structures and identified those having interior β turns. Because of their polar character (irrespective of their side chain types), β turns are found predominantly at the molecular surface. From their data base, Rose et al. found only six protein classes (the pancreatic trypsin family being one) that contain interior turns. It was also found these turns all had an associated set of buried waters that formed a hydrogen-bonding complex with the carbonyl oxygens of the turn residue. Rose et al. suggested that these buried waters function as structural prosthetic groups, supplementing intramolecular hydrogen bonding by providing necessary site-specific polarity.

Figure 13 pictures a region of the trypsin structure containing an interior β turn (194−197) with the corresponding neutron densities for the associated water molecules. The CO group of 196 is buried and has no H-bonding partner on the protein molecule itself. From the standpoint of energetics, absence of an H-bonding partner would be an unfavorable situation. As seen from the density contours, however, there is a water molecule adjacent to the carbonyl which satisfies its H-bonding requirement.

Note that the density contours for this water molecule are highly asymmetric. This occurs because the water is highly ordered and the scattering from its two deuterium atoms is readily observable in the density map. Within the density, the "+" marks the location of the oxygen. Using this as a point of reference, the orientation of the water molecule with respect to its coordinating ligands (196−0, 197−0, 45-OG) can be deduced.

The ability to assign orientation of waters based on experimental data has important implications for understanding the forces involved in protein folding. Using the example above, the orientation of this water is dictated by the electrostatic forces of its coordinating ligands and by local environmental influences. From Figure 13, these H bond geometries are not optimal for any individual group; rather, the situation seems to be one where compro-

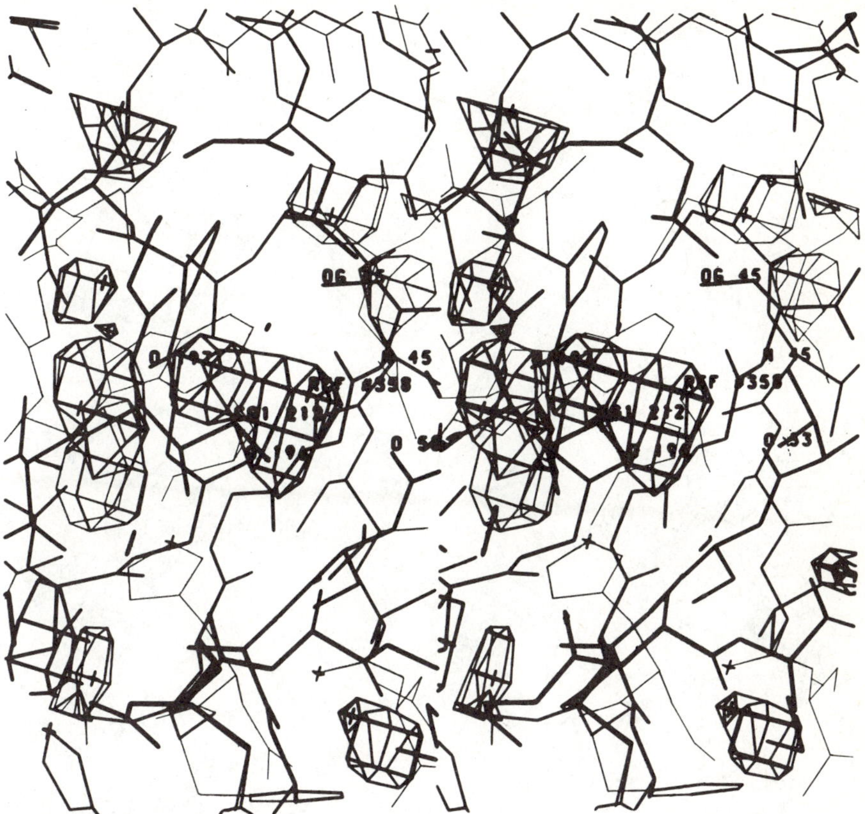

Figure 13. Ordered water (D_2O) bound in an internal cavity. Oxygen position of D_2O marked as + Ref #358.

mises in H bonds are made between the three ligands. If a collection of similar examples of water molecule orientations could be assembled, it might provide an experimental basis for developing and testing theoretical dynamics and energy minimization algorithms.

From X-ray results, there are several seemingly unoccupied cavities in trypsin that are large enough to hold at least one water (M. Connelly, I. Kuntz, unpublished results). In the neutron solvent difference maps (see below), however, there existed interpretable neutron density in these cavity regions (A. Kossiakoff, unpublished results). An example of this is shown in Figure 14, where two isolated solvent molecules are bound in internal pockets of the molecule. It is noteworthy that neither water has the

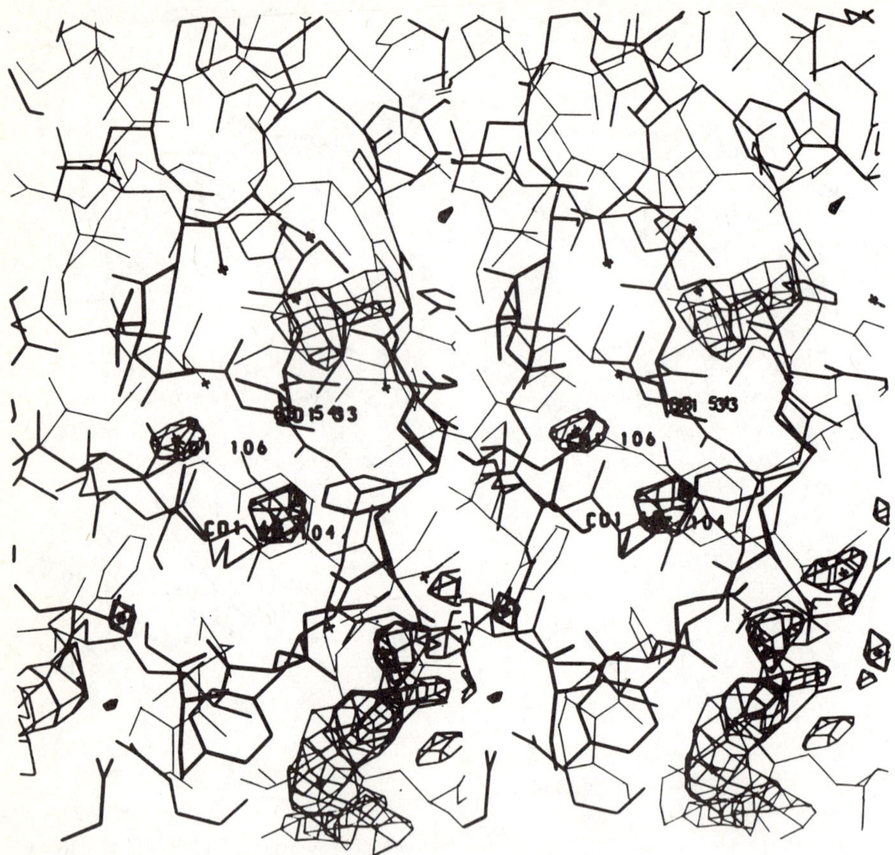

Figure 14. Buried waters in hydrophobic cavities. Because they form no H bonds, these waters are not well ordered and were not identified in the X-ray analyses.

hydrogen-bonding ligands usually associated with protein-solvent interactions. Analysis of the groups that form the interfaces of these water cavities indicated that the waters resided in a predominantly hydrophobic environment. This example is one of several where partially ordered waters are associated with hydrophobic regions. In summary, folding of the polypeptide chain of trypsin into its tertiary structure involves formation of solvent-filled cavities of varying chemical character.

6.2.2. *Exterior Water Structure*

The typical approach for assigning water molecule positions is to accept peaks in the density map that are above a certain signal-to-noise threshold and are positioned at permissible hydrogen-bonding locations relative to groups on the protein surface. While this approach is generally satisfactory for tightly bound waters whose scattering produces large peaks in the density map, it is considerably less definitive and may lead to erroneous assignments in the case of partially ordered waters where the signal-to-noise ratio is poor. The recent development of the neutron solvent difference map method has greatly improved the accuracy of making such water structure assignments (Kossiakoff, 1985b).

The solvent difference map method is based on the fact that the scattering of neutrons by H_2O and D_2O is characteristically quite different; water in its D form (D_2O) has a scattering potential about six times that of H_2O. The method has been shown to be able to locate water molecules that scatter as low as 10% of fully ordered water sites. This is considerably better than can be achieved by even highly refined X-ray structures. In the refined X-ray model of trypsin, about 150 water positions were assigned at a 1.5 Å resolution (J. Finer-Moore and R. Stroud, personal communication). They compared these positions with the densities contained in the neutron solvent difference maps and found an exceptional correspondence (95% with the well-ordered waters). They further observed that the solvent map contained a number of additional density peaks adjacent to the primary sites. These density peaks are likely to be waters forming a second hydration layer at particular points along the protein surface. Assignment of these additional solvent sites resulted in a total of about 300 unique water molecules.

These 300 waters, for the most part, were found to be associated with hydrophilic regions of the protein surface or were part of the second layer of hydration found at a number of sites. An analysis of the surface area of the protein in direct contact with bound water revealed that only about 15−20% of the surface is covered. This means that in trypsin (and probably in most other proteins as well), most of the surface interfaces with water having the properties of bulk solvent.

7. STRUCTURE-FUNCTION INVESTIGATIONS USING GENETIC ENGINEERING

The compilation of this chapter coincides with the emergence of new methods of genetic engineering that will allow substitution of arbitrary resides at specific sites in the primary sequence of proteins. The power of this approach in allowing insight into the functional importance of an individual or group of residues is obvious. The "site-directed mutagenesis" approach will not only alter the design of experiments, but will also produce more definitive interpretations in a shorter period of time. Two examples of possible applications to structure-function studies using genetic engineering, are discussed below. While interpretation of these experiments might appear straightforward, the discussion will show that in practice they are not. The choice of these particular examples was made to illustrate some of the problems that must be considered in these types of experiments.

7.1. Residue 102 (Asp to Asn)

A primary factor in lowering the energy of the activated complex is the electrostatic potential energy derived from the charge interaction, Asp$^-$, His$^+$, TI$^-$. A possible experiment aimed at quantifying the effect of this interaction on the rate of catalysis would be substitution of a neutral Asn for the charged Asp. From the standpoint of solely examining the three catalytic groups, kinetic experiments could be devised to assess the change. However, a simplified approach ignoring the effects of other groups in the local environment is incorrect. Even though the two groups are stereochemically similar, their modes of H bonding are different; Asp has two H bond acceptors, while Asn has one acceptor and one donor. Figure 15 shows that in the environment of Asp-102, H-bonding modes are important. In order for proton H2 to be transferred to the imidazole during catalysis, the nitrogen, NE2, has to be unprotonated. This requires oxygen OD1 of the Asn to be the H-bond ligand to the imidazole and the ND2 group to be directed at the amide peptide gorups of residues 56 and 57. This latter interaction is unfavorable.

The more likely conformation of the Asn side chain would probably be one where the ND2 group H bonds to the free electron pair of the imidazole ND1 (Fig. 15, bottom), and the OD$_1$ H bonds to the 56 and 57 peptide groups. In this model, NE2 is the protonated nitrogen—a situation making proton transfer from the Ser 195 OG impossible. The result is an inactive enzyme. The point is that an incorrect interpretation could arise by ascribing the enzyme's inactivity to the absence of a charged group at position 102, rather than the effect of a different H-bonding sequence between the Asn and His compared to that between Asp and His. As of this writing, this particular experiment is being undertaken with trypsin (C. Craik,

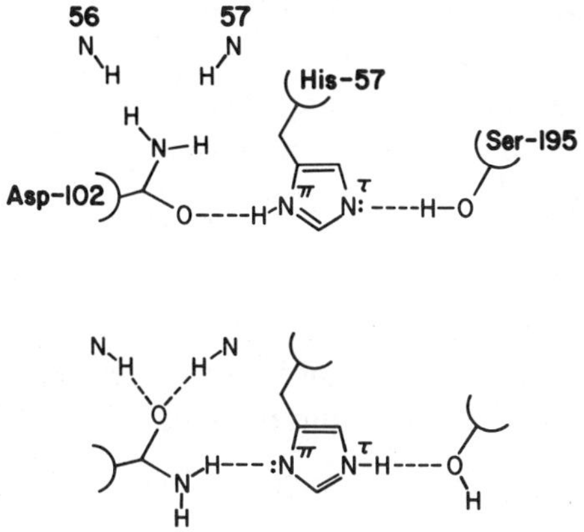

Figure 15. Two possible H-bonding orientations for the Asn-His interaction. Interaction with Ser 214 OG is not shown. H2 is the Ser-195 hydroxyl proton, in this figure NE2 is $N\tau$ and ND1 is $N\pi$.

personal communication) and subtilisin (J. Wells and S. Power, personal communication).

7.2. Residue 189 (Asp to Glu)

Kinetic evidence indicates that trypsin has a 25:1 preference for arginine over lysine substrates (Fersht and Requera, 1971). This may partly be the Arg side chain making a direct salt bridge interaction with the Asp 189 carboxylate, while the lysine interaction is mediated by a water molecule. An interesting experiment would be to replace the Asp with a Glu. This would make a direct Glu-Lys interaction possible and perhaps produce an enzyme completely specific for lysine.

An important caveat to this logic is that the Asp side chain was oriented by H-bonds; a similar type of orientation would not be possible in the Glu mutant with the result that the side chain could be at least somewhat disordered. How much such a situation would affect binding is hard to assess, but it is a factor which cannot be ignored.

The above two experimental proposals point out that genetic engineering experiments have to be well thought through to consider the affects of local induced structure change along side the chemical change. The approach of site-directed mutagenesis, however, will have a major impact on the field of structural biology and will be a primary tool which, coupled with crystallography and other biophysical methods, can better illuminate structure/function relationships in proteins.

ACKNOWLEDGMENTS

I would like to express my thanks to Drs. Robert Stroud and Alex Kossiakoff for their suggestions during preparation of the manuscript.

REFERENCES

Alagona, G., Scrocco, E., and Tomasi, J. (1975). *J. Am. Chem. Soc.* **97**, 6976.

Alden, R. A., Wright, C. S., and Kraut, J. (1970). *Phil. Trans. R. Soc. London,* **B257**, 119.

Allen, L. C. (1981). *Ann. NY Acad. Sci.* **67**, 383.

Bachovchin, W. W., and Roberts, J. D. (1978). *J. Am. Chem. Soc. 100*, 8041.

Bender, M. L., and Kezdy, F. J. (1964). *J. Am. Chem. Soc.* **86**, 3704.

Birktoft, J. J., Blow, D. M., Henderson, R., and Steitz, T. A. (1970). *Phil. Trans. Roy. Soc. London* **B257**, 67.

Blow, D. M. (1976). *Acc. Chem. Res.* **9**, 145.

Blow, D. M., Birktoft, J. J., and Hartley, B. S. (1969). *Nature* **221**, 337.

Bode, W., and Schwager, P. (1975). *J. Mol. Biol.* **98**, 693.

Bode, W., Schwager, P., and Huber, R. (1978). *J. Mol. Biol.* **118**, 99.

Bruice, T. C., and Benkovic, S. J. (1964). *J. Am. Chem. Soc.* **86**, 418.

Burgi, H. B., Duntz, J. D., Lehn, J. M., and Wipft, G. (1974). *Tetrahedron* **30**, 1563.

Caplow, M. J. (1969). *J. Am. Chem. Soc.* **91**, 3639.

Chambers, J. L., and Stroud, R. M. (1977a). *Acta Crystallog.* **B33**, 1824.

Chambers, J. L., and Stroud, R. M. (1977b). Protein Data Bank, Brookhaven National Laboratory.

Cohen, G. H., Silverton, E. W., Matthews, B. W., Braxton, H., and Davies, D. R. (1969). *J. Mol. Biol.* **44**, 129.

Delbaere, L. T., Hutcheon, W. L., James, M. N., and Theissen, W. E. (1975). *Nature* **257**, 758.

Delpuech, J. J., and Serratrice, G. (1975). *Mol. Phys.* **29**, 849.

Desmeules, P. (1980). Ph.D. Thesis, Princeton University.

Edsall, J. T., and McKenzie, A. A. (1983). *Adv. Biophys.* **10**, 37.

Elrod, J. P., Gandour, R. D., Hogg, J. L., Kise, M., Maggiora, G. M., Schowen, R. L., and VenKatasubban, K. S. (1975). *Symp. Faraday Soc.* **10**, 145.

Fahrney, D. E., and Gold, A. M. (1963). *J. Am. Chem. Soc.* **85**, 997.

Fastrez, J., and Fersht, A. R. (1973). *Biochemistry* **12**, 1067.

Fehlhammer, H., and Bode, W. (1975). *J. Mol. Biol.* **98**, 683.

Fehlhammer, H., Bode, W., and Huber, R. (1977). *J. Mol. Biol.* **111**, 415.

Fersht, A. R., and Requena (1971). *J. Am. Chem. Soc.* **93**, 7079.

Finney, J. L. (1979). In *Water: A Comprehensive Treatise,* F. Franks, Ed., Vol. 6, Plenum Press, New York, p. 47.

Freer, S. T., Kraut, J., Robertus, J. D., Wright, H. T., and Xuong, Ng H. (1970). *Biochemistry* **9**, 1997.

Greer, J. (1981). *J. Mol. Biol.* **153**, 1027.

Henderson, R. (1970). *J. Mol. Biol.* **54**, 341.

Henderson, R. (1971). *Biochem. J.* **124**, 13.

Hirono, H., Nakagana, S., and Umeyama, H. (1985). *Proc. Natl. Acad. Sci. USA* (in press).

Huber, R., and Bode, W. (1978). *Acc. Chem Res.* **11**, 114.

Huber, R., Kukla, D., Bode, W., Schwager, P., Bartels, K., Deisenhofer, J., and Steigemann, W. (1974). *J. Mol. Biol.* **89**, 73.

Hunkapiller, M. W., Smallcombe, S. H., Whitaker, D. R., and Richards, J. H. (1973). *Biochemistry* **12**, 4732.

Hunkapiller, M. W., Forgac, M. D., and Richards, J. H. (1976). *Biochemistry* **15**, 5581.

Inagami, T., and Mitsuda, H. (1964a). *J. Biol. Chem.* **239**, 1388.

Inagami, T., and Mitsuda, H. (1964b). *J. Biol. Chem.* **239**, 1395.

Inward, P. M., and Jencks, W. P. (1965). *J. Biol. Chem.* **240**, 1986.

James, M. N., Sielecki, A. R., Brayer, G. D., Delbaere, L. T., and Bauer, C. A. (1980). *J. Mol. Biol.* **144**, 43.

James, M., Brayer, G. D., Delbaere, L. T., Sielecki, A. R., and Gertler, A. (1980b). *J. Mol. Biol.* **139**, 423.

Jansen, E. F., Nutting, M. D., and Balls, A. K. (1949). *J. Biol. Chem.* **179**, 201.

Johnson, C. H., and Knowles, J. R. (1966). *Biochemistry* **101**, 56.

Keil, B. (1971). In *The Enzymes* P. B. Boyer, Ed., Vol 3. Academic Press, New York and London, p. 249.

Kerns, R. C., and Allen, L. C. (1978). *J. Am. Chem. Soc.* **100**, 6587.

Kleier, D. A., Scheiner, S., and Lipscomb, W. H. (1976). *Int. J. Quantum Chem. Quantum Biol. Symp.* **3**, 161.

Koeppe, R. E., and Stroud, R. M. (1976). *Biochemistry* **15**, 3450.

Kollman, P. A., and Hayes, D. M. (1981). *J. Am. Chem. Soc.* **103**, 2955.

Komiyama, M., and Bender, M. L. (1979). *Proc. Natl. Acad. Sci. USA* **76**, 557.

Kossiakoff, A. A. (1982). *Nature* **296**, 713.

Kossiakoff, A. A. (1983). *Ann. Rev. Biophys. Bioeng.* **12**, 159.

Kossiakoff, A. A. (1984). In *Neutrons in Biology, Basic Life Sciences*, B. P. Schoen-born, Ed. Vol. 27. Plenum Press, New York, p. 281.

Kossiakoff, A. A. (1985a). *Ann. Rev. Biochem.* C. C. Richardson, Ed. 54, 1195–1227.

Kossiakoff, A. A. (1985b). *Methods in Enzymology: Biomembranes* L. Packer Ed. **127**, 329.

Kossiakoff, A. A., and Spencer, S. A. (1980). *Nature* **288**, 414.

Kossiakoff, A. A., and Spencer, S. A. (1981). *Biochemistry* **20**, 6462.

Kossiakoff, A. A., Chambers, J. L., Kay, L. M., and Stroud, R. M. (1977). *Biochemistry* **16**, 654.

Kraut, J. (1977). *Ann. Rev. Biochem.* **46**, 331.

Kraut, J., Robertus, J. D., Birktoft, J. J., Alden, R. A., Wilcox, P. E., and Powers, J. C. (1971). *Cold Spring Harbor Symp. Qual. Biol.* **36**, 117.

Krieger, M., Kay, L. M., and Dickerson, R. E. (1974). *J. Mol. Biol.* **83**, 209.

Laskowski, M. J., and Sealock, R. W. (1971). In *The Enzymes,* Vol. 3 P. B. Boyer, Ed. Academic Press, New York, p. 376.

Markley, J. L., and Porubean, M. A. (1976). *J. Mol. Biol.* **102**, 487.

Martinek, K., Savin, Y. V., and Berezin, I. V. (1971). *Biokhimiya (Moscow)* **36**, 806.

Matthews, D. A., Alden, R. A., Birktoft, J. J., Freer, S. T., and Kraut, J. (1977). *J. Biol. Chem.* **252**, 8875.

Morgan, P. H., Robinson, N. C., Walsh, K. A., and Neurath, H. (1972). *Proc. Natl. Acad. Sci. USA* **69**, 3312.

Nakagawa, S., and Umeyama, H. (1984). *J. Mol. Biol.* **179**, 103.

Nakagawa, S., Umeyama H., and Kudo, T. (1980). *Chem. Pharm. Bull.* **28**, 1342.

Neurath, H., and Hartleg, B. S. (1959). *J. Cell Comp. Physiol.* **54**, Suppl. 1, 179.

O'Leary, M. H., and Kluetz, M. D. (1972). *J. Am. Chem. Soc.* **94**, 3585.

Ong, E. B., Shaw, E., and Schoellmann, G. (1964). *J. Am. Chem. Soc.* **86**, 1271.

Poland, D., and Scheraga, H. A. (1967). *Biochemistry* **6**, 3791.

Pollock, E., Hogg, J. L., and Schowen, R. L. (1973). *J. Am. Chem. Soc.* **95**, 968.

Porubean, M. A., Neves, D. E., Ransch, S. K., and Markley, J. C. (1978). *Biochemistry* **17**, 4640.

Porubean, M. A., Westler, W. M., Ibanez, I. B., and Markley, J. L. (1979). *Biochemistry* **18**, 4108.

Richarz, R. Tschesche, H., and Wuthrich, K. (1980). *Biochemistry* **19**, 5711.

Robertus, J. D., Alden, R. A., Birktoft, J. J., Kraut, J., Powers, J. C., and Wilcox, P. E. (1972). *Biochemistry* **11**, 2439.

Robillard, G., and Shulman, R. G. (1972). *J. Mol. Biol.* **71**, 507.

Robillard, G., and Shulman, R. G. (1974a). *J. Mol. Biol.* **86**, 519.

Robillard, G., and Shulman, R. G. (1974b). *J. Mol. Biol.* **86**, 541.

Robinson, N. C., Neurath, H., and Walsh, K. A. (1973). *Biochemistry* **12**, 420.

Rose, G. D., Young, W. B., and Gierasch, L. M. (1983). *Nature* **304**, 654.

Ruhlmann, A., Kukla, D., Schwager, P., Bartels, K., and Huber, R. (1973). *J. Mol. Biol.* **77**, 417.

Satterthwait, A. C., and Jencks, W. P. (1974a). *J. Am. Chem. Soc.* **96**, 7018.

Satterthwait, A. C., and Jencks, W. P. (1974b). *J. Am. Chem. Soc.* **96**, 7031.

Schechter, F., and Berger, A. (1967). *Biochem. Biophys. Res. Commun.* **27**, 157.

Scheiner, S., and Lipscomb, W. N. (1976). *Proc. Natl. Acad. Sci. USA* **73**, 432.

Scheiner, S., Kleier, D. A., and Lipscomb, W. N. (1975). *Proc. Natl. Acad. Sci. USA* **72**, 2606.

Schuster, P. (1970). *Theoret. Chem. Acta* **19**, 212.

Segal, D. M., Cohen, G. H., Davies, D. R., Powers, J. C., and Wilcox, P. E. (1971). *Cold Spring Harbor Symp. Qual. Biol.* **36**, 85.

Shaw, E., Mares-Guia, M., and Cohn W. (1965). *Biochemistry* **4**, 2219.

Shotton, D. M., and Watson, H. C. (1970). *Nature* **225**, 311.

Shotton, D. M., White, N. J., and Watson, H. C. (1971). *Cold Spring Harbor Symp. Qual. Biol.* **36**, 91.

Steitz, T. A., and Shulman, R. G. (1982). *Ann. Rev. Biophys. Bioeng.* **11**, 419.

Steitz, T. A., Henderson, R., and Blow, D. M. (1969). *J. Mol. Biol.* **46**, 337.

Stroud, R. M. (1974). *Sci. Am.* **231**, 74.

Stroud, R. M., Kay, L. M., and Dickerson, R. E. (1971). *Cold Spring Harbor Symp. Qual. Biol.* **36**, 125.

Stroud, R. M., Kay, L. M., and Dickerson, R. E. (1974). *J. Mol. Biol.* **83**, 185.

Stroud, R. M., Krieger, M., Koeepe, R. E., Kossiakoff, A. A., and Chambers, J. L. (1975). *Proteases and Biological Control.* p. 13. Cold Spring Harbor Laboratory, Cold Spring Harbor, New York.

Sweet, R. M., Wright, H. T., Janin, J., Chothia, C. H., and Blow, D. M., (1974). *Biochemistry* **13**, 4212.

Tulinsky, A., Vandlen, R. L., Morimoto, C. N., Mani, N. V., and Wright, L. H., (1973) *Biochemistry* **12**, 4185.

Umeyama, H., Nakagawa, S., and Kudo, T. (1981). *J. Mol. Biol.* **150**, 409.

Vehar, G. A., Kohr, W. J., Bennett, W. F., Pennica, D., Ward, C. A., Keck, R., Harkins, R. N., and Collen, D. (1984). *Biotechnology* **2**, 1051.

Wang, J. H. (1970). *Proc. Natl. Acad. Sci. USA* **66**, 874.

Warshel, A. (1978). *Proc. Natl. Acad. Sci. USA* **75**, 5250.

Warshel, A. (1981). *Acc. Chem. Res.* **14**, 284.

Warshel, A., and Levitt, M. (1976). *J. Mol. Biol.* **103**, 227.

Weiner, S. (1984). Ph.D. Thesis, University of California, San Francisco.

Weiner, S. J., Singh, N. C., and Kollman, P. A. (1985). *J. Am. Chem. Soc.* (in press).

Wright, H. T. (1973a). *J. Mol. Biol.* **79**, 1.

Wright, H. T. (1973b). *J. Mol. Biol.* **79**, 13.

8

Aspartic Proteinases and Their Catalytic Pathway

MICHAEL N. G. JAMES
ANITA R. SIELECKI
Medical Research Council of Canada
Group in Protein Structure and Function
Department of Biochemistry
University of Alberta
Edmonton, Alberta, Canada

CONTENTS

1. **INTRODUCTION**

2. **MOLECULAR PROPERTIES OF ASPARTIC PROTEINASES**
   2.1. Primary Structures
   2.2. Results From Biochemical and Kinetic Studies

3. **TERTIARY STRUCTURES OF ASPARTIC PROTEINASES**
   3.1. The Aspartic Proteinase Fold
   3.2. The Refinement of Penicillopepsin and Its Inhibitor Complexes
   3.3. Active Site Architecture
   3.4. Binding Interactions with Inhibitor and Substrate Molecules
   3.5. Proposed Productive Substrate Binding to Aspartic Proteinases
   3.6. Proposed Catalytic Hydrolysis Pathway for Aspartic Proteinases

4. **COMPARISON TO THE SERINE PROTEINASE FAMILY**

5. **COMPARATIVE MOLECULAR MODELING: RENIN**

6. **CONCLUDING REMARKS**

REFERENCES

1. INTRODUCTION

There are four classes of enzymes that catalyze the hydrolysis of peptide bonds: the serine proteinases, the thiol or cysteine proteinases, the metalloproteinases, and the aspartic proteinases (Hartley, 1960). The latter class is thus named because its members have two aspartic acid residues in their active sites that are intimately involved with the catalytic hydrolysis of oligopeptide substrates (Clement, 1973). This class of enzymes has also been referred to as acid or carboxyl proteinases. The present chapter will review the known structural data pertaining to their catalytic mechanism.

Members of the aspartic proteinase family include the much studied gastric proteinases (Foltmann, 1981), for example, pepsin A (EC3.4.23.1), gastricsin (EC3.4.23.3), and chymosin (EC3.4.23.4). The latter is isolated from the stomach of neonatal mammals and is widely used in the production of cheese. Other members of this enzyme family have important roles in the control and regulation of a variety of physiological functions. Certain mammalian lysosomal enzymes, cathepsin D and E, are implicated in protein turnover (Barrett, 1977). The renin-angiotensin system is paramount in the regulation of blood pressure (Ondetti and Cushman, 1982). Renin is found not only in the kidney, produced by the juxtaglomerular cells, but also in many other tissues such as the submaxillary gland and brain and in the amniotic fluid.

Aspartic proteinases are also produced by molds. Due to a declining supply of chymosin (in a complex mixture called rennet), some of these mold enzymes are being actively studied as possible substitutes for chymosin in the cheese-making industry. The three mold enzymes that will figure prominently in this review are penicillopepsin from *Penicillium janthinellum* (Sodek and Hofmann, 1970), *Rhizopus chinensis* pepsin (Fukumoto et al., 1967), and *Endothia parasitica* pepsin (see Hagemeyer et al., 1968; also, general review by Hofmann, 1974).

In view of the many roles that aspartic proteinases have in a variety of biological processes, it is important to understand the chemical mechanism whereby these enzymes catalyze the hydrolysis of peptide bonds. Some of them, involved with general protein degradation such as the gastric proteinases, have a broad specificity. Others, such as renin, exhibit an exquisite specificity. In the hydrolysis of the N-terminal peptide of angiotensinogen, renin cleaves a single bond—between Leu 10-Leu 11—thus releasing the decapeptide angiotensin I. A knowledge of the structural principles that confer such specificity could aid in the rational development of selective inhibitors of renin.

In all, the three-dimensional molecular structures of four aspartyl proteinases are known. The structure of penicillopepsin, originally determined at 2.8 Å resolution (Hsu et al., 1977a, 1977b; James et al., 1977), has now been fully refined at 1.8 Å resolution (James and Sielecki, 1983). The structure of porcine pepsin A was first described at 2.7 Å resolution (Andreeva et al., 1977–1979), and has now been partially refined at 2 Å resolution (Andreeva et al., 1984). Progress in the studies of rhizopuspepsin and endothiapepsin has been hampered by the lack of complete amino acid data. Their structures at medium resolution have been described (Subramanian et al., 1977a, 1977b; Jenkins et al., 1977; Blundell et al., 1979), although the assignment of the primary sequence of parts of the molecules based on the crystallographic data alone may still be ambiguous. In addition, a model of murine submaxillary renin based on the 2.1 Å structure of endothiapepsin has been proposed (Blundell et al., 1983), and work is in progress on the determination of the crystal structure of this important regulatory enzyme (Navia et al., 1984).

The mode of inhibitor binding to some of these enzymes has also been investigated crystallographically. Pepstatin (Umezawa et al., 1970) is an extremely potent inhibitor of aspartyl proteinases (Workman and Burkitt, 1979). Its sequence is Iva-Val-Val-Sta-Ala-Sta, where Iva is an isovaleryl group and statine (Sta) is an unusual amino acid [(4S,3S)-4-amino-3-hydroxy-6-methylheptanoic acid]. It has been suggested that part of the powerful inhibitory properties of pepstatin is due to the presence of the central Sta residue that could mimic the tetrahedral transition state of a good substrate (Marciniszyn et al., 1976; Marshall, 1976). Bott et al. (1982) have reported the binding of pepstatin to rhizopuspepsin at 2.5 Å resolution. Pepstatin analogs bound to penicillopepsin have been studied at 1.8 Å resolution (James et al., 1982, 1983, 1985), leading to a new proposal for the catalytic pathway of peptide hydrolysis (James and Sielecki, 1985). Before analyzing how the structural results have influenced the understanding of the catalytic action, we shall discuss the general molecular and chemical properties of the aspartic proteinases.

2. MOLECULAR PROPERTIES OF ASPARTIC PROTEINASES

All known aspartic proteinases from vertebrates are biosynthetically produced as inactive precursors or proenzymes (zymogens). Zymogen precursors of the fungal enzymes have not been detected thus far. The active molecules consist of approximately 330 amino acid residues, with a molecular weight of ~35,000. In the vertebrate enzymes, this active form results from limited proteolysis that removes an N-terminal fragment of approximately 45 additional amino acids (Kassell and Kay, 1973; Tang, 1976). The reason for the lack of proteolytic activity of the zymogen form is not at all understood in molecular terms. Evidence for both a unimolecular and a bimolecular mechanism of activation has been reviewed (Tang, 1976).

2.1. Primary Structures

The first complete amino acid sequence determination of an aspartyl proteinase was that of porcine pepsin A (Tang et al., 1973; Moravek and Kostka, 1974; Sepulveda et al., 1975). This gastric proteinase is made of a single polypeptide chain of 327 residues with three intramolecular disulfide bridges. There is an overwhelming preponderance of acidic carboxyl groups over the groups with a pK_a in the basic range (i.e., 43 aspartic and glutamic acid residues, compared to only 2 arginine, 1 lysine, and 1 histidine residues). This amino acid composition results in a low pK_I for this enzyme. The single phosphoserine residue in pepsin is not involved with its mechanism of action (Perlmann, 1952). Chemical evidence has implicated two aspartic acid residues in the catalytic function of pepsin (Knowles, 1970; Fruton, 1970). The sequence determination confirmed the presence of these two groups in segments that have extraordinary sequence homology: Asp-Thr-Gly-Ser in the N-terminal portion of the molecule, and Asp-Thr-Gly-Thr in the C-terminal part of pepsin. This fact suggested a gene duplication event in the evolution of the aspartic proteinases. Strong support for this proposal (Tang et al., 1978) had to await the determination of a tertiary structure.

The only published complete sequence of a mold aspartic proteinase is that of penicillopepsin. The chemical sequence was first determined in 16 major fragments (Cunningham et al., 1976). These fragments were then arranged in their proper order by fitting the amino acids to the 2.8 Å electron density map of penicillopepsin (Hsu et al., 1977a, 1977b). Difficult regions were clarified by renewed isolation of specific peptides and their chemical sequencing. The final amino acid sequence is reported in the description of the 1.8 Å resolution structure of penicillopepsin (James and Sielecki, 1983). Regions of the molecule that have had the sequence alignment with pepsin (Hsu et al., 1977a, 1977b) revised as a result of new chemical data (personal communication, T. Hofmann), or new structural information resulting from the refinement of penicillopepsin, are reported in Table 1. This alignment is not based on a detailed structural comparison, since the only pepsin coordinates available at the time this study was carried out corresponded to the initial, unrefined 2.7 Å resolution model. It is therefore stressed that the alignment is still tentative, and it is presented here primarily to assist in the subsequent discussions.

A great deal of interest has focused on the published sequences for mouse submaxillary gland renin (Misono et al., 1982; Panthier et al., 1982). These two studies were highly complementary; the former used traditional amino acid chemical-sequencing methods and the latter the new, rapid technique of nucleotide sequencing. The structural model for mouse submaxillary gland renin that has been constructed using interactive computer graphics (Blundell et al., 1983) exemplifies the keen interest in this molecule.

Table 1. Portions of the Sequence Alignment of Penicillopepsin (Penpep) and Porcine Pepsin A (Pepsin)[a]

```
        20                            28
Penpep  P  V  T  I  G  —  —  G  T  T  L
Pepsin  T  I  G  I  G  T  P  A  Q  D  F
        17                            27


        206                          213         216
Penpep  D  G  F  —  —  —  S  G  I  A  D  T  G  T
Pepsin  A  C  S  G  G  C  Q  A  I  V  D  T  G  T
        205 └________________┘            215  ↙ 218


        247                                                      264
Penpep  F  D  C  S  T  —  —  N  L  P  D  F  S  V  S  I  S  G  Y  T
Pepsin  I  S  C  S  S  I  D  S  L  P  D  I  V  F  T  I  —  —  D  G
        248                                                      265


        289              294           295                 304
Penpep  S  N  S  G  I  G  —  —  —  —  —  F  S  I  F  G  D  I  F  L  K
Pepsin  G  M  D  V  P  T  S  S  G  E  L  W  I  L  G  D  V  F  I  R
        289                                                      308
```

[a]The numbers above the sequences refer to the sequential penicillopepsin numbering; those below, are the pepsin sequence numbers. This tentative revised alignment is based on the 2.7 Å resolution model of porcine pepsin (Andreeva et al., 1977–1979) and the refined penicillopepsin structure with the final sequence data (James and Sielecki, 1983).

2.2. Results From Biochemical and Kinetic Studies

A great many chemical and kinetic analyses of the aspartic proteinase mechanism were carried out prior to the tertiary structural determinations in 1977. Several excellent reviews of these studies have been published by Fruton (1970, 1976), Knowles (1970), Clement (1973), and Hofmann (1974). In this section, only some of the more relevant biochemical findings will be summarized, as they have important consequences for the proposal of a mechanism based on the structural results. There is general agreement in the interpretations of many of the biochemical observations. However, some of the more esoteric results have produced conflicting views of the pathway. It is also true that many of the structural results are only interpretable in light of the biochemical deductions.

Evidence for the involvement of two carboxyl groups in the peptide-bond breaking step has come from studies of the pH dependence of pepsin action (Denburg et al., 1968; Cornish-Bowden and Knowles, 1969; Clement et al., 1970). This pH dependence of the kinetic parameters also suggests pK_a values of ~1.2 and 4.7 for those residues (see review by Clement, 1973), although it is not established unequivocally which carboxyl group has the lower pK_a. Chemical modification studies of pepsin by active site-directed reagents such as diazo compounds (Delpierre and Fruton, 1966; Rajagopalan et al., 1966) and epoxides (Hartsuck and Tang, 1972), as well as inhibition of the enzyme by EPNP [1,2-epoxy-3-(p-nitrophenoxy)propane] (Chen and Tang, 1972), were also consistent with a reaction with two carboxyl groups. Completion of the sequence of pepsin led to the proposal that the two carboxyl groups were those of Asp 33(32)* and Asp 213(215) (Tang et al., 1973). Similar studies on the inhibition of penicillopepsin by diazo compounds were also carried out (Sodek and Hofmann, 1971); later it was shown crystallographically that EPNP indeed reacts with both carboxyl groups of Asp 33(32) and Asp 213(215) in pepicillopepsin (Hsu et al, 1977b; James et al., 1977).

---

*In the present discussion, the sequential amino acid residue numbering of penicillopepsin will be used. In addition, for clarity, the sequence numbers of equivalent residues in porcine pepsin (Hsu et al., 1977a, 1977b) will be included in parentheses. A new tertiary structural alignment based on the refined structures of penicillopepsin and pepsin has not been done, so this equivalence is only tentative.

A dominant feature of aspartic proteinase-catalyzed hydrolysis of peptide bonds is the pronounced dependence of k_{cat} on secondary binding from both the amino and carboxyl sides of the scissile bond (Fruton, 1970, 1976). From studies with synthetic substrates, the subsite specificities of pepsin favor large hydrophobic residues at the P_1 and P'_1 positions of a substrate[+] (Powers et al., 1977). Extending the length of the polypeptide chain of a substrate towards either the amino or carboxy termini, bring about large increases of k_{cat} with relatively little variation in the K_m value (Fruton, 1970). These results, obtained first for pepsin, have also been confirmed with other aspartic proteinases, particularly penicillopepsin (Hofmann and Hodges, 1982). The obvious implication of these experiments was that aspartic proteinases would have an extended binding cleft sufficient to accommodate peptides of up to seven amino acid residues (Sampath-Kumar and Fruton, 1974; Wang and Hofmann, 1976, 1977). A more subtle interpretation of those same results in order to explain the behavior of k_{cat} and K_m was to suggest that conformational change and strain could be important contributions to the catalytic mechanism of peptide hydrolysis (Fruton, 1976). In this respect, it was proposed that a significant portion of the binding energy in the enzyme-substrate complex formation would be used to reduce the activation energy of the bond-breaking step (Jencks, 1975). The popular view of this effect in molecular terms, involves the imposition of strain on the substrate by the enzyme to produce a distortion towards the transition state. Thus, as the substrates become longer, more of the binding energy could be used to lower the activation energy for hydrolysis, resulting in an enhanced k_{cat} value. This strain would be exerted through the substrate.

Nevertheless, there is experimental evidence suggesting that the enhanced k_{cat} is the result of strain transmitted through the enzyme (an induced-fit hypothesis), rather than through the substrate. Penicillopepsin does not cleave the tripeptide Leu-Gly-Leu, but its presence in an hydrolysis mixture increases the k_{cat} value 10-fold for the dipeptide substrate L-leucyl-L-tyrosyl amide (Wang et al., 1974). Additionally, circular dichroism spectra have been interpreted to indicate a detectable conformational

---

+ The nomenclature of Schechter and Berger (1967) is used. Amino acid side chain positions on the substrate are denoted as $P_1...P_n$ towards the amino terminus and $P'_1...P'_n$ towards the carboxy terminus, with the origin of the numbering at the scissile bond. The corresponding binding subsites on the enzyme are denoted $S_1...S_n$ and $S'_1...S'_n$.

change in the enzyme when the activator tripeptide binds. This conformational change could be similar to that observed crystallographically upon binding of pepstatin analog inhibitors to penicillopepsin (James et al., 1982). Conformational flexibility has also been invoked to rationalize some of the fluorescence experiments with pepsin (Sachdev et al., 1975; Sachdev and Fruton, 1975).

A basic group has long been implicated in the catalytic mechanism of aspartic proteinases (Kitson and Knowles, 1971). Treatment of porcine pepsin with phenylglyoxal results in approximately 50% inactivation, as measured with hemoglobin as substrate. Inactivation of pepsin by butane-2,3-dione was also attributed to reaction with a basic group identified as Arg 316 of this enzyme (Huang and Tang, 1972). It was shown later that butane-2,3-dione inactivation is due rather to a photosensitized modification of tyrosine and tryptophan residues (Gripon and Hofmann, 1981). This same work verified that modification of one or more lysine residues by diethylpyrocarbonate leads to inactivation of penicillopepsin, pointing again to the role of a basic group in catalysis. The only positions in the sequences of penicillopepsin and porcine pepsin occupied by homologous basic residues correspond to Lys 304 and Arg 308, respectively (Hsu et al., 1977a, 1977b), suggesting a similar role for these groups in the corresponding enzymes. The molecular basis for such a role in catalysis, though, is not conclusive.

A central question that still haunts some of those involved with the biochemistry of reactions catalyzed by aspartic proteinases is whether there exists a covalently attached intermediate on the hydrolytic pathway. The involvement of a covalently attached acyl enzyme intermediate in the hydrolytic pathway of serine proteinases was first suggested by Hartley and Kilby (1954) to explain the burst kinetics observed with α-chymotrypsin and certain chromophoric substrates. In the absence of structural data, it was logical to speculate that covalent intermediates would also occur on the pathways of aspartic proteinases. Evidence for this has always been indirect and based primarily on the unusual property of transpeptidation, which is a reaction catalyzed by aspartic proteinases at pH values close to neutrality ($\sim$4.0−6.0).

Transpeptidation reactions occur via a covalently attached amino enzyme intermediate, and they involve the transfer of a C-terminal amino acid (-N-R′ in scheme 1 below) to a new acceptor group (R″-COO⁻), resulting in the formation of R″-CO-NH-R′ (Neumann et al., 1959; Fruton et al.,

1961). Observations on the nature of peptide products in other transpeptidation reactions catalyzed by pepsin and penicillopepsin (Takahashi et al., 1974; Takahashi and Hofmann, 1975; Newmark and Knowles, 1975), in which the N-terminal amino acid of a substrate is transferred to a new acceptor group, were interpreted as occurring via a covalently attached acyl enzyme intermediate (an anhydride). The initial steps in the formation of the covalent acyl enzyme intermediate are shown in scheme 2 below. For both schemes, 1 and 2, the formation of the initial Michaelis complex is implied and not shown explicitly.

The several proposals for a covalent amino enzyme intermediate have been reviewed by Knowles (1970) and Clement (1973). Since it is accepted that carboxyl groups are the catalytically important groups in the aspartic proteinases, then a covalently bound amino intermediate requires the formation of an amide bond from the amino portion of the substrate to the enzyme (scheme 1). This reaction involves the proposal of a four-center exchange reaction (Delpierre and Fruton, 1965).

$$\text{(1)}$$

Aside from the steric restrictions at the active site that this hypothesis imposes, there are many implausible components to such a suggestion (Silver and Stoddard, 1972). One of these is the breakdown of the amino enzyme during hydrolysis. A molecule of water would be required, and it would have to attack the carbonyl-carbon atom of the amide intermediate.

Although this mechanism might explain amino transfer reactions via exchange of the $RCOO^-$ group, the acyl transfer transpeptidation would require prior hydrolysis and exchange of the amino moiety, while the acyl moiety remained presumably noncovalently bound to the enzyme. It is unlikely that this sequence of events could occur. The steric objections to such a proposal have been discussed (James et al., 1981).

It has been shown that amino transfer and acyl transfer reactions with pepsin and penicillopepsin occur in the same reaction vessel under the appropriate conditions (Takahashi and Hofmann, 1975; Newmark and Knowles, 1975; Wang and Hofmann, 1976, 1977). The acyl component of such a mechanism must involve the initial nucleophilic attack by the carboxylate of one of the active site aspartic acid residues, as in the amino enzyme mechanism above. But then, rather than the complicated four-center reaction to form the amino enzyme (Delpierre and Fruton, 1965), the tetrahedral intermediate (hereinafter referred to as TI) would collapse to the acid anhydride acyl enzyme following protonation of the leaving-group nitrogen atom.

$$
\begin{array}{ccc}
\text{E-C} \overset{\displaystyle O}{\underset{\displaystyle O}{\diagup}} \ominus + \text{R-C-N-R}'
\quad\rightleftharpoons\quad
\left[\text{E-C} \overset{\displaystyle O}{\diagup} \;\; \begin{array}{c} O^{\ominus} \\ | \\ \text{O-C-N-R}' \\ | \;\; H \\ R \end{array} \right]
\end{array}
$$

$$ -H^+ \;\updownarrow\; +H^+ \tag{2} $$

$$
\left[\text{E-C} \overset{\displaystyle O}{\diagup} \;\; \begin{array}{c} O \\ \| \\ \text{O-C-R} \end{array} \right] + \; H_2N\text{-}R'
$$

The steric problems involved with this proposed pathway have also been discussed (James et al., 1981). Breakdown of the anhydride could take place via two possible routes in which water (or an activated OH^- ion) would attack either the carbonyl carbon of the substrate or the enzyme. The latter is not possible on steric grounds. This reaction pathway would accommodate acyl transfer via exchange of the amino moiety. Amino transfer requires the unlikely *prior* hydrolysis of the anhydride, while the amino portion of the substrate remained bound. It has been suggested that whether an amino enzyme intermediate or an acyl enzyme intermediate occurs is a function of the nature of the substrate (Takahashi and Hofmann, 1975). This proposal would imply an even more complex mechanism.

Additional evidence for the existence of a covalently attached acyl enzyme intermediate came from experiments in which pepsin catalyzed the ^{18}O exchange between a virtual substrate and water (Sharon et al., 1962). Nevertheless, Knowles (1970) put forward an alternative scheme that accounted for the ^{18}O exchange without the involvement of a covalent intermediate. More recently, additional ^{18}O isotope exchange experiments have also been interpreted in terms of a mechanism that does not involve covalently attached intermediates (Antonov et al., 1978, 1981). Many attempts have been made to trap *any* covalent intermediate that might occur on the catalytic pathway. None of these attempts has succeeded (Fruton, 1976). Cryoenzymatic experiments aimed at the detection of a burst reaction or a trapped intermediate for pepsin (Dunn and Fink, 1984) and penicillopepsin (Hofmann and Fink, 1984) have also failed. Quite correctly, the authors conclude that their negative results are not a definite proof for or against the existence of a covalent intermediate, although conditions of the experiments (in aqueous methanol at $-60°C$) would strongly enhance the probability of detecting such a species, if it existed.

In an extremely insightful analysis of the catalytic mechanism of peptide bond hydrolysis by pepsin, Fruton (1976) first suggested that there may be no covalent intermediates involved in aspartyl proteinase catalysis. He then outlined the sequence of events that would most likely take place during hydrolysis. He invoked conformational change of the extended active site region of pepsin in order to explain the acyl or amino transfer reactions and ordered release of products. All of these hypotheses were advanced in the absence of structural information.

What follows is a description of the tertiary structures available to date. In this description, it will be seen that the catalytic pathway is most consistent with no covalent intermediates; the conformational change observed when specific inhibitors bind to penicillopepsin is also entirely consistent with Fruton's suggestions to explain acyl and amino transfer reactions.

3. TERTIARY STRUCTURES OF ASPARTIC PROTEINASES

The first reports of the medium-resolution structures of aspartic proteinases took place in November 1976 at the University of Oklahoma, during a meeting organized by Dr. J. Tang. During that conference, it became clear that the three mold enzymes, penicillopepsin, endothiapepsin, and rhizopuspepsin, all had extremely similar polypeptide chain conformations (Subramanian et al., 1977a, 1977b; Jenkins et al., 1977; Hsu et al., 1977b). Structural comparisons of the topologically equivalent α-carbon atoms of penicillopepsin and endothiapepsin (approximately 200 pairs from over 320 total residues) have a root-mean-square (r.m.s.) deviation of 1.35 Å (James et al., 1981). The structures of penicillopepsin and rhizopuspepsin seem to be more similar, as indicated by an analogous comparison of α-carbon atom coordinates: an r.m.s. deviation of 1.2 Å for approximately 260 α-carbon atom pairs (Bott and Davies, 1983). These values are not definitive, since they correspond to the unrefined medium-resolution coordinates. In any case, the mold enzymes seem to have highly conserved tertiary structures.

The structure of porcine pepsin reported at the same meeting had problems caused by closely packed molecules and required a reinterpretation of the electron density map (Andreeva et al., 1977). It is now confirmed that the tertiary structure of pepsin is very similar to those of the mold aspartic proteinases (Andreeva et al., 1984). The crystal data of the four enzymes are summarized in Table 2.

The earlier medium-resolution structures were determined by the multiple isomorphous replacement (MIR) method. Phases deduced in this way, while adequate for interpreting an initial molecular structure, are not sufficiently accurate to describe the details of hydrogen-bonding interactions, solvent structure, or how inhibitors or substrate analogs are bound to the enzyme. It is absolutely necessary to refine the structures of the native enzyme and of substrate/inhibitor complexes at close to atomic resolution in order to be confident of the results.

Mechanistic overinterpretations at lower resolutions easily occur. A mechanism based on the medium-resolution structure of penicillopepsin was advanced that attempted to rationalize the role of a basic residue in catalysis, the observed conformational changes on substrate or activator peptide binding, as well as the enhancement of k_{cat} with increased substrate

Table 2. Crystal Data for Aspartyl Proteinases

	Pepsin	Rhizopuspepsin	Endothiapepsin	Penicillopepsin
Initial multiple isomorphous replacement structure determinations				
Space group	$P2_1$	$P2_12_12_1$	$P2_1$	C2
Cell dimensions				
a (Å)	54.7	60.3	53.6	97.4
b (Å)	36.3	60.7	74.1	46.6
c (Å)	73.5	107.0	45.7	65.4
β (°)	103.8	90.0	110.0	115.4
Z (mol/unit cell)	2	4	2	4
Crystallization	Water-ethanol	20 mM	2.7M	2.5M
conditions		$Ca(OAc)_2$	$(NH_4)_2SO_4$	Li_2SO_4
pH	2	6.0	4.5−6.3	4.4
Data collection	Film/diffractometer	Diffractometer	Diffractometer	Diffractometer
Resolution (Å)	2.7	2.5	3.0	2.8
Number of reflections	6500	10,405	—	6127
Mean figure of merit	0.67	0.77	0.70	0.90
Current published data				
Resolution (Å)	2.0[a]	1.8[b]	2.1[c]	1.8[d]
Number of reflections	13,768	—	—	21,962
R factor[e]	0.292	—	0.16	0.136

[a]Andreeva et al., 1984.
[b]Bott and Davies, 1983.
[c]Pearl and Blundell, 1984.
[d]James and Sielecki, 1983.
[e]The crystallographic R factor is defined as $\Sigma\|F_o|-|F_c\|/\Sigma|F_o|$, where $|F_o|$ and $|F_c|$ are the observed and calculated structure factor amplitudes, respectively.

chain length (James et al., 1977, 1981). This proposed pathway involved an extended charge relay scheme with residues Lys 304(308), Asp 14(11), Asp 300(304), His 159(157), and the peptide bond between Thr 214(216) and Gly 215(217). The refinement of penicillopepsin brought to light a number of features that provided evidence against this early proposal. There are several water molecules in the immediate vicinity of, and hydrogen bonded to, the NH_3^+ group of Lys 304(308) and to the carboxyl side chain of Asp 300(304). The presence of these water molecules indicates that the influence originally attributed to these charged groups diminishes, since they are solvated. In addition, not all of the key residues invoked in the extended charge relay are conserved in all aspartyl proteinases. The recently determined amino acid sequence of mouse renin (Misono et al., 1982; Panthier et al., 1982) has established that Asp 14(11) and Asp 300(304) are replaced in that enzyme by asparagine and alanine residues, respectively. As we shall show, it is possible to advance other satisfactory mechanisms that do not involve covalent catalysis and that take into account all the structural data now available.

3.1. The Aspartic Proteinase Fold

The aspartic proteinases are rather asymmetric globular proteins of rough dimensions 65 × 49 × 39 Å and have an extensive hydrophobic interior. They may be classified as essentially all β-sheet proteins, and have the general shape of a croissant. An extended, deep substrate-binding groove is prominently located at the junction of the N- and C-terminal domains (Figs. 1 and 2). This feature was immediately evident in the medium-resolution structures, as had already been predicted from the work of Fruton (1970, 1976). The binding cleft is ~25−30 Å in length, which is sufficient to accommodate seven or eight amino acids in an extended β-sheet conformation.

The aspartic proteinases are composed of two structurally similar domains (Tang et al., 1978). The 147 α-carbon atoms of the C-terminal domain can be rotated and translated in a least-squares minimization procedure to best fit the 176 α-carbon atoms of the N-terminal domain. When this is done using the coordinates of the refined penicillopepsin structure, there are 70 topologically equivalent pairs of α-carbon atoms with an r.m.s. deviation of 1.91 Å (James and Sielecki, 1983). Of these, 15 (21%) correspond to chemically identical residues. An approximate two-fold symmetry axis re-

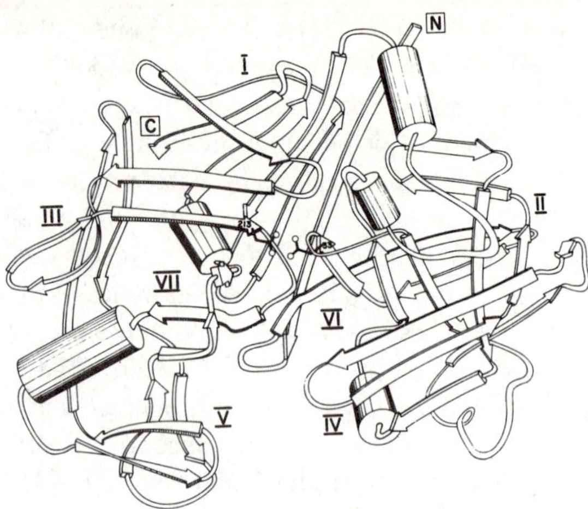

Figure 1. Diagrammatic representation of penicillopepsin (after James and Sielecki, 1983). The arrows represent individual strands of polypeptide chain that are involved in β-sheet conformation. The cylindrical barrels represent portions of polypeptide chain that adopt helical conformations. The N- and C-termini of the polypeptide chain are labeled N and C, respectively. The roman numerals stand for the sheets that are represented separately in Figures 3–5. The approximate positions of Asp 33(32) and Asp 213(215) are shown in sheets VI and VII, respectively.

lates the 70 pairs of atoms. It passes between the two active site aspartic acid residues and involves a rotation of 177° and a translation of 1.2 Å to superimpose the equivalent α-carbon atoms of the N- and C-terminal domains. The molecular symmetry is most striking for the residues in close proximity to Asp 33(32) and Asp 213(215). In penicillopepsin, when the two strands Ile 211(213) to Thr 216(218), and Ile 297(301) to Gly 299(303) are fitted to Asn 31(30) to Ser 36(35) and Leu 121(120) to Gly 123(122), respectively, the r.m.s. deviation for the 46 common atom pairs is 0.49 Å.

There is general consensus that the aspartic proteinases have *intra*domain approximate twofold symmetry (Andreeva et al., 1978; Andreeva and Guschina, 1979; Blundell et al., 1979) in addition to the *inter*domain (intramolecular) approximate twofold symmetry (Tang et al., 1978). Thus, each domain is composed of two structurally similar portions related by a local dyad. Within the N-terminal domain of endothiapepsin, 22 pairs of α-carbon atoms are equivalent with an r.m.s. deviation of 1.92 Å, whereas in the C-terminal domain, 17 pairs are equivalent with a corresponding value of

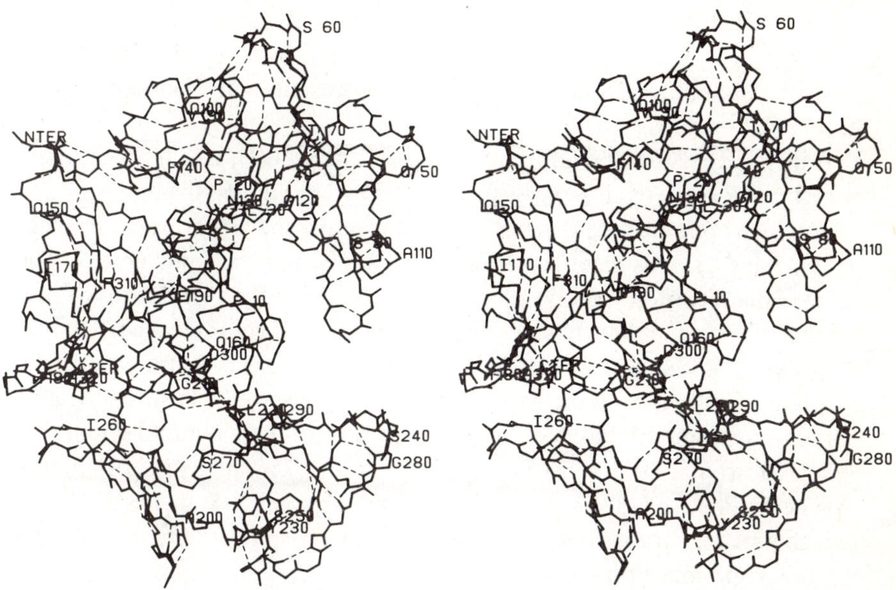

Figure 2. Stereo drawing of the atoms of the polypeptide chain (N, C$^\alpha$, C, O) of penicillopepsin. This figure is oriented approximately 90° from that in Figure 1. Every tenth residue has been labeled with the residue type in the one-letter code and the sequence number of penicillopepsin. The dashed lines represent the hydrogen bonds of the main chain (i.e., >N-H⋯O=C<). The several β-sheet regions are readily discerned from this figure. Upon inclusion of the side chain atoms, the interior of this shell-like structure develops into the closely packed hydrophobic core characteristic of the aspartic proteinases.

2.31 Å (Blundell et al., 1979). These authors have suggested that such results are consistent with the hypothesis that the model of evolution for the aspartic proteinases is a quadruple gene product with four copies of an ancestral β-sheet unit of approximately 45 residues. Various mutation events would be required to add the loops found in the modern proteinases. Alternatively, since there is little sequence homology, the evolutionary history of the domains may be the result of strong convergent pressure. As in most evolutionary arguments based on amino acid sequencing or tertiary structural data, a definitive answer is seldom possible.

The three-dimensional fold of aspartic proteinases has been discussed by several authors (Tang et al., 1978; Blundell et al., 1979; Clothia and Janin, 1982; James and Sielecki, 1983; Andreeva et al., 1984). Making use of the refined structure of penicillopepsin to describe more reliably the intramolecular hydrogen-bonding interactions (James and Sielecki, 1983), it is possible to arrive at another slightly different way of presenting the elaborate structural relationships.

The large, 6-stranded antiparallel β-pleated sheet (I), which is a common structural feature of all aspartic proteinases, is situated on the opposite side of the hydrophobic core that surrounds the active site aspartic acid residues (Figs. 1 and 2). This sheet I is represented diagrammatically in Figure 3. The interdomain twofold symmetry axis passes between the two central strands of this sheet, such that the two hairpin loops Gln 150(148) to Phe 169(169) and Lys 304(308) to Ala 323(327) are related. There is a stretch of polypeptide chain from the turn at Gly 168(168) to Ser 178(178), connecting strand j to strand k, that does not obey the interdomain twofold symmetry. The N-terminal strand (a) is related to strand k, and two short helical segments Phe 140(138) to Lys 145(143) and Asp 300(304) to Lys 304(308), which precede strands i and q, respectively, are also related to one another by the same interdomain dyad.

The two lobes of the croissant-shaped molecule are most conveniently described by the orthogonal packing of β-pleated sheets, as suggested by Chothia and Janin (1982). Figures 4 and 5 show diagrammatically how these sheets are hydrogen bonded, and when properly folded, how they form the N-terminal and C-terminal domains (see Fig. 2 for a stereo representation of the whole polypeptide chain conformation). The N-terminal domain consists of a three-layer orthogonal β packing (Figs. 4a and 4b). Sheet VI is in the middle; sheet II folds backwards to pack against sheet VI, and sheet IV folds forward in front of the middle sheet to make a three-layer sandwich. The middle sheet (VI) is made up of 7 strands, from which strands c_1 and h_1 and strands d and g are parallel pairs. Strand c_1 carries the catalytically important Asp 33(32).

Figures 4 and 5 have been drawn so that the topological symmetry and the differences between the two domains are emphasized. The C-terminal domain has only two orthogonally packed sheets, VII and III. Sheet III folds backwards at a diagonal junction with sheet VII. Sheet III, although much more extensive, is topologically equivalent to sheet II of the N-terminal domain. Sheet V is not topologically related to sheet IV. In the structural analysis presented in the paper by Tang et al. (1978), it is shown that the "flap" region, Ser 72(72) to Ser 82(82), has an insertion [Tyr 75(75) to Ser 80(80)] over the corresponding region in the C-terminal domain, Ala

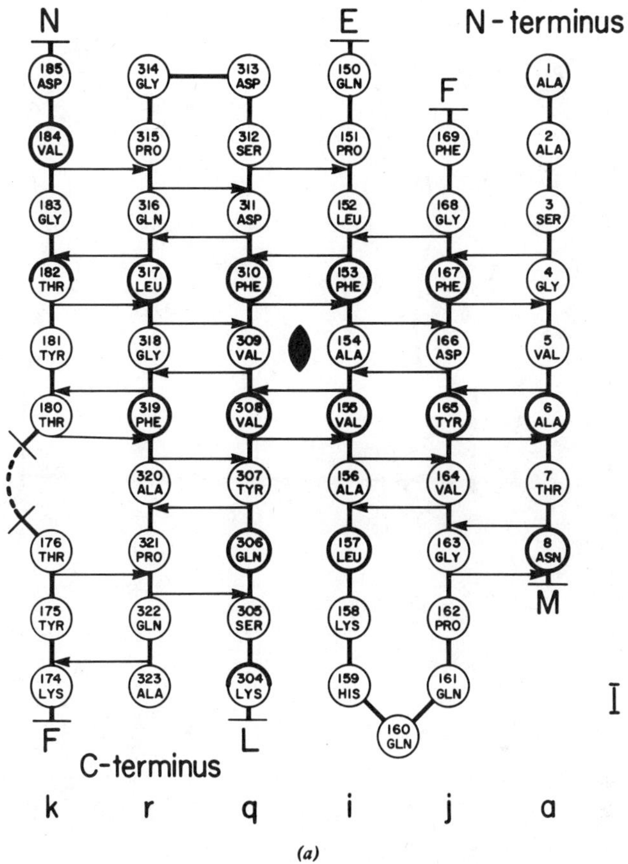

Figure 3. (*a*) A diagrammatic representation of the large 6-stranded antiparallel β-sheet of penicillopepsin. The strand nomenclature is taken from James and Sielecki (1983) and is represented by the lowercase letters below the strands. The connectivities are indicated by uppercase letters at the ends of the strands (i.e. strand i [E, at residue Gln 150] is connected to strand h_1 of sheet VI in Fig. 4*a* [E, at residue Leu 124]). Where there is a break in the residue number sequence, the connecting residues do not belong to any of the β-sheet systems. Hydrogen bonds within the sheet are denoted by arrows drawn from donor N-H to acceptor C=O. The interdomain (intramolecular) twofold axis is shown between residues Ala 154(154) and Val 309(313), and is approximately perpendicular to this sheet. Those residues that are encircled with a thick line are the residues that contribute to the hydrophobic core of penicillopepsin. Those with only a thick line on the top (e.g. Lys 304(308)) are so denoted because they are on the edge of the nonpolar core, with some of their polar atoms accessible to solvent. (*b*) A stereographic drawing of the α-carbon atoms that constitute the residues of sheet I. The connecting segment between Phe 169(169) and Lys 174(174) (from F in strand k, to F in strand j of Fig. 3*a*) is located behind the sheet. The hydrophobic surface of sheet I is facing the viewer.

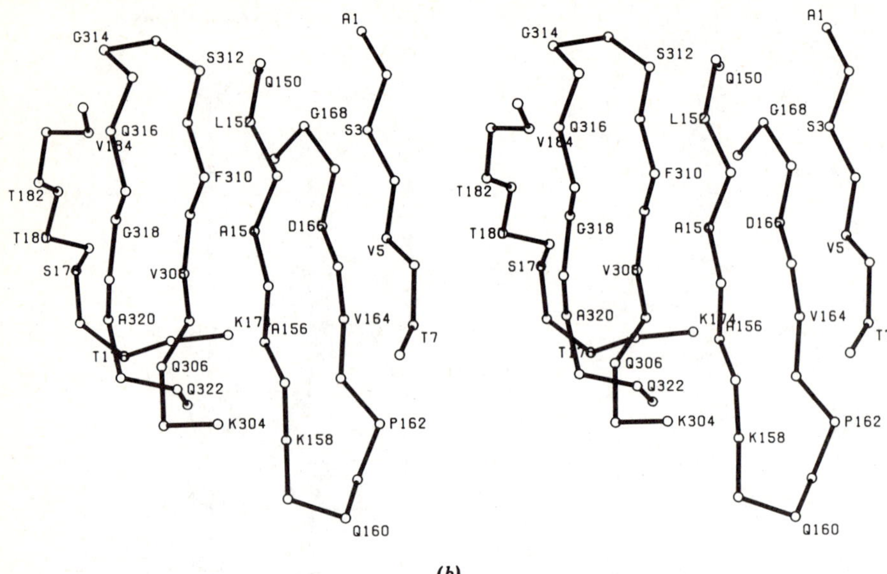

(*b*)

Figure 3. (*b*) *(continued)*

Figure 4. (*a*) The three sheets represented diagrammatically here, II, IV, and VI, constitute a three-layer sandwich in which the strands of adjacent sheets are approximately orthogonal to one another (as defined by Chothia and Janin, 1982). The central or middle sheet is sheet VI. Sheet II folds *backwards* at the junction of residues 20–21 and 28–29 to become the bottom layer. Sheet IV is folded *forwards* in front of sheet VI along the diagonal line to form the topmost sheet of the sandwich. As in Figure 3*a*, strand connectivities are indicated by uppercase letters. Those residues encircled with a thick line are the residues forming the intersheet contacts between the middle and bottom sheets (VI and II); those with squares surrounding the residues numbers are the intersheet contacts of sheet VI and IV. The intradomain symmetry axis runs between Trp 40(39) and Leu 121(120) in sheet VI and between Thr 22(21) and Thr 91(90) in sheet II. (*b*) The stereo diagram of the α-carbon atom coordinates used to define sheets VI (middle), II (bottom), and IV (top) of the triple-decker sandwich. The "flap" of the aspartic proteinases is the loop Trp 71(71) to Gly 83(82). It is the movement of the residues at the tip of the flap, 76, 77, 78, that is important in the binding of inhibitors, and presumably substrates, to penicillopepsin. The topology of these three strands is discussed by Chothia and Janin (1982).

432

VI
E
C
f₁
B
e
A
c₁
d
124 LEU
99 GLY
g
100 GLN
88 ASP
64 LYS
33 ASP
123 GLY
38 ASP
101 ALA
87 THR
65 GLU
32 PHE
122 LEU
39 LEU
102 VAL
86 PHE
66 LEU
31 ASN
121 LEU
40 TRP
103 GLN
85 VAL
M
19 THR
30 LEU
120 GLY
41 VAL
104 ALA
84 ASN
71 TRP
20 PRO
29 ASN
119 ASP
42 PHE
105 ALA
83 GLY
72 SER
II
D
A
106 GLN
82 SER
94 GLY
93 GLY
h₁
107 GLN
73 ILE
95 VAL
92 VAL
21 VAL
28 LEU
108 ILE
81 ALA
96 THR
91 THR
22 THR
27 THR
109 SER
80 SER
74 SER
97 ALA
90 VAL
23 ILE
26 THR
D
79 SER
75 TYR
98 HIS
89 SER
24 GLY
25 GLY
g₁
78 GLY
76 GLY
C
B
77 ASP
e₁
f₃
f₂
b
c
IV
f

(a)

D33 G94 G123 T96 Q100 H98
V92 V90 L66
Q76 L39 S72 D88 G68
T119 V21 V102 F86
L12 I23
G78 S80 V4 G25 S82 N84
N29 T27 A104
D119 Q106
I108

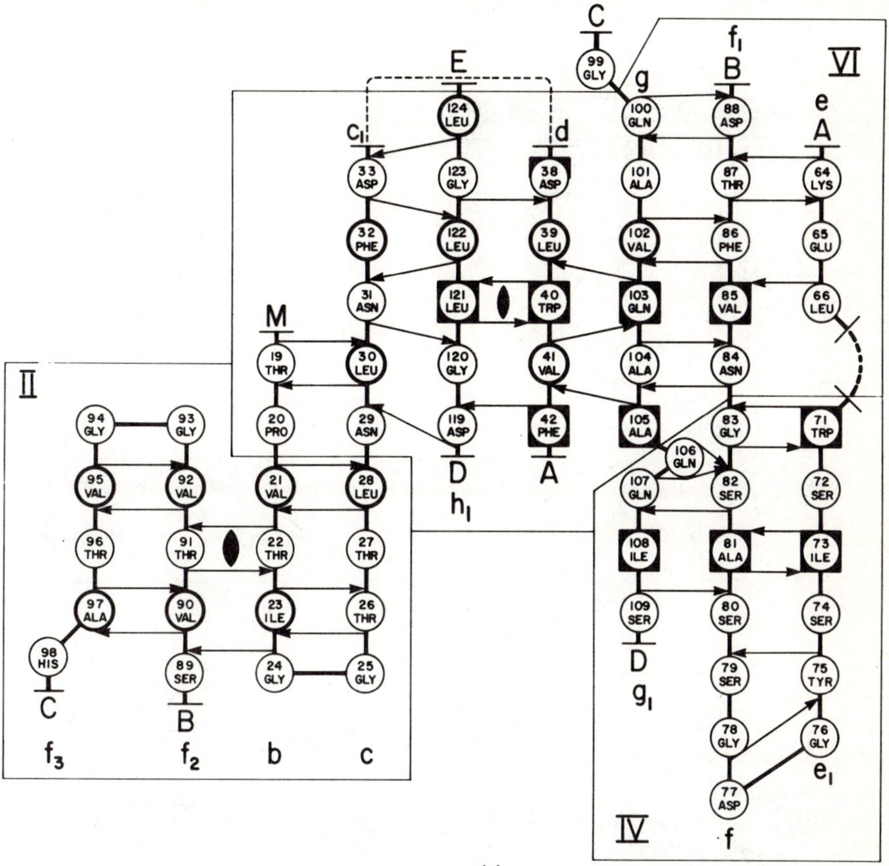

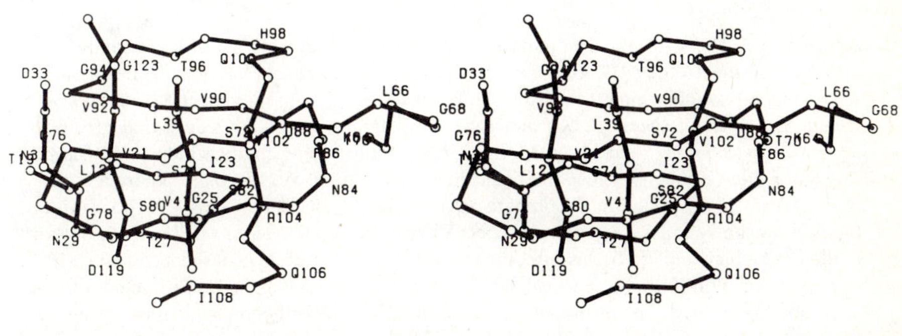

(b)

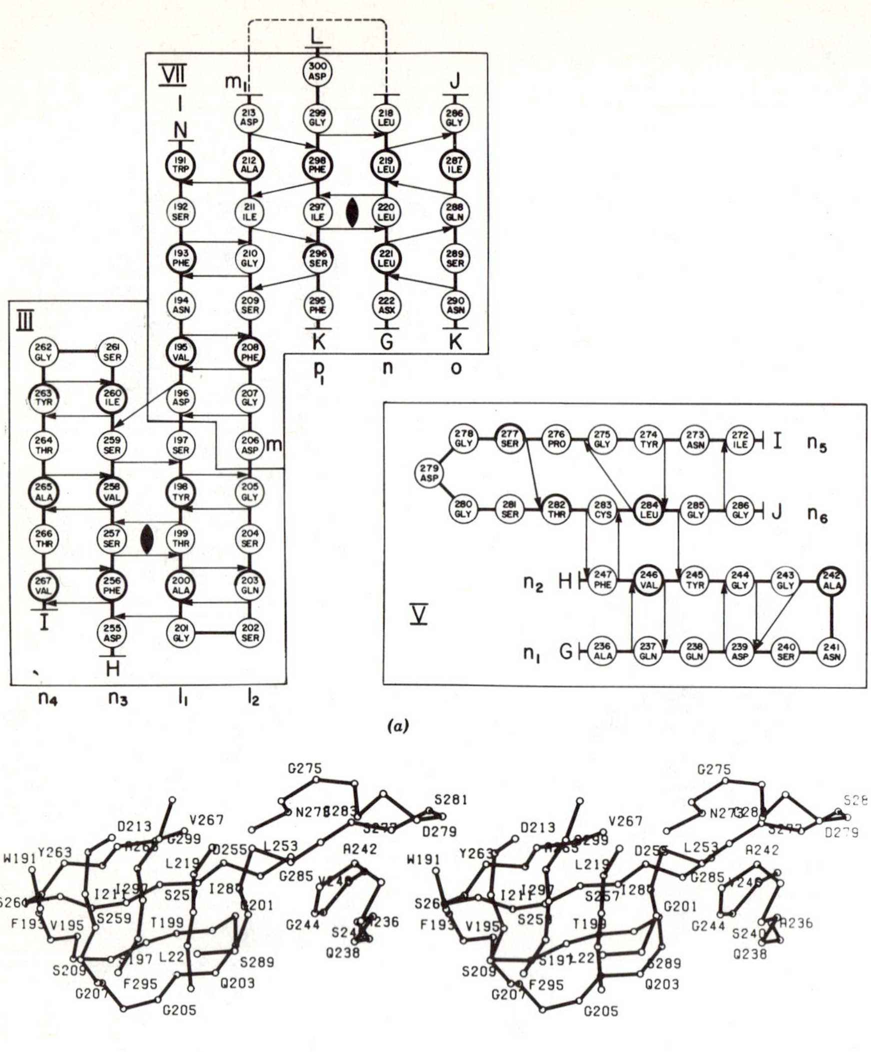

Figure 5. (*a*) The C-terminal domain of penicillopepsin has only a two-layer sandwich of orthogonal β sheets. Sheet III folds backwards to pack against sheet VII, and the intradomain, approximate twofold symmetry axis lies between strands p₁ and n of sheet VII and between strands l₁ and n₃ of sheet III. As in the previous figure, internal residues that are in the intersheet space are marked with thick-line circles. Strand connectivity is indicated by uppercase letters, and strand nomenclature by the lowercase letters. The two pairs of parallel strands in sheet VII are m₁,p₁ and n,o. Comparison of the arrangement of sheets VII and III in this figure shows how similar it is to that of sheets VI and II shown in Figure 4*a*. Sheet V is also part of the C-terminal domain. In this case, the thick-line circles indicate intrasheet contacts. (*b*) A stereographic plot of the three sheets III, V, and VII, of part (*a*). The α-carbon coordinates alone are represented. The strands in sheet III and VII are not orthogonal, but make an average angle of ∼67° to one another. This can be compared to sheets VI and II of the N-terminal domain, where the interstrand angle is 71° (Chothia and Janin, 1982).

239(240) to Gly 243(244). Similarly, the C-terminal domain has a large insertion, Gly 269(270) to Gly 285(285), over the corresponding region in the N-terminal domain between strands f_3 and g. The resulting twisted sheet V of penicillopepsin is in a different location from that expected when the interdomain two-fold symmetry relation is applied to sheet IV of the N-terminal lobe. The detailed comparison of these two domains has been reported in an earlier publication (James and Sielecki, 1983).

It should be noted that the intradomain dyad of the N-terminal domain (Blundell et al., 1979) projects between strands h_1 and d of sheet VI and between strands b and f_2 of sheet II (Fig. 4a). Similarly, the intradomain dyad of the C-terminal domain passes between strands p_1 and n of sheet VII and between strands n_3 and l_1 of sheet III (Fig. 5a). The central two strands in sheets VI and VII are antiparallel.

3.2.   The Refinement of Penicillopepsin and its Inhibitor Complexes

At present, the only published, fully refined structure of an aspartyl proteinase in its native conformation and its inhibitor complexes is that of penicillopepsin (James and Sielecki, 1983; James et al., 1982, 1983, 1985).

The amount of crystallographic experimental data that is possible to obtain from macromolecules is limited by the degree of order in the crystalline samples. This implies that the ratio of available observations to the parameters that should be refined (coordinates and, at least, a temperature factor for each atom in the macromolecule and its associated solvent) is very low. A logical solution is the incorporation into the refinement procedure of other types of available information, such as the expected stereochemistry of individual amino acids (as obtained from small molecule structures), planarity of peptide bonds, minimum distances for van der Waals contacts for each atom type, interatomic energy potentials, etc. One of the resulting procedures, the restrained least-squares refinement (Hendrickson and Konnert, 1980), is now widely used by the crystallographic community. Table 3 gives a summary of the Hendrickson-Konnert refinement statistics for penicillopepsin and two of its complexes, so that the reader may assess the quality of these crystal structures independently.

The refinments of the native molecule and of penicillopepsin inhibited by the pepstatin analog Iva-Val-Val-StaOEt are considered complete. Data up to 1.8 Å resolution have been used in the refinements, and the resulting R

Table 3. Crystallographic Refinement Statistics for Penicillopepsin and Its Complexes with Inhibitors[a]

Parameters	Native Penicillopepsin	Penicillopepsin + Iva-V-V-StaOEt[a]	Penicillopepsin + Iva-V-V-LystaOEt[a]
Space group	C2	C2	C2
Unit cell			
a (Å)	97.38	97.24	97.57
b (Å)	46.61	46.50	46.47
c (Å)	65.42	65.65	66.39
β (°)	115.39	115.34	116.15
Resolution range (Å)	8.0−1.8	8.0−1.8	8.0−1.8
Reflections used $[I \geqslant \sigma(I)]$[b]	21,962	21,197	22,494
Total atoms	2366[d]	2401	2402
No. solvents	319	322	170
No. cycles	86	21	15
R factor	0.136	0.131	0.149
rms Δ[c] (bond distances, Å)	0.014	0.015	0.012
rms Δ (angle distances, Å)	0.038	0.038	0.036
rms Δ (planar groups, Å)	0.020	0.017	0.016
rms Δ (chiral centers, Å^3)	0.10	0.15	0.15
rms Δ (ω of peptide bonds, °)	3.0	3.3	3.0
Mean $\||F_o|-|F_c|\|$ (e)	45.3	44.6	50.3

[a]The two inhibitors are pepstatin analogs Iva-Val-Val-StaOEt and Iva-Val-Val-LystaOEt. Lysta is the lysyl side chain analog of statine.

[b]The data used for refinement were all those reflections with intensity (I) greater than the standard deviation of its measurement, $\sigma(I)$.

[c]rms Δ refers to the root mean square deviations of the refined stereochemical parameters from the ideal values derived from the crystal structure analyses of the amino acids and peptides (Sielecki et al., 1979).

[d]The discrepancy between this number and that previously reported (2363) is due to the revision of the chemical sequence that modified the identification of residues at positions 251−252 from Ser-Ser to Thr-Asn (T. Hofmann, personal communication).

factors (0.136 and 0.131) are sufficiently low to indicate that further refinement would not significantly improve the accuracy of the atomic parameters. The r.m.s. deviations of distances and angles from expected ideal values derived from small molecule crystal structures are also low, and indicate that the resulting model has excellent stereochemical geometry.

At present, the refinement of the complex of penicillopepsin with a second statine-based inhibitor, Iva-Val-Val-LystaOEt, is not complete; the solvent model has not been fully defined. However, the structure is pertinent, and the R factor of 0.149 is low enough to discuss the results at this interim stage of refinement. The conformation of the inhibitor (except for the ethyl group of the ester) is unambiguous, and the interactions that it makes with penicillopepsin are clear. The current status of this refinement is given in comparison to the native enzyme and the other inhibitor complex in Table 3.

The importance of refinement of the atomic parameters of bound inhibitors or substrate analogs in the active sites of enzymes can be illustrated with the results obtained for the complex of penicillopepsin with Iva-Val-Val-StaOEt (James et al., 1982, 1983). A difference electron density map was computed with coefficients $|F_I|$ - $|F_N|$ and calculated phases, α_N, from the final refined model of native penicillopepsin at cycle 86 ($|F_I|$ and $|F_N|$ are the structure factor amplitudes measured from the inhibitor complexed and native crystals, respectively). This map was used to derive an initial model of the inhibitor, Iva-Val-Val-StaOEt, as shown in Figure 6a. Two main sources of error affect, in particular, such difference maps: (1) the phases used for the computation are those of the native enzyme and are only approximately correct for the crystal of the complex; (2) the inhibitor displaces solvent molecules from the surface of the enzyme, thus introducing distortions in the difference electron density.

The initial model was subsequently subjected to restrained least-squares refinement (Hendrickson and Konnert, 1980) and reinterpreted from the density of an "omit" map with coefficients ($|F_I|$ - $|F_c|$), and phases α_c. The calculated structure factor amplitudes $|F_c|$ and the associated phases α_c were based on the refined model of the complex, with the omission of the atoms of the Iva and ester groups of the inhibitor. Such "omit" difference maps are extremely useful when trying to evaluate the correctness of suspected regions of the model (James et al., 1983). The final refined molecular structure of the inhibitor based on that reinterpretation, with the electron density of the map computed after cycle 21 of the refinement, is shown in Figure 6b. The initial model and the final refined position of the pepstatin analog are compared in Figure 6c. It can be seen that there are large differences in the region of the ethyl ester group between the two models. The r.m.s. difference between corresponding atomic positions of the inhibitor is 0.90 Å. The atoms comprising the statine residue alone have shifted by an average of 0.85 Å as a result of the refinement and the reinterpretation of the electron density maps obtained throughout that process. Even though the initial coordinates were derived from a high-quality difference electron density map at 1.8 Å resolution (Figure 6a), the refinement has resulted in

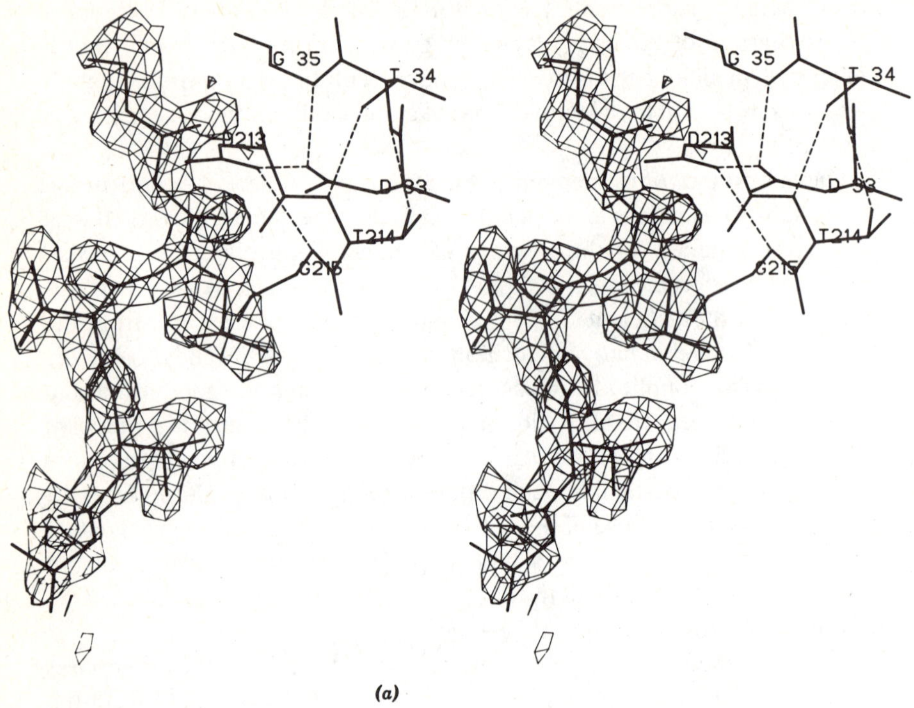

Figure 6. (*a*) Difference electron density map of Iva-Val-Val-StaOEt bound to penicillopepsin (after James et al., 1982). The phases used were the computed phases of the native penicillopepsin molecule, α_N, and the coefficients were the differences between structure factor amplitudes $|F_I| - |F_N|$, of the complexed and the native penicillopepsin crystals. The contour level is 0.18 eÅ^{-3}. A portion of the native active site with the two aspartyl residues is shown. The initial model of the fitted inhibitor is superposed into the density. Dashed lines represent possible hydrogen bonds. (*b*) A view of the final electron density in the region of the pepstatin analog inhibitor at the conclusion of the least-squares refinement of the complex (James et al., 1983). The view is similar to that shown in (*a*). The model represents the final refined coordinates. The contour surface is drawn at 0.40 eÅ^{-3}. The coefficients for the computation of this map were $2|F_I| - |F_c|$ and phases α_c, based on the refined structure of the complex. (*c*) Comparison of the initial coordinates (dashed lines) of Iva-Val-Val-StaOEt deduced from the map shown in (*a*), with the final refined coordinates from cycle 21 (solid lines), as shown in (*b*) (after James et al., 1983). The slight conformational change of the active site residues is also indicated (full thin line, native penicillopepsin; full thick line, penicillopepsin in the complex with the pepstatin analog). The r.m.s. distance between corresponding atoms of the inhibitor is 0.90 Å. The largest change is in the conformation of the ethyl ester that, after refinement, best fits a high-energy nonplanar conformation (see James et al., 1983).

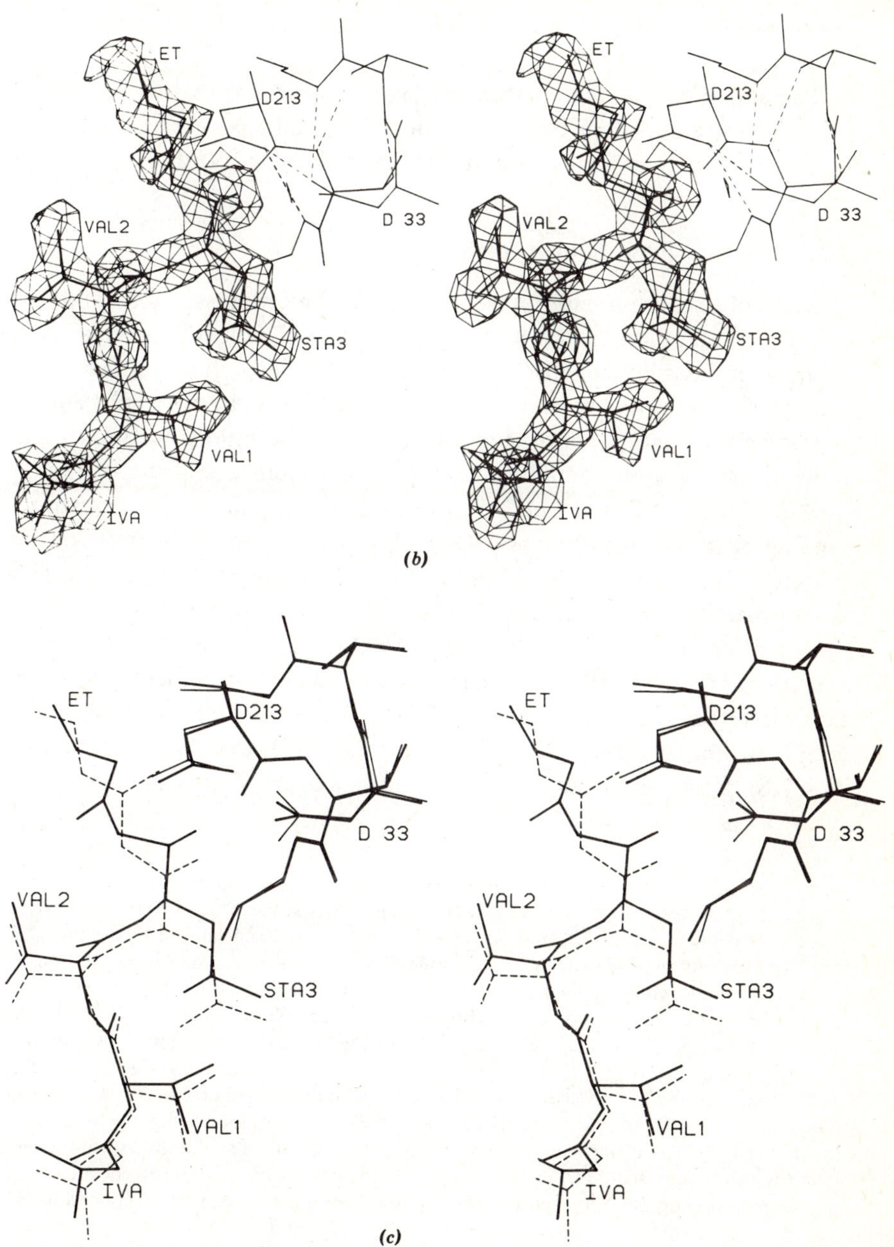

(b)

(c)

relatively large atomic movements relative to the catalytic groups on penicillopepsin.

The above results demonstrate that molecular conformations of inhibitors or substrate analogs deduced from difference maps, even at relatively high resolution, can have quite inaccurate molecular coordinates.

3.3. Active Site Architecture

In view of the arguments presented in the preceding section, and our obvious greater familiarity with the penicillopepsin molecule, we shall center most of the following discussion on the higher resolution, refined structures available for that molecule. Nevertheless, due to the homology among aspartyl proteinases, a majority of the structural descriptions and interpretations of those results should be valid for other members of the family.

Asp 33(32) and Asp 213(215) are located centrally in the substrate-binding cleft of penicillopepsin (Fig. 7a). The interdomain dyad passes between these two amino acids and relates them and the residues in their close proximity. In the left part of Figure 7a, the two aspartic acid residues are surrounded by an extensive hydrophobic patch. This feature is conserved in other aspartyl proteinases. The residues of penicillopepsin involved are : Leu 284(284), Tyr 274(275), Leu 218(220), Leu 220(222), Ile 293(297), Phe 295(299), Ile 297(301), Ile 211(213), Phe 190(189), and Ile 129(128). Residues displayed on the right side of this figure that are impor-

Figure 7. (*a*) A stereo optic view of the residues that comprise the active site and binding cleft of penicillopepsin. Asp 33(32) and Asp 213(215) are located centrally in this figure. The one-letter code for the amino acids is used, along with the sequential numbering for residues in penicillopepsin. Residues of importance have been highlighted with a thicker line for the bonds in the side chains. The hydrophobic-binding patch is made up of residues Leu 284(284), Tyr 274(275), Leu 218(220), Leu 220(222), Ile 293(297), Phe 295(299), Ile 297(301), Ile 211(213), Phe 190(189), Ile 129(128). A portion of the flap, from Ser 72(72) to Ser 80(80), is shown. (*b*) A stereo optic view of the active site of penicillopepsin with the refined positions of neighboring solvent molecules (represented by circles). This is the same view of the active site as shown in (*a*), but the residues of the flap have been omitted for clarity. Possible hydrogen-bonding interactions are represented by dashed lines. Note the paucity of ordered solvent sites in the regions surrounding the side chains of the exposed hydrophobic residues mentioned in (*a*). Waters are assigned the one-letter code ''O,'' whereas ''Z'' identifies solvents associated with a symmetry-related protein molecule.

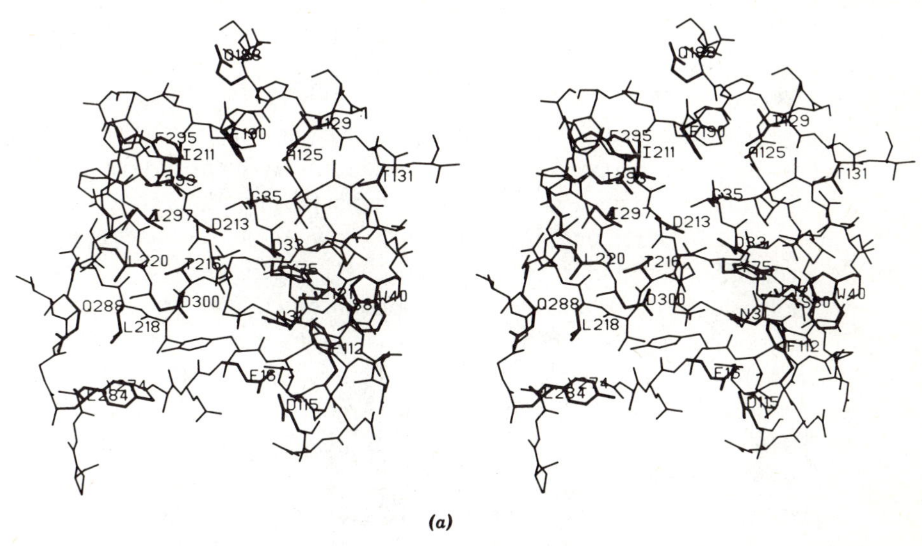

(a)

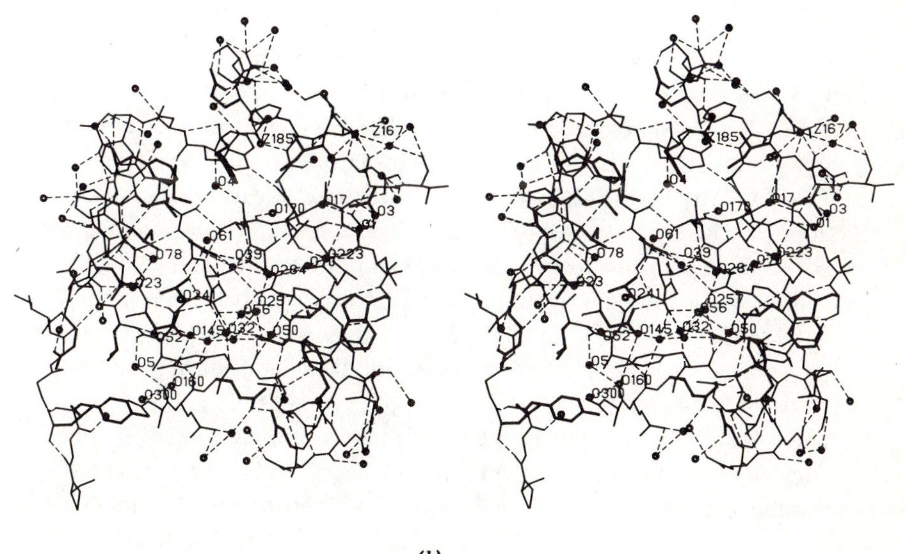

(b)

tant in substrate-binding interactions include, Asn 31(30), Phe 112(111), Leu 121(120), in addition to residues on the "inside" of the "flap," Tyr 75(75) and Asp 77(77).

Solvent molecules in the active site region of penicillopepsin have been identified during the progress of the 1.8 Å resolution refinement (Fig. 7b). The ordinal number of these molecules is such that lower numbers represent the more highly ordered ones in the structure (low temperature factor and high occupancy sites; James and Sielecki, 1983). There are few ordered water molecules in the region of the hydrophobic patch, and they all make hydrogen-bonded interactions with the polar atoms of the main chain. Several of the solvent molecules depicted in Figure 7b occupy sites that have been identified as substrate-binding pockets. Among them, O39, O52, O61, O145, O160, O241, O257, O284, and O300 are displaced when specific inhibitors bind to penicillopepsin.

Pepsin crystals are grown from ethanol-water solutions. The initial refinement of this structure has shown bound ethanol molecules in the active site region (Andreeva et al., 1984). Ethanol is an inhibitor of pepsin, and the identification of these tightly bound solvents may explain the inhibited state of the enzyme in the crystalline form used for the structure determination. Further refinement is in progress in order to define the active site region of pepsin as accurately as possible.

A possible hydrogen-bonding scheme at the active site of penicillopepsin in its native conformation is illustrated in Figure 8. Contact distances less than 3.5 Å are summarized in Table 4. There is an extensive hydrogen-bonding network, with the segments Asp 33(32) to Ser 36(35) and Asp 213(215) to Thr 216(218) in type-I β bends (Crawford et al., 1973). In addition, each aspartate carboxyl group is in a special conformation, with one of its oxygen atoms forming a hydrogen bond to the main chain NH of the amino acid located two residues further along the polypeptide chain [Gly 35(34)N...Asp 33(32)$O^{\delta 2}$ and Gly 215(217)N...Asp 213(215)$O^{\delta 1}$]. These special interactions of aspartate or asparagine side chains have been termed Asx turns (Rees et al., 1983).

The scheme for the hydrogen bonding between the two active site carboxyl groups themselves is not definite. As described in a previous section, pK_a values of ~1.2 and ~4.7 have been assigned to these residues. An analogy with the conformation of the maleic acid molecule has been made to explain such unusual pK_a's (Hsu et al., 1977a, 1977b). More recently, it has

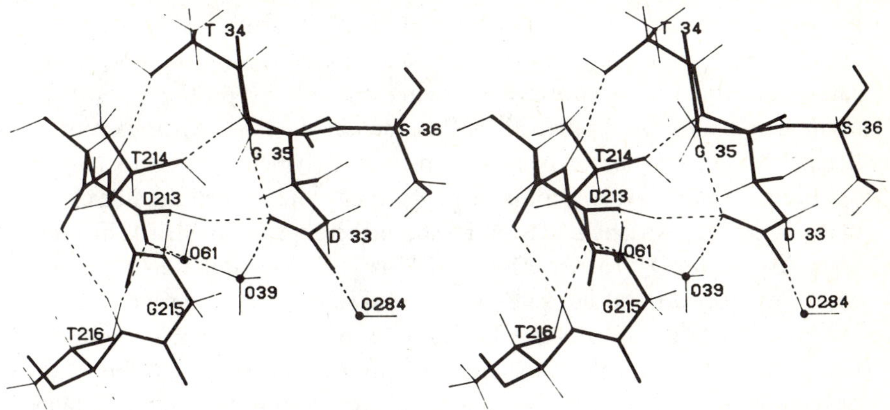

Figure 8. A stereoscopic view of possible proton positions and hydrogen-bonding interactions (contact distances < 3.4 Å) in the active site of native penicillopepsin. Solvent site O39 is represented here as a positively charged NH_4^+ ion, with sites O61 and O284 as water molecules. The proton shared by Asp 33(32) and Asp 213(215) is shown localized on the carboxyl group of Asp 213(215), so that the carboxylate of Asp 33(32) has the negative charge (James and Sielecki, 1983). Other possible schemes are shown diagrammatically in Figures $9a-f$. Hydrogen bonding is indicated by dashed lines from the hydrogen atoms to the acceptor. Contact distances less than 3.5 Å are listed in Table 4.

Table 4. Possible Hydrogen Bonding and Contact Distances in the Active Site of Native Penicillopepsin[a]

Groups	Distance (Å)	Groups	Distance (Å)
Thr 34 NH......Thr 214 $O^{\gamma 1}$	2.89	Thr 214 NH......Thr 34 $O^{\gamma 1}$	2.86
Thr 34 $O^{\gamma 1}$H...Ala 212 O	2.80	Thr 214 $O^{\gamma 1}$H...Phe 32 O	2.75
Gly 35 NH......Asp 33 $O^{\delta 2}$	2.76	Gly 215 NH......Asp 213 $O^{\delta 1}$	2.93
Ser 36 O^{γ}H....Asp 33 $O^{\delta 1}$	3.43	Thr 216 $O^{\gamma 1}$H...Asp 213 $O^{\delta 2}$	2.75
Ser 36 NH......Asp 33 O	3.39	Thr 216 NH......Asp 213 O	3.22
Asp 33 $O^{\delta 2}$ ..... Asp 213 $O^{\delta 1}$	2.87	—	
Asp 33 $O^{\delta 2}$ ..... O 39	2.62	Asp 213 $O^{\delta 1}$ ..... O 39	2.83
Asp 33 $O^{\delta 1}$ ..... O 39	3.15	Asp 213 $O^{\delta 2}$ ..... O 39	2.86
Asp 33 $O^{\delta 1}$ ..... O 257	3.11	—	
Asp 33 $O^{\delta 1}$ ..... O 284	2.73	Asp 213 $O^{\delta 2}$ ..... O 61	2.89

[a]The table has been constructed to emphasize the symmetry at the active site. Note that the actual hydrogen bonds will also depend upon the distribution of protons at the active site.

Hydrogen bonding distances are always given from donor to acceptor atom, whether or not the proton is included in the notation.

been suggested that two other highly conserved aspartic acid residues, Asp 14(11) and Asp 300(304), might have a role in lowering the pK_a of the catalytically important carboxyl groups in those aspartyl proteinases with an optimal activity at lower pH values (Blundell et al., 1983). In any event, at the pH of the crystals of penicillopepsin, 4.4, which is close to the pH optimum for the enzyme (Hofmann et al., 1984), only one of the side chains of the active site aspartic acid residues should be deprotonated. The distance from Asp 33(32)$O^{\delta 2}$ to Asp 213(215)$O^{\delta 1}$ is 2.87 Å, and is thus consistent with a hydrogen bond between these groups, with a proton localized on either carboxyl group or shared equally by both. However, it should be noted that when a proton is shared equally by two carboxyl groups, the oxygen-oxygen contact distance is usually ~0.3 Å shorter than the distance we observe here (James and Williams, 1974).

It should be emphasized that even in a well-refined structure, the positions of hydrogen atoms cannot be determined from X-ray crystallographic measurements at 1.8 Å resolution. However, it is straightforward to position protons on main chain nitrogen atoms or other planar groups by assuming a standard configuration. Even the orientation of hydroxyl groups of residues such as Thr 34(33), Thr 214(216), and Thr 216(218) (Fig. 8) can be inferred from the interactions in which they are involved. The elucidation of the charge distribution at the active site carboxyl groups, however, must be obtained by other means.

The unambiguous identification of the chemical nature of the peak denoted O39 is problematic as well. It was refined, as all other solvents in the penicillopepsin structure, as an oxygen atom (8 electrons; temperature factor, $B = 17.4$ Å^2; occupancy $= 0.92$). However, this peak could instead correspond to an NH_4^+ ion, as the native penicillopepsin crystals were obtained initially from $(NH_4)_2SO_4$ and then transferred to Li_2SO_4. O39 is not a Li^+ ion, as the ionic radius of lithium is far too small for the contact distances observed for O39 (Table 4). Also, it is unlikely to be a Na^+ ion, as it does not have the correct ligand coordination (i.e., $Na^+...O$ distances of ~2.4 Å). In order to examine whether it is consistent with a positive ion site in penicillopepsin, crystals were soaked in a thallium acetate solution, and the resulting structure was solved by conventional methods. A strong electron density peak in the corresponding 2.5 Å difference map showed that Tl^+ indeed binds at this site (James and Sielecki, 1985). Thus, at present, we interpret O39 as an NH_4^+ ion (Fig. 8). O39 could also be a hydronium ion (H_3O^+) stabilized by the net negative charge at the two aspartic acid carboxyl groups (Fig. 9a). In this case, a strong electrophilic group would be present in the native structure to protonate the carbonyl oxygen atom of a substrate peptide bond.

In the representation of the catalytic site of Figures 8 and 9a, the net negative charge has been localized on the carboxyl group of Asp 33(32). This assumption is more consistent with the observed stereochemistry at the active site of penicillopepsin (James and Sielecki, 1983). However, the proton could reside on the carboxylate of Asp 33(32), implying that Asp 213(215) would carry the net negative charge (Fig. 9b). Such a situation would not be consistent with the productive mode of substrate binding to be discussed in a subsequent section.

A solvent site equivalent to O39 of penicillopepsin is present in the refined structure of endothiapepsin (Pearl and Blundell, 1984). These authors propose hydrogen-bonding schemes equivalent to those depicted in Figures 9a and 9b. In addition, they suggest alternative schemes making minor contributions in which the O39 site is occupied by an hydroxyl ion (Figs. 9c and d). Whereas this interpretation would provide an effective nucleophilic group to attack the carbonyl-carbon atom of a substrate when forming the TI, it is not compatible with the pK_a's of carboxyl groups and H_2O. The pK_a of a water molecule is ~15, which is 10 orders of magnitude higher than that of a carboxyl group. A very special environment would be required to create a hydroxyl group at this position. Moreover, the location of the protons on the aspartyl side chains, as proposed, leads to unacceptable close contacts between these hydrogen atoms. Alternative conformations that maintain the interpretation of O39 as an hydroxyl group, but alleviate those contacts, are depicted in Figures 9e and 9f.

The active site of rhizopuspepsin has been described only in terms of the pepstatin-bound enzyme (Bott et al., 1982). It is not known whether there is a peak in the native electron density map that would correspond to O39 of penicillopepsin. The active site of porcine pepsin has been described at 2 Å resolution (Andreeva et al., 1984). As mentioned, there are three ethanol molecules in the proximity of the two aspartate residues. Asp 32 and Asp 215 in porcine pepsin seem to adopt very different conformations from those observed for the catalytic groups in pencillopepsin (see Fig. 10 in Andreeva et al., 1984). The hydrogen-bonded environments are not conserved, the two carboxyl groups do not form a hydrogen bond, nor is there a solvent peak corresponding to O39. It is not possible to determine yet whether these observations are due to the presence of ethanol molecules bound at the

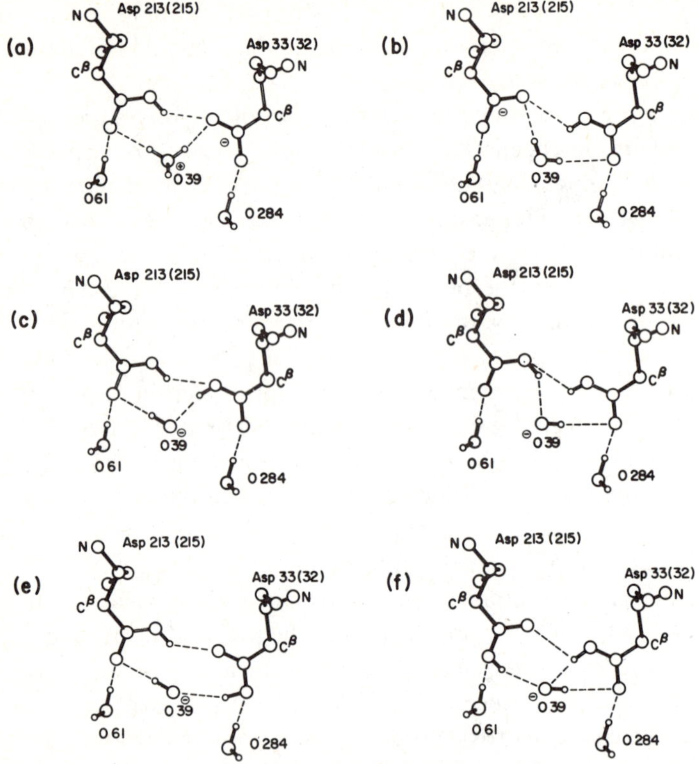

Figure 9. Several possible hydrogen-bonding schemes for the two catalytic aspartic residues and the adjacent solvent molecules, O39, O61, and O284. These diagrammatic representations have been based on the refined atomic positions of the nonhydrogen atoms of penicillopepsin; protons have been positioned using O-H bond lengths of 0.98 Å; H-O-H bond angles of 105° for H_2O molecules, and ~115° for the X-O-H angle of carboxyl groups. (*a*) The active site aspartic acid residues with the protons and charge distribution as shown in the stereo representation of Figure 8. The solvent O39 is shown here as a hydronium ion, H_3O^+, and the negative charge is localized on Asp 33(32). This charge distribution implies a nucleophilic role for water O284 (James and Sielecki, 1985). (*b*) An alternative to the hydrogen-bonding distribution shown in (*a*). Here the net negative charge is localized on Asp 213(215), implying that O61 would act as the nucleophile in a general base attack on the carbonyl-carbon atom of a substrate. For this to be the case, a different orientation of the scissile bond of the substrate would be required, in which the S_1-P_1 and S_1'-P_1' binding interactions would not be as those observed experimentally with specific pepstatin-based inhibitors. This alternative is therefore less likely than (*a*) (see Fig. 3 in James and Sielecki, 1985). O39 is represented in this panel as a water molecule to emphasize the fact that the possible hydrogen-bonding interactions proposed are not dependent on its nature. (*c* and *d*) Two of the possible hydrogen-bonding interactions at the active site of endothiapepsin after Pearl and Blundell (1984). These schemes were proposed in order

active site or to the incomplete state of the refinement of the pepsin structure.

For a number of enzyme systems, it has been shown that substrate-binding sites have isotropic thermal parameters (B factors) that are higher than the average values for the whole molecule (Sielecki et al., 1979; Artymiuk et al., 1979; James et al., 1980). In penicillopepsin, segments of polypeptide chain that have been associated with substrate-binding regions have B factors much greater than the whole main chain average of 12.8 Å^2 (James and Sielecki, 1983). On the other hand, the regions of the molecule supporting the catalytic aspartyl groups have very low B factors (4−7 Å^2 for main chain atoms), indicating a relatively rigid conformation. The intricate hydrogen-bonding scheme described in Figure 8 is therefore reflected in the very low isotropic temperature factors obtained during the refinement of penicillopepsin for most of the atoms depicted in that figure (James and Sielecki, 1983). This suggests that with the exception of the "flap," the groups that form the active site are ideally positioned for the catalytic event, since the relative rigidity provided by the hydrogen-bonding network would make them resist large conformational changes during catalysis.

3.4. Binding Interactions with Inhibitor and Substrate Molecules

Speculations on the catalytic mechanism of any enzyme can only be placed on a firm ground in light of its three-dimensional structure (Fruton, 1976). In this respect, the description of the enzyme in complex with a good substrate is the most important configuration necessary for the deduction of its chemical mechanism. This ideal is probably unattainable since substrates are converted into products much too rapidly to be described with a crystallo-

---

to advance the idea that O39 could act as the nucleophilic group in the catalytic mechanism. The protons have been located on adjacent oxygen atoms of the Asp 33(32) and Asp 213(215) groups. This leads to unacceptable close contacts between hydrogen atoms (< 1.5 Å). Unfavorable dipole moments would make these configurations unlikely. (*e* and *f*). Alternative configurations for the generation of an hydroxyl group at O39. These alternatives to (*c*) and (*d*) have favorable dipoles and acceptable van der Waals contacts. Despite the possible canonical forms shown here, a nucleophilic role for solvent O39 is unlikely, as discussed in the text.

graphic experiment. Data on possible binding modes can be obtained from crystal structures of inhibitors bound to the active site of the enzyme. Such information is indirect insofar as these molecules are inhibitors, rather than substrates, and it must be assumed that the binding modes are similar. In the present section, the structures of a number of enzyme-inhibitor complexes will be reviewed.

3.4.1. Penicillopepsin

The synthetic epoxide 1,2-epoxy-3-(*p*-nitrophenoxy)propane (EPNP) is an effective inhibitor of aspartic proteinases (Chen and Tang, 1972). A 2.8 Å resolution difference electron density map of EPNP bound to penicillopepsin showed that this molecule reacts covalently with both aspartic acid residues Asp 33(32) and Asp 213(215) (James et al., 1977, 1981). Also evident on this early medium-resolution difference map (James et al., 1981) was a peak of negative density corresponding to the displacement of solvent peak O39 (Fig. 8) on reaction of Asp 33(32) and Asp 213(215) with EPNP. A third feature was a pair of adjacent positive and negative regions of electron density in the vicinity of Tyr 75(75). This was taken to indicate that a conformational change of the side chain of Tyr 75(75) had taken place. The phenolic ring moved in a direction that suggested a possible proton donor role for Tyr 75(75) in a fashion analogous to that originally proposed for Tyr 248 of carboxypeptidase A. Such a role for Tyr 75(75) has been withdrawn in light of the results of later inhibitor-binding studies with penicillopepsin carried out at higher resolution (James et al., 1982).

The crystallographic study at medium resolution of the complex with the pentapeptide Ac-Pro-Ala-Pro-Ala-PheOH showed the direction of binding that an extended polypeptide substrate would take (James et al., 1981). The electron density distribution of the 2.8 Å resolution difference map was not clearly interpretable except for the C-terminal Ala-PheOH portion of the peptide. No conformational changes in the region of the "flap" were observed on the binding of this pentapeptide.

Based on the results obtained with these two compounds, a possible substrate binding mode to penicillopepsin was proposed (James, 1980). From the point of view of deriving a possible productive substrate-binding mode, however, the most important structural study to date of a competitive inhibitor with penicillopepsin was that with Iva-Val-Val-StaOEt (James et al., 1982, 1983). This is a synthetic compound (Rich et al., 1980; Rich and

Sun, 1980) based on pepstatin, which is a potent natural inhibitor of aspartyl proteinases.

The initial difference electron density map and mode of binding of this peptide deducted from it is shown in Figure 6*a*. In addition, there was clear indication of a major conformational change in the position of the "flap" (the β loop from residues 72−82; Fig. 10). The importance of refinement of this complex at 1.8 Å resolution has already been discussed in Section 3.2.

A second peptide of the same type has been synthesized (Rich et al., 1983). In this inhibitor, the statine residue was replaced by a lysyl (Lysta) analog of statine, DAHOA [(3S,4S)-4,8-diamino-3-hydroxyoctanoic acid]. It binds to penicillopepsin with a K_i value of 4×10^{-10} M, as compared to 4.7×10^{-8} M for Iva-Val-Val-StaOEt. The rationale behind the incor-

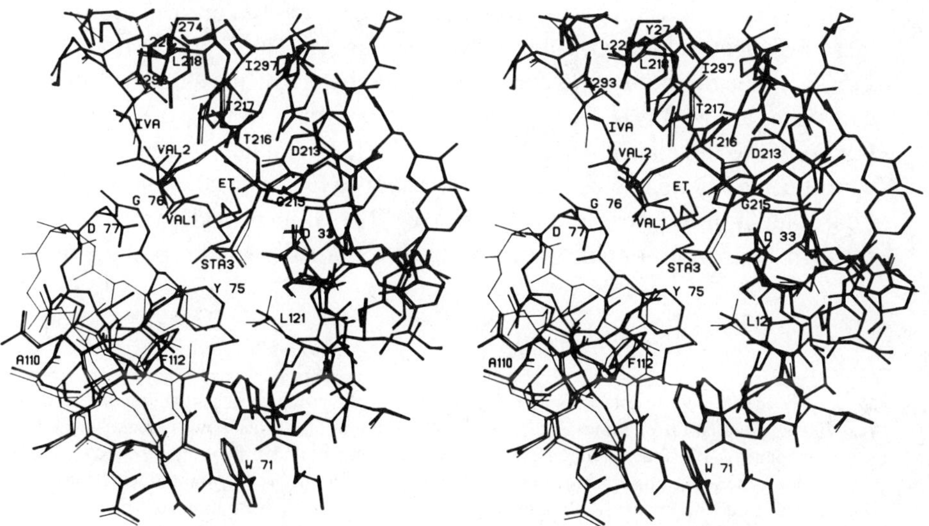

Figure 10. Conformational changes in the active site of penicillopepsin induced upon binding the inhibitor Iva-Val-Val-StaOEt (after James et al., 1983). Thick lines represent the refined structures of the enzyme and inhibitor in the crystals of the complex; thin lines represent the atomic positions of native penicillopepsin. The segment of polypeptide chain that moves furthest is that from Tyr 75(75) to Asp 77(77), with the atoms of Gly 76(76) moving an average of 4 Å from their refined positions in the native enzyme. Residues from Ser 109(108) to Gln 113(112) also move in concert with those of the flap, but by a much smaller amount. Several individual residues appear to have changed their conformations as well: Ile 73(73), Leu 121(120), Ile 293(293), and Ile 297(301).

poration of the Lysta moiety was to obtain direct experimental evidence for a cation-binding site in penicillopepsin. Fungal aspartyl proteases have the ability to activate trypsinogen by the cleavage of a single bond, Lys-Ile. None of the mammalian aspartic proteases exhibits such lysine specificity, but rather show preference for hydrophobic residues at the P_1 position. It is therefore of interest to characterize further the nature of the residues that constitute the S_1-binding site. The difference electron density map computed with complexed and native penicillopepsin data showed conclusively that the cation-binding site consists of Asp 77(77) and Ser 79(79) on the interior face of the "flap" (James et al., 1985). The molecular structure of this complex has been partially refined at 1.8 Å resolution. The present R value is 0.149 for the data from 8.0−1.8 Å resolution (Table 3).

The binding mode of these two pepstatin analogs is similar (Figs. 11a and 11b). Table 5 lists the hydrogen-bonding interactions between the two inhibitors and penicillopepsin. Both adopt an extended β conformation and form hydrogen bonds with the enzyme surface at Thr 217(219) and Gly 215(217). Furthermore, the conformational change of the "flap" in both complexes allows for the formation of hydrogen bonds from Asp 77(77) and Gly 76(76) to the main chain NH and CO of the P_2 residue (Figs. 11a, 11b), contributing to an additional binding energy (James et al., 1983). This "flap" and its movement upon binding (Fig. 10) would provide stabilization through hydrogen-bonding interactions to both sides of the polypeptide

---

Figure 11. (*a*) A stereo optic view of the active site of the complex of penicillopepsin with Iva-Val-Val-StaOEt (after James et al., 1983). The orientation of this view is similar to that shown in Figure 7. Possible hydrogen-bonding interactions, detailed in Table 5, are depicted by dashed lines. Residues important for the binding of this inhibitor are highlighted with thicker lines and labeled with the one-letter code and sequence number. Solvent molecules have been omitted for clarity. (*b*) A similar view to that in (*a*), but for the inhibitor Iva-Val-Val-LystaOEt bound at the active site of penicillopepsin. χ_2 for the lysyl side chain of the lysta residue is currently −50°, thus in an alternative staggered conformation to the fully extended form at $\chi_2 \cong 180°$. This conformation brings the ε-NH_3^+ group into hydrogen-bonding distance with the carboxylate of Asp 77(77) and with O^γ of Ser 79(79). A third hydrogen bond to a water molecule, O87 is also present. Other solvent molecules have been omitted for clarity. Comparison of the remainder of the inhibitor molecule with that of Iva-Val-Val-StaOEt in (*a*) shows that the other interactions are almost identical (with the possible exception of the less ordered ester group). The additional electrostatic and hydrogen-bonding interactions of the lysyl side chain enhance the K_i for this inhibitor by a factor of ~100.

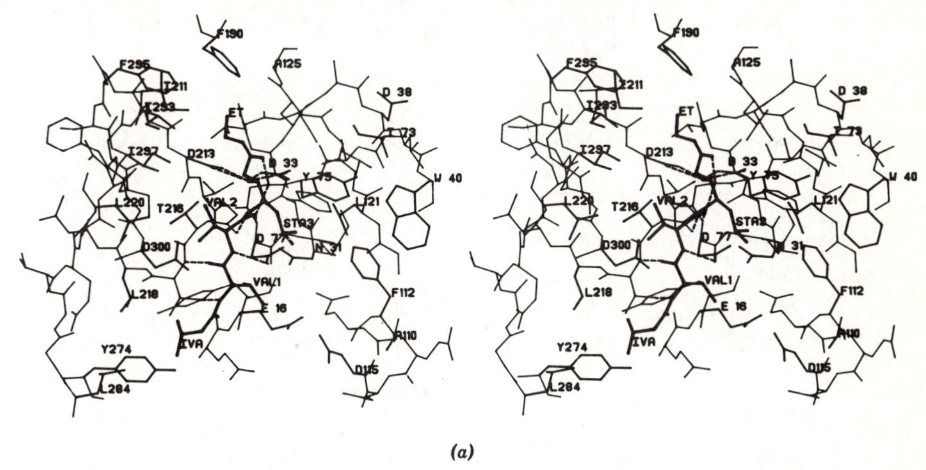

(a)

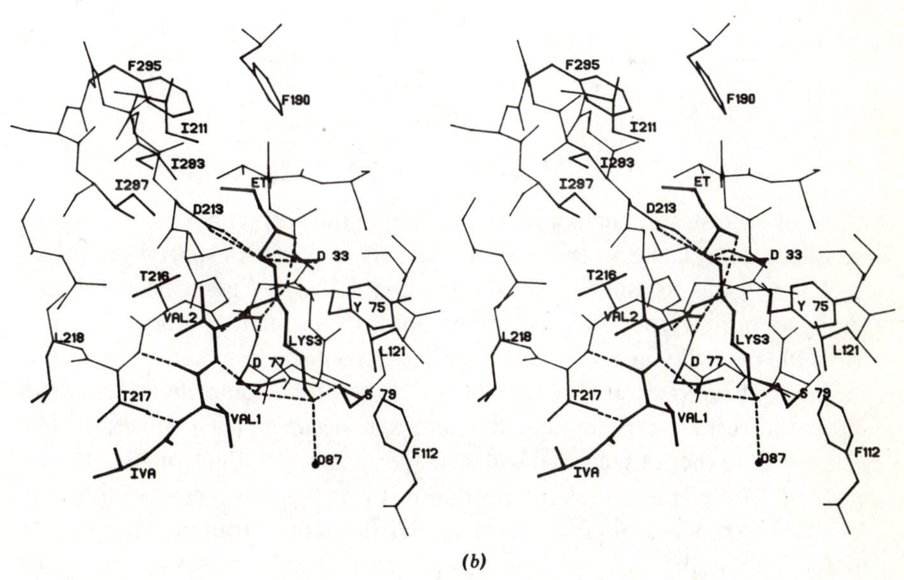

(b)

Table 5. Hydrogen–Bonding Interactions Between Penicillopepsin and Pepstatin Analog–Type Inhibitors

Penicillopepsin Residues	Inhibitor Residues		Distances (Å) Iva-Val-Val-StaOEt	Iva-Val-Val-LystaOEt
Thr 217 $O^{\gamma 1}$	H-N Val		2.84	2.89
Thr 217 N-H	O=C Val	P_3	3.03	2.94
Wat 13 O-H[a]	O=C Val		2.80	2.82
Asp 77 $O^{\delta 1}$	H-N Val		2.71	2.97
Gly 76 N-H	O=C Val	P_2	3.21	3.25
Asp 77 N-H	O=C Val		3.24	3.17
Gly 215 C=O	H-N Sta/Lysta		3.07	3.07
Asp 33 $O^{\delta 2}$			2.50	2.46
Asp 33 $O^{\delta 1}$	H-O-C(3) Sta/Lysta	P_1	3.31	3.31
Asp 213 $O^{\delta 1}$			2.89	2.94
Asp 213 $O^{\delta 2}$			2.60	2.68
Asp 77 $O^{\delta 1}$	H-N$^{\zeta}$ Lysta		—	2.65
Ser 79 O^{γ}	H-N$^{\zeta}$ Lysta	P_1	—	2.95
Wat 87 O	H-N$^{\zeta}$ Lysta		—	2.90
Gly 76 N-H	O=C Ester		3.24	3.47
Wat 282 O-H	O=C Ester		2.88	—

[a]This solvent, in identical positions in both complexes, is within 0.2 Å of O52 in native penicillopepsin (Fig. 7b).

chain of a substrate, and contribute as well to the correct positioning of the scissile bond relative to the two catalytically important aspartyl groups.

The subsites S_4 and $S_1{}'$, on the binding surface of penicillopepsin are easily identified from these crystallographic experiments. Table 6 summarizes the penicillopepsin residues that make contacts less than 4.1 Å to the residues of the pepstatin analogs. Clearly, it will be somewhat dependent upon the residue type of the substrate that occupies each subsite of the enzyme as to the contacts that will be made at each site; but from the data of Table 6, it is evident that the residue in P_1 makes the largest number of contacts to residues of penicillopepsin. Furthermore, from the interactions described for the ester group, it is possible to infer that the residue in position $P_1{}'$ would also interact with a large number of residues on the

Table 6. Penicillopepsin Residues in Contact with Pepstatin Analog–Type Inhibitors

Pepstatin Analog	Penicillopepsin Residues Forming Contacts < 4.1 Å to:	
	Iva-Val-Val-StaOEt	Iva-Val-Val-LystaOEt
P$_4$ Iva	Glu 15(12), Thr 217 (219), Leu 218(220), Leu 284(284)	Glu 15(12), Thr 217(219), Leu 218(220), Tyr 274(275), Leu 284(284)
P$_3$ Val	Glu 15(12), Asp 77(77), Gly 215(217), Thr 216(218), Thr 217(219)	Glu 15(12), Asp 77(77), Gly 215(217), Thr 216(218), Thr 217(219)
P$_2$ Val	Tyr 75(75), Gly 76(76), Asp 77(77), Thr 216(218), Leu 220(222), Ile 297(301)	Tyr 75(75), Gly 76(76), Asp 77(77), Thr 216(218)
P$_1$ Sta/Lysta	Asn 31(30), Asp 33(32), Tyr 75(75), Asp 77(77), Ser 79(79), Leu 121(120), Asp 213(215), Gly 215(217), Thr 216(218)	Asp 33(32), Gly 35(34), Tyr 75(75), Asp 77(77), Ser 79(79), Phe 112(111), Leu 121(120), Asp 213(215), Gly 215(217), Thr 216(217)
P$_1'$ ester	Gly 35(34), Thr 75(75), Gly 76(76), Phe 190(189), Ile 211(213), Asp 213(215), Thr 216(218)	Gly 35(34), Tyr 75(75), Gly 76(76), Asp 213(215), Thr 216(218)

enzyme, confirming the observation that the two residues on either side of the scissile peptide are the most important for effective catalysis (Fruton, 1976).

Iva-Val-Val-LystaOEt binds ~100 times more tightly to penicillopepsin than does Iva-Val-Val-StaOEt (Rich et al., 1983). Since the only difference is its lysyl side chain, as compared to the leucyl side chain of the latter, the electrostatic interactions of the ϵ-NH$_3{}^+$ group must be the major contributors to this ~100-fold factor. Figure 11*b* shows the hydrogen-bonding interactions -N$^\zeta$H...Asp 77(77)O$^{\delta 1}$, $-$N$^\zeta$H...Ser 79(79)O$^\gamma$, and -N$^\zeta$H...Wat 87. The tighter binding is also reflected in changed *B* factors for the residues of the "flap".

The mean *B* factors of two segments in the substrate-binding region of native penicillopepsin are shown in Figure 12*a*; segment Ile 73(73) to Ser 80(80), the "flap" region, and Ile 108(107) to Asn 118(117). Figures 12*b* and 12*c* show the changes in the average *B* factors for these same residues of penicillopepsin in the complex with Iva-Val-Val-StaOEt and Iva-Val-Val-LystaOEt, respectively. The region Ile 108(107) to Asn 118(117) behaves in a predictable manner, in that the mean *B* factors of both main chain and side chains are reduced (by $5-8$ Å^2) in the presence of either inhibitor. Both

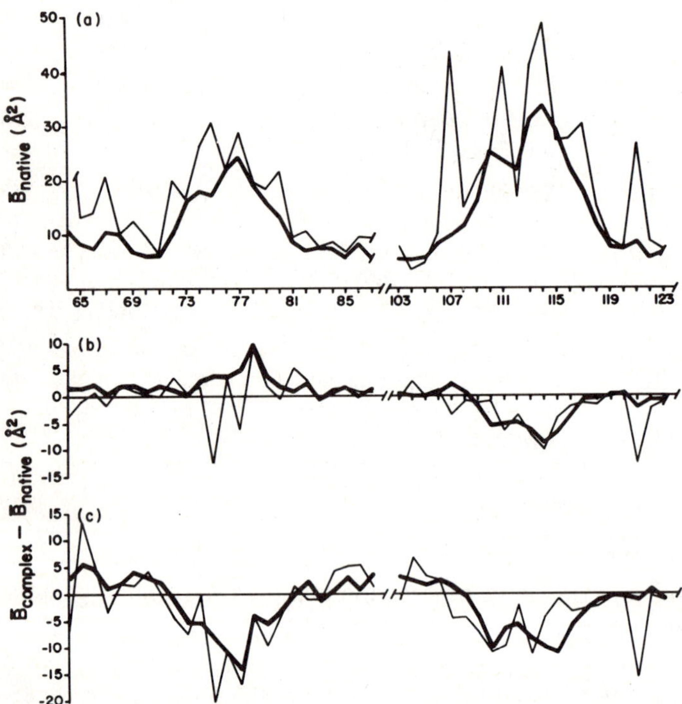

Figure 12. (*a*) A plot of the refined isotropic *B* factors for two regions of the native penicillopepsin molecule, residues Lys 64(63) to Thr 87(86) and residues Gln 103(102) to Gly 123(122). The thick line in all three panels represents the mean *B* factor for the atoms of the main chain of that residue (N, C$^\alpha$, C, O); the thin line refers to the mean *B* factor of the atoms of the corresponding side chain. The overall average *B* factor for the 2366 nonhydrogen atoms of penicillopepsin is 14.5 Å^2 (12.8 Å^2 for atoms of the main chain) (data from James and Sielecki, 1983). (*b*) A plot of the difference in the mean *B* factors for atoms of the same regions of penicillopepsin in the complex with Iva-Val-Val-StaOEt ($\bar{B}_{\text{complex}} - \bar{B}_{\text{native}}$). (*c*) The same as (*b*), but for the complex with Iva-Val-Val-LystaOEt at cycle 15 of the crystallographic refinement.

inhibitors reduce, upon binding, the mean B factor of the side chain of Leu 121(120) considerably, as a result of the contacts of the leucyl group with the statine or Lysta moieties (see Figs. 11a and 11b).

Residues of the "flap" exhibit a different change in mean B factor depending on the nature of the inhibitor bound. As expected, the mean B factors of the atoms of the main chain for residues Ser 74(74) to Ser 79(79) are dramatically reduced in the complex with Iva-Val-Val-LystaOEt over those for the same residues in the native enzyme. However, in the complex with the statine inhibitor, the mean B factors of these same residues are increased over those for the native enzyme. The explanation for this apparently anomalous behavior resides in the presence of a crystallographic twofold symmetry axis at ½, y, ½ in the crystals of penicillopepsin (James and Sielecki, 1983). In the conformation of the native molecule, residues in the "flap" region of two neighboring molecules come into close contact due to the location of this axis, allowing for the formation of two strong intermolecular hydrogen bonds and several other stabilizing nonbonded interactions. These are disrupted upon inhibitor binding due to the concomitant movement of the flap towards the inhibitor molecules. The interactions that the residues of the flap in their new positions make with the statine inhibitor seem to be weaker, since the result is an apparently greater conformational flexibility for this region than that exhibited in the native structure, where the intermolecular interactions across the crystallographic symmetry axis can take place.

Obviously, in the complex with Iva-Val-Val-LystaOEt the additional electrostatic and hydrogen-bonding interactions of the lysyl side chain with Asp 77(77) and Ser 79(79) are more than enough to compensate for the loss of the intermolecular contacts across the symmetry axis in the native conformation, as indicated by much reduced isotropic B factors for the main chain atoms of the flap. It is of interest to note that the side chains of Tyr 75(75) and Asp 77(77) have significantly reduced B factors in both complexes, as compared with those of the native enzyme.

3.4.2. *Rhizopuspepsin*

The naturally occurring inhibitor pepstatin has been bound to rhizopuspepsin and the complex analyzed in a 2.5 Å difference electron density map (Bott et al., 1982; Bott and Davies, 1983). The portion corresponding to the first four residues, Iva-Val-Val-Sta, were easily identified in the difference density map. There was more ambiguity in the fitting of the C-terminal Ala-StaOH groups. The observed binding and hydrogen-bonding interactions are very similar to those reported for the pepstatin fragment complexed with penicillopepsin. The flap position in the crystals of native rhizopuspepsin is in a more closed conformation when compared to native penicillopepsin. Therefore, it is not surprising that the conformational change observed in that enzyme upon binding of pepstatin is smaller—an r.m.s. movement of 1.2 Å for residues in positions 71–77—in the pepsin numbering scheme (Bott and Davies, 1983).

The residues that form the S_1 binding site in rhizopuspepsin are given as Phe 75, Gly 76, Asp 77, Phe 111, and Leu 120. The residues forming the S_1' pocket are Ile 213, Ile 301, and Tyr 189. These residues have the numbering scheme of porcine pepsin. It should be noted that the homologous residue to Tyr 75(75) has been interpreted as a phenylalanine in rhizopuspepsin. Likewise, the residue at position 77, originally thought to be a leucine (Bott et al., 1982), is currently reported as an aspartic acid (Bott and Davies, 1983). Clearly, the completion of the chemical sequencing will be extremely useful in clarifying the nature of these important residues.

3.4.3.   Porcine Pepsin

In a recent publication, the crystal structure of the dipeptide Phe-TyrI$_2$-OMe bound to pepsin has been described (Andreeva et al., 1984). Due to disorder in the crystals of the complex, this work was limited to 3 Å resolution. A positive peak of electron density on the difference map was interpreted in terms of the molecular structure of the dipeptide. It should be pointed out that the assignment of atomic positions at such resolution is, of necessity, tentative, and only the general arrangement of the molecule has been given. The diiodotyrosine residue is considered to be in contact with Ile 213, Leu 301, and Leu 299 in the pepsin numbering. Not all of these contacts are easily discernible in Figure 12 of Andreeva et al. (1984), but it is stated that the binding site coincides with what has been identified as the S_1' site in penicillopepsin. The phenylalanine ring is described as being in contact with Val 184, Trp 190, Ile 192, Val 214, and Val 321 (pepsin numbering). This pocket is quite different from the S_1 site described for the fungal enzymes, and small movements of several residues are invoked in order to accommodate the bulky phenyl side chain. Indeed, the homologous residues of penicillopepsin are well into the interior of the closely packed hydrophobic core of the molecule, and a similar binding could not be achieved in this structure without a serious rearrangement of several β strands. In addition, the positioning of the N-terminal phenylalanine, with its charged amino group buried in the hydrophobic interior, makes it very unlikely that this interpretation will stand the test of higher resolution and atomic refinement.

It has been pointed out that the S_1' binding pocket of pepsin may also involve residues of a second flap that are not present in the penicillopepsin molecule (James et al., 1982; Andreeva et al., 1984). There is a four-residue insertion in pepsin, from Ser 295 to Glu 298 (Table 1), that could make important interactions with the P_n' residues of an oligopeptide substrate.

3.5. Proposed Productive Substrate Binding to Aspartic Proteinases

The inhibitor-binding studies described in the preceding section for penicillopepsin, rhizopuspepsin, and porcine pepsin place severe restrictions on possible binding modes of good substrates. These studies have identified the location of the S_1 and S_1' binding sites. In agreement with these structural studies, the binding region for P_1 and P_1' side chains has been proposed in endothiapepsin (Blundell et al., 1983). The structures of pepstatin-based inhibitors bound to penicillopepsin and rhizopuspepsin provide the detailed interactions for the acyl portion of substrates. The binding sites for P_4, P_3, P_2, and P_1 residues are clearly defined from the refined structures of Iva-Val-Val-StaOEt and Iva-Val-Val-LystaOEt bound to penicillopepsin. As these are refined structures at 1.8 Å resolution (James et al., 1983, 1985), the hydrogen-bonding interactions (Table 5) and the nonbonded contacts (Table 6) give reliable guidance on the positioning of models of good substrates. These experimentally determined structures were used as basic "templates" for the model-building studies of substrates in the active site of penicillopepsin with the use of interactive computer graphics (Hofmann et al., 1984). Manipulation of the molecular model was done by rotations about single bonds, or global translations or rotations only. Inter- and intramolecular nonbonded contacts that were too close were monitored by a "bump check" feature of the program M3, designed and implemented by Colin Broughton (Sielecki et al., 1982) on the MMS-X interactive graphics system (Barry et al., 1976).

The best substrate found so far for penicillopepsin is the hexapeptide Ac-Ala-Ala-Lys-pNO$_2$-Phe-Ala-Ala-NH$_2$ (Hofmann and Hodges, 1982). Based on this sequence, a shorter tetrapeptide, Ac-Ala-Lys-Tyr-Ala-NH$_2$, was used for the molecular modeling to penicillopepsin. The pentapeptide Val-Val-Leu-Phe-Leu was used by Bott et al. (1982) to explore substrate binding to rhizopuspepsin. However, the TI they present includes only the description of the interactions of the P_1 Leu and the P_1' Phe residues. Upon close examination of the figures and tables in their paper, several excessively short nonbonded contacts are discernible in the ES complex of the model proposed by these authors. It is hard to judge how extensive the movements would have to be to relieve those contacts to acceptable van der Waals distances while maintaining the same overall proposed model.

The deduced productive binding mode for the substrate Ac-Ala-Lys-Tyr-Ala-NH$_2$ to penicillopepsin is presented in Figure 13. The hydrogen-bonding interactions for the Ac-Ala-Lys portion of the substrate are taken from the refined positions of the residues in the analogous P_1 and P_2 positions of the inhibitor Iva-Val-Val-LystaOEt. The ionic interaction be-

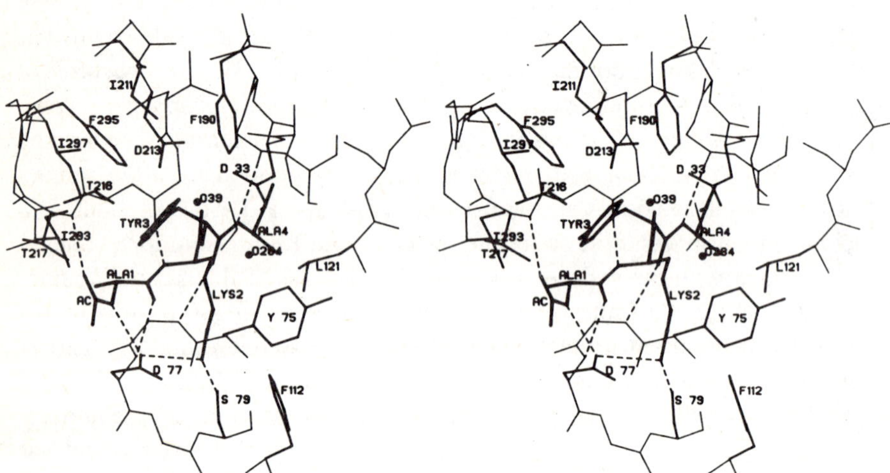

Figure 13. A stereo optic representation of a hypothetical model of the substrate Ac-Ala-Lys-Tyr-Ala-NH$_2$ bound in a productive mode in the active site of penicillopepsin (after Hofmann et al., 1984). The binding interactions portrayed in this figure are based on the pepstatin analog interactions for the acyl portion (Ac-Ala-Lys-) of the substrate. The corresponding amino portion was built by interactive computer modeling. In this representation, the positions of O39 and O284, as observed in the native structure, are given for reference. It is obvious that the solvent molecule O39 would be displaced by the carbonyl-oxygen atom of the substrate. Solvent O284 is positioned ideally for nucleophilic attack on the substrate carbonyl-carbon atom.

tween P_1 Lys and Asp 77(77) and the hydrogen bond from P_1 Lys NH to Gly 215(217)O are preserved (Figs. 11*b* and 13). The binding interactions for the P_1' Tyr and P_2' Ala are based solely on molecular model building. Hydrophobic interactions to the phenolic ring of P_1' tyrosine are made by Phe 190(189), Ile 211(213), Ile 293(293), Phe 295(299), and Ile 297(301). The hydrogen-bonding interactions suggested by this model building are from Gly 76(76)NH to P_1' Tyr CO and from P_2' Ala NH to Gly 35(34)CO. Similar hydrogen-bonding interactions from rhizopuspepsin to their hypothetical substrate have been described by Bott et al. (1982), and were also proposed in an earlier enzyme-substrate model for penicillopepsin (James, 1980).

It is worth noting several additional points regarding this enzyme-substrate productive-binding proposal. Among the interactions described in the preceding paragraph, there are two important hydrogen bonds for the proposed positioning of the substrate that involve the peptide amide groups from the P_1 and P_2' residues to the carbonyl oxygen atoms of Gly 215(217) and Gly 35(34), respectively. It is possible to draw an interesting analogy with the serine proteinase family, where equivalent hydrogen bonds have been proposed from the P_1 and P_2' NH groups of a good substrate to the carbonyl-oxygen atoms of two residues of the enzymes, in particular to the conserved Ser 214 and to residue 41 in the chymotrypsin numbering (James et al., 1980). These interactions have been observed experimentally in the X-ray structures of the complexes of serine proteinases and their inhibitors (Huber and Bode, 1978; Fujinaga et al., 1982; Read et al., 1983). On the proposed catalytic pathway, the hydrogen bond from the P_1 NH group to Ser 214O shortens (strengthens) as the carbonyl-carbon atom of the substrate approaches the TI (Robertus et al., 1972). Similarly, in the binding mode depicted in Figure 13, the distances P_1 Lys NH to Gly 215(213)O and P_2' Ala NH to Gly 35(34)O are long ($\sim$3.2 Å), but could shorten (strengthen) as the substrate approached a TI, thus providing additional stabilization energy for the transition state.

In the productive substrate-binding model shown in Figure 13, the carbonyl-oxygen atom of the scissile bond points in towards the two aspartic acid residues Asp 33(32) and Asp 213(215). In doing so, it comes within 0.6 Å of the observed position of O39 in the native enzyme structure (Fig. 14*a*), clearly displacing this solvent molecule. In addition, solvent O284 is within 0.2 Å of the ideal position defined for an attacking nucleophile relative to the carbonyl-carbon atom (Bürgi et al., 1973). The identification of the

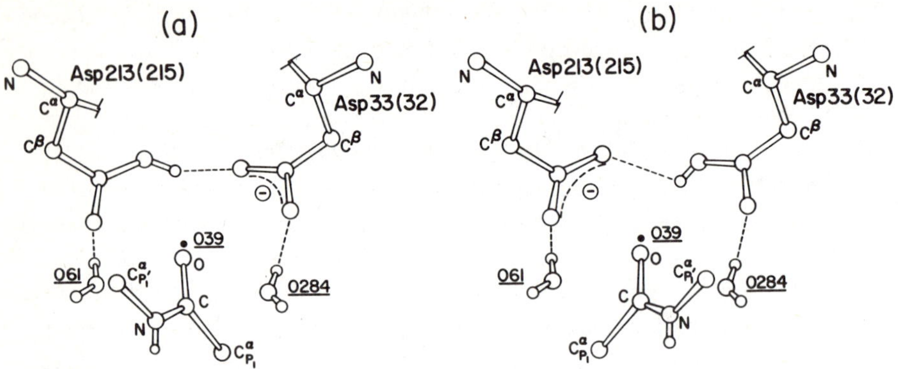

Figure 14. Two alternatives for the charged state of the active site aspartic acid residues in the E-S Michaelis complex. (*a*) The charge is shown localized on the Asp 33(32) carboxylate, and solvent molecules O39 and O61 would be displaced by the substrate in this version of the Michaelis complex. This is the favored mode of interaction, as the P_1-S_1 and P_1'-S_1' binding interactions are those deduced from the bound position of Iva-Val-Val-StaOEt. Water O284 is the nucleophilic group in this proposal. (*b*) Similar to (*a*), but with the scissile peptide plane oriented such that O61 would be the nucleophilic group and the net negative charge would be on Asp 213(215). This possibility is sterically less favorable than that in (*a*), as it would require different P_1-S_1 and P_1'-S_1' interactions from those observed experimentally with the pepstatin analog inhibitors.

incipient nucleophile is thus determined by the proposed substrate-binding mode. Figure 14*b* depicts an alternative asymmetric interaction that the scissile bond could make with Asp 33(32) and Asp 213(215). With this orientation of the substrate, O61 would be appropriately positioned for nucleophilic attack. However, the interactions that the P_1 and P_1' residues make with the enzyme in this binding mode are different from those observed experimentally in the structures of the pepstatin analog complexes. This direction of approach is therefore a less likely possibility. Other binding approaches for the carbonyl bond have been explored (Hofmann et al., 1984) and have been rejected on steric grounds.

In a mechanism that does not involve covalently attached intermediates, there is still a third candidate for the attacking nucleophile in the event that precedes the formation of the TI. This is solvent O39. Pearl and Blundell (1984) have suggested that the equivalent solvent site in the endothiapepsin structure could be interpreted also as an hydroxyl ion (see Figs. 9*c*–9*f*), assigning to it the possible role of the nucleophile. We have discussed in Section 3.3 some of the arguments against such an assignment. Besides, it is unlikely that this site, which favors positively charged ions in penicillopepsin, would support an hydroxide required for nucleophilic attack of the peptide carbonyl-carbon atom. Of necessity, the approach of the peptide plane of the substrate would be different from that proposed and shown in Figures 13 and 14*a*. Both are approximately orthogonal to one another. In the alternative proposal of Pearl and Blundell (1984), the plane of the scissile peptide bond is approximately perpendicular to the interdomain pseudo twofold symmetry axis located between the two aspartic acid side chains (Fig. 15). For such a direction of approach, there is no electrophilic group to polarize the carbonyl bond and generate a partially positively charged carbon atom of the peptide. Nor is there an oxyanion-binding site that would stabilize the negative charge developing on the carbonyl oxygen as the nucleophilic attack proceeds. It is also proposed by those authors that the extended binding site of the aspartyl proteinases could provide sufficient binding energy to compensate for a distortion of the peptide plane of the substrate from planarity. Similar strain mechanisms have been described in the case of bacterial serine proteinases, where rate enhancements of a similar order have been measured (Bauer, 1978). However, these enzymes have a well-defined oxyanion-binding site that stabi-

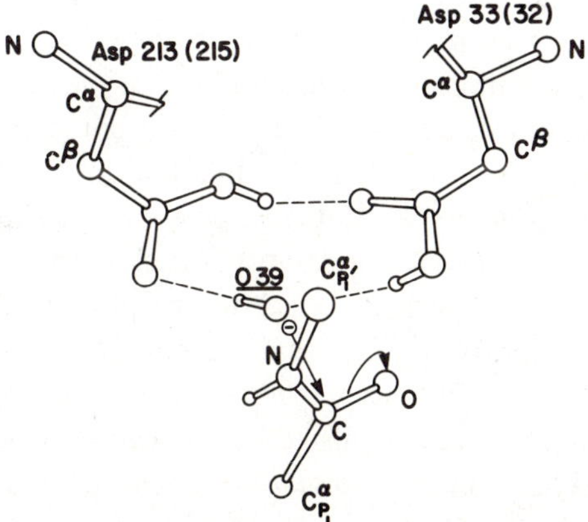

Figure 15. A diagrammatic interpretation of the proposal that O39 is the nucleophilic group facilitating the formation of the tetrahedral intermediate (Pearl and Blundell, 1984). This proposal has no electrophilic component, and the developing charge on the substrate carbonyl-oxygen atom would have no apparent positively charged group or favorable dipole to stabilize it.

lizes the negative charge as it develops on the carbonyl-oxygen atom of a substrate (James et al., 1980).

In summary, the productive binding mode we favor is the one shown diagrammatically in Figure 14a, in which O39 and O61 would be displaced by an incoming substrate. The negative charge at the active site would be localized on Asp 33(32), with O284 being assigned the role of the attacking nucleophile.

3.6. Proposed Catalytic Hydrolysis Pathway for Aspartic Proteinases

It has been pointed out (Drenth, 1980) that all proteolytic enzymes have three essential components required for catalytic hydrolysis of amides or esters: (1) an electrophile, which will enhance the polarization of the carbonyl bond to assist in the nucleophilic attack and stabilize the negative charge that develops on the carbonyl oxygen atom of the intermediate; (2) a nucleophilic center, which will attack the carbonyl-carbon atom to form a TI; and (3) a proton donor, especially important in the hydrolysis of amides, which will make the nitrogen a good leaving group in the breakdown of the TI. In the case of the serine proteinases, the role for each of these has been well established because the hydrolytic cleavage involves a covalently attached acyl enzyme intermediate. In these systems, Ser 195 is the neucleophilic center, the two peptide NH groups of Gly 193 and Ser 195 form the electrophilic-binding site for the carbonyl-oxygen atom of the substrate, and the imidazolium side chain of His 57 is the source of the proton for the leaving-group nitrogen (Kraut, 1977). In addition, the binding interactions at the substrate binding sites S_n to S_n' position the substrate optimally so that hydrolysis of the scissile bond between residues $P_1 - P_1'$ may take place.

As already discussed in Section 2.2, the weight of evidence is against the existence of a covalently attached acyl or amino intermediate in the catalytic pathway of aspartyl proteinases. Indirect evidence supporting a pathway that involves a noncovalent TI also comes from the strong inhibitory nature of pepstatin and its analogs. The several crystal structures of the complexes of pepstatin or its synthetic derivatives with aspartic proteinases all show clearly that the tetrahedral carbon-hydroxyl of the statine residue is bound in a noncovalent mode. This suggests that the TI in the hydrolytic pathway of a good substrate should also be a noncovalently bound species. Certainly, the structural data have produced a consensus on this point, and all mechanisms that have been advanced have not involved a covalently attached intermediate (James, 1980; James et al., 1977, 1981; Bott et al., 1982; Pearl and Blundell, 1984).

In the following two sections, the formation and breakdown of this noncovalently bound TI is examined with particular attention to the stereochemistry of the reaction steps.

3.6.1. *Formation of the Tetrahedral Intermediate*

It was discussed in the previous section how selection of the electrophile, nucleophile, and proton donor in the catalytic hydrolysis of a peptide bond by aspartic proteinases is strongly dependent upon the interpretation of the productive binding mode of a good substrate. If we accept the proposal for the productive binding mode shown in Figures. 13 and 14*a*, we must still select the possible electrophile among, at least, two candidates. The first one is the proton shared by Asp 33(32) and Asp 213(215) (James et al., 1977, 1981; Bott et al., 1982; James and Sielecki, 1985). The protonation of the carbonyl-oxygen atom of the substrate by this component would give the hydrolytic mechanism of aspartic proteinases a predominantly electrophilic character, as advocated by Knowles (1970). Alternatively, if solvent site O39 is interpreted as an hydronium ion, as proposed in Section 3.3., it could represent the electrophilic component of the mechanism, and acid catalysis could be a possible pathway. In the absence of definitive experimental data as to the nature of O39, we shall describe a catalytic pathway based on the first alternative.

Catalysis, in that case, would be initiated with the protonation of the carbonyl-oxygen atom of the substrate by the proton shared between the two catalytic carboxyl groups. Possibly concerted with the protonation event are the electronic rearrangements necessary for the nucleophilic attack of O284 assisted by the general base: the $-COO^-$ of Asp 33(32). The development of the positive charge on the carbonyl-carbon atom of the substrate after protonation of the carbonyl-oxygen atom would also favor the nucleophilic attack by O284 as an OH^- ion. The position of O284 is indicated in Figure 13, ~2.8−2.9 Å from the *si* face of the peptide bond. The O284-C direction makes an angle of ~110° to the C=O direction, which is in agreement with the specifications of Bürgi et al. (1973). One of the protons of water O284 is transferred to the Asp 33(32) carboxylate group as the C-O284 distance shortens during the formation of the TI.

The protonation of the carbonyl-oxygen atom disrupts the delocalization

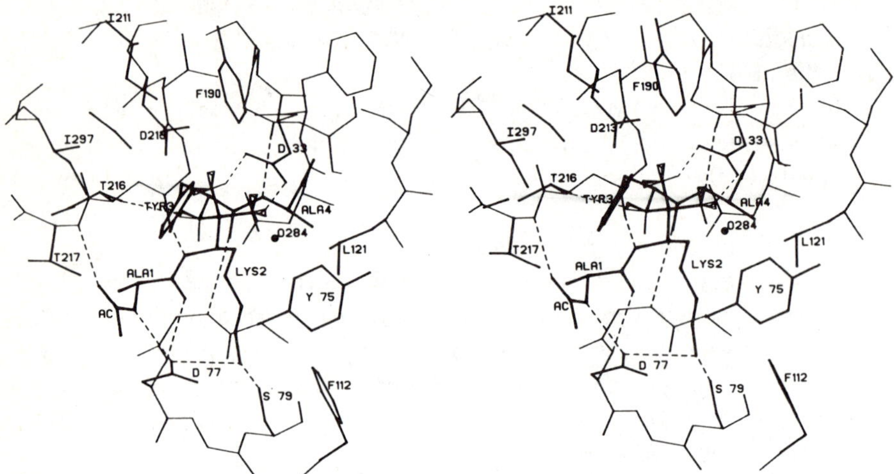

Figure 16. A stereoscopic view of the proposed tetrahedral intermediate bound in the active site of penicillopepsin. This intermediate would result from the general base-assisted nucleophilic attack of an OH⁻ ion (O284) on the carbonyl-carbon atom of the substrate following the electrophilic protonation of the carbonyl-oxygen atom of the scissile bond. The lone-pair electrons (depicted as elongated pyramids) on the three heteroatoms are shown, with those on the tetrahedral nitrogen and the protonated carbonyl oxygen of the substrate oriented antiperiplanar to the direction of bond formation. Stabilizing hydrogen-bonding interactions between the tetrahedral intermediate and the enzyme are indicated by dashed lines.

of the π electrons in the peptide bond. Both the carbonyl-carbon atom and the nitrogen atom become pyramidal as the nucleophilic attack by O284 proceeds. The stereochemistry of these initial events has been examined by molecular modeling of a proposed TI (Fig. 16) (James and Sielecki, 1985). The formation and breakdown of this intermediate is consistent with the principles of the stereoelectronic effect in amide hydrolysis (Deslongchamps, 1975). This stereoelectronic hypothesis postulates that the departure of the leaving group is facilitated if the lone-pair electrons on the heteroatoms of the TI are antiperiplanar to that group. Thus, during formation of a bond by nucleophilic attack of O284 on the carbonyl-carbon atom of the scissile peptide, the two heteroatoms, N and O, bonded to the carbonyl-carbon atoms should each have lone-pair electrons oriented antiperiplanar to the bond being formed. This stereochemical configuration is shown in Figure 16. (The original position of the proposed nucleophile, O284, is also indicated in the figure.)

Several favorable electrostatic interactions provide stabilization of this TI. In the conformer shown in Figure 16, the two hydroxyl groups (a *gem* diol on the carbonyl-carbon atom) are hydrogen bonded to the protonated carboxyl group of Asp 33(32). Asp 33(32) is a donor of a hydrogen bond to the hydroxyl-oxygen atom that arises from O284 (protons are not shown on the enzyme in Figure 16). The other oxygen atom of the carboxyl group of Asp 33(32) is a hydrogen bond acceptor from the hydroxyl group on the carbonyl-carbon that resulted from protonation of the carbonyl-oxygen atom of the substrate peptide. This interaction is reminiscent of a carboxylic acid dimer interaction, although it clearly cannot be described electronically in such a way.

Another electrostatic interaction stabilizing the TI is that between the lone-pair electrons on the pyramidal-nitrogen atom and the hydroxyl group of Thr 216(218). In native penicillopepsin and in the structures of its two pepstatin analog complexes, the Thr 216(218) $O^{\gamma 1}H$ forms a hydrogen bond to $O^{\delta 1}$ of Asp 213(215). In the present TI, this hydrogen bond could be bifurcated to both Asp 213(215) and the lone-pair electrons on the nitrogen atom of the substrate.

In the proposal by Bott et al. (1982), the electrophilic component is also the proton shared by the carboxyl groups of Asp 213(215) and Asp 33(32). The nucleophile is suggested to be a water molecule held by hydrogen bonds to the carboxyl of Asp 33(32), the carbonyl-oxygen atom of Gly 35(34), and possibly the O^{γ} of Ser 36(35). The position of such a water was inferred from model building of the TI based on the statine component of pepstatin bound to rhizopuspepsin. It has not been identified in the structure of the native enzyme. The difference between the TI proposed by Bott et al. (1982) and the conformation reported here (Fig. 16) have important consequences in the second step of proton transfer to the leaving group.

3.6.2. Breakdown of the Tetrahedral Intermediate

The TI depicted in Figure 16 can be analyzed in terms of the requirements for cleavage of the C-N bond to yield products. There are two necessary prerequisites for cleavage of this bond. First, the leaving-group nitrogen atom must receive a second proton to form a good leaving group ($-NH_2$). Second, according to the stereoelectronic effect hypothesis (Deslongchamps, 1975), there should be lone-pair orbitals on the two oxygen atoms bonded to the tetrahedral carbon that are oriented antiperiplanar to the C-N bond.

In light of the stereochemical distribution of groups in the TI-enzyme complexes, there are two possible sources for the second proton. First, from bulk solvent, especially in the cases of those aspartic proteinases that have pH optima in the strongly acidic region. In the conformation necessary for formation of the TI (Fig. 16), the lone-pair orbital on the nitrogen atom points towards the surface of the enzyme, and protonation from bulk solvent could be restricted. If this would be the case, it is also possible to postulate an inversion of configuration of the nitrogen atom, as suggested by Bizzozero and Dutler (1981), for the serine proteinases. Such inversion would result in the lone-pair orbital on the nitrogen pointing towards the bulk solvent and therefore being more readily accessible for protonation.

Alternatively, proton donation to the leaving-group nitrogen may originate from the innermost -OH group of the TI, with the proton being transferred possibly via the carboxylate of Asp 213(215). If this mechanism of transfer would occur, the nitrogen atom should *not* undergo an inversion of configuration, but the hydroxyl proton must adopt an alternative conformation attainable by a rotation about the C-O bond, placing the hydroxyl proton and lone-pair electrons on the nitrogen of the leaving group in a *syn* conformation. The energetic differences between these two alternatives is difficult to assess; at this stage, it is arbitrary as to how to make the choice of one over the other. Both pathways would have favorable aspects.

The second requirement necessary to conform to the dictates of the stereoelectronic theory is that the two lone-pair orbitals on the oxygen atoms of the TI should be oriented antiperiplanar to the C-N bond. The required conformation is shown in Figure 17 and attained by single bond rotations that should imply relatively small rotational energy barriers.

Similar proposals for protonation of the leaving group have been suggested for the rhizopuspepsin (Bott et al., 1982). Either the proton comes from bulk solvent or originates from the TI itself. In the latter event, the authors suggest that the carbonyl of Gly 35(34) could facilitate the shift of protons from Asp 33(32) to the hydroxide and from the hydroxide to the amide nitrogen. This is a very improbable role for the carbonyl-oxygen atom of Gly 35(34), and is quite unlikely in light of the refined three-dimensional structure of penicillopepsin and its complexes. Examination of Figures 16 and 17 clearly show that the carbonyl oxygen of Gly 35(34) is in no position to facilitate such a proton transfer. The role that we ascribe to this carbonyl of Gly 35(34) is to function as a hydrogen bond acceptor for the NH of a P_2' residue of the substrate.

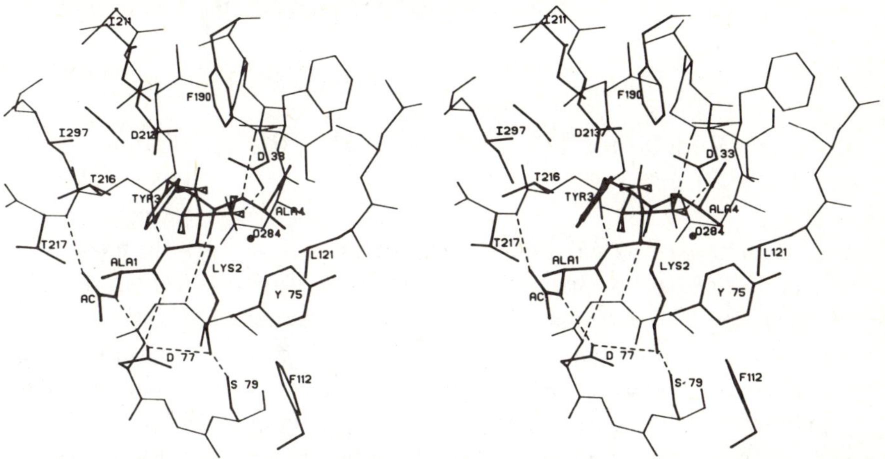

Figure 17. A stereo view of the proposed second tetrahedral intermediate on the catalytic pathway of aspartic proteinases. Two differences to the tetrahedral intermediate shown in Figure 16 are evident. The configuration at the nitrogen is inverted so that the lone-pair electrons are more accessible to bulk solvent allowing for the protonation of the leaving group nitrogen. Second, the rotomers for the C-OH groups are arranged with lone-pair orbitals on the oxygen atoms antiperiplanar to the C-N bond, thus facilitating cleavage (Deslongchamp, 1975).

Proton transfers figure prominently in all of the mechanisms that have been presented above. There is, at present, certainly from the structural data alone, no way to define a rate-determining proton transfer step. In order to make mechanistic interpretations, it would also be of paramount importance to know proton positions in the enzyme-substrate complex. X-ray crystal structure analysis of proteins and enzymes does not provide data on the location of hydrogen atoms. They can only be inferred from the nonhydrogen atom positions or with the use of high-resolution neutron diffraction (Schoenborn, 1975; Kossiakoff and Spencer, 1981; Phillips and Schoenborn, 1981). Therefore, it is necessary to turn to the chemical data for help in this regard. Unfortunately, for pepsin, there is an apparent absence of solvent isotope effect on the maximum rate of catalytic hydrolysis (Clement and Snyder, 1966). Although this is an interesting observation, there are serious problems in basing mechanistic deductions on the effects of D_2O on the catalytic rates of enzymes (Knowles, 1970). It is clear that one of the major difficulties in defining the catalytic pathway of aspartic proteinases is the lack of a covalently attached intermediate, be it an acyl or an amino enzyme.

4.   COMPARISON TO THE SERINE PROTEINASE FAMILY

The mechanism of action of the serine proteinases is perhaps the most fully understood enzyme mechanism. The reader is referred to excellent reviews by Kraut (1977), Huber and Bode (1978), and Bizzozero and Dutler (1981). Extensive structural studies on the bacterial serine proteinases confirm that they share a catalytic pathway similar to those of the mammalian enzymes (James et al., 1980; Fujinaga et al., 1982; Read et al., 1983).

There are many structural and chemical differences between the serine and aspartic proteinase families. The serine enzymes have pH optima in the alkaline range 7.0−10.0, the pH optima for the aspartic proteinases are mainly in the acid range 1.0−7.0. The groups at the active site of these enzyme families are clearly quite different, as are their overall tertiary structures. However, as both enzyme families catalyze the hydrolysis of peptide bonds, there must be structural and chemical features that are analogous, which is indicative of the convergent evolutionary pressures that have given rise to enzymes that carry out similar functions on the same substrates.

There are structural similarities on a gross level, in that the residues directly involved with the catalytic event are located at the junction of two domains. Thus, Asp 102, His 57, and Ser 195 are located at the junction of the N- and C-terminal domains of the serine proteinases in a manner similar to the disposition of Asp 33(32) and Asp 213(215) that we have discussed. The binding modes of inhibitors, as determined by X-ray crystallographic analyses, provide many interesting analogies. Both families bind polypeptide chains in an extended conformation, with the location of the side chains of adjacent amino acids of the substrate peptides alternating to either side along the polypeptide chain. Both enzyme families have well-defined subsites for the binding of these side chains, thereby conferring specificity and determining how the scissile bond occupies the region next to the electrophile, the nucleophile, and the proton donor.

In Figure 18, the proposed productive substrate-binding modes for both enzyme families are represented. It can be seen that many of the interactions

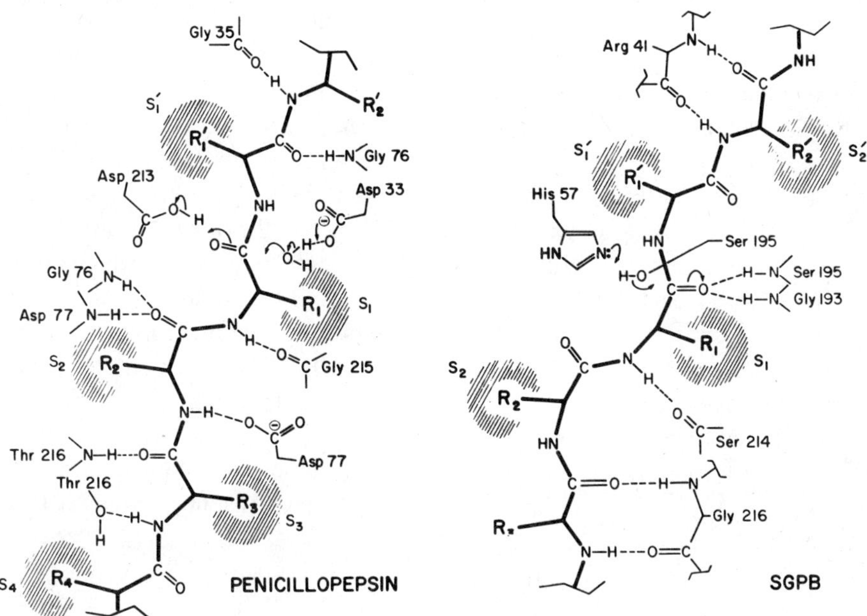

Figure 18. A diagrammatic representation of the binding of substrates in the active site of penicillopepsin, compared to the substrate-binding mode in the serine proteinase SGPB (after James, 1980). The two important hydrogen bonds from the N-H at P_1 and P_2' of the substrate are probably not fully formed in the Michaelis complex, but provide additional stabilization for the transition state. They can only be formed optimally in the tetrahedral intermediate. The hydrogen-bonding interactions at residue P_2 of the aspartic proteinases have no counterparts as they originate in the unique feature of the flap that surrounds the substrate in the E-S complex. Residue numbers refer to the sequential numbering of penicillopepsin. For SGPB, they are based on the α-chymotrypsin numbering according to its sequence alignment with the bacterial serine proteases (Fujinaga et al., 1985).

are similar in both families. For the serine proteinases, the electrophilic center consists of the two hydrogen-bonding interactions to the carbonyl-oxygen atom from Gly 193NH and Ser 195NH. The nucleophile is the Ser 195 alkoxide anion after transferring its proton to the imidazole ring of His 57. The proton donor for the leaving-group nitrogen is His 57. The nucleophilic attack is on the *re* face of the scissile peptide bond. It can be seen in the productive substrate-binding mode shown in Figure 18 that for aspartic proteinases, the nucleophilic attack on the carbonyl-carbon atom is on the *si* face (see also Fig. 17).

It is also of interest to compare the TI generated by the nucleophilic attack of Ser 195 on the scissile peptide to that of the noncovalently bound intermediate in Figure 16. Binding studies of oligopeptide products and a covalently attached aldehyde inhibitor with *Streptomyces griseus* Protease A have led to a proposal for a TI model for the serine proteases (James et al., 1980). This structure is shown in Figure 19. The TI is covalently attached to O^γ of Ser 195. The lone-pair orbitals on O^γ and on the oxygen atom in the oxyanion-binding site are shown in staggered conformations, in which both oxygen atoms each have one lone-pair orbital antiperiplanar to the C-N bond. This configuration would lead to the breakdown of the intermediate into the acyl enzyme and the leaving-group $-NH_2$. The lone-pair electrons on the nitrogen atom are shown in a favorable orientation to accept a proton from the imidazolium group of His 57.

The important hydrogen bond that helps to stabilize the TI from the NH of P_1 Phe to the CO of Ser 214 is formed. This is comparable to the

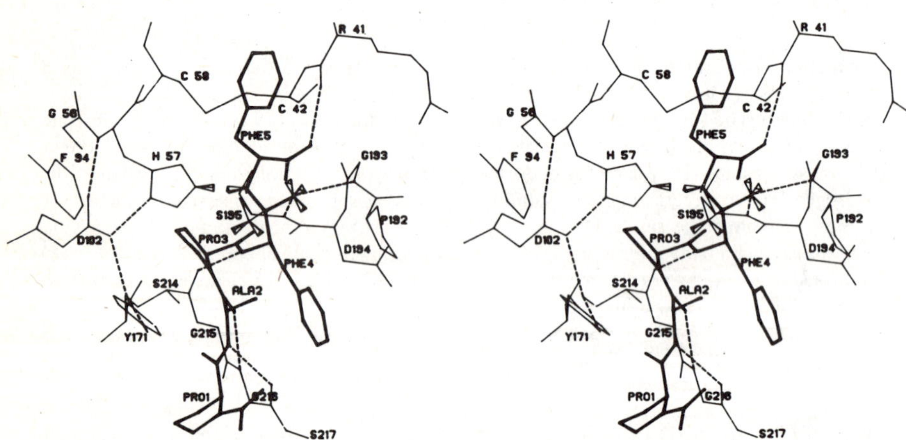

Figure 19. A stereo optic view of the active site of the bacterial serine proteinase SGPA with a proposed model of a tetrahedral intermediate. The hypothetical substrate Ac-Pro-Ala-Pro-Phe-Phe displayed in this representation was derived from the refined coordinates of a tetrapeptide aldehyde inhibitor Ac-Pro-Ala-Pro-PheH bound to SGPA (James et al., 1980). The C-terminal phenylalanine of the substrate was constructed based on the conformation of the P_1' residue of the third domain of the turkey ovomucoid inhibitor bound to SGPB (Fujinaga et al., 1982; Read et al., 1983). Lone-pair electrons on Ser 195 O^γ and on the oxyanion of the intermediate (depicted as elongated pyramids) are antiperiplanar to the C-N bond. This is required by the theory of stereoelectronic control to facilitate hydrolytic cleavage of the scissile bond, following protonation of the leaving-group nitrogen by His 57.

shortening of the hydrogen bond from the P_1 NH to the CO of Gly 215(217) proposed in the case of the aspartic proteinases, as discussed in Section 3.5.

5. COMPARATIVE MOLECULAR MODELING: RENIN

An intense interest in the design of effective renin inhibitors for the regulation of blood pressure has encouraged several groups to construct molecular models of the renin enzyme on the basis of its sequence homology with other aspartic proteinases and the known tertiary structures of the several fungal enzymes (Blundell et al., 1983; Carlson et al., 1983; Sibanda et al., 1984). Other modeling studies have been done, but are not yet published (D. Davies, private communication; this laboratory). All these molecular models of human or mouse renin have been based on the tertiary structures of endothiapepsin, rhizopuspepsin, or penicillopepsin. This is a severe limitation, since there is no evidence for such a high tertiary structural homology among the mammalian and fungal enzymes.

Most authors have emphasized that, at present, the only way of producing an accurate model of a protein molecule is by the solution of its X-ray crystal structure. Experience with the serine proteinase family certainly confirms this fact, especially upon comparison of the sequences and known tertiary structures of the bacterial vs. the mammalian enzymes (McLachlan and Shotton, 1971; Jurášek et al., 1976, Delbaere et al., 1979; Read et al., 1984). For regions of little or no sequence homology, it is presently impossible to predict how a polypeptide could fold, and only general conformations can be suggested (Read et al., 1984). In regions of insertions or deletions, major conformational differences of main chain and side chains can result. Completely misleading speculation on possible enzyme-substrate interactions can be made if these insertions or deletions are in binding regions neighboring the active site.

Several authors have also attempted to improve the predicted structures by minimizing the total energy of the proposed models in order to relieve impossibly close contacts, unfavorable conformations, etc. The fact that the resulting model has a lower energy state is, of course, no guarantee of being closer to the true structure. This is impressively illustrated by the recent study of Novotný et al. (1984). Two completely unrelated proteins, hemerythrin and an immunoglobin VL domain, both with 113 amino acids, were each used as a model of the other. All of the amino acid side chains of the all helical protein hemerythrin were replaced by the side chains of the all β protein immunoglobin. The resultant structures were then subjected to energy minimization. The final values of the total energy for the incorrect models were not significantly different from those of the correct structures. Simple overall potential energy values cannot be used to judge the validity of a predicted structure. Other criteria have to be examined (Novotńy et al., 1984). On the one hand, it is true that this represents an extreme of two proteins with no sequence homology and, of course, no tertiary structural homology; and it is clear that one would not select either of these structures to model the other. However, on the other hand, even in this case that is blatantly incorrect, the energy minimization criterion was unable to differentiate the correct from the incorrect model.

Keeping in mind all of these objections and cautionary caveats, Figure 20 represents a model of the active site of human kidney renin that was constructed from the human renin sequence (Imai et al., 1983) based on the tertiary structure of penicillopepsin. A decapeptide fragment of horse

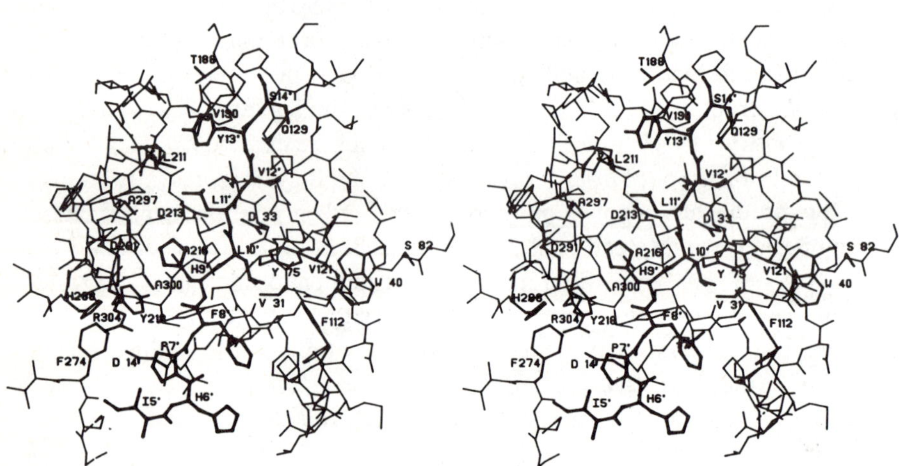

Figure 20. A stereo view of the active site region of a hypothetical model of human kidney renin. This model was constructed from the amino acid sequence of the human enzyme by comparative model building based on the refined structure of penicillopepsin. The amino acid numbering shown refers to the sequential numbering of penicillopepsin. The hypothetical substrate-binding position of a fragment of horse angiotensinogen (Skeggs et al., 1957), from Ile 5 to Ser 14, was based on the pepstatin analog binding shown in Figures 11*a* and 11*b*. The substrate residues displayed correspond to positions P_6 to P_4'.

angiotensinogen (Skeggs et al., 1957) is shown in the proposed binding site. The procedure used is, of necessity, similar for all such constructions, and has been outlined by Blundell et al. (1983) and Sibanda et al. (1984). The model of the angiotensinogen substrate, from P_4 to P_1', was based on the structure of the pepstatin analogs bound in the active site of penicillopepsin (James et al., 1983). Analogous constructions have been recently published for a human renin model based on the endothiapepsin structure, with a portion of the human angiotensinogen molecule (P_4 to P_3' residues) bound (Sibanda et al., 1984). In this case, the suggested mode of binding for residues P_4 to P_1 was deduced by analogy with the inhibitors bound to rhizopuspepsin (Bott et al., 1982) and penicillopepsin (James et al., 1982). Substrate residues P_1' to P_3' were fitted into the remaining half of the cleft. Both substrate-modeling studies assume, therefore, that the productive binding mode for human kidney renin is going to be the same as that deduced for the fungal enzymes penicillopepsin and rhizopuspepsin. Thus, the substrates are shown in an essentially fully extended conformation for the polypeptide chain (Fig. 20). Neither of these studies has attempted to account for how such a conformation would explain the very unusual kinetic data that implicate residues seven to eight amino acids removed from the scissile bond in a marked rate-enhancing effect (Skeggs et al., 1980).

It has been suggested that the requirement for a minimum length of peptide substrate indicates that conformational factors might be responsible for the remarkable specificity of renin (Oliveira et al., 1977). These same workers proposed a β-turn conformation for the fragment His-Pro-Phe-His at positions 6−9 of the N-terminal segment of angiotensinogen. However, a hairpin β bend at positions 6−9 of the substrate cannot be attained if the conformation of the human renin molecule is similar to the fungal enzymes in the region of the "flap." With the absence of structural data for either the angiotensinogen peptide or an alternative conformation for the residues of the "flap," there is no basis to attempt to model build those regions so that such a β bend in the substrate could be accommodated in the active site of renin.

In spite of the relatively highly conserved hydrophobic cores between the fungal enzymes and the hypothetical renin models, there are enough sequence differences in segments close to the active site and substrate-binding regions to suggest that there may be significant conformational differences among these molecules. Blundell et al. (1983) have pointed out that Asp 300(304) of the fungal enzymes and the mammalian gastric enzymes is mutated to an alanine in the renins. They have suggested that the carboxyl group of Asp 300(304) may contribute to lowering the pK_a of the two catalytic aspartic acids in enzymes with a low pH optimum, such as pepsin, although they do not specify how this would be attained. It is tacitly assumed that the renin active site, that is, Asp 33(32) to Ser 36(35) and Asp 213(215) to Thr 216(218), will have the same conformation at its pH optimum of ~7.0 as do the fungal enzymes at the pH optima of ~4 or lower (Hofmann et al., 1984). In human renin, one of the stabilizing hydrogen bonds to Asp 213(215) is lost, as Thr 216(218) has been mutated to an alanine (Imai et al., 1983). Sibanda et al. (1984) suggest that the resulting slight asymmetry at the active site may contribute to a small change of the properties of the catalytic residues, but they fail to say which properties. They do, however, exclude an influence on the different pH optimum between renin and other aspartic proteinases on the basis that in mouse submaxillary and kidney renins, residue 216(218) is a serine and could therefore form an equivalent hydrogen bond with Asp 213(215).

There are other regions of the hypothetical substrate-binding site of renin (Fig. 20) that will be very strongly influenced by amino acid insertions/ deletions or mutations relative to the known fungal enzyme structures. The conformation of the "flap" region [shown from Ile 73(73) to Ser 82(82) in Fig. 20] is influenced by a deletion of one residue at position 79(79), relative to penicillopepsin and a change of sequence from Gly 76(76) and Asp 77(77) to serine and threonine, respectively. This could have a profound influence on the substrate-binding interactions at P_3, P_2, and P_1 (Fig. 20). Similarly, problems with sequence alignment in the segment from 104(103) to 118(117), where the homology between different aspartic proteinases is very low, make structure prediction in this important S_1-binding region very tentative (Sibanda et al., 1984). Third, the tetrapeptide insertion at position 294(294) (Table 1) may have the greatest influence on possible enzyme-substrate interactions in human kidney renin. On the other hand, renin has the same number of residues as pepsin in this segment. However, the human renin sequence Pro-Pro-Pro-Thr-Gly-Pro-Thr, from position 293(293) to 295(299), makes it unlikely that the structure will resemble this region of pepsin, let alone be predicted from the fungal enzyme structures. It can be seen in Figure 20 that this segment of renin could have a strong influence on substrate-binding subsites P_2, $P_1{}'$, and $P_3{}'$.

It is not yet possible to account, in a quantitative manner, for the effects that amino acid changes will have either on protein structure or on various enzyme functions such as pH stability, pH optima, specificity, etc. This knowledge will be extremely important in the design of biological catalysts for specific chemical reactions. Besides, although it appears to be a trivial point whether O39 or O284 is the nucleophilic group for the catalytic pathway, it is of paramount importance in the design of the stereochemistry of a catalytic surface. The direction of nucleophilic attack must be known or at least anticipated.

This discussion has been intended to emphasize the very fundamental problems in comparative molecular modeling. It is not intended to dissuade its practitioners, because certainly in the absence of an enzyme-substrate complex structure for the system in which one is interested, this approach is by far the most rewarding. In the design of an inhibitor, a great many possible chemical syntheses can be rejected even on a crude structural model of the enzyme; many new syntheses can be suggested as well (Boger, 1983).

6. CONCLUDING REMARKS

We have reviewed structural features of the aspartyl proteinases that suggest a pathway for the hydrolysis of oligopeptide substrates. In our view, the most plausible pathway does not involve a covalently attached intermediate. Models of the productive binding modes of substrates to penicillopepsin and to the homologous fungal enzyme rhizopuspepsin have been constructed on the basis of noncovalently bound inhibitors containing statine in place of the scissile bond. These substrate-binding models implicate the proton shared by the carboxyl groups of Asp 33(32) and Asp 213(215) as the electrophilic group. The attacking nucleophile is a water molecule bound in the vicinity of Asp 33(32) (O284 in the solvent numbering of penicillopepsin). The proton donor is not well established, but it is likely to come either from bulk solvent or from the innermost hydroxyl group of the TI in a transfer facilitated by the carboxyl group of Asp 213 (215). The stereochemistry of the groups at the active site throughout this pathway is consistent with the principles of the stereoelectronic effect.

There are several alternatives to the steps outlined above for the aspartic proteinase mechanism. The noncovalent character of the intermediates hamper the unambiguous characterization of the pathway. Knowledge of the chemical nature of solvent O39 would rule out some of these alternatives. Neutron diffraction experiments on sufficiently large crystals of one of the aspartyl proteinases could help in the definitive identification of this site. In addition, the hydrogen bonding at the active site of the native enzyme and its inhibited complexes could be definitively established. The success of such experiments is dependent upon the ability to grow crystals of over 1 mm^3 in size and diffracting to high resolution.

Many possible binding modes of substrate molecules have been examined using the facilities of interactive computer graphics. These studies have led to a relatively small number of alternatives, and it is likely that the true mechanism of hydrolysis is to be found among those discussed herein. Comparison of the most probable productive binding modes of substrates to penicillopepsin with the known virtual substrate and protein inhibitior binding modes to serine proteinases shows a number of structural similarities. These, in turn, suggest that the catalytic pathways for these two enzyme families have common features. The most obvious differences stem from the very different nature of the groups at the active sites that are responsible for the covalency changes experienced by the substrate as hydrolysis proceeds.

ACKNOWLEDGMENTS

Dr. Theo Hofmann has provided purified penicillopepsin and much encouragement over the several years of our collaboration. Dr. Dan Rich has provided the synthetic pepstatin analog inhibitors that we have bound to penicillopepsin. Koto Hayakawa has grown excellent single crystals during the course of this work. We especially thank Masao Fujinaga for his very critical reading of this manuscript and for his many suggestions. The Medical Research Council of Canada has generously supported research in the authors' laboratory during the period in which this work was carried out.

REFERENCES

Andreeva, N. S., and Gustchina, A. E. (1979). *Biochem. Biophys. Res. Commun.* **87**, 32–42.

Andreeva, N. S., Gustchina, A. E., Fedorov, A. A., Shutzkever, N. E., and Volnova, T. V. (1977). In *Acid Proteases, Structure, Function, and Biology,* J. Tang, Ed. Plenum Press, New York, pp. 23–31.

Andreeva, N. S., Fedorov, A. A., Gustchina, A. E., Riskulov, R. R., Shutzkever,

N. E., and Safro, M. G. (1978). *Mol. Biol. Engl. Transl. Mol. Biol. (Mosc)* **12**, 704–716.

Andreeva, N. S., Zdanov, A. S., Gustchina, A. E., and Fedorov, A. A. (1984). *J. Biol. Chem.* **259**, 11353–11365.

Antonov, V. K., Ginodman, L. M., Kapitannikov, Y. V., Barshevskaya, T. N., Gurova, A. G., and Rumsh, L. D. (1978). *FEBS Lett.* **88**, 87–90.

Antonov, V. K., Ginodman, L. M., Rumsh, L. D., Kapitannikov, Y. V., Barshevskaya, T. N., Yavashev, L. P., Gurova, A. G., and Volkova, L. I. (1981). *Eur. J. Biochem.* **117**, 195–200.

Artymiuk, P. J., Blake, C. C. F., Grace, D. E. P., Oatley, S. J., Phillips, D. C., and Sternberg, M. J. E. (1979). *Nature* **280**, 563–568.

Barrett, A. J. (1977). In *Proteinases in Mammalian Cells and Tissues*, A. J. Barrett, Ed. North-Holland, Amsterdam, pp. 209–248.

Barry, C. D., Molnar, C. E., and Rosenberger, F. U. (1976). *Technical Memorandum No. 229*, Computer Systems Laboratory, Washington University, St. Louis. MO.

Bauer, C.-A. (1978). *Biochemistry* **17**, 375–380.

Bizzozero, S. A., and Dutler, H. (1981). *Bioorg. Chem.* **10**, 46–62.

Blundell, T. L., Sewell, B. T., and McLachlan, A. D. (1979). *Biochim. Biophys. Acta* **580**, 24–31.

Blundell, T., Sibanda, B. L., and Pearl, L. (1983). *Nature* **304**, 273–275.

Boger, J. (1983). In *Peptides: Structure and Function*, V. Hruby, and D. Rich, Eds. Proceedings of the 8th American Peptide Symposium, Pierce Chemical Company, Rockford, IL, pp. 569–587.

Bott, R. R. and Davies, D. R. (1983). In *Peptides: Structure and Function*, V. Hruby, and D. Rich, Eds. Proceedings of the 8th American Peptide Symposium, Pierce Chemical Company, Rockford, IL, pp. 531–540.

Bott, R., Subramanian, E., and Davies, D. R. (1982). *Biochemistry* **21**, 6956–6962.

Bürgi, H. B., Dunitz, J. D., and Shefter, E. (1973). *J. Am. Chem. Soc.* **95**, 5065–5067.

Carlson, W., Haber, E., Feldman, R., and Karplus, M. (1983). In *Peptides: Structure and Function*, V. Hruby, and D. Rich, Eds. Proceedings of the 8th American Peptide Symposium, Pierce Chemical Company, Rockford, IL, pp. 821–824.

Chen, K. C. S., and Tang, J. (1972). *J. Biol. Chem.* **247**, 2566–2574.

Chothia, C., and Janin, J. (1982). *Biochemistry* **21**, 3955–3965.

Clement, G. E. (1973). In *Progress in Bioorganic Chemistry*, E. T. Kaiser, and F. J. Kezdy, Eds. Vol. 2. John Wiley & Sons, New York, pp. 177–238.

Clement, G. E., and Snyder, S. L. (1966). *J. Am. Chem. Soc.* **88**, 5338–5339.

Clement, G. E., Rooney, J., Zahkeim, D., and Eastman, J. (1970). *J. Am. Chem. Soc.* **92**, 186–189.

Cornish-Bowden, A. J., and Knowles, J. R. (1969). *Biochem. J.* **113**, 353–362.

Crawford, J. L., Lipscomb, W. N., and Schellman, C. G. (1973). *Proc. Natl. Acad. Sci. USA* **70**, 538–542.

Cunningham, A., Wang, H.-M., Jones, S. R., Kurosky, A., Rao, L., Harris, C. I., Rhee, S. H., and Hofmann, T. (1976). *Can. J. Biochem.* **54**, 902–914.

Delbaere, L. T. J., Brayer, G. D., and James, M. N. G. (1979). *Can. J. Biochem.* **57**, 135–144.

Delpierre, G. R., and Fruton, J. S. (1965). *Proc. Natl. Acad. Sci. USA* **54**, 1161–1167.

Delpierre, G. R., and Fruton, J. S. (1966). *Proc. Natl. Acad. Sci. USA* **56**, 1817–1822.

Denburg, J. L., Nelson, R., and Silver, M.S. (1968). *J. Am. Chem. Soc.* **90**, 479–486.

Deslongchamps, P. (1975). *Tetrahedron* **31**, 2463–2490.

Drenth, J. (1980). *Recl. Trav. Chim. Pays-Bas* **99**, 185–190.

Dunn, B. M., and Fink, A. L. (1984). *Biochemistry* **23**, 5241–5247.

Foltmann, B. (1981). *Essays in Biochemistry* **17**, 52–84.

Fruton, J. S. (1970). *Adv. Enzymol.* **33**, 401–443.

Fruton, J. S. (1976). *Adv. Enzymol.* **44**, 1–36.

Fruton, J. S. (1977). *Proc. Am. Philos. Soc.* **121**, 309–315.

Fruton, J. S., Fujii, S., and Knappenberger, M. H. (1961). *Proc. Natl. Acad. Sci USA* **47**, 759–761.

Fujinaga, M., Read, R. J., Sielecki, A. R., Ardelt, W., Laskowski, M., Jr., and James, M. N. G. (1982). *Proc. Natl. Acad. Sci. USA* **79**, 4868–4872.

Fujinaga, M., Delbaere, L. T. J., Brayer, G. D., and James, M. N. G. (1985). J. Mol. Biol. **184**, 479–502.

Fukumoto, J., Tsuru, D., and Yamamoto, T. (1967). *Agr. Biol. Chem.* **31**, 710–717.

Gripon, J.-C., and Hofmann, T. (1981). *Biochem. J.* **193**, 55–65.

Hagemeyer, K., Fawwal, I., and Whitaker, J. R. (1968). *J. Dairy Sci.* **51**, 1916–1930.

Hartley, B. S. (1960). *Ann. Rev. Biochem.* **29**, 45–72.

Hartley, B. S., and Kilby, B. A. (1954). *Biochem. J.* **56**, 288–297.

Hartsuck, J. A., and Tang, J. (1972). *J. Biol. Chem.* **247**, 2575–2580.

Hendrickson, W. A., and Konnert, J. H. (1980). In *Computing in Crystallography*, R. Diamond, S. Ramaseshan, and K. Venkatesan, Eds. Indian Academy of Sciences, Internatl. Union of Crystallography, Bangalore, India, pp. 13.01–13.23.

Hofmann, T. (1974). *Adv. Chem. Ser.* **136**, 146–185.

Hofmann, T., and Hodges, R. S. (1982). *Biochem. J.* **203**, 603–610.

Hofmann, T., and Fink, A. L. (1984). *Biochemistry* **23**, 5247–5256.

Hofmann, T., Hodges, R. S., and James, M. N. G. (1984). *Biochemistry* **23**, 635–643.

Hsu, I.-N., Delbaere, L. T. J., James, M. N. G., and Hofmann, T. (1977a). *Nature (London)* **266**, 140–145.

Hsu, I.-N., Delbaere, L. T. J., James, M. N. G., and Hofmann, T. (1977b). In *Acid Proteases, Structure, Function, and Biology*, J. Tang, Ed. Plenum Press, New York, pp. 61–81.

Huang, W.-Y., and Tang, J. (1972). *J. Biol. Chem.* **247**, 2704–2710.

Huber, R., and Bode, W. (1978). *Acc. Chem. Res.* **11**, 114–121.

Imai, T., Miyazaki, H., Hirose, S., Hori, H., Hayashi, T., Kageyama, R., Ohkubo, H., Nakanishi, S., and Marukami, K. (1983). *Proc. Natl. Acad. Sci. USA* **80**, 7405–7409.

James, M. N. G. (1980). *Can. J. Biochem.* **58**, 251–271.

James, M. N. G., and Williams, G. J. B. (1974). *Acta Cryst.* **B30**, 1249–1257.

James, M. N. G., and Sielecki, A. R. (1983). *J. Mol. Biol.* **163**, 299–361.

James, M. N. G., and Sielecki, A. R. (1985). *Biochemistry,* **24**, 3701–3713.

James, M. N. G., Hsu, I.-N., and Delbaere, L. T. J. (1977). *Nature (London)* **267**, 808–813.

James, M. N. G., Sielecki, A. R., Brayer, G. D., Delbaere, L. T. J., and Bauer, C.-A. (1980). *J. Mol. Biol.* **144**, 43–88.

James, M. N. G., Hsu, I.-N., Hofmann, T., and Sielecki, A. R. (1981). In *Structural Studies on Molecules of Biological Interest,* G. Dodson, J. P. Glusker, and D. Sayre, Eds. Clarendon Press, Oxford, pp. 350–389.

James, M. N. G., Sielecki, A. R., Salituro, F., Rich, D. H., and Hofmann, T. (1982). *Proc. Natl. Acad. Sci. USA* **79**, 6137–6141.

James, M. N. G., Sielecki, A. R., and Moult, J. (1983). In *Peptides: Structure and Function,* V. Hruby, and D. Rich, Eds. Proceedings of the 8th American Peptide Symposium, Pierce Chemical Company, Rockford, IL, pp. 521–530.

James, M. N. G., Sielecki, A. R., and Hofmann, T. (1985). In *Aspartic Proteinases and their Inhibitors,* V. Kostka, Ed. Proceedings of the FEBS Advanced Course No. 84/07, Walter de Gruyter, Berlin, pp. 163–177.

Jencks, W. P. (1975). *Adv. Enzymol.* **43**, 219–410.

Jenkins, J. A., Tickle, I., Sewell, T., Ungaretti, L., Wollmer, A., and Blundell, T. (1977). In *Acid Proteases, Structure, Function, and Biology*, J. Tang, Ed. Plenum Press, New York, pp. 43–60.

Jurášek, L., Olafson, R. W., Johnson, P., and Smillie, L. B. (1976). In *Proteolysis and Physiological Regulation,* D. W. Ribbons, and K. Brew, Eds. Vol. 11. Miami Winter Symposia, Academic Press, New York, pp. 93–123.

Kassell, B., and Kay, J. (1973). *Science* **180**, 1022–1027.

Kitson, T. M. and Knowles, J. R. (1971). *FEBS Lett.* **16**, 337–338.

Knowles, J. R., (1970). *Phil. Trans. Roy. Soc.* **B257**, 135–146.

Kossiakoff, A. A., and Spencer, S. A. (1981). *Biochemistry* **20**, 6462–6474.

Kraut, J. (1977). *Ann. Rev. Biochem.* **46**, 331–358.

Marcinszyn, J., Jr., Hartsuck, J. A., and Tang, J. (1976). *J. Biol. Chem.* **251**, 7088–7094.

Marshall, G. R. (1976). *Federation Proc.* **35**, 2494–2501.

McLachlan, A. D., and Shotton, D. M. (1971). *Nature (London)* **229**, 202–205.

Misono, K. S., Chang, J.-J., and Inagami, T. (1982). *Proc. Natl. Acad. Sci. USA* **79**, 4858–4862.

Moravek, L., and Kostka, V. (1974). *FEBS Lett.* **43**, 207–212.

Navia, M. A., Springer, J. P., Poe, M., Boger, J., and Hoogsteen, K. (1984). *J. Biol. Chem.* **259**, 12714–12717.

Neumann, H., Levine, J., Berger, A., and Katchalski, E. (1959). *Biochem. J.* **73**, 33–41.

Newmark, A., and Knowles, J. R. (1975). *J. Am. Chem. Soc.* **97**, 3557–3559.

Novotńy, J., Bruccoleri, R. E., and Newell, J. (1984). *J. Mol. Biol.* **177**, 567–573.

Oliveira, M. C. F., Juliano, L., and Paiva, A. C. M. (1977). *Biochemistry* **16**, 2606–2611.

Ondetti, M. A., and Cushman, D. W. (1982). *Ann. Rev. Biochem.* **51**, 283–308.

Panthier, J.-J., Foote, S., Chambraud, B., Strosberg, A. D., Corvol, P., and Rougeon, F. (1982). *Nature* **298**, 90–92.

Pearl, L., and Blundell, T. (1984). *FEBS Lett.* **174**, 96–101.

Perlmann, G. E. (1952). *J. Am. Chem. Soc.* **74**, 6308–6309.

Phillips, S. E. V., and Schoenborn, B. P. (1981). *Nature* **292**, 81–82.

Powers, J. C., Harley, A. D., and Myers, D. V. (1977). In *Acid Proteases, Structure, Function, and Biology,* J. Tang, Ed. Plenum Press, New York, pp. 141–157.

Rajagopalan, T. G., Stein, W. H., and Moore, S., (1966). *J. Biol. Chem.* **241**, 4295–4297.

Read, R. J., Fujinaga, M., Sielecki, A. R., and James, M. N. G. (1983). *Biochemistry* **22**, 4420–4433.

Read, R. J., Brayer, G. D., Jurášek, L., and James, M. N. G. (1984). *Biochemistry* **23**, 6570–6575.

Rees, D. C., Lewis, M., and Lipscomb, W. N. (1983). *J. Mol. Biol.* **168**, 367–387.

Rich, D. H., and Sun, E. T. O. (1980). *Biochem. Pharmacol.* **29**, 2205–2212.

Rich, D. H., Sun, E. T. O., and Ulm, E. (1980). *J. Med. Chem.* **23**, 27–33.

Rich, D. H., Salituro, F. G., and Holladay, M. W. (1983). In *Peptides: Structure and Function,* V. Hruby, and D. Rich, Eds. Proceedings of the 8th American Peptide Symposium, Pierce Chemical Company, Rockford, IL, pp. 511–520.

Robertus, J. D., Kraut, J., Alden, R. A., and Birktoft, J. J. (1972). *Biochemistry* **11**, 4293–4303.

Sachdev, G. P., and Fruton, J. S. (1975). *Proc. Natl. Acad. Sci. USA* **72**, 3424–3427.

Sachdev, G. P., Brownstein, A. D., and Fruton, J. S. (1975). *J. Biol. Chem.* **250**, 501–507.

Sampath-Kumar, P. S., and Fruton, J. S. (1974). *Proc. Natl. Acad. Sci. USA* **71**, 1070–1072.

Schechter, I., and Berger, A. (1967). *Biochem. Biophys. Res. Commun.* **27**, 157–162.

Schoenborn, B. P., (1975). *Brookhaven Symp. Biol.* **27**, 1–10.

Sepulveda, P., Marciniszyn, J., Liu, D., and Tang, J. (1975). *J. Biol. Chem.* **250**, 5082–5088.

Sharon, N., Grisaro, V., and Neumann, H. (1962). *Arch. Biochem. Biophys.* **97**, 219–221.

Sibanda, B. L., Blundell, T., Hobart, P. M., Fogliano, M., Bindra, J. S., Dominy, B. W., and Chirgwin, J. M. (1984). *FEBS Lett.* **174**, 102–111.

Sielecki, A. R., Hendrickson, W. A., Broughton, C. G., Delbaere, L. T. J., Brayer, G. D., and James, M. N. G. (1979). *J. Mol. Biol.* **134**, 781–804.

Sielecki, A. R., James, M. N. G., and Broughton, C. G. (1982). In *Computational Crystallography*, D. Sayre, Ed. Clarendon Press, Oxford, pp. 409–419.

Silver, M. S., and Stoddard, M. (1972). *Biochemistry* **11**, 191–200.

Skeggs, L. T., Khan, J. R., Lentz, K., and Shumway, N. P. (1957). *J. Exp. Med.* **106**, 439–453.

Skeggs, L. T., Dorer, F. E., Levine, M., Lentz, K. E., and Kahn, J. R. (1980). In *The Renin-Angiotensin System*, J. A. Johnson, and R. R. Anderson, Eds. Plenum Press, New York, pp. 1–27.

Sodek, J., and Hofmann, T. (1970). *Can. J. Biochem.* **48**, 425–431.

Sodek, J., and Hofmann, T. (1971). *Meth. Enzymol.* **19**, 372–396.

Subramanian, E., Liu, M., Swan, I. D. A., and Davies, D. R. (1977a). In *Acid Proteases, Structure, Function, and Biology*, J. Tang, Ed. Plenum Press, New York, pp. 33–60.

Subramanian, E., Swan, I. D. A., Liu, M., Davies, D. R., Jenkins, J. A., Tickle, I. J., and Blundell, T. L. (1977b). *Proc. Natl. Acad. Sci USA* **74**, 556–559.

Takahashi, M., and Hofmann, T. (1975). *Biochem. J.* **147**, 549–563.

Takahashi, M., Wang, T.-T., and Hofmann, T. (1974). *Biochem. Biophys. Res. Commun.* **57**, 39–46.

Tang, J. (1976). *Trends Biochem. Sci.* **1**, 205–208.

Tang, J., Sepulveda, P., Marciniszyn, J., Chen, K. C. S., Huang, W.-Y., Too N., Liu, D., and Lanier, J. P. (1973). *Proc. Natl. Acad. Sci. USA* **70**, 3437–3439.

Tang, J., James, M. N. G., Hsu, I.-N., Jenkins, J. A., and Blundell, T. L. (1978). *Nature* **271**, 618–621.

Umezawa, H., Aoyagi, T., Morishima, H., Matsuzaki, M., Hamada, H., and Takeuchi, T. (1970). *J. Antibiot.* **23**, 259–262.

Wang, T.-T., and Hofmann, T. (1976). *Biochem. J.* **153**, 691–699.

Wang, T.-T., and Hofmann, T. (1977). *Can. J. Biochem.* **55**, 286–294.

Wang, T.-T., Dorrington, K. J., and Hofmann, T. (1974). *Biochem. Biophys. Res. Commun.* **57**, 865–869.

Workman, R. J., and Burkitt, D. W. (1979). *Arch. Biochem. Biophys.* **194**, 157–164.

Structural Basis
for Catalysis
by Rhodanese

WIM G. J. HOL
Laboratory of Chemical Physics
Department of Chemistry
University of Groningen
Groningen, The Netherlands

CONTENTS

1. THE ENZYME RHODANESE: MECHANISM AND FUNCTION
   1.1.   Reactions Catalyzed
   1.2.   Inhibitors
   1.3.   The pH Dependence
   1.4.   The Function of Rhodanese: A Puzzle for Biochemists

2. THE ENZYME RHODANESE AT THE ATOMIC LEVEL
   2.1.   Molecular Weight and Amino Acid Sequence
   2.2.   The Three-Dimensional Structure of Sulfur-Rhodanese
   2.3.   The Localization of Inhibitors
   2.4.   The Three-Dimensional Structure of Sulfur-Free
          Rhodanese
   2.5.   The Structure of the Complex of Sulfur-Rhodanese with
          Thiosulfate

3. THE CATALYTIC MECHANISM

4. EVOLUTION OF RHODANESE

5. CONCLUSIONS

REFERENCES

1. THE ENZYME RHODANESE: MECHANISM AND FUNCTION

Rhodanese is a challenge to biophysicists and a puzzle to biochemists. Despite its wide distribution in nature and 50 years of investigations, its "true" function remains a puzzle (Westley, 1973, 1977, 1980; Sörbo, 1975; Westley et al., 1983). The challenge is to reconcile its three-dimensional structure and amino acid sequence: the polypeptide chain is folded into two domains that have a very similar structure, but nevertheless display very little sequence homology (Ploegman et al., 1978a).

This review will focus on the intriguing puzzle and imposing challenge raised by this enzyme. We will start, however, with a summary of the properties of rhodanese as revealed by kinetic and chemical investigations.

1.1. Reactions Catalyzed

Lang (1933) reported that certain tissues of higher animals contained an enzyme that catalyzed the following reaction:

$$CN^- + S_2O_3^{2-} \longrightarrow SCN^- + SO_3^{2-} \tag{1}$$

Further investigations, mainly by the groups of Sörbo and Westley, have shown that the enzyme catalyzes the transfer of divalent sulfur from a donor molecule (DS) to an acceptor (A) to form an acceptor-sulfur molecule (AS) according to the general scheme

$$DS + A \longleftrightarrow AS + D.$$

Rhodanese can use alternatives for thiosulfate as sulfur donors. Especially active are thiosulfonates and persulfides. Alternatives for cyanide as an acceptor are sulfite, sulfinates, and certain thiol compounds. A summary of these compounds is given in Table 1. As can be seen, the substrate specificity is quite broad. There is some specificity regarding thiol compounds, as Sörbo (1962) reported that mercaptoethanol and thioglycolate were weakly active as acceptors, whereas cysteamine and glutathione were inactive.

Light was shed on the catalytic mechanism by the elegant isotope exchange studies of Westley and Nakamoto (1962). It appeared from their studies that bovine liver rhodanese carried out reaction 1 in two steps:

$$S_2O_3^{2-} + Rhod \longleftrightarrow SO_3^{2-} + Rhod\text{-}S \tag{3a}$$

$$Rhod\text{-}S + CN^- \longleftrightarrow Rhod + SCN^- \tag{3b}$$

where "Rhod-S" is "sulfur-rhodanese" containing an extra sulfur atom, and "Rhod" is the "sulfur-free" enzyme.

Cannella, et al. (1975a) have provided evidence that instead of thiosulfate, selenosulfate can act as a substrate. In that case, a selenium-rhodanese intermediate is formed, and rhodanese acts as a selenium transferase. As little is known about this reaction, we will focus in the following sections on the sulfur-transferring properties of the enzyme rhodanese.

Table 1. Compounds Active as Rhodanese Substrates

Compound	Structure	References
Acceptors		
Cyanide	CN^-	Lang, 1933
Sulfite	SO_3^{2-}	Sörbo, 1957
Sulfinates	RSO_2^-	Sörbo, 1962
Thiols	RSH	Sörbo, 1962
		Villarejo & Westley, 1963
Donors		
Thiosulfate	SO_3S^{2-}	Lang, 1933
Thiosulfonates	RSO_2S^-	Sörbo, 1953a, 1953c, 1962
		Mintel & Westley, 1966a
		Eriksson & Sörbo, 1967
		Westley & Heyse, 1971
Organic persulfide	$RSSH$	Villarejo & Westley, 1963
Inorganic persulfide	SS_n^-	Sörbo, 1960
Thiocystine	$RSSSR$	Szczepkowski & Wood, 1967
		Abdolrasulnia & Wood, 1979

1.2. Inhibitors

Important information concerning the active site region came from detailed kinetic studies, and specifically from the discovery of three major classes of inhibitors. These studies have been extensively discussed in two excellent reviews (Westley, 1973; Sörbo, 1975), so that only general points have to be presented here.

The first class of inhibitors consists of sulfhydryl reagents (Saunders and Himwich, 1950; Sörbo, 1951; Horowitz and Criscimagna, 1982). Wang and Volini (1968) demonstrated unequivocally that: (1) enzymatic activity is lost when one sulfhydryl group is alkylated or oxidized to a mixed disulfide; (2) thiosulfate could prevent these reactions; and (3) prolonged treatment with thiosulfate fully reversed the inactivation caused by disulfide formation, but not that caused by alkylation. All of these experiments showed that "sulfur-free" rhodanese contains an SH group that is crucial for a proper functioning of the enzyme. In sulfur-rhodanese, this sulfhydryl group is somehow masked.

The second class of inhibitors consists of anions. Davidson and Westley (1965) reported a decrease of velocity at low substrate concentrations in a solution of high ionic strength. Wang and Volini (1973) described ion pair formation of rhodanese not only with the substrate and product anions thiosulfate, cyanide, sulfite, and thiocyanate, but also with such diverse anionic species as acetate, formate, sulfate, and glycinate. Oi (1975) observed inhibition by di- and tricarboxylic acids as well as α-keto acids. Volini et al. (1978b) investigated rhodanese inhibition by copper, nickel, and zinc cyanide complexes. A particularly interesting example of the importance of the cationic region near the active site was provided recently by Horowitz and Criscimagna (1982): sulfur-free rhodanese is alkylated rapidly by iodoacetic acid, but not by iodoacetamide. It is likely, as the authors suggest, that the carboxylate group of the acid interacts with the positive charges of the active site, bringing the reactive part of the molecule close to the essential sulfhydryl group. Iodoacetamide lacks the favorable electrostatic interaction, and this could explain why it reacts much slower with the essential cysteine. The difference in apparent reactivity of these two compounds might be enhanced by the 160-fold excess of phosphate with respect to both alkylating agents used during these experiments. Phosphate ions would occupy the anion-binding site near the essential cysteine and might be displaced more rapidly by iodoacetic acid than by iodoacetamide.

Mintel and Westley (1966b) carried out a series of detailed kinetic experiments, varying ionic strength as well as the dielectric constant of the solution. They concluded that the active site contained *two* positive charges. Chemical evidence for the involvement of one or more lysine residues in the catalytic mechanism was provided by studying the inactivation by pyridoxal 5'-phosphate (Cannella et al., 1975b). Chemical modification by phenylglyoxal (Weng et al., 1978b) implicated an arginine residue. These studies turned out to be correct when the three-dimensional structure of the enzyme became known.

The third class of inhibitors is formed by aromatic compounds. Many of these inhibit rhodanese competitively with respect to thiosulfate. One of these inhibitors is the pyridyl pyridinium ion (Davidson and Westley, 1965). Wang and Volini (1968) observed that aromatic anions, such as benzenesulfonate, are much stronger competitive inhibitors than aliphatic anions. If we also keep in mind that destruction of a single tryptophane residue in rhodanese with *N*-bromo succinimide (Davidson and Westley, 1965) inactivates the enzyme, then it can be quite safely concluded (Westley, 1973) that the active site region contains a hydrophobic region that comprises at least one tryptophane side chain.

Thus, from kinetic and chemical studies, three major features of the active site of bovine liver rhodanese were deduced: (1) an essential sulfhydryl residue; (2) a cationic region with two positive charges; and (3) a hydrophobic region.

1.3.   The pH Dependence

Schlesinger and Westley (1974) carried out a detailed investigation of the sulfur transfer catalyzed by rhodanese. The binding of thiosulfate was shown to be dependent upon a pK' of 9.9, with the protonated enzyme binding the substrate much stronger than the deprotonated form. Such a pK value would correspond well with that of a lysine side chain. The authors also observed that the formation of the sulfur-substituted enzyme is dependent upon a pK' of 6.5, which they ascribe to that of the essential cysteine sulfhydryl.

The reaction of sulfur-rhodanese with cyanide appears to be dependent on pK' values of 5.9 and 9.4. The second-order rate constants for the reactions of sulfur-rhodanese with CN^- are $8.9 \times 10^8 \ M^{-1} \ sec^{-1}$, $2 \times 10^7 \ M^{-1} \ sec^{-1}$ and $< 10^4 \ M^{-1} \ sec^{-1}$, going from the most protonated to the least protonated form. The pK' of 9.4 would again agree well with that of a lysine residue, but the pK' value of 5.9 is much higher than would be expected for a persulfide, according to the authors. What should be pointed out is the very high rate constant for the reaction with cyanide at low pH values—it is close to the diffusion-controlled maximum value.

1.4.   The Function of Rhodanese: A Puzzle for Biochemists

As noted in the opening paragraph of this chapter, rhodanese occurs in numerous living organisms. It is observed in plants, bacteria, fungi, and animals (e.g., see Sörbo, 1975 and Westley, 1973 for references). In mammals, the highest concentrations are found in liver and kidney (Lang, 1933; Reinwein, 1961). In the liver of birds and mammals, rhodanese occurs in the mitochondrial matrix, as shown by cell fractionation studies (De Duve et al., 1955; Dudek et al., 1980). However, in fish, amphibia, and reptilia, the majority of rhodanese was found in the cytosol (Dudek, et al, 1980). Numerous suggestions have been made as to the crucial importance of this enzyme for living organisms (Westley, 1973, 1977, 1980; Sörbo, 1975; Westley, et al., 1983), but the issue has not yet really been settled. Three of the more frequently made proposals are the following.

(1) *Rhodanese is essential for cyanide detoxification.* This is suggested by the catalyzed reaction 1. Also, the location, in the mitochondria of the liver in mammals, makes some sense because cyanide exerts its primary biochemical effect by reversible inhibition of cytochrome oxidase. Moreover, cyanide intake for which a detoxifying mechanism would have evolved enters mostly by means of cyanogenic glycosides in plants, such as cassava, commonly used for food. Therefore, the presence of high concentrations of rhodanese in the liver correlates if detoxification is the primary function of the enzyme.

Westley (1980) has made several objections against an easy acceptance of detoxification being the physiological role of rhodanese. One of the points made was that the central nervous system is the major site of attack of cyanide poisoning, and the rhodanese content of nerve tissue in general is low compared with liver and kidney. This argument has to be used with care, of course, as Westley pointed out, because rhodanese seems to be well located to cope with cyanide poisoning caused by food intake, as mentioned above.

Another publication questioning the importance of rhodanese for cyanide detoxication has appeared recently (Long and Brattsten, 1982). Observing that, in insects, there is no apparent correlation between rhodanese activity and accentuated feeding on highly cyanogenic foliage, the authors conclude that rhodanese does not play a prominent role in the detoxification of dietary cyanide. As this conclusion is based on studies of insects, it may be worthwhile to mention quite different suggestions based on studies of higher animals. Dudek et al. (1980) discovered that in the process of vertebrate evolution, rhodanese in the liver greatly increased in activity and underwent translocation from the soluble cytoplasm into the mitochondrial compartment. This might be related to the organization of the electron transport chain and to increased importance of oxidative phosphorylation in higher animals (Dudek et al., 1980), and hence to the need for greater protection.

(2) *Rhodanese is involved in the formation of iron-sulfur centers in iron-sulfur proteins.* This has been investigated by reconstitution experiments on ferredoxins (Finazzi-Agrò et al., 1971; Bonomi et al., 1977a; Pagani et al., 1982), succinate dehydrogenase (Pagani et al., 1975; Bonomi et al., 1977b), NADH dehydrogenase (Pagani and Galante, 1983), and xanthine oxidase (Nishino et al., 1985). Although the experimental data undoubtedly demon-

strate "sulfur insertase" activity of rhodanese, these observations are hard to reconcile with the uneven distribution of rhodanese in the various organs of higher animals. Nevertheless, it might be that the sulfur restoration capability is a secondary function of rhodanese at those places where it is already present for carring out its detoxifying activity.

(3) *Rhodanese is essential for maintaining an adequate level of sulfane sulfur in the organism* (Sulfane sulfur is, for example, the "outer" sulfur of thiosulfate persulfides and thiosulfonates, and is present also in the eight-membered ring of elemental sulfur.) This role of rhodanese has been advocated (Westley, 1977, 1980; Westley, et al., 1983) and, in fact, includes the possibility that rhodanese indirectly contributes to cyanide detoxification and (re)construction of iron-sulfur complexes by ensuring a sufficient concentration of sulfane sulfur. Other enzymes involved in maintaining the pool of sulfane sulfur are 3-mercapto pyruvate sulfurtransferase and thiosulfate reductase, while serum albumin is the proposed carrier of sulfane sulfur from its site of formation, for example, the liver, to the peripheral tissues (Westley, 1977, 1980; Westley et al., 1983). These proposals are highly interesting, but given the scope of this article, the reader is referred to Westley's papers for further information.

2. THE ENZYME RHODANESE AT THE ATOMIC LEVEL

Having described some catalytic properties and the questions regarding the physiological role of rhodanese, we now turn to its molecular properties, which are known down to the atomic level. The amino acid sequence of bovine liver rhodanese has been determined by Heinrikson and collaborators in Chicago, while the Groningen protein crystallography group elucidated the three-dimensional structure of sulfur-rhodanese and of a number of rhodanese-inhibitor complexes.

2.1. Molecular Weight and Amino Acid Sequence

Heinrikson and collaborators (Weng et al., 1978a; Russell et al., 1978) established the complete amino acid sequence, which is depicted in Figure 1. The molecular weight turned out to be 32,900, which is in good agreement with the early work of Sörbo (1953b), who obtained a value of 37,000 by

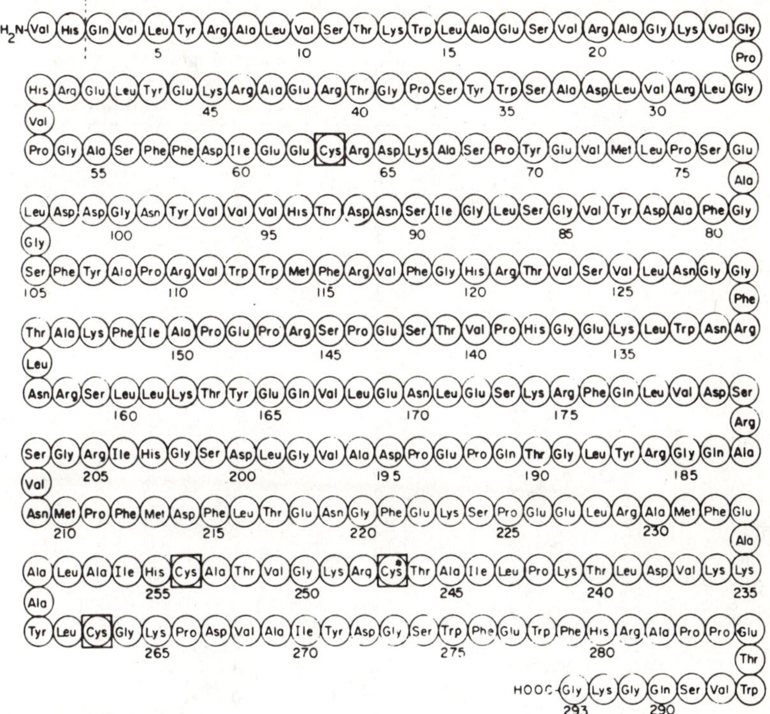

Figure 1. The complete amino acid sequence of bovine liver rhodanese as determined by Heinrikson and collaborators. Cysteines are boxed, and the essential Cys 247 is indicated with an asterisk. (Reprinted by permission from Ploegman et al., 1978a. Copyright Macmillan Journals Limited.)

ultra centrifugation, and with several other investigations (Trumpower et al., 1974; Ellis and Woodward, 1975; Russell et al., 1975; Bergsma et al., 1975). Bovine liver rhodanese contains 293 residues with four cysteine residues, of which Cys 247 is crucial for the catalytic activity. The amino acid composition is shown in Table 2. The sequence was readily correlated with the electron density map obtained by crystallographic investigations (Ploegman et al., 1978a, 1978b).

The bovine liver enzyme, so far, is the only rhodanese whose amino acid sequence has been determined. Rhodanese from vervet monkey liver was recently reported to have a molecular weight of 37,000 (Janse van Rensburg and Schabort, 1984a), and is also probably closely related to the bovine

Table 2. The Amino Acid Composition of Bovine Liver Rhodanese

	Total	Domain I Total[a]	Domain II Total[b]	Homologous Parts[a]		Identities[d]
				Domain I	Domain II	
Glycine	25	14	11	11	7	3
Alanine	23	10	11	8	9	1
Valine	25	17	8	13	7	1
Leucine	25	11	13	9	12	2
Isoleucine	7	2	4	2	4	—
Cysteine	4	1	3	1	3	—
Methionine	5	2	3	1	3	—
Histidine	8	5	3	3	3	—
Phenylalanine	15	7	7	7	6	—
Tyrosine	11	7	4	4	4	1
Tryptophane	8	5	3	4	3	1
Proline	18	7	8	4	3	—
Serine	21	12	8	10	7	2
Threonine	13	5	7	4	5	—
Asparagine	8	4	3	4	2	—
Glutamine	6	1	5	0	3	—
Aspartic acid	14	7	7	4	5	1
Glutamic acid	22	9	11	7	8	1
Lysine	15	5	9	4	8	—
Arginine	20	11	7	9	7	—
Totals	293	142	135	109	109	14

[a]Domain I comprises residues 1–142.
[b]Domain II comprises residues 159–293.
[c]Homologous parts as aligned in Figure 7.
[d]The "boxed" identical pairs in Figure 7.

enzyme in view of similar kinetic characteristics (Janse van Rensburg and Schabort, 1984b).

2.2. The Three-Dimensional Structure of Sulfur-Rhodanese

2.2.1. Structure Determination

Crystals of the bovine liver enzyme were obtained from ~$2M$ ammonium sulfate solutions, pH 7.3 and 1 mM in $Na_2S_2O_3$ (Drenth and Smit, 1971). The space group was C2 with cell dimensions: a = 156 Å, b = 49 Å, c = 42.2 Å, and β = 98.5°. There is one molecule of ~33,000 daltons per asymmetric unit.

With heavy atom derivatives obtained by soaking in 2 mM para-chloromercury (II) benzensulfonic acid (PCMS) and 5 mM $K_2Pt(CN)_4$, a 3.9 Å electron density map was obtained that allowed a chain tracing (Smit et al., 1974) that later turned out to be essentially correct. With an additional uranyl acetate derivative, Bergsma et al. (1975) extended the resolution to 3 Å and firmly established that bovine liver rhodanese consists of a single polypeptide chain folded into two domains of very similar three-dimensional structure.

In 1978, the amino acid sequence became known and a 2.5 Å resolution electron density map was calculated, for which six heavy atom derivatives were used for phase determination: 2 mM PCMS, 5 mM $K_2Pt(CN)_4$, ~2mM $UO_2(Ac)_2$, 10 mM $NaReO_4$, 5 mM $NaAu(CN)_2$, and 0.5 mM 2,6 dichloro-mercury-4-nitrophenol (Ploegman et al., 1978a, 1978b). The resultant electron density distribution was of excellent quality. The entire polypeptide chain was clearly visible, with the exception of the first and last two residues, which are probably mobile or disordered, or both. It was clear that the structure of sulfur-rhodanese had been solved, as density for the extra sulfur atom was present near the S^γ of Cys 247 (Fig. 2). Subsequently, data from sulfur-rhodanese crystals were collected to 2.1 Å resolution and refined by restrained least-squares methods to an R factor of 19.1% (Lijk et al., unpublished results).

2.2.2. The Structure of Bovine Liver Rhodanese

The 293 residues of this mitochondrial sulfurtransferase are folded into a compact ellipsoid, with approximate dimensions of 60 × 50 × 40 Å. The molecule consists of two domains of equal size that are folded very similarly and are related by a pseudo twofold axis. A loop of 16 residues, fully exposed to the solvent, connects the two domains. Domain I consists of residues 1−142 and domain II of residues 159−293. A highly schematic representation of the course of the polypeptide chain is presented in Figure 3, which also indicates the secondary structure elements. A stereo picture, viewed along the pseudo dyad, is given in Figure 4.

The core of each domain is a 5-stranded parallel pleated sheet flanked by two helices on one side and three on the other. Two of the helices in each domain run antiparallel to the neighboring β strands, giving rise to right-

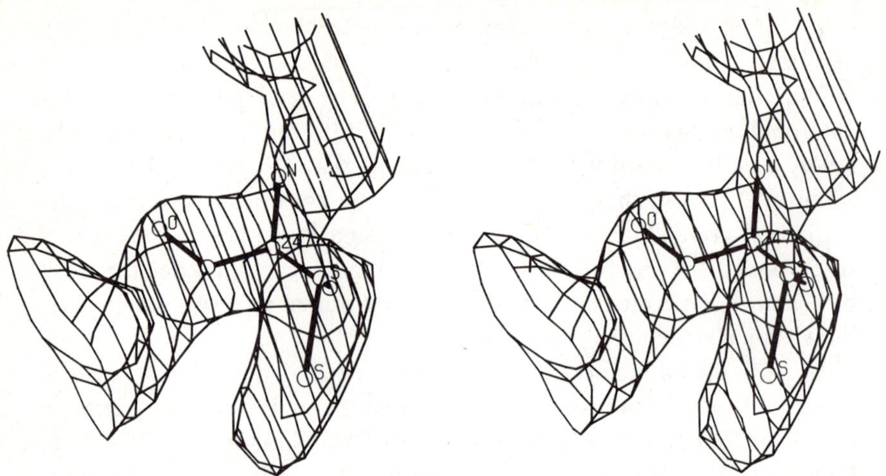

Figure 2. The electron density of Cys 247 in the 2.5 Å multiple isomorphous electron density map, clearly showing that in sulfur-rhodanese, the extra sulfur atom occurs as a persulfide (Reprinted by permission from Ploegman et al., 1978b. Copyright Academic Press, Inc., London.)

handed $\beta\alpha\beta$ units. In total, rhodanese thus contains four such folding units. This can be seen in Figures 3 and 4, and also in the beautiful drawings of the two rhodanese domains by Jane Richardson (Fig. 5).

A listing of the positions of the α helices and β strands are given in Table 3. An analysis of turns has also been carried out (Ploegman et al., 1978a). There are 19 turns, each composed of four residues, where the polypeptide chain reverses its course by almost 180°. All turns occur at the surface of the molecule, except those beginning at residues 105 and 271.

In summary, the X-ray analysis revealed a well-defined polypeptide chain with two domains of the α-β type. It also established that the structure solved is that of sulfur-rhodanese. We will look first at the relationship of the two domains and, second, at the active site region.

2.2.3. *The Two Domains: A Challenge for Biophysicists*

The structures of the two domains are very similar indeed, as is immediately obvious from Figure 6. For 117 C^α pairs, a deviation of 1.95 Å (r.m.s.) was obtained, whereas the total numbers of C^α atoms in domains I and II are 142 and 135, respectively (Ploegman et al., 1978b). In view of this close structural relationship, one might have expected an obvious correspondence in

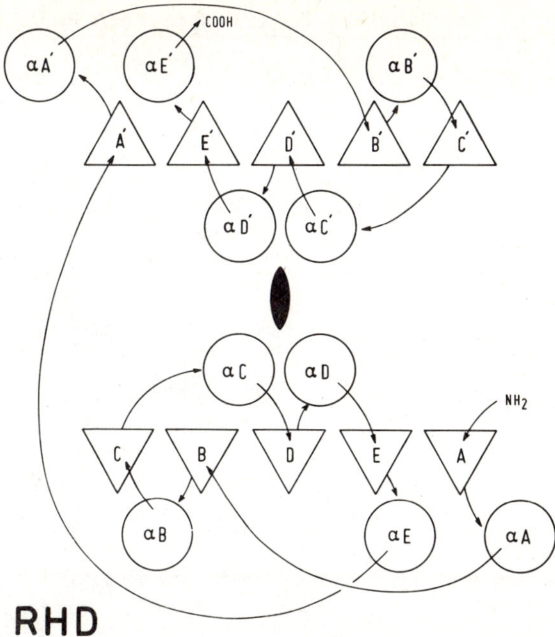

Figure 3. Schematic representation of the tertiary structure of bovine liver rhodanese. α helices are indicated with circles, β strands with triangles. Lower part, domain I; upper part, domain II. The symbol ● represents the pseudo two-fold rotation axis relating the two domains.

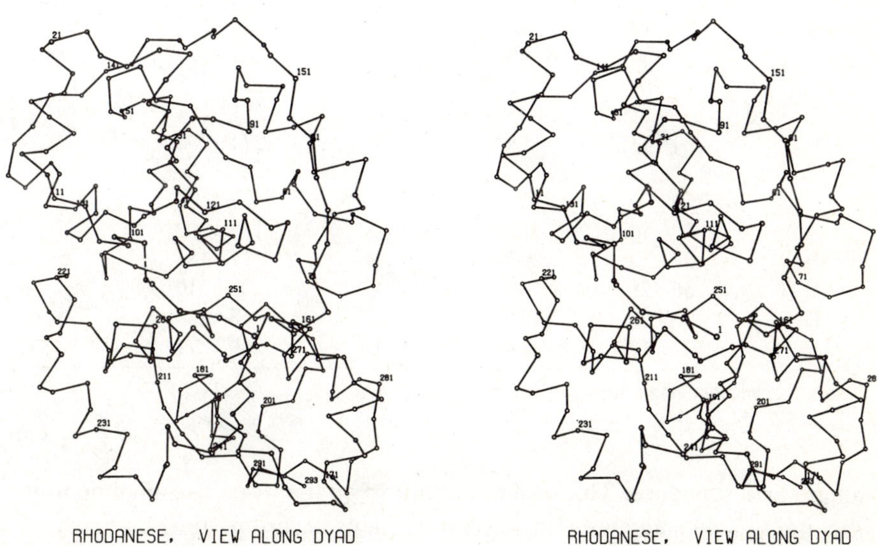

Figure 4. The course of the polypeptide chain of bovine liver rhodanese in stereo. Upper half, domain I; lower half, domain II. (Reproduced by permission, from Ploegman et al., 1978b. Copyright Academic Press, Inc., Ltd., London.)

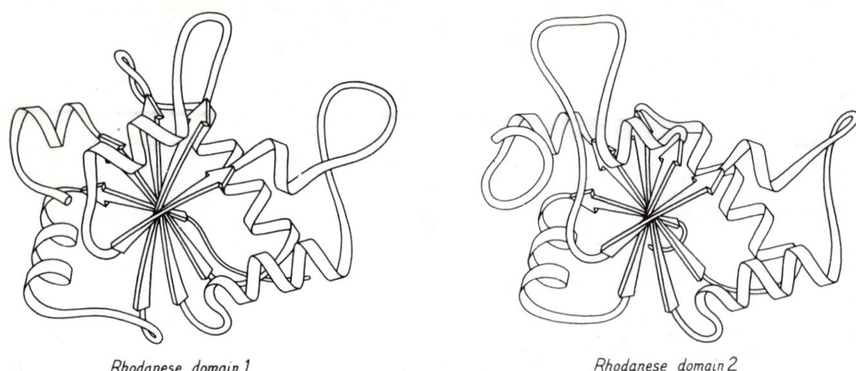

Figure 5. The two domains of rhodanese in similar orientation side by side. Arrows indicate β strands. (Reproduced by permission from Richardson, 1981. Copyright Academic Press, Inc., New York.)

Table 3. Secondary Structure Elements in Bovine Liver Rhodanese

Identifier	Domain I Residues	No. of Residues	Identifier	Domain II Residues	No. of Residues
		Helices			
A	11−22	12	A′	163−174	12
B	42−50	9	B′	183−189	7
C	76−87	12	C′	224−235	12
D	107−119	13	D′	251−264	14
E	129−137	9	E′	274−282	9
		β Strands			
A	8−10	3	A′	160−162	3
B	27−33	7	B′	177−181	5
C	56−58	3	C′	208−210	3
D	94−98	5	D′	242−246	5
E	122−127	6	E′	269−271	3

Source: From Ploegman et al. (1978b)

amino acid sequence. This turned out not to be the case. The alignment of the amino acid sequences of the two domains according to the structural superposition is shown in Figure 7. There are only 14 identical residues at equivalent positions, which is a mere 13%. The minimum base change per

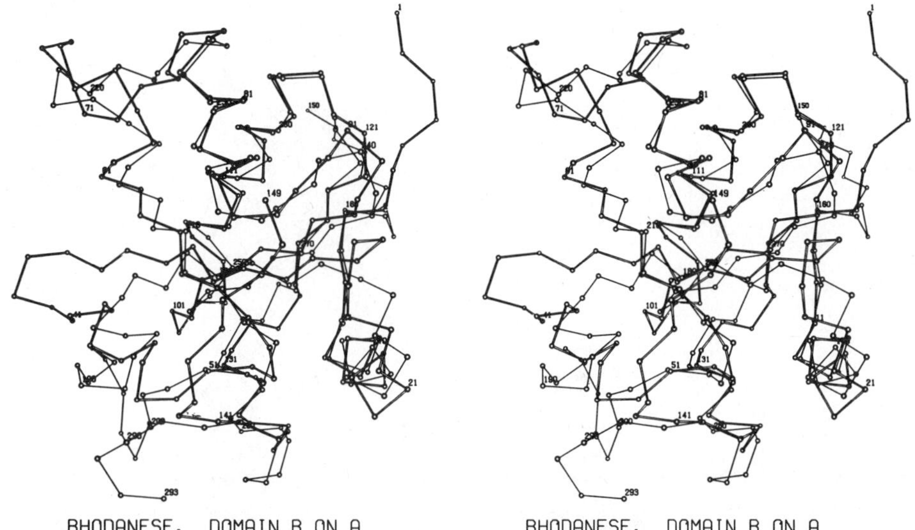

Figure 6. Stereo picture of the C^α atoms of domain I and domain II superimposed. Single bonds connect atoms of domain II, double lines connect those of domain I. (Reproduced by permission from Ploegman et al., 1978b. Copyright Academic Press, Inc., Ltd., London.)

codon between equivalent residues is 1.27, which is close to random. The amino acid composition of the two domains and their homologous parts are given in Table 2. It appears that 3 of 11 glycines in domain II are conserved in domain I. The dissimilarity of the sequence in three dimensions is shown in Figure 8, where a cluster of hydrophobic residues in one domain is superimposed onto the related cluster in the other domain. Although the two equivalent tryptophanes (Trp 133 and Trp 278) have their ring systems virtually superimposed, the impression is one of great disversity in the nature and size of side chains.

It is because of this dissimilarity that rhodanese is a challenge for theoretical biophysicists: what are the key features in the two sequences that cause these two halves of the molecule to fold in such a strikingly similar manner? Ploegman et al. (1978b) were unable to find an answer to this question. Keim et al. (1981) have carried out a detailed statistical analysis of the amino acid sequence. These authors conclude that there is just a statistically significant similarity in amino acid sequence of the structurally equivalent

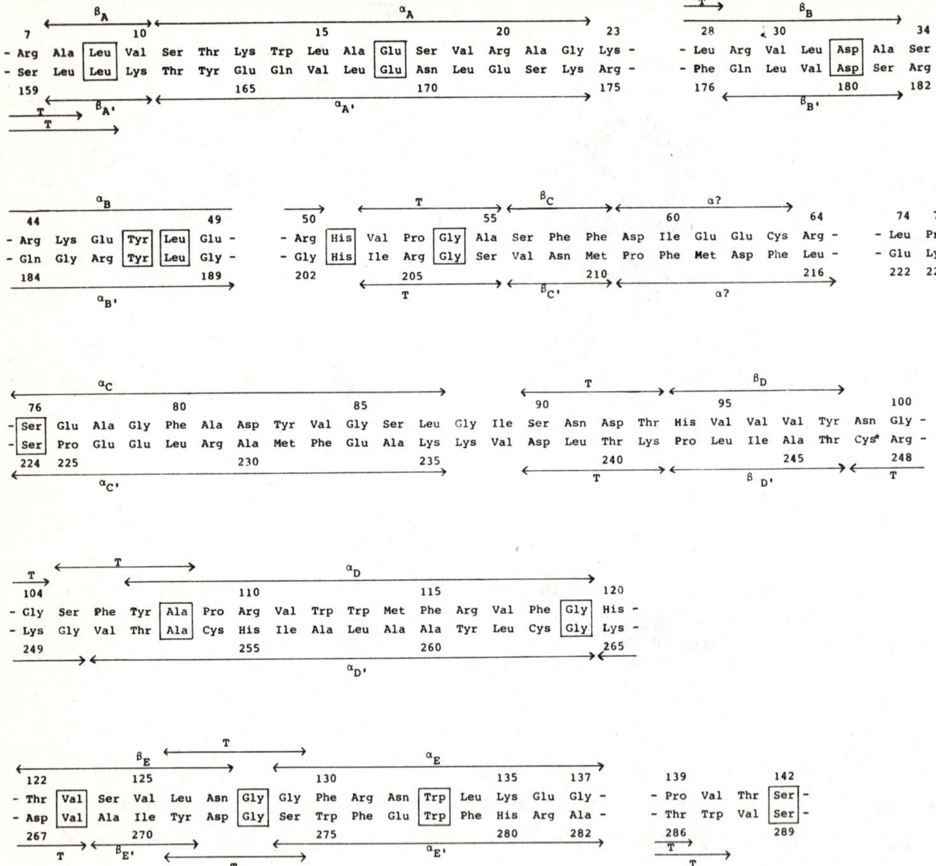

Figure 7. Alignment of the amino acid sequences of the two domains of bovine liver rhodanese according to the results of the tertiary structure superposition. Secondary structure elements, as defined in Table 2 and depicted schematically in Figure 3, are indicated. Identical residues occupying equivalent positions are boxed.

parts of the two domains. However, if science claims a deep understanding of protein structures, then it should arrive at the structural similarity of the two domains starting from the amino acid sequence *only*.

2.2.4. *The Active Site*

The active site is located near the interface of the two domains. The extra sulfur atom in sulfur-rhodanese is clearly visible in the electron density map

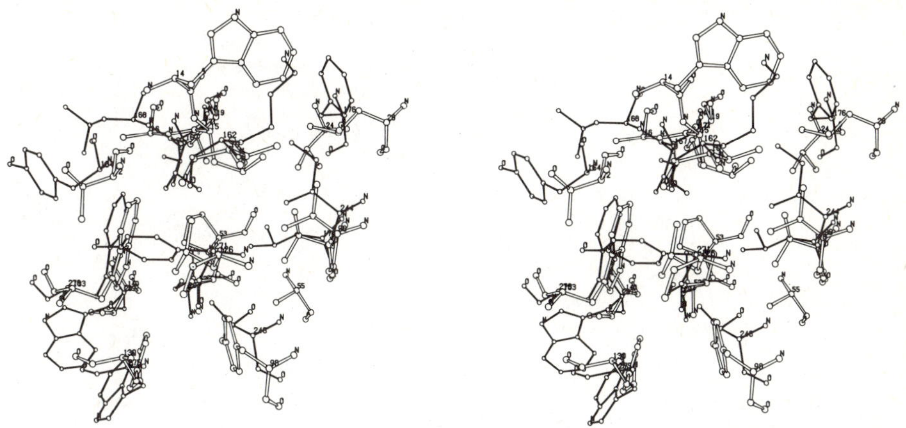

Figure 8. Superposition of hydrophobic clusters in domains I and II. Atoms from domain I are connected by double bonds and those from domain II by single bonds. The rotational and translation operations used were obtained by superimposing equivalent C^α atoms. (Reproduced by permission from Ploegman et al. 1978b. Copyright Academic Press Inc., Ltd., London.)

(Fig. 2), and apparently forms a persulfide bond with S^γ of Cys 247. This is in excellent agreement with fluorescence studies of Finazzi-Agrò et al. (1972). Definite proof of the density in Figure 2 being that of a cysteine-persulfide was obtained by treating sulfur-rhodanese crystals with cyanide; the density for the extra sulfur atom disappeared (Ploegman et al., 1979; Section 2.4.1).

Many investigators have noted that the sulfur-rhodanese intermediate of the catalytic cycle is very stable, provided that no sulfur acceptors are present in the solution. Why would this persulfide have such unusual properties? The three-dimensional structure suggests that this is due to a large number of hydrogen bonds, most of which occur between peptide NH and sulfur atoms (Table 4). Although it may be that the energy gained by forming a single N-H...S or OH...S hydrogen bond is rather small, a total of up to seven such bonds may lead to a considerable increase of persulfide stability.

The essential thiol group is located at the bottom of a rather narrow pocket. Two positively charged residues, Lys 249 and Arg 186, are located at the entrance of this pocket (Fig. 9). The lysine residue does not interact with nearby charges. Arg 186, however, is interacting strongly with the

Table 4. Hydrogen Bonds with the Persulfide of Cys 247 in Sulfur Rhodanese

Involving the "Persulfide Sulfur Atom"

Residue	Atom	Distance (Å)
Arg 248	Peptide N	3.2
Lys 249	Peptide N	3.0
Val 251	Peptide N	2.9
Thr 252	Peptide N	3.6
Thr 252	$O^{\gamma1}$	3.6

Involving the S^{γ} Atom

Gly 250	Peptide N	2.9
Ser 274	O^{γ}	3.1

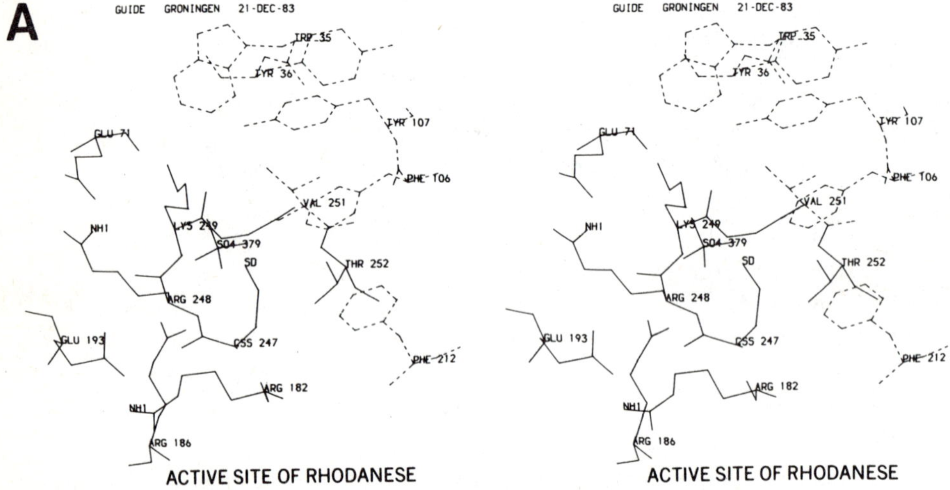

Figure 9. The active site of bovine liver rhodanese. Cys 247 is the essential cysteine shown with an extra S^{δ} (indicated as "SD") sulfur atom in persulfide linkage to S^{γ}, as observed in the structure of sulfur rhodanese. This persulfide is stabilized by a large number of hydrogen bonds (Table 4). Residues Trp 35, Phe 106, Tyr 107, and Phe 212 form part of the hydrophobic area near the surface of the active site. These residues are shown with dotted lines. Lys 249 and Arg 186 provide two positive charges at the entrance of the active site pocket. Arg 186 is involved in an extensive salt bridge network shown also in Figure 10. "SO₄ 379" is the sulfate ion bound at the active-site entrance. (Reprinted, by permission from Lijk et al., 1984. Copyright European Federation of Biochemical Societies.)

carboxyl group of Glu 193, which, in its turn, is interacting with two more guanidinium groups. In fact, these residues form part of a quite remarkable salt bridge network, which is shown schematicaly in Figure 10. This figure also shows that not a trace of this network is present in domain I. We will return to the importance of Lys 249 and Arg 186 when discussing the mode of action of several inhibitors (Section 2.3) and the catalytic mechanism (Section 3). It should be pointed out, however, that these crystallographic results are in perfect agreement with the deduction of two positive charges near the active site by Mintel and Westley (1966b), and with chemical modifications studies implicating a lysine (Cannella et al., 1975b) and an arginine residue (Weng et al., 1978b) as essential for catalytic activity.

While "half" of the active site pocket may be dominated by the hydrophilic residues just described; the other "half" is entirely different in nature and consists of a number of hydrophobic residues in close proximity to each other: Trp 35, Phe 106, Tyr 107, Phe 212, and Val 251 (Fig. 9). This is in good agreement with the results of kinetic investigations (Section 1.2), and it seems likely that reaction with Trp 35 is the crucial event of inactivation of rhodanese by N-bromosuccinimide (Davidson and Westley, 1965). The function of this hydrophobic region for carrying out the best-studied reaction, that of thiosulfate with cyanide, is entirely unclear, as both of these substrates as well as the resultant products are very hydrophilic anions.

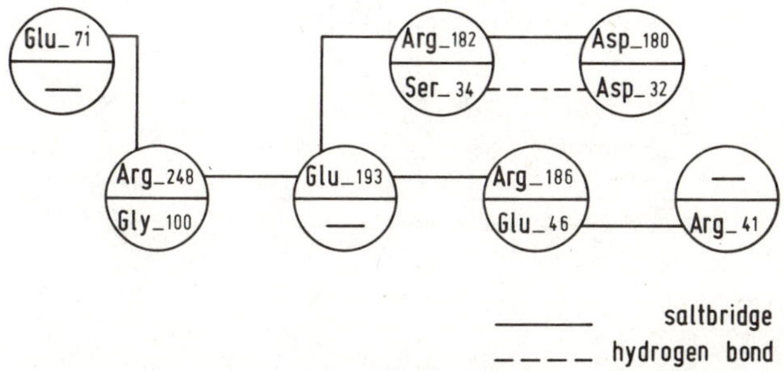

Figure 10. Schematic representation of the salt bridge network at the active site of bovine liver rhodanese (upper half of the circles). In the lower half of each circle, the equivalent residues in the other domain are indicated. A hyphen means that no equivalent residue exists. It is obvious that none of the ionic interactions of the salt bridge network has a counterpart in the other domain.

The essential cysteine is situated close to the N-terminus of the long helix $\alpha_{D'}$. It is also the N-terminus of helix $\alpha_{E'}$, which is elongated by the peptide units of residues 247−249 (Fig. 11). It is well known that the α helix carries a considerable dipole moment (Wada, 1976), which can be represented by placing half a positive unit charge near the N-terminus of the helix and half a negative unit charge near the C-terminus (Hol et al., 1978). Consequently, the essential cysteine is close to the N-termini of both helices, and it is likely that this is a major contributing factor to the unusual low pK of 6.5 of the essential Cys 247 (See Section 1.3). It is unlikely that charged residues contribute to the lowering of the pK of this sulfhydryl group, as the nearest charge is that of the buried Asp 180 at a distance of 6 Å from the S^γ atom (Ploegman et al., 1979).

Having described some general molecular features and the active site, we will now look first at crystallographic studies of complexes of bovine liver rhodanese with inhibitors, and then turn to the catalytic mechanism.

2.3. The Localization of Inhibitors

As discussed in Section 1.2, numerous inhibitors of rhodanese have been discovered, and we would like to have detailed structural information about

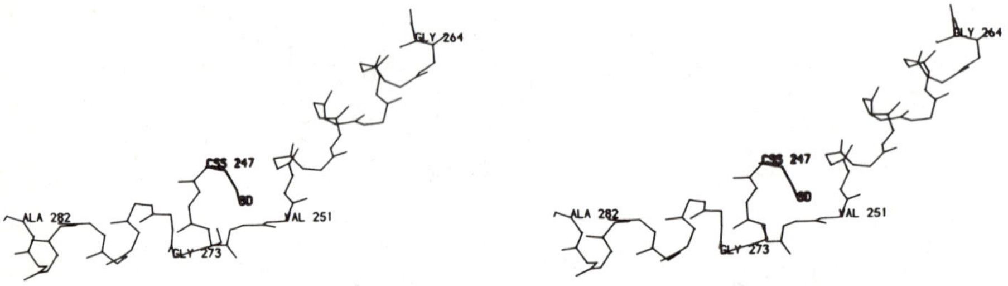

Figure 11. Two helices pointing with their N-termini towards the essential sulfhydryl group of Cys 247. Residues 251−264 are from helix $\alpha_{D'}$, while helix $\alpha_{E'}$ (residues 273−282) is "extended" by the peptide units 247−249. For clarity, only main chain atoms are shown, except for Cys 247. (Reprinted, by permission from Lijk et al., 1984. Copyright European Federation of Biochemical Societies.)

the mode of action for each of these. This is, of course, not the current situation, but we shall see that a few cases are now well understood.

2.3.1. Metalcyanides

Volini et al. (1978b) reported the inhibition of rhodanese by several metalcyanides, and this has been followed up by crystallographic investigations (Lijk et al., 1983). In fact, $Au(CN)_2^-$ and $Pt(CN)_4^{2-}$ had served as heavy atom derivatives in the course of the structure determination (see Section 2.2.1). X-ray data were collected for two additional complexes: $Ni(CN)_4^{2-}$ and $Zn(CN)_4^{2-}$. Pertinent data for the four rhodanese metalcyanide complexes are provided in Table 5.

It appears that all four metal complexes bind to rhodanese in a characteristic manner at the active site entrance, close to the positive charges of Lys 249 and Arg 186, thereby explaining the mode of inhibition by these compounds. The position of the metalcyanide near the active site entrance was virtually identical for all four metals, and was also the same for sulfur-rhodanese and sulfur-free rhodanese. Additional binding sites occur for the gold and zinc complexes that can be explained by the different affinities of these metals for other ligands than cyanide (Table 5; Lijk et al., 1983). It is unlikely, however, that these additional sites are of importance for the inhibition observed.

The difference electron density of the gold cyanide complex (Fig. 12) turned out to be intriguing, as it showed an additional gold atom bound deep in the active site pocket, forming a complex with S^γ of Cys 247. Evidently, the extra sulfur atom of sulfur-rhodanese had been removed, although in this case the crystals had not been treated with cyanide prior to adding $Au(CN)_2^-$. A likely explanation is that: (1) cyanide was liberated when $Au(CN)_2^-$ formed complexes with two nonessential sulfhydryl groups, those of Cys 63 and Cys 263; (2) these cyanide ions removed the persulfide sulfur from Cys 247; and (3) $Au(CN)_2^\theta$ then reacted with the thiol group of Cys 247, forming a covalent Au-S bond. It should be pointed out that the presence of even two gold ion complexes in the active site does not lead to significant conformational changes of the enzyme.

2.3.2. Sulfate and Selenate

Many anions are inhibitors of bovine liver rhodanese (Section 1.2), and the mode of inhibition of metalcyanides has been described in the previous section. One of the other inhibitors is sulfate, and as the rhodanese crystals

Table 5. Metal–Cyanide Binding to Bovine Liver Rhodanese

	Added to Crystals of[a]	No. of Crystals Used for Data Collection	No. of Independent Measurements	Resolution (Å)	Major Binding Sites[b]
Na[Au(CN)$_2$], 5 mM	Sulfur-rhodanese	11	4094	3.2	"Arg 186 & Lys 249," Cys 247, Cys 63 and Cys 263
K$_2$[Pt(CN)$_4$], 10 mM	Sulfur-rhodanese	9	2963	3.9	"Arg 186 & Lys 249"
K$_2$[Ni(CN)$_4$], 10 mM	Sulfur-free Rhodanese	6	9443	2.5	"Arg 186 & Lys 249"
K$_2$[Zn(CN)$_4$], 10 mM	Sulfur-free Rhodanese	5	2043	4.5	"Arg 186 & Lys 249," Asp 101, His 203

From Lijk et al. (1984).
[a]Crystals of sulfur-free rhodanese had been prepared by soaking sulfur-rhodanese crystals for 2 × 30 min. in a solution of 50 mM KCN, 2.0 M(NH$_4$)$_2$SO$_4$ followed by 2 ×30 min. in 2.0 M(NH$_4$)$_2$SO$_4$
[b]"Arg 186 & Lys 249" indicates the common position at the entrance of the active site pocket, that is, the "general anion-binding site."

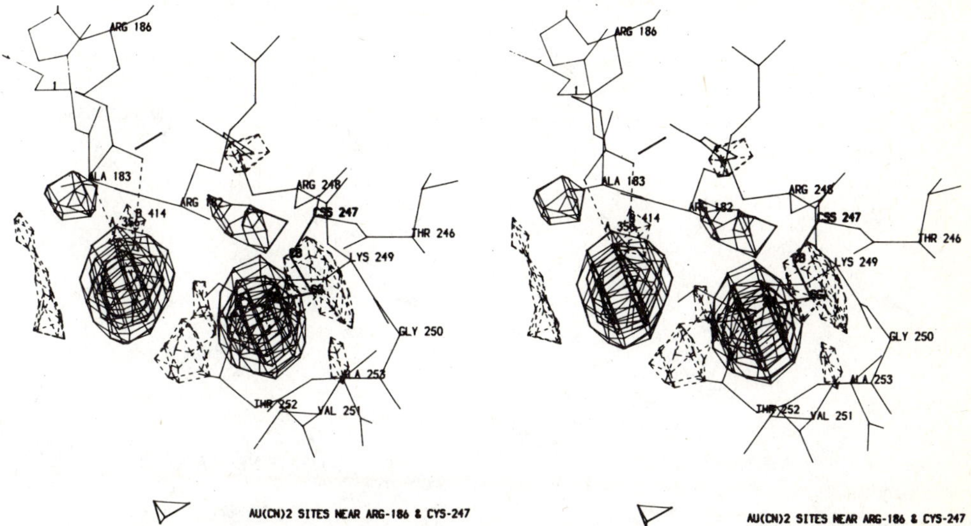

Figure 12. The binding of Au(CN)$_2^-$ in the active site region of bovine liver rhodanese. The difference electron density is shown between crystals of sulfur rhodanese soaked in a 5mM Na[Au(CN)$_2$] and of "native" sulfur rhodanese. One gold atom has evidently displaced the extra sulfur atom bound to the S$^\gamma$ of Cys 247. A second gold complex is situated at the entrance to the active site pocket, interacting electrostatically with Arg 186 and Lys 249.

were obtained from 2 M ammonium sulfate solutions, it was possible to determine the major sulfate-binding sites. The procedure followed was that of sulfate-selenate exchange pioneered by Tulinsky and Wright (1973). To this end, sulfur-rhodanese crystals were transferred to 1.8 M (NH$_4$)$_2$SeO$_4$ solutions, pH 7.1, and X-ray data collected to a resolution of 2.9 Å. As the refined 2.1 Å structure of sulfur-rhodanese in sulfate was available, accurate difference electron density maps could be calculated (Lijk et al., 1984). It appeared that both sulfate and selenate bind at two positions. One position occurs near Arg 29 (Fig. 13). Out of 20 arginines, four do not extend their side chains freely into solution. Three of these (182, 186, and 248) are involved in the salt bridge network depicted in Figure 10. The fourth, Arg 29, is unique in that it is buried up to the C$^\zeta$ atom in the hydrophobic core of the first domain. Together with the main chain NH of Glu 148, the guanidinium group of this arginine apparently creates an excellent sulfate and selenate binding site.

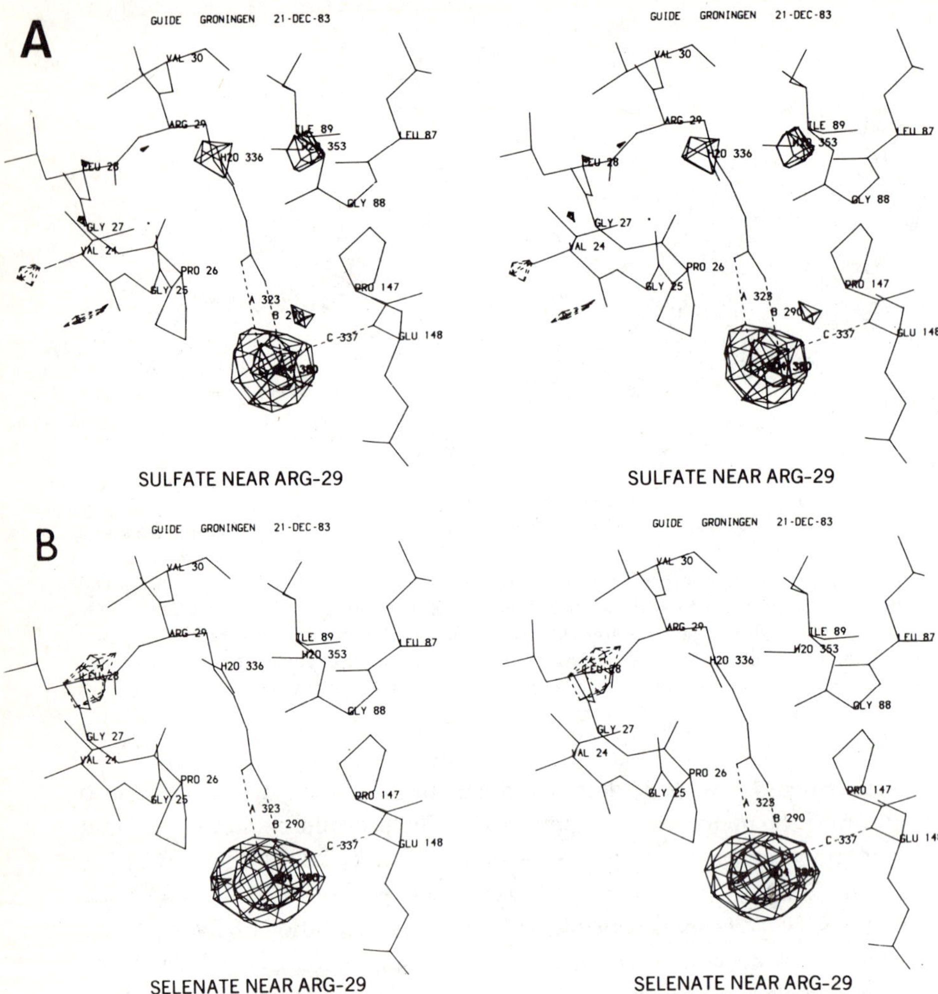

Figure 13. Sulfate and selenate binding near Arg 29 at the position indicated "SO$_4$ 380." The position of the sulfur atom is the result of crystallographic refinement. The distance of a number of sulfate oxygens to nearest protein atoms are indicated in pm along the dotted lines labeled A, B, and C. In Figures 13–16, solid lines represent positive contour levels, while dotted lines are negative ones. (*a*) The 2.1 Å resolution difference electron density calculated with coefficients ($|F_{obs,SO_4}| - |F_{calc}|$) and calculated phases. From the calculated structure factors were excluded: ions, water molecules, and the extra S$_8$ atom bound in persulfide linkage to Cys 247. The contour levels are at −22, +22, and +45 arbitrary units. (*b*) The 2.9 Å resolution difference electron density calculated with coefficients ($|F_{obs,SeO_4}| - |F_{calc}|$). Contour levels are at −19, +19, and +38 arbitrary units. (Reprinted by permission from Lijk et al., 1984. Copyright European Federation of Biochemical Societies.)

506

A second selenate-binding site could clearly be established close to Arg 186 and Lys 249 (Fig. 14). Comparison of the electron density levels in Figures 13b and 14b shows that the occupancies of selenate are the same at the two ion sites. For sulfate, this is not the case; the position near Arg 186 is clearly less well occupied than the position near Arg 29 (Fig. 14a vs. Fig. 13a). Apparently the small difference in size between SO_4 and SeO_4 (S-O distance, 1.49 Å, Udalova and Pinsker, 1963; Se-O distance, 1.62 Å, Kruglik et al., 1973) is sufficient to cause a difference in occupancies. The distances from the sulfate peak in Figure 14a to neighboring protein atoms are long and correspond better with those expected for a sulfate ion than for a water molecule. Henceforth, we conclude that the inhibition observed by sulfate (Wang and Volini, 1973) is due to sulfate binding, albeit weak, to this position in rhodanese. These crystallographic observations also suggest that selenate might be a somewhat better inhibitor of rhodanese than sulfate.

As both metalcyanides and sulfate bind essentially to the same position in the active site region of rhodanese, it is reasonable to assume that this position is a "general anion-binding site," where numerous other anions bind with enough affinity to decrease the speed of the enzymatic reaction.

2.4. The Three-Dimensional Structure of Sulfur–Free Rhodanese

2.4.1. No Evidence for a Conformational Change From X-ray Crystallography

The crystal structure described so far was that of sulfur-rhodanese. In order to unravel fully the details of the catalytic mechanism, it is necessary to also know the three-dimensional structure of sulfur-free rhodanese. Crystallization experiments, unfortunately, have never given crystals of sulfur-free rhodanese that were suitable for an X-ray crystallographic investigation. Therefore, the only way information could be obtained about the tertiary structure of the sulfur-free enzyme was to treat sulfur-rhodanese crystals with cyanide and to see what the effect on the three-dimensional structures would be.

The results of this approach (Ploegman et al., 1979; Lijk et al., 1984) are as follows (Fig. 15): (1) the persulfide sulfur clearly disappears upon treatment with cyanide; (2) at a distance of ~3 Å from the S^γ of Cys 247, a solvent molecule occupies a position in the sulfur-free enzyme that partially over-

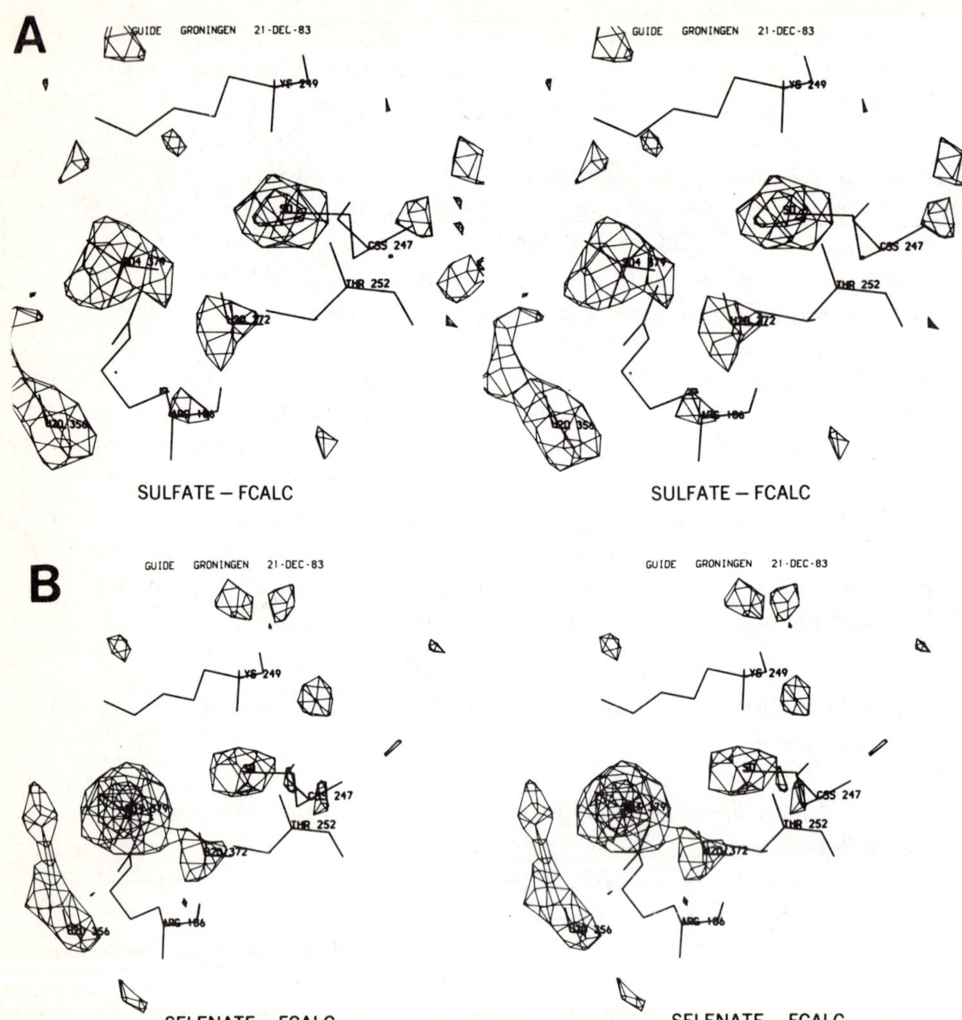

Figure 14. Binding of sulfate and selenate (indicated with "SO₄ 379") near Arg 186 at the active site entrance. (*a*) 2.1 Å difference electron density calculated with coefficients ($|F_{obs,SO_4}| - |F_{calc}|$) and calculated phases. Contour levels are the same as in Figure 13*a*. As explained in the legend to Figure 13*a*, the calculated structure factors did not include the extra sulfur atom (SD) bound to S$^\gamma$ of Cys 247. Henceforth, density for this atom shows up in this difference map. Comparison of the "SO₄ 379 density" with the "SO₄ 380 density" in Figure 13*a* shows that the occupancy of the sulfate site near Arg 186 is less than near Arg 29. (*b*) 2.9 Å difference electron density with coefficients ($|F_{obs,SeO_4}| - |F_{calc}|$) and calculated phases. Contour levels are the same as in Figure 13*b*. The selenate is bound at the "position SO₄ 379" with the same occupancy as at "position 380" (Fig. 13*b*). (Reprinted by permission from Lijk et al., 1984. Copyright European Federation of Biochemical Societies.)

508

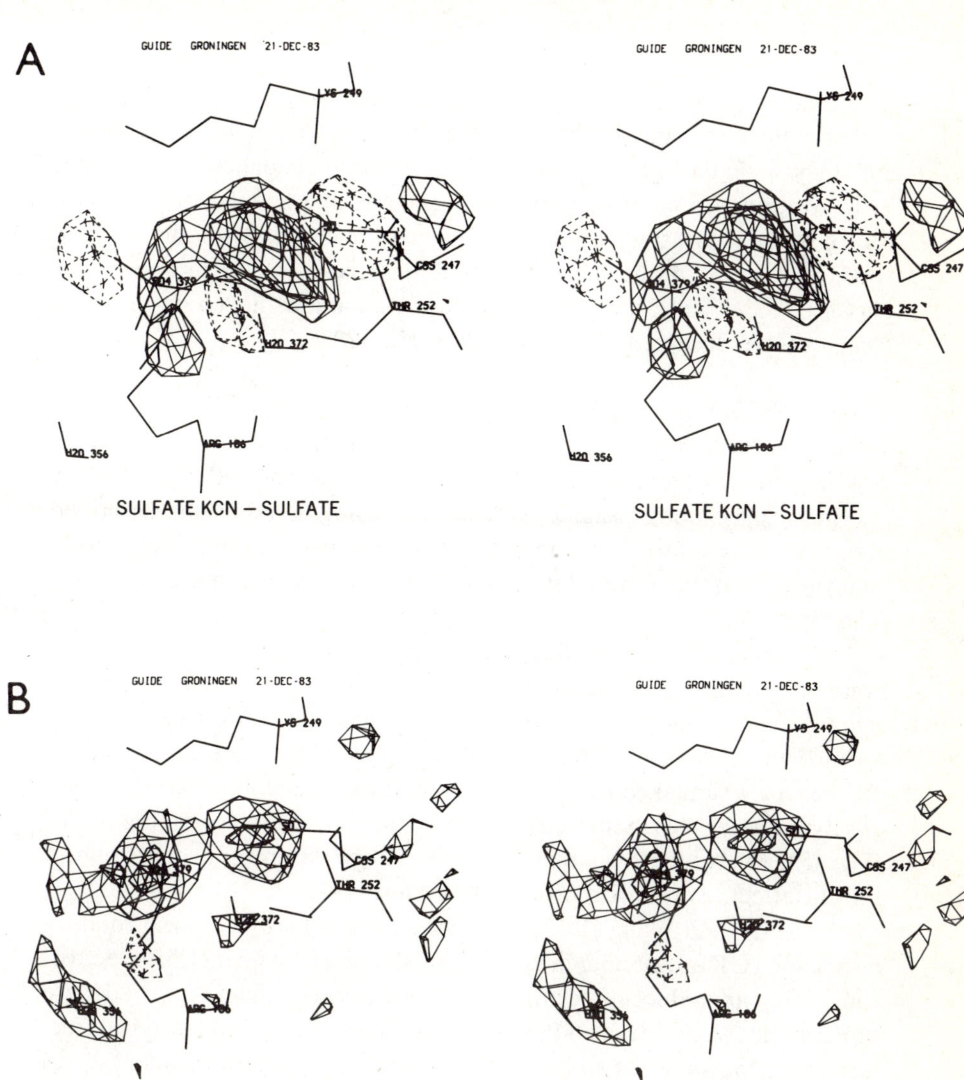

Figure 15. The active site in sulfur-free rhodanese. The position of the atoms shown are the same as in Figures 14*a* and 14*b*. (*a*) The 2.6 Å difference electron density calculated with coefficients ($|F_{obs,SO_4,KCN}| - |F_{obs,SO_4}|$) and calculated phases. Contour levels are at -5, $+5$, and $+10$. (*b*) The 2.6 Å difference electron density calculated with coefficients ($|F_{obs,SO_4,KCN}| - |F_{calc}|$) and calculated phases. As in Figures $13-16$, the persulfide sulfur was not included in the calculations. Contour levels are at -19, $+19$, and $+38$. (Reprinted by permission from Lijk et al., 1984. Copyright European Federation of Biochemical Societies.)

laps with the position of the extra sulfur in sulfur-rhodanese; (3) a weakly bound sulfate ion occupies virtually the same "SO$_4$ 379" position as in sulfur-rhodanese; and (4) selenate-sulfate exchange studies (Lijk et al., 1984) show that selenate also binds in sulfur-free rhodanese at the "SO$_4$ 379" position. An important additional point is that no evidence exists on the basis of these studies that a conformational change of the enzyme rhodanese occurs. It may also be pointed out that in the binding studies of Ni(CN)$_4^{2-}$ and Zn(CN)$_4^{2-}$ on sulfur-free rhodanese, no conformational changes were observed (Table 5; Lijk et al., 1983).

2.4.2.   *Studies in Solution Suggesting a Conformational Change*

During the last 17 years, several authors have proposed that significant conformational differences exist between rhodanese and sulfur-rhodanese. As these suggestions disagree with the crystallographic results described above, we may try to obtain an impression of how solid the evidence for a conformational change in solution is. Investigations to be considered are the following:

1.   Leininger and Westley (1968) concluded that the very large nonelectrostatic entropy term upon substrate binding "probably indicates a significant conformational change in the protein on binding of S$_2$O$_3^{2-}$ to the enzyme." Wang and Volini (1973) also reached, on the basis of entropy differences, a similar conclusion. In view of the considerable rearrangement of solvent molecules in the active site upon removal of the persulfide sulfur (Fig. 15), one wonders whether the large entropy differences could not be a solvent effect, as was also proposed by Smit (1973).

2.   Volini and Wang (1973) concluded, on the basis of optical rotatory dispersion (ORD), circular dichroism (CD), and ultraviolet (UV) spectroscopy, "that upon thiosulfate binding and sulfite ion discharge the rhodanese protein undergoes a transition from disordered coil-type to a β-type structure involving some 20 amino acid residues." As in sulfur-rhodanese, all β-type structures occur entirely in the center of the two domains (Figs. 2, 5, and 6); it is hard to see how much a "β ↔ disorder transition" could occur without, for example, also leading to unfolding of α helices. Nevertheless, the recorded spectra of the two enzyme forms shows differences in several respects.

3.   Jarabak and Westley (1974a, 1974b) report different rate constants, k_{+3}, in the reaction of cyanide with sulfur-rhodanese for different sulfur donors. They suggest that the enzyme appears to "remember" the structure

of the substrate from which it received the transferable sulfur atom for a significant period after the first product, and suggest: "This phenomenon is probably related to the high conformational mobility previously observed." Intriguing as this observation is, it does not give clear insight into the extent of the conformational difference proposed.

4. Solvent perturbation difference spectra (Guido et al., 1976) indicate that some tyrosine and tryptophane residues are buried as rhodanese is converted into sulfur-rhodanese. This does not necessarily indicate conformational changes, as the extra sulfur atom would almost certainly affect the accessibility of Trp 35 and Tyr 107 in the active site hydrophobic region (Fig. 9).

5. Horowitz and Patel (1980) report differences in solubility of rhodanese and sulfur rhodanese. Although this would indeed be consistent with the idea of a conformational difference between the two enzyme forms, our understanding of protein solubility is so limited that no firm conclusions can be based upon this observation. It cannot be excluded that the mere presence of the extra sulfur atom plus the rearrangement of solvent molecules in the active site (Fig. 15) are the causes of the differences in solubility.

6. A similar argument holds for the fluorescence studies of Wasylewski and Horowitz (1982). Sulfur-rhodanese might indeed be more stable than rhodanese, but this by no means proves that this is the result of "changes in the interactions between the structural domains into which rhodanese is folded." The presence of a persulfide making numerous hydrogen bonds with the protein (Table 4) may well be sufficient for the observed stability difference.

7. The temperature dependences of the proton relaxation time T_1 of rhodanese and sulfur-rhodanese are not the same (Blicharska et al., 1982). This is not very hard proof for conformational differences for reasons similar to those given under (5) and (6) above. It may also be pointed out that rhodanese was studied in 5 mM KCN and sulfur-rhodanese in 3 mM $Na_2S_2O_3$; and it is possible that the difference in *cation* is partly the reason for the differences in relaxation times (see Section 5). Changes in the mobilities of bound water molecules, suggested as an underlying cause by the authors, are entirely in agreement with the solvent rearrangements shown in Figure 15.

8. A study of the apolar interaction of rhodanese with octyl-substituted agarose gel (Horowitz and Criscimagna, 1983) revealed that rhodanese binds better than sulfur-rhodanese. This might, of course, be due to a

difference in conformation, but it could also be explained by greater accessibility of the hydrophobic region in rhodanese than in sulfur-rhodanese due to the presence of the extra sulfur atoms in the latter. As long as no detailed insight is obtained into the mode of binding of the octyl chain to rhodanese or sulfur-rhodanese, conclusions are deemed to be somewhat speculative.

In summary, one might say that the evidence for significant conformation changes when going from rhodanese to sulfur-rhodanese could have been stronger. On the other hand, the crystallographic evidence for *no* conformational difference is also not as solid as one would like. First, crystal contacts may prevent a conformational change to take place. Second, the high ammonium sulfate concentrations in the crystals may "lock" rhodanese in a conformation similar to that of sulfur-rhodanese. The studies on thiosulfate binding to sulfur-rhodanese—the subject of the next section—offer a possible answer to the dispute described in this section.

2.5. The Structure of the Complex of Sulfur–Rhodanese with Thiosulfate

The occurrence of a dead-end complex of sulfur-rhodanese with thiosulfate during the catalytic cycle of the enzyme (Schlesinger and Westley, 1974) led to an investigation by X-ray diffraction techniques (Lijk et al., 1984). Thiosulfate appears to occupy the "general anion binding site" in sulfur-rhodanese with a preferred orientation of its sulfur-sulfur bond in the direction of the persulfide of Cys 247 (Fig. 16). Thus, in this dead-end complex, four sulfurs occur close to each other with a distance of approximately 3 Å between the outer sulfur of thiosulfate and the extra sulfur of sulfur-rhodanese. This could be explained by a $S^-1...H\text{-}S$ hydrogen bond, although in solid hydrogen sulfide, S...H-S bonds of 3.9 Å are observed (Harada and Kitamura, 1964). The negative charge on one of the sulfur atoms may be the cause of the short sulfur-sulfur hydrogen bond in this case. The negatively charged sulfur would probably be the thiosulfate sulfur because of its interaction with the guanidinium group of Arg 186. The persulfide sulfur does not interact with nearby positive charges, and is therefore more likely to be protonated. It is obvious that binding of thiosulfate at this site hinders sulfur acceptors in reaching the essential Cys 247.

The thiosulfate-sulfur-rhodanese complex did reveal a minor conformational change with respect to both sulfur-rhodanese and rhodanese. The

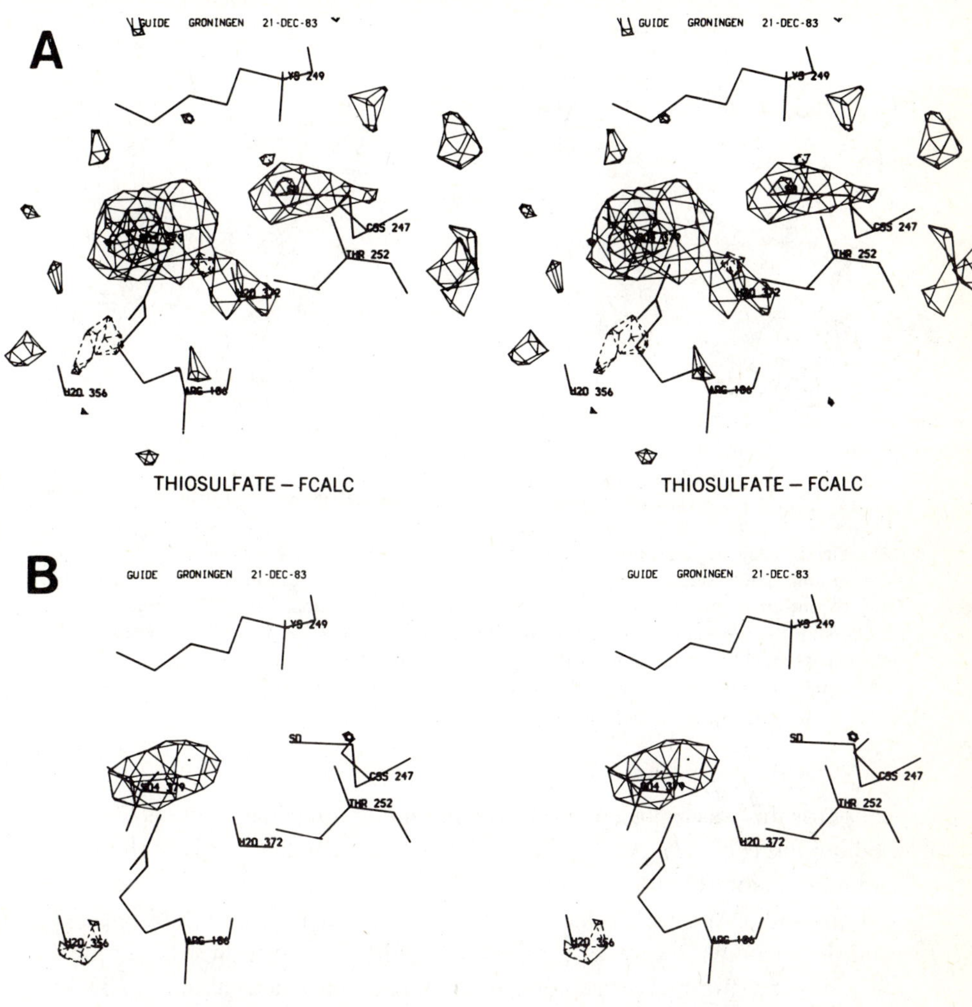

Figure 16. Thiosulfate binding near the active site of sulfur-rhodanese. The positions of the atoms shown are the same as in Figures 14 and 15. (*a*) The 2.35 Å resolution difference electron density with coefficients ($|F_{obs,S_2O_3}| - |F_{calc}|$) and calculated phases. The thiosulfate clearly occupies the same position as sulfate and selenate (Figs. 14 and 15). The occupancy of thiosulfate is the same as at the anion-binding site near Arg 29. (*b*) The 2.35 Å ($|F_{obs,S_2O_3}| - |F_{obs,SO_4}|$) difference density. Contour levels are drawn at -15, $+15$, $+30$. This difference density indicates that the outer sulfur atom of the thiosulfate ion points in the direction of S^δ-247. A second contour level just indicates the position of the outer sulfur atom of the bound thiosulfate. (Reproduced by permission from Lijk et al., 1984. Copyright European Federation of Biochemical Societies.)

513

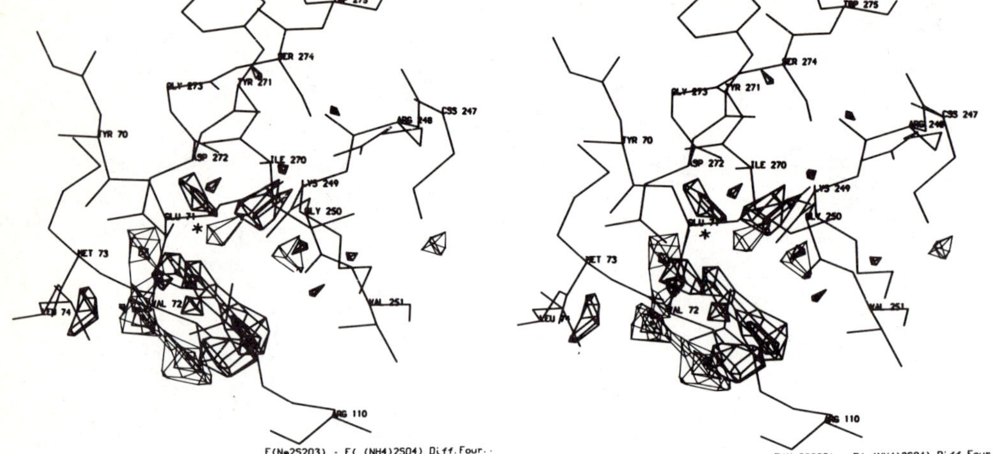

Figure 17. Cation binding near Asp 272 as revealed by the $(|F_{obs.S_2O_3}| - |F_{obs.SO_4}|)$ difference electron density. The position of the ion as obtained by the refinement of sulfur-rhodanese in 2 *M* ammonium sulfate is indicated by an asterisk. Light contour levels are negative and heavy contours represent positive regions in the difference map. It is clear that Val 72 undergoes a conformational change involving its side chain. A second ligand, the carbonyl oxygen of Lys 249, also moves towards the cation-binding site when the larger ammonium ion is replaced by a sodium ion. The third protein ligand, a carboxylate oxygen of Asp 272 does not appear to change its position significantly.

cause for this was replacement of an ammonium ion (Ploegman, 1977) by a sodium ion (Fig. 17). As the sodium ion is 0.5 Å smaller in radius than the ammonium ion (Cotton and Wilkinson, 1980), the ligands surrounding this cation-binding site are displaced. Two of the protein ligands, $O^{\delta 1}$ of Asp 272 and the carbonyl oxygen of Lys 249 change little in position, but the third ligand,—the carbonyl oxygen of Val 72—moves and causes a rotation of the side chain of this valine residue. Other minor differences occur concomitantly.

These crystallographic observations suggest that the conformational changes ascribed by various investigators to differences between rhodanese and sulfur-rhodanese, *might* actually be the result of structural differences in rhodanese-*cation* complexes.

3.   THE CATALYTIC MECHANISM

Knowing the three-dimensional structure of sulfur-rhodanese, assuming only minor structural differences between sulfur-rhodanese and rhodanese, and realizing the importance of the "general anion binding site" near Arg 186 and Lys 249, a proposal can now be made for the catalytic mechanism of rhodanese (Fig. 18). The catalytic efficiency of rhodanese is obtained by a purposeful arrangement of the following elements:

1. Two positive charges, those of Lys 249 and Arg 186, at the entrance of the active site pocket. The tasks of these charges may be threefold. First, masking the negative charge of the S^- of Cys 247 for the approaching negatively charged substrates (Figs. 18*a* and 18*f*). Second, providing a well-defined binding site for these substrates (Figs. 14 and 15). Third, polarizing the sulfur-sulfur bond in thiosulfate such that its outer sulfur atom becomes less negatively charged so that it can more easily be attacked by the S^- of Cys 247.

2. A set of N-H and O-H hydrogen bond donors (Table 4). These may not only give rise to an unusually stable persulfide, but may also play an important role in stabilizing transition-state intermediates (Figs. 18*c* and 18*h*).

3. The N-termini of the α helix dipole $\alpha_D{}'$ (Fig. 11) and, perhaps, also of $\alpha_E{}'$. These probably lower the pK of the essential Cys 247, and by doing so create a more powerful thiophile.

4. The sulfhydryl of Cys 247; the actual sulfur-transferring moiety of the enzyme rhodanese.

No function in this scheme is envisaged for the hydrophobic region of the active site (Fig. 9; Section 2). However, this region is very likely involved in the binding of substrates such as the aromatic and aliphatic thiosulfonates. The hydrocarbon groups of these substrates probably make contact with the hydrophobic part of the active site, while the thiosulfonate moiety occupies the "general anion binding site" and follows a similar catalytic cycle as outined in Figure 18.

4. EVOLUTION OF RHODANESE

The three-dimensional structure of rhodanese revealed that this protein consists of two domains of a very similar fold and very dissimilar amino acid sequence (Section 2.2.3). Moreover, the two domains are related by a pseudo twofold axis that deviates only marginally from a perfect twofold axis (Ploegman et al., 1978b). These results suggest that rhodanese is the product of gene duplication, which initially formed a dimer of two identical subunits. At a later stage, the genes of the two monomers fused, and the amino acid

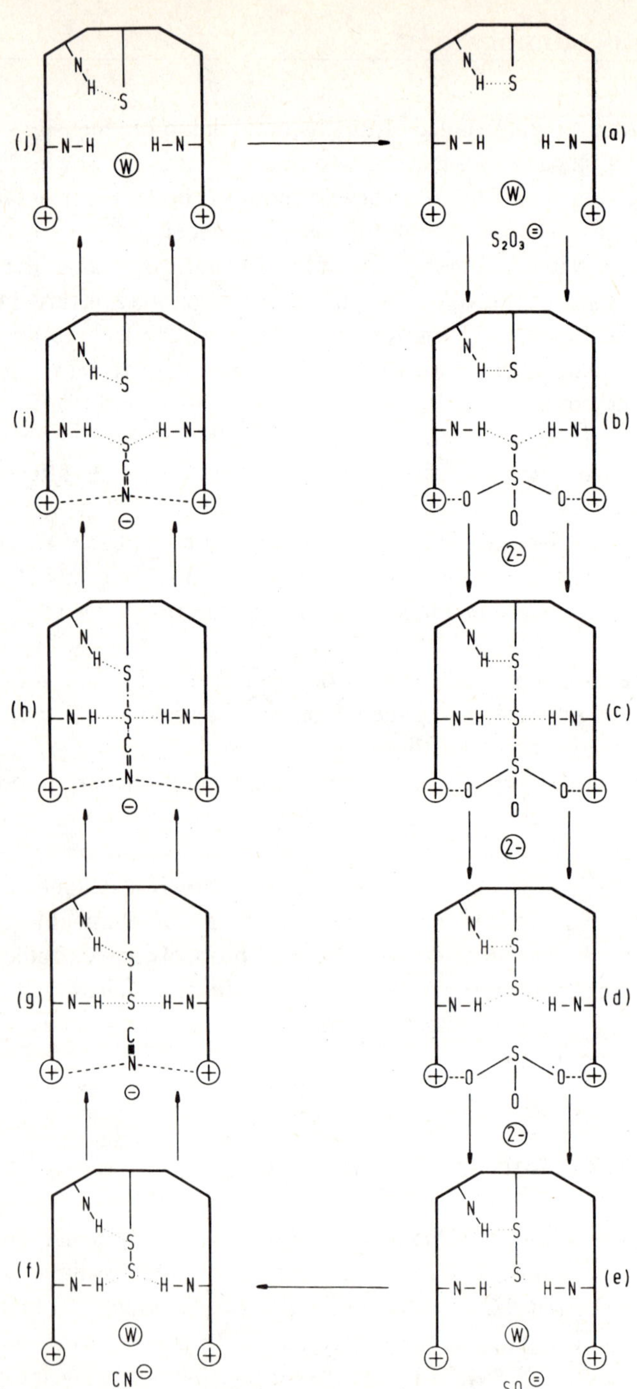

(j)
(i)
(h)
(g)
(f)
CN
(a)
S₂O₃
(b)
2-
(c)
2-
(d)
2-
(e)
SO₃

sequences deviated more and more until the present situation was reached where any sequence homology is difficult, if not impossible, to detect. Alternatively, the gene duplication and fusion may have occurred in a single step.

In a small number of other enzymes with known three-dimensional structure, similar events have probably taken place. In two of those cases, the aspartyl proteases and the trypsin-like serine proteases, the interface of both domains created a marvelous binding region for long substrate molecules (Tang et al., 1978; McLachlan, 1979; Holm et al., 1984). In another case, glutathione reductase, both domains have the same functionally important dinucleotide-binding fold (Schulz, 1980). In rhodanese, the active site does occur close to the interface of both domains, but all crucial residues for the reaction between $S_2O_3^=$ and CN^- are provided by the second domain. Indeed, Volini et al. (1978a) have reported results that can be explained by assuming that bovine liver rhodanese fragments, with molecular weights down to 18,000, are still catalytically active. If this is true why then would it have been advantageous to create a double domain rhodanese molecule? Possible answers to this question are:

1. It increased the stability of the molecule.

2. It allowed free alterations of the first domain, aimed at targeting it to the correct organelle, while the second domain continued to function normally.

3. Thiosulfate and cyanide are not the "true" substrates of rhodanese in the living cell. The true substrates are larger, and the additional first domain would allow the creation of an extensive binding site.

Figure 18. Simplified diagram of the catalytic mechanism of rhodanese. The three N-H groups shown represent the hydrogen bond donors listed in Table 4. The two positive charges are those of Arg 186 and Lys 249. (*a*) Thiosulfate approaching the active site. (*b*) Thiosulfate bound at the "general anion-binding site," which is the presumed Michaelis-Menten complex. (*c*) Intermediate bond breaking and formation step, or steps. (*d*) Sulfur-rhodanese with sulfite on the "general anion-binding site." (*e*) Sulfur-rhodanese releasing sulfite. (*f*) Cyanide approaching the active site of sulfur-rhodanese. (*g*) A possible Michaelis-Menten complex that, however, has not been detected kinetically. (*h*) Intermediate bond breaking and formation step, or steps. (*i*) Sulfur-free rhodanese with thiocyanate bound. (*j*) Sulfur-free rhodanese ready for the next catalytic cycle.

The last possibility might be linked with a function of rhodanese as a "sulfur insertase" for certain specific iron-sulfur proteins. As the possible functions of rhodanese have already been discussed in Section 1.4, it is probably best to refrain here from further speculations regarding the function and evolution of this enzyme.

5. CONCLUSIONS

Since its discovery by Lang in 1933, rhodanese has been subject to numerous studies ranging from physiology to biophysics. In particular, the enzyme from bovine liver has been characterized in great detail. Our knowledge may be summarized as follows:

1. The enzyme occurs widely distributed in nature.
2. When evolution progressed, liver rhodanese became more and more concentrated in mitochondria.
3. The enzyme catalyzes the transfer of a single sulfur atom between a sulfur donor and a sulfur acceptor.
4. The bovine liver enzyme consists of 293 residues folded into two domains of highly similar structure and very dissimilar amino acid sequence.
5. The active site contains an essential sulfhydryl group, a set of hydrogen bond donors, an active site α helix dipole, and two crucial positive charges at its entrance.
6. A hydrophobic region near the active site is probably involved in the binding of aromatic and aliphatic moieties of substrates and inhibitors.
7. A cationic-binding site accommodates different radii of cations bound by minor conformational alterations.

Major remaining questions are:

1. What is the physiological importance of the enzyme?
2. Why do the two dissimilar sequences of the two domains adopt such similar folds?

3. Can the suggested conformational differences between rhodanese and sulfur-rhodanese in solution be the result of conformational changes induced by different cations?

4. What would be the evolutionary advantage of gene duplication for this enzyme?

It will be highly interesting to see if future investigations will be able to answer these questions.

ACKNOWLEDGMENTS

It is a great pleasure to thank the persons who actually did the crystallographic studies on rhodanese in our laboratory over the years. The initial investigations of Drs. Jan Derk Smit and Hans Jansonius were followed by those of Drs. Jan Harm Ploegman, Jack Bergsma, Hein te Riele, and Leo Lijk. Crucial for all the X-ray investigations was the excellent maintenance and guidance of the equipment by Kor H. Kalk. We are indebted to the excellent technical assistance of Gé Drent and Chris Torfs. Roelof Kloosterman prepared numerous drawings, while Roeline Hogenkamp assisted greatly in the preparation of the text. I would like to thank, in particular, Professor J. Drenth for his continuous interest and stimulating discussions.

The investigations were supported by the Dutch Foundation for Chemical Research (SON) with financial aid from the Dutch Organization for the Advancement of Pure Research (ZWO). All calculations were performed on the successive Cybers of the Groningen University Computer Center.

REFERENCES

Abdolrasulnia, R., and Wood, J. L. (1979). *Biochim. Biophys. Acta* **567**, 135–143.

Bergsma, J., Hol, W. G. J., Jansonius, J. N., Kalk, K. H., Ploegman, J. H., and Smit, J. D. G. (1975). *J. Mol. Biol.* **98**, 637–643.

Blicharska, B., Koloczek, H., and Wasylewski, Z. (1982). *Biochim. Biophys. Acta* **708**, 326–329.

Bonomi, F., Pagani, S., and Cerletti, O. (1977a). *FEBS Lett.* **84**, 149–152.

Bonomi, F., Pagani, S., Cerletti, P., and Cannella, C. (1977b). *Eur. J. Biochem.* **72**, 17−24.

Cannella, C., Pecci, L., Finazzi-Agrò, A., Federici, G., Pensa, B., and Cavallini, D. (1975a) *Eur. J. Biochem.* **55**, 285−289.

Cannella, C., Pecci, L., Costra, M., Pensa, B., and Cavallini, D. (1975b). *Eur. J. Biochem.* **56**, 283−287.

Cotton, F. A., and Wilkinson, G. (1980). *Advanced Inorganic Chemistry.* Wiley-Interscience, New York, p. 14.

Davidson, B., and Westley, J. (1965). *J. Biol. Chem.* **240**, 4463−4469.

De Duve, C., Pressman, B. C., Gianetto, R., Wattiaux, R., and Appelmans, F. (1955). *Biochem. J.* **60**, 604−617.

Drenth, J., and Smit, J. D. G. (1971). *Biochem. Biophys. Res. Comm.* **45**, 1320−1322.

Dudek, M., Frendo, J., and Koj, A. (1980). *Comp. Biochem. Physiol.* **65B**, 383−386.

Ellis, L. M., and Woodward, C. K. (1975). *Biochim. Biophys. Acta* **379**, 385−396.

Eriksson, B., and Sörbo, B. H. (1967). *Acta Chem. Skand.* **21**, 958−960.

Finazzi, Agrò, A., Cannella, C., Graziani, M. T., and Cavallini, D. (1971). *FEBS Lett.* **16**, 172−174.

Finazzi, Agrò, A., Federici, G., Giovagnoli, C., Cannella, C., and Cavallini, D. (1972). *Eur. J. Biochem.* **28**, 89−93.

Guido, K., Baillie, R. D., and Horowitz, P. M. (1976). *Biochim. Biophys. Acta* **427**, 600−607.

Harada, J., and Kitamura, N. (1964). *J. Phys. Soc. Japan* **19**, 328−343.

Hol, W. G. J., Van Duijnen, P. Th., and Berendsen, H. J. C. (1978). *Nature* **273**, 443−446.

Holm, I., Ollo, R., Panthier, J.-J., and Rougeon, F. (1984). *EMBO J.* **3**, 557−562.

Horowitz, P. M., and Patel, K. (1980). *Biochem. Biophys. Res. Comm.* **94**, 419−423.

Horowitz, P. M., and Criscimagna, N. L. (1982). *Biochim. Biophys. Acta* **702**, 173−177.

Horowitz, P. M., and Criscimagna, N. L. (1983). *Biochem. Biophys. Res. Comm.* **111**, 595−601.

Janse van Rensburg, L., and Schabort, J. C. (1984a). *Int. J. Biochem.* **16**, 539−546.

Janse van Rensburg, L., and Schabort, J. C. (1984b). *Int. J. Biochem.* **16**, 547−551.

Jarabak, R., and Westley, J. (1974a). *Biochemistry* **13**, 3237−3239.

Jarabak, R., and Westley, J. (1974b). *Biochemistry* **13**, 3240−3243.

Keim, P., Heinrikson, R. L., and Fitch, W. M. (1981). *J. Mol. Biol.* **151**, 179−197.

Kruglik, A. I., Simonov, V. I., and Yuzvak, V. I. (1973). *Kristallografiya* **18**, 287−292.

Lang, K. (1933). *Biochem. Z.* **259**, 243–250.

Leininger, K. R., and Westley, J. (1968). *J. Biol. Chem.* **243**, 1892–1899.

Lijk, L. J., Kalk, K. H., Brandenburg, N. P., and Hol, W. G. J. (1983). *Biochemistry* **22**, 2952–2957.

Lijk, L. J., Torfs, C. A., Kalk, K. H., De Maeyer, M. C. H., and Hol, W. G. J. (1984). *Eur. J. Biochem.* **142**, 399–408.

Long, K. Y., and Brattsten, L. B. (1982). *Insect Biochem.* **12**, 367–375.

McLachlan, A. D. (1979). *J. Mol. Biol.* **128**, 49–79.

Mintel, R., and Westley, J. (1966a). *J. Biol. Chem.* **241**, 3381–3385.

Mintel, R., and Westley, J. (1966b). *J. Biol. Chem.* **241**, 3386–3389.

Nishino, T., Usami, C., and Tsushima, K. (1983). *Proc. Natl. Acad. Sci. USA* **80**, 1826–1829.

Oi, S. (1975). *J. Biochem.* **78**, 825–834.

Pagani, S., and Galante, Y. M. (1983). *Biochim. Biophys. Acta* **742**, 278–284.

Pagani, S., Cannella, C., Cerletti, R., and Pecci, L. (1975). *FEBS Lett.* **51**, 112–115.

Pagani, S., Bonomi, F., and Cerletti, P. (1982). *Biochim. Biophys. Acta* **700**, 154–164.

Ploegman, J. H. (1977). Ph.D. Thesis, University of Groningen, p. 75.

Ploegman, J. H., Drent, G., Kalk, K. H., Hol, W. G. J., Heinrikson, R. L., Keim, P. S., Weng, L., and Russell, J. (1978a) *Nature* **273**, 124–129.

Ploegman, J. H., Drent, G., Kalk, K. H., and Hol, W. G. J. (1978b). *J. Mol. Biol.* **123**, 557–594.

Ploegman, J. H., Drent, G., Kalk, K. H., and Hol, W. G. J. (1979). *J. Mol. Biol.* **127**, 149–162.

Reinwein, D. (1961). *Hoppe-Seyler's Z. Physiol. Chem.* **326**, 94–101.

Richardson, J. S. (1981). *Adv. Prot. Chem.* **34**, 168–339.

Russell, J., Weng, L., Keim, P. S., and Heinrikson, R. L. (1975). *Biochem. Biophys. Res. Commun.* **64**, 1090–1097.

Russell, J., Weng, L., Keim, P. S., and Heinrikson, R. L. (1978). *J. Biol. Chem.* **253**, 8102–8108.

Saunders, J. P., and Himwich, W. A. (1950). *Am. J. Physiol.* **163**, 404–409.

Schlesinger, P., and Westley, J. (1974). *J. Biol. Chem.* **249**, 780–788.

Schulz, G. E. (1980). *J. Mol. Biol.* **138**, 335–347.

Smit, J. D. G. (1973). Ph.D. Thesis, University of Groningen, p. 31.

Smit, J. D. G., Ploegman, J. H., Pierrot, M., Kalk, K. H., Jansonius, J. N., and Drenth, J. (1974). *Isr. J. Chem.* **12**, 287–304.

Sörbo, B. H. (1951) *Acta Chem. Skand.* **5**, 724–734.

Sörbo, B. H. (1953a). *Acta Chem. Skand.* **7**, 32–37.

Sörbo, B. H. (1953b). *Acta Chem. Skand.* **7**, 1129–1136.

Sörbo, B. H. (1953c). *Acta Chem. Skand.* **7**, 1137−1145.

Sörbo, B. H. (1957). *Acta Chem. Skand.* **11**, 628−633.

Sörbo, B. H. (1960). *Biochim. Biophys. Acta* **38**, 349−351.

Sörbo, B. H. (1962). *Acta Chem. Skand.* **16**, 243−245.

Sörbo, B. H. (1975). In *Metabolic Pathways*, D. M. Greenberg, Ed., 3rd ed., Vol. 7. Academic Press, New York, pp. 433−456.

Szczepkowski, T. W., and Wood, J. L. (1967). *Biochim Biophys. Acta* **139**, 469−478.

Tang, J., James, M. N. G., Hsu, I. N., Jenkins, J. A., and Blundell, T. L. (1978). *Nature* **271**, 618−621.

Trumpower, B. L., Katki, A., and Horowitz, P. (1974). *Biochem. Biophys. Res. Commun.* **57**, 532−538.

Tulinsky, A., and Wright, L. H. (1973). *J. Mol. Biol.* **81**, 47−56.

Udalova, V. V., and Pinsker, Z. G. (1963). *Kristallografiya* **8**, 538−547.

Villarejo, M., and Westley, J. (1963). *J. Biol. Chem.* **238**, 4016−4020.

Volini, M., and Wang, S.-F. (1973). *J. Biol. Chem.* **248**, 7386−7391.

Volini, M., Craven, D., and Ogata, K. (1978a). *J. Biol. Chem.* **253**, 7591−7594.

Volini, M., Van Sweringen, B., and Chen, F.-S. (1978b). *Arch. Bioch. Biophys.* **191**, 205−215.

Wada, A. (1976). *Adv. Biophys.* **9**, 1−63.

Wang, S.-F., Volini, M. (1968). *J. Biol. Chem.* **243**, 5465−5470.

Wang, S.-F., and Volini, M. (1973). *J. Biol. Chem.* **248**, 7367−7385.

Wasylewski, Z., and Horowitz, P. M. (1982). *Biochim. Biophys. Acta* **701**, 12−18.

Weng, L., Russell, J., and Heinrikson, R. L. (1978a). *J. Biol. Chem.* **253**, 8093−8101.

Weng, L., Heinrikson, R. L., and Westley, J. (1978b). *J. Biol. Chem.* **253**, 8109−8119.

Westley, J. (1973). *Adv. Enzymol.* **39**, 327−368.

Westley, J. (1977). In *Bioinorganic Chemistry*, E. E. Van Tamelen, Ed., Vol. 1. Academic Press, New York, pp. 371−390.

Westley, J. (1980). In *Enzymatic Basis of Detoxication*, W. D. Jakoby, Ed., Vol. 2. Academic Press, New York, pp. 245−262.

Westley, J., and Nakamoto, T. (1962). *J. Biol. Chem.* **237**, 547−549.

Westley, J., Adler, H., Westley, L., and Nishida, C. (1983). *Fund. Appl. Toxicol.* **3**, 377−382.

Westley, J., and Heyse, D. (1971). *J. Biol. Chem.* **246**, 1468−1474.

Catalysis By Seleno Glutathione Peroxidase

10

RUDOLF LADENSTEIN
OTTO EPP
Max-Planck-Institut für Biochemie Martiasried West Germany

CONTENTS

1. ENZYMOLOGY OF GLUTATHIONE PEROXIDASE
  1.1. Biological Function and Relation to Selenium Biochemistry
  1.2. Substrate Specificity and Glutathione Binding
  1.3. Kinetic Mechanism and Inhibition Studies

2. PRIMARY STRUCTURE
  2.1. Chemical Sequence Analyses of the Bovine Erythrocyte and the Rat Liver Enzyme
  2.2. Comparison with the Sequence Predicted From X-ray Analysis

3. SECONDARY AND TERTIARY STRUCTURE
  3.1. The Folding of the Peptide Chain
  3.2. Comparison with the Chain Folding of Bacteriophage-T_4-Thioredoxin
  3.3. The Structure of the Active Center
  3.4. Structured Solvent
  3.5. A Preliminary Description of GSH Peroxidase Catalysis in Molecular Terms

4. QUATERNARY STRUCTURE AND MOLECULAR SYMMETRY

5. SEVERAL QUESTIONS REMAIN OPEN

REFERENCES

---

1. ENZYMOLOGY OF GLUTATHIONE PEROXIDASE

1.1. Biological Function and Relation to Selenium Biochemistry

The redox chemistry of oxygen is complicated; and many oxygen derivatives are highly reactive and toxic to living organisms. Nevertheless, nature frequently uses many of the intermediates of oxygen metabolism. Both heme and nonheme proteins are involved in the manipulation and biochemical conversion of oxygen derivatives (Fig. 1). In the case of hydrogen peroxide, a powerful oxidizing agent, organisms have developed at least four ways to handle this toxic, yet necessary, molecule in distinct compartments of the cell. The first line of defense seems to be provided by superoxide dismutase catalyzing the reaction

$$2\,O_2^- + 2\,H^+ \longrightarrow O_2 + H_2O_2. \tag{1}$$

Seleno glutathione (GSH) peroxidase and the heme protein, catalase, decompose H_2O_2 to oxygen and water at essentially diffusion-controlled rates. In the liver, for instance, the catalase reaction is restricted to the peroxisomal space, whereas GSH peroxidase fulfills its task in the cytosolic and mitochondrial spaces. Heme peroxidases, on the other hand, use the oxidizing power of hydrogen peroxide to catalyze selective one-electron oxidations of organic substrates, usually phenols and amino phenols. In the last line of defense, which is required only if a polyunsaturated membrane

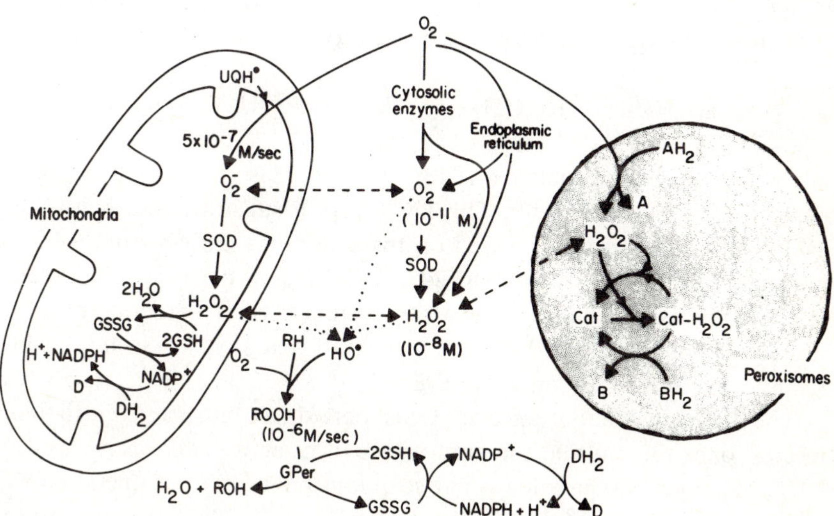

Figure 1. Sources and sinks for oxygen reduction products in the mitochondrial, cytosolic, and peroxisomal spaces. Abbreviations: UQH•, ubiquinone radical; SOD, superoxide dismutase; Cat, catalase; GPer, GSH peroxidase. The concentrations and formation rates of some metabolites are indicated. (Adapted with permission from Chance et al., 1978.)

lipid has been peroxidized, GSH peroxidase seems to be the sole enzyme preventing further propagation of a radical chain reaction that would lead to lipid peroxidation and severe disturbances of membrane function.

GSH peroxidase (EC 1.11.1.9) was discovered by G.C. Mills in 1957 as an enzyme catalyzing the reduction of H_2O_2 by GSH,

$$2\ GSH + H_2O_2 \longrightarrow GSSG + 2\ H_2O; \qquad (2)$$

and thereby protecting the hemoglobin in red blood cells from oxidative destruction (Mills, 1957). In 1968, a mitochondrial protein that was able to prevent the GSH-induced swelling of mitochondria, and was therefore named the contraction factor, was identified as GSH peroxidase (Neubert et al., 1962). This observation could be interpreted only when it had become evident that "high-amplitude swelling" of mitochondria resulted from peroxidative destruction of membrane phospholipids (Hunter et al., 1964). In an in vitro system containing isolated mitochondria, it could be demonstrated that GSH peroxidase was able to prevent lipid peroxidation (Flohe and Zimmermann, 1970). In fact, GSH peroxidase was shown not only to react with H_2O_2, but also with hydroperoxides derived from unsaturated fatty acids (Little and O'Brien, 1968: Christophersen, 1968), nucleic acids (Christophersen, 1969), and other essential biomolecules (Flohe, 1982). Therefore equation 2 may be generalized to

$$2\ GSH + ROOH \longrightarrow GSSG + ROH + H_2O, \qquad (3)$$

introducing a much broader role for the peroxidase in detoxification and protection of biomembranes. This low acceptor-substrate specificity is unusual among peroxidases. Based on these findings, an essential role was postulated for GSH peroxidase (Flohe, 1971) as part of the cellular defense system comprising catalase, superoxide dismutases, and α-tocopherol, which protects organisms from the consequences of oxidative stress and enables them to live in the presence of oxygen.

The biological importance of GSH peroxidase attracted much more interest once the unusual chemical nature of its active site was realized. In 1972, evidence was provided that selenium might play a functional role within the active site of this enzyme (Rotruck et al., 1972). This hypothesis was confirmed by neutron activation analysis that revealed stoichiometric amounts of selenium in crystalline GSH peroxidase (Flohe et al., 1973). This discovery opened up a new field of biochemical research centering on the catalytic function of selenium in enzymes. Furthermore, some light was cast on the nature of the selenium requirement of mammals first described in 1957 (Schwarz and Foltz, 1957). In vertebrates, GSH peroxidase still remains the only selenoenzyme with an established catalytic function. In bacteria, at least four selenoenzymes have now been discovered (Stadtman, 1984).

Another selenium-independent GSH peroxidase activity was found in 1976 (Lawrence and Burk, 1976). Unlike the selenium-dependent enzyme, the 47,000 dalton protein exhibits negligible activity with H_2O_2 as substrate, but seems to be rather active with organic hydroperoxides (Prohaska, 1980). The presence of this enzyme in many tissues and its relation to the glutathione-S-transferases require reevaluation of the role of GSH peroxidases in hydroperoxide metabolism and complicates the understanding of the metabolic interplay of enzymes involved in detoxification. The biological relevance of seleno GSH peroxidase has been pointed out in several reviews. (For more information concerning the function of this enzyme in hydroperoxide metabolism, the reader is referred to the following review articles: Flohe, 1979, 1982; Wendel, 1980,) Some important molecular and functional parameters of seleno GSH peroxidase from bovine red blood cells are summarized in Table 1.

Table 1. Molecular and Functional Parameters of Seleno GSH Peroxidase From Bovine Red Blood Cells

Molecular weight	$M = 84,000$ (tetramer)
Subunits	$M = 21,000$ (identical)
Monomer radius	$R = 19$ Å
Molecular symmetry	222
Isoelectric point	pH (I) $= 5.8$
pH optimum	pH $= 8.8$
Temperature optimum	$T = 42°C$
Prosthetic groups	1 selenocysteine/monomer
Crystal space group	C2, monoclinic
Unit cell constants	a $= 90.4$ Å, b $= 109.5$ Å, c $= 58.2$ Å, $\beta = 99°$
Subunits per asymmetric unit	dimer, $M = 42,000$
Packing density	$V_m = 3.4$ Å^3/dalton

Source: Data extracted from Flohe et al. (1971a, 1971b, 1971c); Ladenstein and Wendel (1976); Ladenstein et al. (1979).

1.2. Substrate Specificity and Glutathione Binding

GSH peroxidase catalyzes the reduction of a broad array of hydroperoxides to the corresponding alcohols. The only efficient physiological donor substrate appears to be GSH. The hydroperoxides accepted as substrates include H_2O_2, ethyl hydroperoxide, *t*-butyl-hydroperoxide, cumene hydroperoxide, thymine hydroperoxide, hydroperoxides of unsaturated fatty acids, and the corresponding esters, hydroperoxides of steroids, nucleic acids, and prostaglandin G_2 and the primary intermediate of prostaglandin biosynthesis (Flohe et al., 1976; Flohe, 1982). It is thus tempting to speculate that the enzyme reacts nonspecifically with every hydroperoxide, unless the reaction with the hydroperoxy group is sterically unfavorable. Dialkyl peroxides, as well as cyclic peroxides, are apparently not metabolized.

Specificity studies with various thiol compounds indicate that both carboxyl groups of the GSH molecule contribute to substrate binding. A dramatic decrease of the enzymatic activity is observed if the γ-Glu residue of GSH is substituted by a β-Asp or N-acetyl residue, or if the glycine residue is replaced by a methoxy or amide group (Table 2).

For the bovine red blood cell enzyme, the existence of a typical ES complex in the presence of excess GSH (enzyme/GSH = 1:10) could not be demonstrated by binding studies (Flohe et al., 1971c). In another attempt, substrate binding to the oxidized enzyme was studied by cochromatography with labeled GSH (Epp et al., 1983). The experiment was performed at low GSH concentration with ^{3}H-GSH in an equimolar ratio, with respect to the number of active sites of enzyme. Although this binding experiment was carried out under nonequilibrium conditions, a considerable amount of the

Table 2. Characteristic Examples, Showing the Dependence of the Activity of Seleno GSH Peroxidase on Structural Variations of the GSH Molecule

GSH Analog		Enzymatic Activity (%)
γ-Glu-Cys-Gly	GSH	100.0
β-Asp-Cys-Gly		7.6
Cys-Gly	Variations of γ-Glu	6.8
N-Ac-Cys-Gly		2.7
γ-Glu-Cys-OMe		26.0
γ-Glu-Cys-NH$_2$	Variations of Gly	1.4

Source: From Flohe et al. (1971b).

labeled GSH (about 10% of the total) remained bound to the enzyme and could be removed by dialysis in a subsequent step. The experiment demonstrates that GSH is bound noncovalently to the peroxidase, presumably by weak interactions. It seems to be more of an orientation process mediated by electrostatic forces, inducing the proper binding geometry of the GSH molecule at the active site, than the formation of a stable ES complex in the usual sense of an enzyme-substrate complex.

It has also been demonstrated that GSH binds covalently to peroxidase, but the observed binding stoichiometries appear to be confusing. Under oxidizing conditions in solution, GSH peroxidase seems to contain one covalently bound GSH molecule per subunit, presumably via a selenosulfide linkage (E-Se-SG) (Kraus and Ganther, 1980). In a labeling experiment with crystalline GSH peroxidase, binding of 2 mol GSH/mol tetramer was reproducibly observed (Epp. et al., 1983). Because extensive washing of the crystals in phosphate buffer prior to the scintillation measurements did not remove the bound radioactivity, it was concluded that under oxidizing conditions, GSH formed covalent bonds with the crystalline enzyme.

The crystallographic analysis of GSH binding (Epp. et al., 1983) by difference Fourier methods could be carried out only with great difficulties, because the observed binding stoichiometry in the crystalline state would allow, on the average, only a maximum occupancy of 0.5 GSH per binding site. The clearest, although still ill-defined, density features supporting the binding interaction of GSH at the selenocysteine residue were observed in a difference electron density map between GSH reduced crystals and the oxidized enzyme (Fig. 2). Because the resulting difference density appeared to be discontinuous and poorly defined, the structure of the ES complex was not refined crystallographically. Instead, a hypothetical model of GSH binding was proposed, based on an interpretation of the difference density giving the most probable fit (Fig. 3).

In the course of binding, the γ-Glu carboxyl group of the GSH molecule is presumably fixed by a salt bridge to Arg 177. The C-terminal Gly carboxyl group, which because of steric considerations would point into the solvent space, seems to form an additional salt bridge with Arg 50, which is a part of helix α_1. The existence of the arginine residues corresponding to Arg 50 and Arg 177 was also confirmed by chemical sequence analysis (Guenzler et al., 1984). Within the three-dimensional structure, they can be found in the neighborhood of selenocysteine residue 45 and may provide the positive charges previously indicated by substrate specificity studies (Flohe et al.,

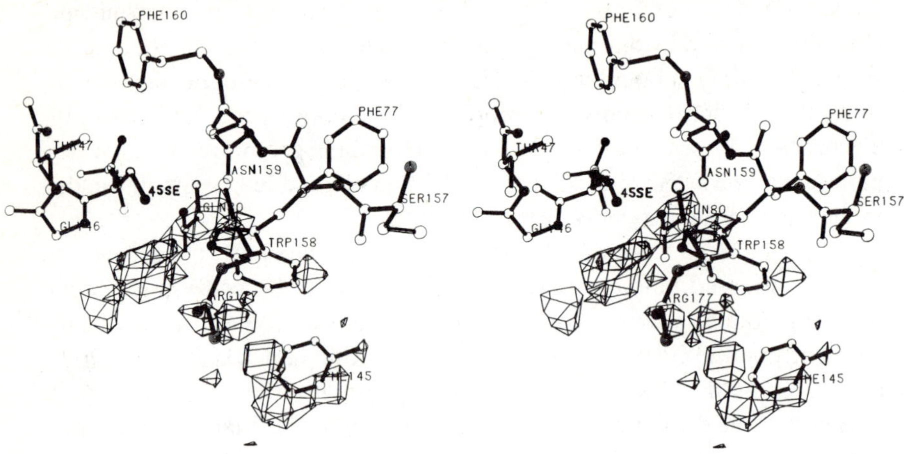

Figure 2. Stereo drawing of the active center of reduced GSH peroxidase, overlaid with the positive difference electron density calculated between reduced and oxidized enzyme. Notice that in the region of the Se-S bond, the electron density corresponding to the two oxygen atoms of the seleninic acid is subtracted; contour level, 2.2 σ.

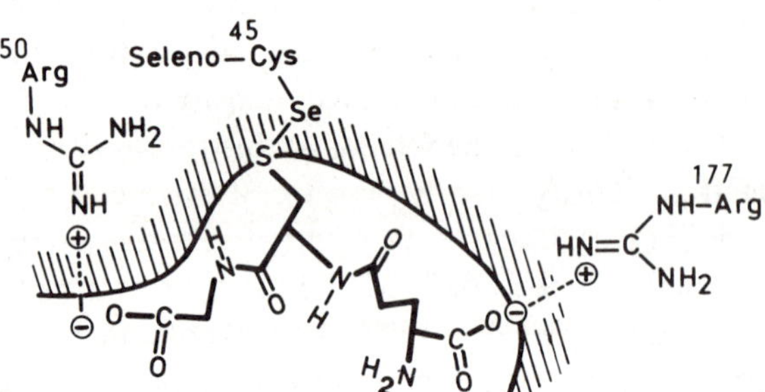

Figure 3. Proposed model of GSH-binding interaction at the active sites of GSH peroxidase. The γ-Glu-carboxyl group and the C-terminal Gly-carboxyl group of the GSH molecule are fixed by salt bridges to Arg 177 and Arg 50, respectively. (Adapted with permission from Epp et al., 1983.)

1971b). These induce proper orientation of the carboxylate groups of the relatively flexible tripeptide glutathione at the substrate-binding site, and force the thiol group to be directed towards the selenolate anion of the catalytic site. The disposition of the functional residues of the peroxidase and their presumed counterparts on the GSH molecule are accommodated by the hypothetical binding model.

Recently, the inactivation of yeast glyoxalase I by arginine-specific reagents has been described (Schasteen and Reed, 1983). The authors have provided experimental evidence in favor of the involvement of arginine residues in GSH binding to this enzyme. From these results and the X-ray studies of GSH binding to GSH peroxidase, arginyl residues can also be proposed as anionic recognition sites for GSH in other GSH-binding enzymes.

The conformations of four GSH derivatives bound at the active site of human red blood cell glyoxalase I have been determined by nuclear magnetic resonance measurements and by subsequent computer model building studies using a distance geometry approach (Rosevear et al., 1984). Extended Y-shaped conformations were detected for each of the bound GSH derivatives, and were similar to the X-ray structure of GSH (Wright, 1958), to the theoretically calculated GSH conformation (Rosevear et al., 1984), and to the hypothetical binding model of GSH at the active site of GSH peroxidase. The similarities suggest that the extended Y-shaped form of the GSH molecule is a low-energy conformation.

1.3. Kinetic Mechanism and Inhibition Studies

The dependence of the initial rate of reaction on the concentrations of the substrates can be described by a rate equation analogous to that established for other peroxidases (Dalziel, 1957).

$$v = \frac{(E_o)}{d(\text{ROOH})/dt} = \frac{\phi_1}{(\text{ROOH})} + \frac{\phi_2}{(\text{GSH})} . \tag{1}$$

The kinetic constants ϕ_1 and ϕ_2 are defined by the following relations:

$$\phi_1 = \frac{1}{k_{+1}}, \ \phi_2 = \frac{1}{k_{+2}} + \frac{1}{k_{+3}} , \tag{2}$$

where k_{+1} describes the reaction of the reduced enzyme with the hydroperoxide, k_{+2} and k_{+3} represent the rate constants of the reaction of the oxidized enzyme species with GSH, and (E_o) stands for the total enzyme

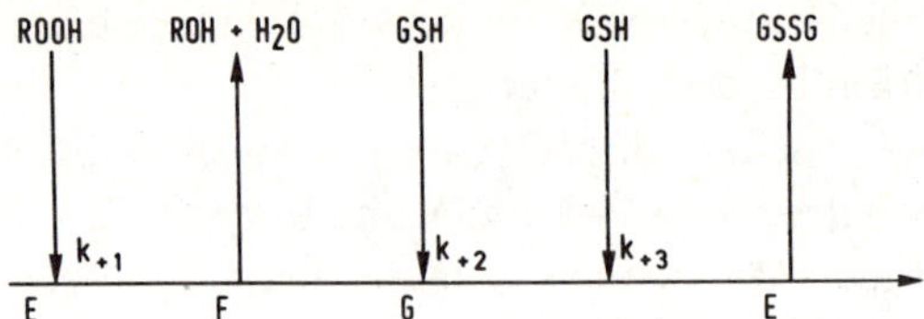

Figure 4. Ping-pong mechanism describing the reaction catalyzed by GSH peroxidase in Cleland's notation. Abbreviations: E, reduced enzyme; F, G, oxidized enzyme species; k_i, rate constants.

concentration. Equation 1 describes a double displacement mechanism, which may be written in Clelands terms as follows.

The absence of a kinetic constant ϕ_o implies that no enzyme-substrate complexes are formed, or at least that the formation of (E · GSH) complexes does not influence the initial velocity, because these complexes do not accumulate even at high GSH levels. The kinetic data for GSH peroxidase from bovine blood (Flohe et al., 1972; Guenzler et al., 1972) and rat liver (Chiu et al., 1975) can be satisfactorily described by equations 1 and 2, and by the reaction scheme shown in Figure 4. The kinetic constants are summarized in Table 3. The first-order rate constant k_{+1} of the oxidation of the reduced enzyme species by the hydroperoxide substrate is on the order of $10^8\ M^{-1}s^{-1}$ (pH, 7.0, 37°C), and it implies a diffusion-controlled reaction. This is in accord with the exposure of the enzyme-bound selenium on the surface of the molecule. The mechanism is best interpreted with the assumption that the enzyme goes through consecutive steps of reduction and

Table 3. Kinetic Constants for GSH Peroxidase From Bovine Erythrocytes at pH 6.7

Substrate	$\phi_1 \cdot 10^{-8}\ (M \cdot sec)$	$\phi_2 \cdot 10^{-6}\ (M \cdot sec)$
Hydrogen peroxide	1.7	2.2
Ethyl hydroperoxide	3.3	2.2
Cumene hydroperoxide	7.8	2.2
t-butyl-hydroperoxide	13.5	2.2

$K_m$app (H$_2$O$_2$) = 0.008 mM, (GSH) = 2.5 mM
$K_m$app (GSH) = 0.130 mM, (H$_2$O$_2$) = 0.001 mM

Source: From Flohe et al. (1972).

oxidation during catalysis. Flohe and coworkers pointed out that several characteristic features of this kinetic mechanism deserve mention:

1. Within the concentration ranges considered, the peroxidase showed no saturation with respect to GSH; thus only apparent K_M values could be determined.

2. The kinetic mechanism implies the existence of three enzyme species, exhibiting different redox states that may be assigned as E-Se⁻, E-SeOH, and E-SeSG.

Oxidized GSH peroxidase can be inactivated and depleted of its selenium after prolonged exposure to high concentrations of potassium cyanide in solution, as well as in the crystalline state (Ladenstein et al., 1979; Kraus and Ganther, 1980; Epp et al., 1983). Selenium depletion led to the identification of four selenium sites in the tetrameric enzyme as the most prominent features in a difference electron density map calculated between selenium-free (DESE) and native (NATI) enzyme crystals. From difference Fourier analysis of the deseleno enzyme (Epp et al., 1983), convincing evidence was provided that the reaction of the selenocysteine residues with cyanide proceeds via a displacement reaction that can be described by

$$R\text{-}CH_2\text{-}Se^- + CN^- \longrightarrow R\text{-}CH_2\text{-}CN + Se^{2-}. \tag{3}$$

Inspection of a difference Fourier map calculated with coefficients $(|F_o^{DESE}| - |F_c^{DESE}|)* \exp i\alpha_c^{DESE}$ revealed a residual density maximum adjacent to the β-carbon of the selenocysteine residue that could be readily interpreted as a bound cyano group. $|F_o|$ and $|F_c|$ represent the observed and calculated structure factor amplitudes of DESE, respectively.

Seleno GSH peroxidase can be inhibited readily by treatment with iodoacetate (Flohe and Guenzler, 1974) or chloroacetate (Wendel et al., 1978) if reduced by GSH or potassium borohydride. The stoichiometry of the binding of labeled haloacetates is shown in Table 4.

Curiously, iodoacetamide does not inactivate the peroxidase. An explanation for the finding that 2 molecules of chloroacetate or iodoacetate are bound to the tetrameric enzyme, concomitant with complete inactivation, will be given in Section 3.5. The control experiment with cyanide-treated enzyme showed that the presence of selenium is required for chloroacetate binding. The binding energy of the Se 3d electrons measured by X-ray

Table 4. Binding Stoichiometry of ^{14}C-labeled Haloacetates to Native and Cyanide-treated GSH Peroxidase of the Reduced Form

Alkylating Reagent	Treatment	Reagent Bound per Tetramer	Selenium Bound per Tetramer
Iodoacetate	—	2.1	3.5
Iodoacetamide	—	0.1	4.0
Chloroacetate	—	2.2	4.0
Chloroacetate	Cyanide	0.1	0.02

Source: From Wendel et al. (1978).

photoelectron spectroscopy indicates some change in the electron orbitals of the Se atom, and also with 56.8 eV, a position between the values for reduced (54.1 eV) and oxidized (58.0 eV) peroxidase (Wendel et al., 1975, 1978). These data demonstrate that the active site selenium can be chemically modified by treatment with haloacetate compounds.

In attempts to titrate the active sites of reduced GSH peroxidase, the inhibition by iodoacetate was used to evaluate the number of active sites (Guenzler, 1974; Flohe and Guenzler, 1974). Stepwise addition of *t*-butyl-hydroperoxide to the GSH-reduced enzyme resulted in a gradual decrease in the inhibition by iodoacetate until full protection of the enzyme was achieved. This point was reached at a molar ratio of 2.2 hydroperoxide/tetrameric enzyme, but the authors discussed four active sites and postulated reoxidation for the enzyme prior to titration.

In a systematic search for effectors of GSH peroxidase, a number of mercaptocarboxylic acids and tertiary mercaptans (Table 5) were found to be strong and specific inhibitors of the selenoenzyme (Chaudiere et al., 1984). The results of kinetic investigations supported the formation of reversible enzyme-inhibitor complexes. The active site selenium seemed to be trapped by rapid binding of the strong inhibitors in competition with GSH. Furthermore, these authors claimed a noncovalent GSH-binding site or the presence of a sulfurcysteine at the active site of the enzyme in order to rationalize their experimental observations. They further assumed that a nearby histidine could modulate the protonation state of the sulfurcysteine thiol group. Unfortunately, a preliminary sequence containing an erroneously assigned active site histidine had been published (Ladenstein et al., 1979). This sequence only served as a starting sequence to initiate crystallo-

Table 5. Inhibition of GSH Peroxidase Activity by Mercaptocarboxylic Acids and Related Thiols

Inhibitor (0.2 mM)	Enzyme Activity[a]	pK_a (SH), 25°C
Mercaptosuccinate	0	10.4
2,3-Dimercaptosuccinate	0	—
2-Mercaptobenzoate	0.8	9.96
β-Mercaptovaline	0.5	10.46
t-Butylmercaptan	0	11.22
Methylmercaptoacetate	0.3	9.85

Source: From Chaudiere et al. (1984).
[a]The enzymatic activity is expressed as fractions of the control at 0.25 mM GSH, 37°C.

graphic refinement of the GSH peroxidase structure and had no biochemical relevance. Within the refined three-dimensional structure, neither a sulfurcysteine nor a histidine residue could be detected in a position near the active site of the enzyme (Epp et al., 1983). This result has recently been confirmed by chemical sequence analysis of bovine erythrocyte GSH peroxidase (Guenzler et al., 1984).

2. PRIMARY STRUCTURE

2.1. Chemical Sequence Analyses of the Bovine Erythrocyte and the Rat Liver Enzyme

Recently, important progress regarding the primary structure of GSH peroxidase from red blood cells has been achieved (Guenzler et al., 1984). The amino acid sequence obtained by chemical analysis represented the first complete determination of the primary structure of a selenoprotein (Table 6). Three different methods were employed for cleavage of the peroxidase: tryptic digestion, treatment with endoproteinase Lys-C, and cleavage by cyanogen bromide. The resulting peptide fragments were analyzed by manual and automated Edman degradation techniques. Until now, only a partial N-terminal sequence of the corresponding rat liver peroxidase was available, and this peptide contains the active site selenocysteine residue (Condell and Tappel, 1982). The sequences of the N-terminal part of rat and bovine enzyme show an expected high degree of homology if properly aligned (Table 6).

536

Table 6. Comparison of Sequence Data of GSH Peroxidase

```
       1                                            10                                           20
I    Ala-Ala-Ala-Leu-Ala-[Ala-Ala-Ala]-Pro-Arg-[Thr-Val-Tyr-Ala-Phe-Ser-Ala-Arg-Pro-Leu-Ala-Gly-Gly-
II                  Gly-[Ala-Val-Ala]-Gln-Ser-[Thr-Val-Tyr-Ala-Phe-Ser-Ala-Arg-Pro-Leu-Ala-Gly-Gly-
III  . . .-. . . .-. . .-. . -. . . .-. . . .-. . .-Ala——————————————————————————Gly————

                          30                                   40
I    Glu-Pro-[Phe-Asn-[Leu-Ser-[Ser -Leu-Arg-Gly-Lys-Val-Leu-Leu-Ile-Glu-Asn-Val-Ala-Ser-Leu-Sec-Gly-
II   Glu-Pro-[Val -Ser-[Leu-Gly-[Ser -Leu-Arg-Gly-Lys-Val-Leu-Leu-Ile-Glu-Asn-Val-Ala-Ser-Leu-Sec-Gly-
III  . . . ————————Ser————————Ala————————————————Gln————————

                   50                                   60
I    Thr-Thr-[Val-[Arg]-Asp-Tyr-Thr-Gln-Met-Asn-Asp-Leu-Gln-Arg-Arg-Leu-Gly-Pro-Arg-Gly-Leu-Val-Val-
II   Thr-Thr-[Thr-[Arg]
III  ————————Thr————Asn————Ser————Gln————————————Gln————

     70                                  80                                 90
I    Leu-Gly-Phe-Pro-Cys-Asn-Gln-Phe-Gly-His-Gln-Glu-Asn-Ala-Lys-Asn-Glu-Glu-Ile-Leu-Asn-Cys-Leu-
III  ————————————————Met————Gln————————————Gln————————————His———

                        100                              110
I    Lys-Tyr-Val-Arg-Pro-Gly-Gly-Gly-Phe-Glu-Pro-Asn-Phe-Met-Leu-Phe-Glu-Lys-Cys-Glu-Val-Asn-Gly-
III  Gln————————————————Gln————————Leu————————Gln——————Met-Lys————
```

                    120   130
I   Glu-Lys -Ala -His -Pro -Leu-Phe -Ala -Phe-Leu-Arg-Glu- Val- Leu-Pro- Thr- Pro- Ser- Asp-Asp-Ala- Thr- Ala-
III Ala-Ser ————————————————————————————Lys ——————————————— Thr ——————————————

          140   150   160
I   Leu-Met-Thr -Asp-Pro -Lys -Phe -Ile -Thr- Trp -Ser- Pro- Val- Cys- Arg-Asn-Asp-Val- Ser- Trp- Asn-Phe- Glu-
III ——Gln————Asn——Gln————————————————————Gln ——Asp ————————————————His-

                    170   180
I   Lys- Phe -Leu-Val -Gly -Pro -Asp-Gly -Val- Pro -Val- Arg-Arg-Tyr- Ser- Arg-Arg-Phe-Leu-Thr-Ile- Asp-Ile-
III ————————————————————Asn——Thr ————————————————————————Asn ——

                    190   198
I   Glu-Pro -Asp-Ile -Glu -Thr -Leu-Leu-Ser- Gln -Gly- Ala- Ser- Ala
III ————Asn——Ser ——————————. . .‾. . . .‾. . .‾. . .‾. . .‾. . .

Source: Adapted with permission from Guenzler et al. (1984).

The complete amino acid sequence of bovine red blood cell GSH peroxidase (I) determined by Guenzler et al. (1984) is aligned to the partial sequence reported for the rat liver enzyme (II) (Condell and Tappel, 1982), and to the sequence of the bovine red blood cell enzyme as predicted from crystallographic refinement (III) (Epp et al., 1983). Numbers give positions within sequence I. Sec indicates selenocysteine. Identical residues in the sequences of bovine and rat enzyme are framed in boxes. Only residues that differ from sequence I are listed in sequence III; identity is indicated by a solid line; residues not detected by X-ray crystallography are marked by · · ·.

2.2. Comparison with the Sequence Predicted From X-ray Analysis

The fragmentation strategy and the alignment of peptides took advantage of the sequence proposed from the crystallographically refined structure of GSH peroxidase (Epp et al., 1983). Guenzler and coworkers have presented a comparison of the chemically determined sequence with that predicted from the X-ray structure analysis (Table 6). In this regard, it should again be mentioned that an earlier sequence proposal derived from a 2.8 Å electron density map proved to be misleading and caused several wrong attempts to explain the catalytic function of this enzyme (see section 1.3). With the refined structure at 2 Å resolution, however, no wrong assignment of the aromatic residues occurred, and an overall correlation in the prediction of roughly 80% was achieved. Most erroneous predictions concerned the amidation of Asp and Glu residues, and it is worthwhile to point out wrong predictions of amino acid residues that had been considered to be involved in subunit interaction, GSH binding, and the regulation of catalytic function (Table 7). Guenzler and coworkers described a distinct difference in overall length between the determined and the predicted sequences. The chemi-

Table 7. Wrong Predictions of Amino Acid Residues by Crystallographic Refinement That Had Been Considered to be Involved in the Function of GSH Peroxidase

Function	Predicted Residue[a]	Chemically Determined Residue[b]
Subunit contacts		
Dimer	Met 69	His 79
	His 81	Cys 91
	Gln 134	Lys 144
Tetramer	Val 121	Thr 131
	Gln 130	Met 140
	Gln 142	Cys 152
	Asp 144	Asn 154
GSH binding at active site	Gln 130	Met 140
GSSG binding	His 81/281	Cys 91/291
Flexible side chain	Met 101	Cys 111

[a]From Epp et al. (1983).
[b]From Guenzler et al. (1984).

cally determined sequence contained an N-terminal extension of nine amino acids and a C-terminal extension of six amino acids. Hence, it is tempting to assume that the N- and C-terminal ends of the peptide backbone are flexible within the crystals and are therefore not apparent by X-ray analysis. On the other hand, the N-terminal nonapeptide, which is almost coincident with the tryptic peptide T1, could have been cleaved off in previous preparations. It has been speculated that its hydrophobic character may give it a signal peptide function or, alternatively, may facilitate binding to membrane structures that require protection from attack by hydroperoxides.

In order to facilitate comparison of sequence segments with structure elements and residue numbers, we have adapted the residue-numbering system published by Guenzler et al. (1984) to the three-dimensional structure of bovine erythrocyte GSH peroxidase (Table 6).

3. SECONDARY AND TERTIARY STRUCTURE

3.1. The Folding of the Peptide Chain

The crystal structure of GSH peroxidase has been analyzed by isomorphous replacement at 2.8 Å resolution phased with two mercury and two platinum derivatives (Ladenstein et al., 1979). The model structure has subsequently been refined at 2 Å resolution, using a joint procedure of restrained crystallographic refinement and energy minimization (Epp et al., 1983).

A schematic drawing of the secondary structure elements of a GSH peroxidase subunit is shown in Figure 5. A subunit is built up from a central core of two parallel (β_1, β_2) and two antiparallel (β_3, β_4) strands of a β-pleated sheet surrounded by four α helices. This pleated sheet structure has a right-handed twist of about 45°. The helices α_1, α_2, and α_4 are placed on one side of the β structure and helix α_3 is on the other side. Twelve clearly defined β turns of types I, II, III and III[1] (Crawford et al., 1973) as well as three one-turn 3_{10} helices could be recognized in the monomer structure.

Figure 6 shows two stereo diagrams of the α-carbon backbone corresponding to a monomer in different orientations. In addition, several side chains that are of functional importance are indicated.

The analysis of the atomic temperature factors allowed some conclusions on the mobility of the structure (Epp et al., 1983). Some strands forming β sheets (β_1, β_4), two of the α helical structures (α_1, α_3), and certain parts of

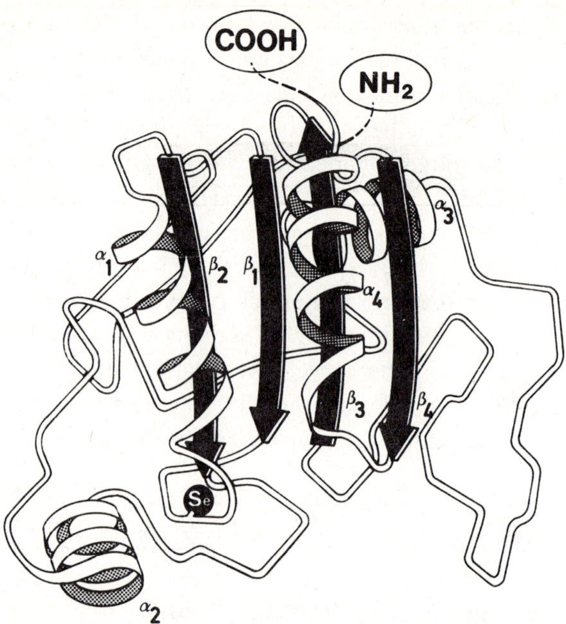

Figure 5. Schematic drawing of the folding pattern of a GSH peroxidase monomer. α_i and β_i indicate α helices and β strands, respectively. Se shows the position of the selenocysteine residue (SeCys 45). (Adapted with permission from Epp et al., 1983.)

the chain involved in subunit contacts especially show significantly lower mean-square displacements than other parts without well-defined secondary structure elements. It is also evident that the mobility of the main chain in the central regions of the α helices is lower than at the ends. N- and C-terminal residues show high temperature factors, indicating that these parts are relatively flexible. The highest values can be found at the positions of a narrow loop (Gly 22, Gly 23, Glu 24) near to the N-terminus and some β turns. These residues are exposed to solvent and, therefore, are less constrained by packing effects. The chain segments around the active site (SeCys 45) are formed by a well-defined secondary structure element: a $\beta\alpha\beta$ structure. They are involved in hydrogen bond interactions and exhibit low mean-square displacements that indicate a relatively rigid arrangement of the peptide chains at the active center region.

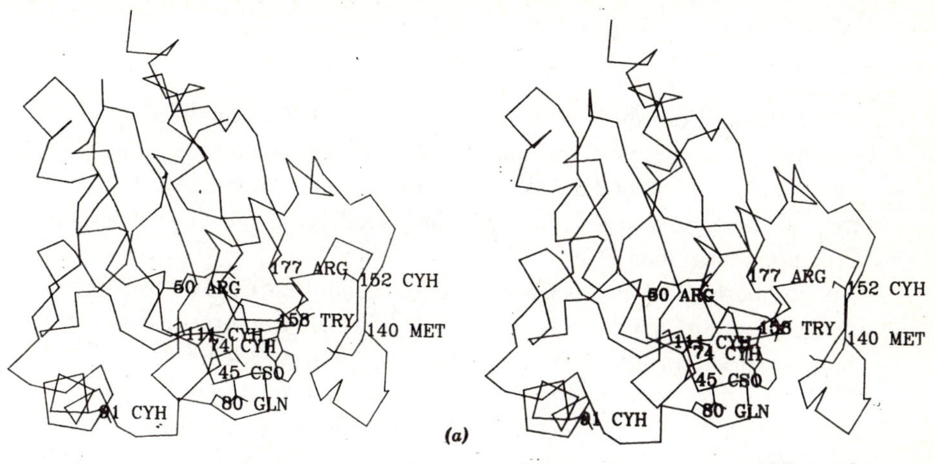

(a)

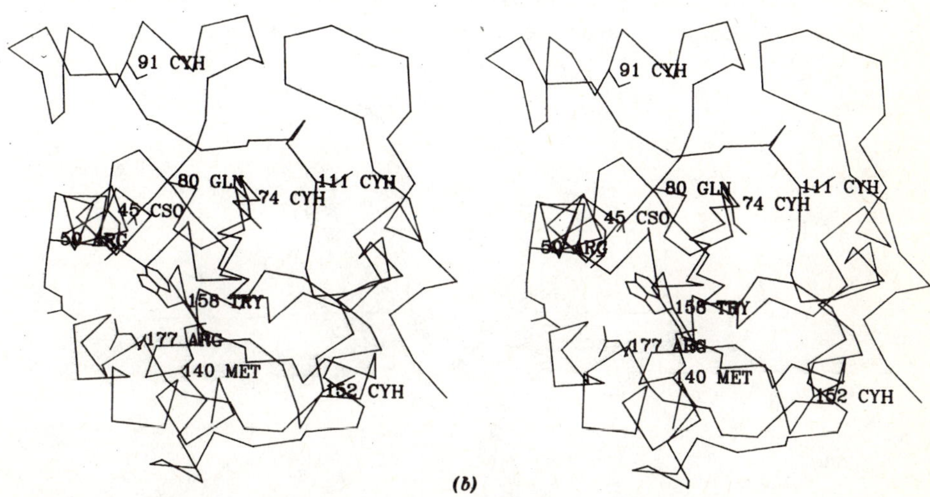

(b)

Figure 6. Stereo diagram of the α-carbon backbone of a GSH peroxidase subunit. Several side chains in the neighborhood of the active site selenocysteine (45 CSO) are shown. (*a*) View along molecular *Q* axis. (*b*) View along crystallographic *y* axis.

3.2.   Comparison with the Chain Folding of Bacteriophage-T$_4$-Thioredoxin

The chemically determined sequence was analyzed by a computer search for homology with other proteins, but no closely related sequences were found, although certain parts of the GSH peroxidase fold are similar to the structures of bacteriophage-T$_4$-thioredoxin (Soederberg et al., 1978) and rhodanese (Ploegman et al., 1978).

A comparison of the GSH peroxidase subunit structure with bacteriophage-T$_4$-thioredoxin revealed large regions of structural resemblance. Thioredoxin shows almost the same folding pattern as the core structure of a GSH peroxidase subunit (Fig. 7) with four strands of β structure and two adjacent α helices. In place of the selenocysteine involved in catalysis, T$_4$-thioredoxin contains a redox-active disulfide at an analogous position of the βαβ unit. The striking similarity of the active regions of these two redox proteins may be the result of convergent evolution or, alternatively, may indicate divergence from a common ancestral redox molecule.

When one considers the T$_4$-thioredoxin fold as a fundamental element of the GSH peroxidase monomer, it seems reasonable to suggest that during protein evolution, two insertions into the backbone of a monomer may have provided additional structure elements—two loops formed by residues

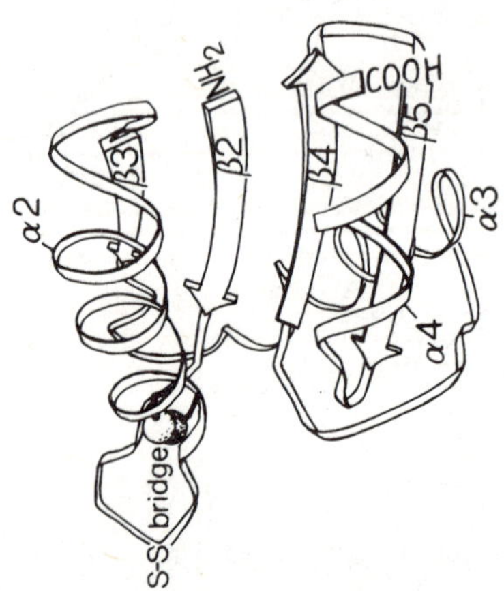

Figure 7. Folding of bacteriophage-T4-thioredoxin. α$_i$ and β$_i$ indicate α helices and β strands. S-S shows the position of the redox active disulfide bridge. (Adapted with permission from Soederberg et al., 1978.)

77–83 and 134–150, and helix α_2 forming contact sites for protein-protein interaction—and may have allowed specific assembly of the oligomeric GSH peroxidase molecule.

3.3. The Structure of the Active Center

The catalytically active selenocysteine residue 45 is located at the N-terminal end of helix α_1, which forms a $\beta\alpha\beta$ substructure together with the two adjacent parallel β strands (β_1, β_2). Similar arrangements of secondary structure are frequently observed at active site regions of enzymes that are folded into α- and β-structural elements. In general, $\beta\alpha\beta$ structures provide a geometrically favorable situation for the occurrence of a binding region near the carboxy ends of two adjacent parallel β strands (Brändén, 1980). Because the active center of GSH peroxidase can be localized in the general vicinity of SeCys 45, where the carboxy ends of two parallel strands (β_1, β_2) meet one another, it is tempting to speculate that this arrangement is of unique importance for catalysis. It has long been known that in the α helix, the alignment of the peptide dipoles parallel to the helix axis gives rise to a macrodipole of considerable strength. For points near the helix N-terminus, the effect of the dipole is equivalent to the effect of half of a positive unit charge (Hol et al., 1978). Hence, one could imagine that the electric field due to the dipole moments of the helices α_1 and α_4 will stabilize the active site selenolate and enhance its nucleophilic reactivity. The charged cosubstrate GSH, itself representing a dipole, may be oriented by this field prior to the electron transfer process.

The active sites of GSH peroxidase are localized in depressions on the molecular surface and are readily accessible. In a space-filling drawing of the molecule (Fig. 8), no prominent clefts or crevices are visible at the active center regions. Exposure of the catalytically active selenocysteine residues at the molecular surface is consistent with the easy access of the substrates and, thus, the high reaction rates of the enzyme.

In the refined electron density map of the oxidized enzyme, no covalent connections of the selenocysteine side chains to ligands in its environment can be observed. Hence, the active centers contain free selenocysteines or the corresponding oxidized derivatives, depending on the functional state of the enzyme. Figure 9 shows the chemical environment of the selenocysteine residue in reduced GSH peroxidase. At physiological pH values, a

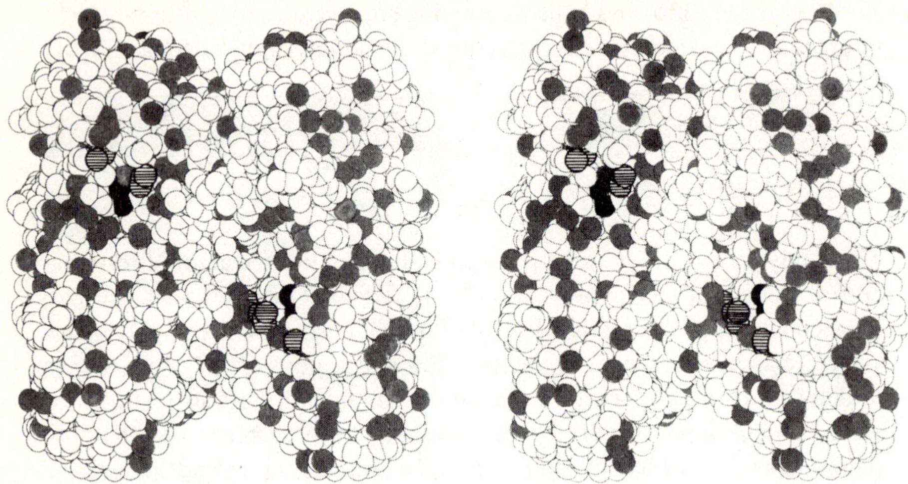

Figure 8. Space-filling drawing of a GSH peroxidase tetramer in the oxidized state (stereo pair). The oxygen atoms corresponding to the seleninic acid group of the enzyme are indicated by black spheres. Bound water molecules are represented by gray spheres, and the residues presumably involved in GSH binding (Arg 50, Arg 177) are marked with small lines. (Adapted from Epp et al., 1983.)

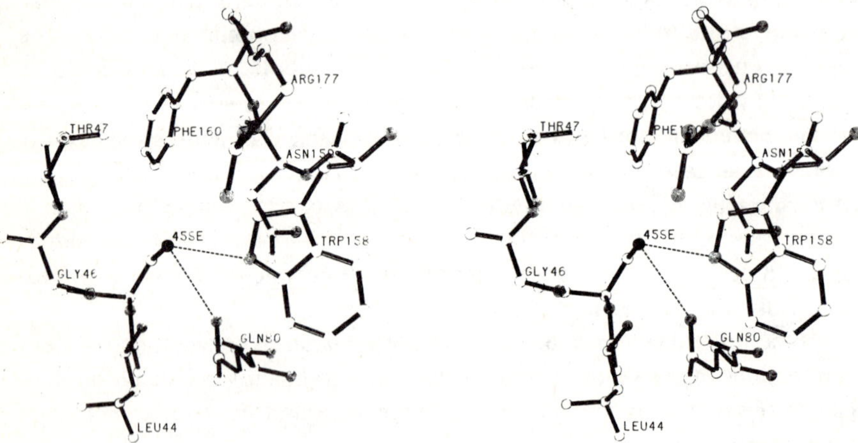

Figure 9. Chemical environment of the selenocysteine residue in the reduced enzyme (stereo pair). (●), position of the selenium atom; (− − − −), hydrogen bonds to 158 Trp-N$^\varepsilon$ and 80 Gln-N$^\varepsilon$. (Adapted with permission from Epp et al., 1983.)

544

selenolate anion, rather than an undissociated free selenol, seems to occur in the reduced enzyme (pK_{SeH} = 5.24; Huber and Criddle, 1967).

Additional evidence is provided by difference Fourier analysis between inhibited enzyme forms and the reduced enzyme (Epp et al., 1983). The observed distances are consistent with hydrogen bond formation of the selenolate anion to 158 Trp-N^ε (d = 3.4 Å) and to 80 Gln-N^ε (d = 3.3 Å). This would certainly result in some stabilization of the active site geometry (see Table 8). A difference Fourier analysis between oxidized and reduced GSH peroxidase has provided evidence for a seleninic acid derivative (E-SeOOH), with pyramidal arrangement of β-carbon, selenium, and oxygen atoms existing in the peroxide-oxidized enzyme form. The seleninic acid derivative could be readily reduced by GSH in the crystalline state to an enzyme form containing the catalytically active selenol, and could also be stabilized through coordination via hydrogen bonds (Fig. 10, Table 8).

The occurrence of selenocysteine residues in the active centers of GSH peroxidase poses an interesting question as to how this unusual amino acid is incorporated into this enzyme. In most cases, a specific post-translational modification of an existing amino acid in a polypeptide chain is responsible for the occurrence of an abnormal amino acid in a protein molecule. A selenocysteine may be formed by addition of selenide or selenoglutathione (G-S-SeH) to a dehydroalanine generated from a specific serine or cysteine.

Table 8. Coordination of the Active Site in Oxidized and Reduced GSH Peroxidase

Functional State	Coordinated Atom	Coordination Partner	Distance (Å)
Oxidized	Oxygen (1)	N^ε-Trp 158	3.2
		N^ε-Gln 80	3.0
		N-Gly 46 main chain	2.9
E-Se-OH (with =O)	Oxygen (2)	O^γ-Thr 47	2.9
		N-Thr 47 main chain	2.9
		O-H_2O bound water	3.0
Reduced E-Se$^-$	Selenolate anion	N^ε-Trp 158	3.4
		N^ε-Gln 80	3.3

Source: From Epp et al. (1983).

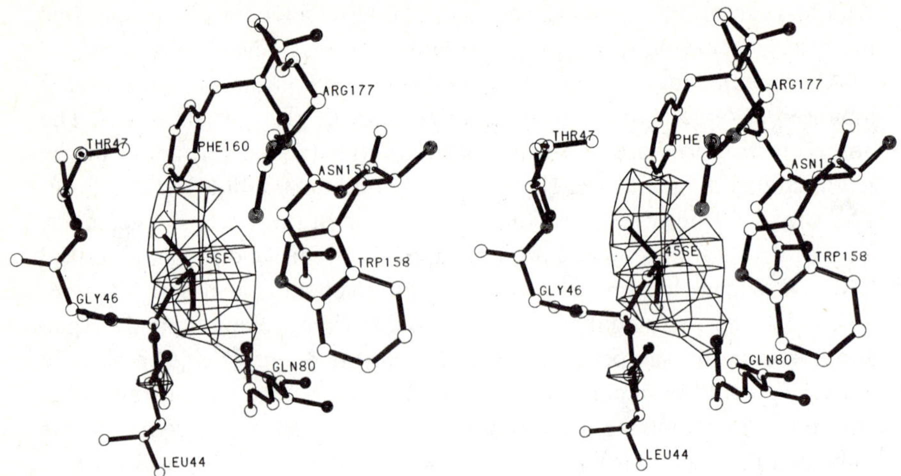

Figure 10. Stereo representation of the environment of the active site in the oxidized enzyme, overlaid with the positive difference electron density calculated between oxidized and reduced enzyme. Contour level, 2.0 σ. (Adapted with permission from Epp et al., 1983.)

In the past few years, however, experimental evidence has accumulated that rat liver contains a form of t-RNA that is aminoacylated specifically with selenocysteine (Hawkes et al., 1982). Although the existence of a specific selenocysteyl t-RNA is consistent with the idea that selenocysteine might be incorporated into selenoproteins during protein synthesis, it does not conclusively prove this point. It remains to be determined if this type of pathway occurs in vivo and, if so, to what extent it could account for the synthesis of known selenocysteine-containing proteins. These results are in contrast to reports of similar experiments in *Escherichia coli* (Young and Kaiser, 1975), where only insignificant differences were observed between the behavior of cysteyl and selenocysteyl t-RNA.

Studies to reconstitute the native enzyme by insertion of seleno and sulfur compounds into selenium-depleted crystals have been unsuccessful thus far (Epp et al., 1983). A possible explanation is the binding of cyanide as proposed in Section 1.3.

3.4. Structured Solvent

During crystallographic refinement, 165 solvent molecules were located in a GSH peroxidase dimer (Epp et al., 1983). All solvent molecules were taken as water molecules; 23 water molecules are internal, and, hence, are constituents of the structure. Five with B factors of $15-30$ Å^2 ($B = 8\,\pi^2 <u>^2$; $<u>$ is the mean-square amplitude of isotropic harmonic displacement) form a chain of water molecules across the local twofold dimer axis at a presumptive GSSG-binding region where the carboxy ends of two parallel β strands (β_1, β_2) and helix α_1 intersect (Fig. 11).

It has been suggested that this site on the dimer axis is involved in GSSG binding. The chain of water molecules would be displaced from the site upon binding of a GSSG molecule. These findings are analogous to the results from the refinement of several serine proteases (James et al., 1980). The observation that the solvent structure of free binding sites due to local charge effects mimics the presence of bound substrates or inhibitors provides supportive evidence for the existence of a binding site that might be composed of residues Gln 80/280, Glu 81/281, Asn 82/282, and Cys 91/291 at the twofold dimer axis (see also Section 3.5). High GSSG concentration does appear to cause inhibition of the enzymatic activity of GSH peroxidase (Wendel, unpublished results).

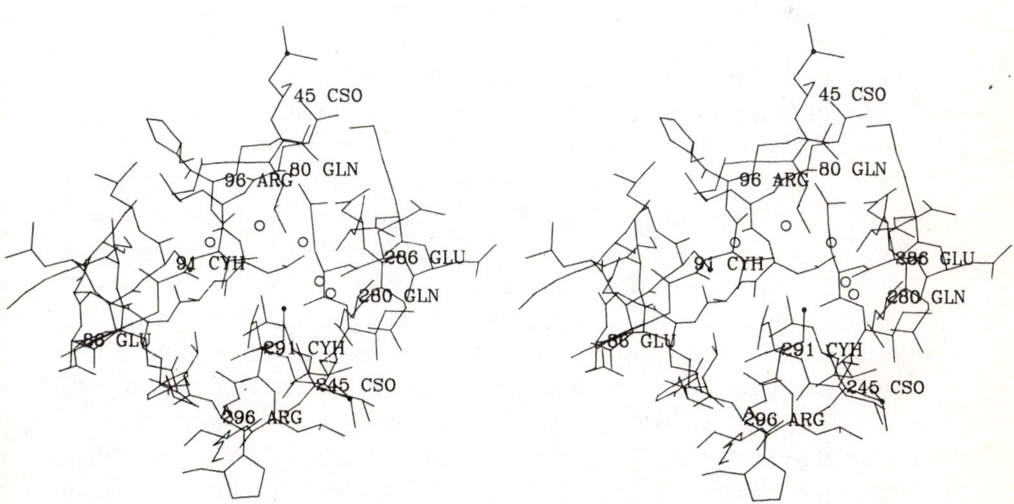

Figure 11. Stereo pair showing the presumptive GSSG-binding region at the local dimer axis. Five water molecules (o) form a solvent chain across the local axis. The sulfur atoms corresponding to the cysteine residues (91 CYH/291 CYH) as well as the selenium atoms of the active sites (45 CSO/245 CSO) are indicated by black spheres (●).

A well-defined water structure is also found near the active site selenocysteine residue 45. The water molecules involved are characterized by B factors in the range $15-25$ Å^2, except for one solvent molecule H_2O 417, which is hydrogen bonded to the main chain carbonyl group of Asn 159 and to a side chain oxygen of the seleninic acid at position 45 (see Table 8). This water molecule shows a low B value of 8.8 Å^2; it would be displaced together with several others upon GSH binding and reduction of the active site.

3.5.   A Preliminary Description of GSH Peroxidase Catalysis in Molecular Terms

In view of the binding data obtained with different inhibitors and GSH, half-site reactivity has been postulated for GSH peroxidase (Epp et al., 1983). Half-site reactivity is related to negative cooperativity, and indicates that constituent protomers of an oligomeric enzyme do not behave as independent entities. This would give a functional meaning to oligomeric structures exhibiting classic Michaelian kinetics. Mechanisms describing the functional properties of half-site enzymes have been proposed (Lazdunski, 1972), and they assume a functional interrelationship between distinct active sites on identical monomers. It is further concluded that these oligomers behave as "polydimers"; indeed, a polydimeric character appears to be a fairly general feature of enzymes with subunit structure.

In crystalline GSH peroxidase, the dimer in an asymmetric unit exhibits local twofold symmetry. This suggests that small structural differences might exist between the monomers. It has been postulated that the tetramers might function as $(\alpha\beta)_2$ entities. Probably, the dimeric structure looses its symmetry by the binding of one GSH molecule. The catalytic events at different active sites of a dimer are presumably controlled by negative cooperativity, suggesting that chemical signals are transferred across the twofold symmetry axis of the dimer. In this hypothetical model, the active sites would react in a flip-flop-type mechanism (Lazdunski, 1972), and would influence each other over a distance of roughly 20 Å. Since the classic work of Monod et al. (1965), the mechanism whereby information might be transmitted from one location to another has been assumed to involve long-range conformational changes in a protein; but as Monod et al. clearly pointed out, there are other alternatives. In particular, one may imagine ways in which changes in dynamic properties of the system might be equally effective in communication between sites (Salemme, 1978; Cooper, 1980). In the dynamic view, the information content of a macromolecule consists not only of its conformation, expressed in terms of the average atomic coordinates, but also of the frequencies and amplitudes of dynamic fluctuations about these positions. It follows that information exchange between sites on a macromolecule might be mediated not only by ligand-induced changes in average atomic positions, but also by changes in the dynamic frequency and amplitude spectrum. It is conceivable that dynamic changes can occur without changing the mean atomic coordinates; in other words, without any conformational change in the conventional sense.

There are several lines of indirect evidence that might indicate small structural rearrangements upon GSH binding. A highly structured ultraviolet (UV) difference spectrum and perturbations in the UV circular dichroism spectrum are observed under the influence of the reducing substrate (Flohe et al., 1971c). A similar behavior could also be observed by difference Fourier analyses of reduced and oxidized enzyme forms (Epp et al., 1983). Hence, small and subtle conformational changes or side chain movements are likely to occur when the oxidation state of the peroxidase is changed. There is no crystallographic evidence, however, indicating the occurrence of large conformational changes upon reduction by GSH.

The catalytic process can be described by the following hypothetical reaction mechanism, which is supported by a broad basis of experimental results:

1. Substrate specificity studies;
2. Analysis of the kinetic mechanism;
3. Inhibitor and substrate-binding data;
4. Biochemical investigations on the nature of the enzyme-bound selenium;
5. X-ray photoelectron spectroscopic studies; and
6. X-ray structure analyses of reduced and oxidized enzyme forms.

The reduced form of the enzyme reacts with a hydroperoxide substrate in a bimolecular reaction. Probably, this reaction proceeds without the formation of a specific ES complex. The reduced enzyme most likely contains a highly dissociated selenol group that gives rise to a selenolate anion ($E\text{-}Se^-$). This negatively charged anion can be stabilized by hydrogen bonding to

N^ε-Trp 158 or N^ε-Gln 80. The interplay between SeCys 45 and Trp 158 seems to be important for catalysis, but is not completely understood at present. Several oxidized forms of the peroxidase may exist, depending on the relative concentrations of reducing and oxidizing substrates. When one assumes physiological substrate concentrations (GSH in the mM range and hydroperoxides in the μM range), the main reaction would certainly shuttle between a selenolate anion and a selenenic acid derivative (E-SeOH) of the active site selenium. In the peroxide-oxidized enzyme (5 mM H_2O_2), however, a seleninic acid derivative (E-SeOOH) could be detected by X-ray crystallography (Epp et al., 1983). However, this form—although it could be reduced to the active site selenol in the crystalline state—is assumed to represent a nonphysiological-oxidized form of the peroxidase. A mixed selenosulfide formed with a nearby cysteine SH group (R-Se-S-R) can be excluded as an alternative, because no sulfur amino acid side chain at a distance from the active site compatible with the formation of covalent bonds could be detected in the molecule. The reduction of the oxidized enzyme is initiated by formation of an (E · GSH) complex. This complex is transformed into a mixed selenosulfide intermediate E-Se-SG. The formation of the complex (E · GSH) seems to be much slower than the intramolecular transformation into the covalent intermediate. In the last step of the mechanism, the second GSH molecule restores the reduced enzyme form.

Conclusions regarding the mechanism and the involvement of additional functional side chains must await further experimental evidence, for example from site-directed mutagenesis. Because we still do not completely understand the catalytic function of this enzyme on a molecular level, we present a hypothetical minimal picture of the assumed catalytic steps (Fig. 12).

It appears likely that catalysis proceeds in a hydrogen bond network between substrates and functional groups, involving a redox shuttle of the selenocysteines without direct participation of sulfurcysteines. Changes in the UV difference spectrum induced by GSH have been interpreted as the participation of a tryptophan in catalysis. The formation of a selenosulfide bond (E-Se-S-G) may be indicated by an absorption maximum (shoulder) at λ = 273nm (Graupe, 1977).

The additional cysteine residues at positions 74, 91, 111, and 152 found by chemical sequence analysis (Guenzler et al., 1984) merit further consideration. A distance calculation (Table 9) revealed that in the crystal structure, they are located too far apart from each other, and from the selenocysteine residue, to form disulfide (-S-S-) or selenosulfide (-Se-S-) bonds within a

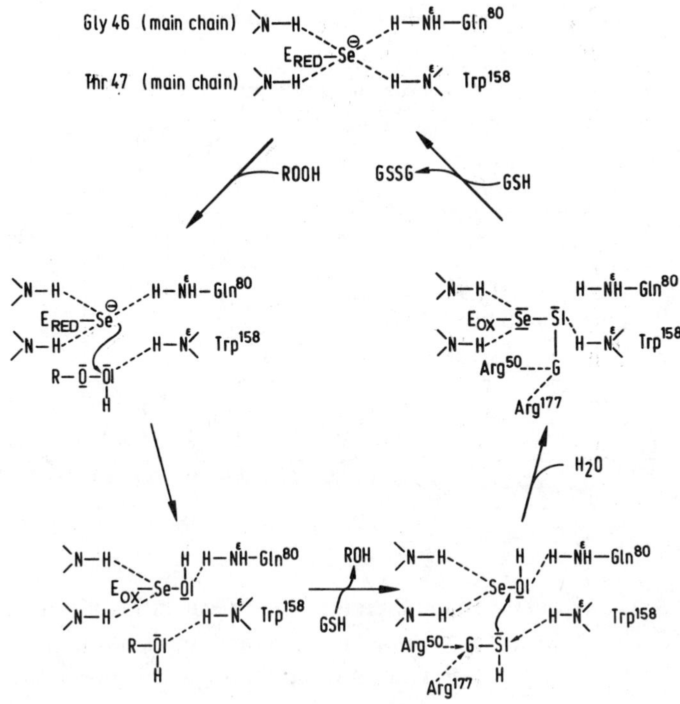

Figure 12. Tentative model of the catalytic process. The occurrence of a short-lived selenenic acid intermediate (E-SeOH) in an oxidized enzyme species is assumed. ($- - - -$), hydrogen bonds and charge interactions presumably formed during catalysis.

monomer or a dimer. This finding, however, does not exclude the possibility that cysteine residues are associated with the catalytic function of the enzyme or even contribute to the regulation of the catalytic activity of GSH peroxidase.

The pair of cysteines 91/291 constitutes a binding site at the intersubunit boundary that can bind GSSG and other thiols such as cysteine-O-methylester (see Fig. 11; Epp et al., 1983). These cysteines can be modified by peroxodisulfate treatment in solution, resulting in a complete loss of the enzymatic activity. Difference Fourier analysis of the corresponding crystalline derivative revealed a density maximum, possibly reflecting oxidation of the SH groups, and stretching symmetrically across the local axis at the positions of the cysteine sulfurs 91/291, nearly at the same position where we assume GSSG binding to occur. Presumably these cysteines, which are

Table 9. Distances (Å) Between Sulfur- and Selenocysteine Residues in a GSH Peroxidase Dimer (S ⟶ S, Se ⟶ Se, Se ⟶ S)

| | Distances of Cysteine-Sulfurs | | | | | Dimer boundary | |
| | | | | | | | |
d (Å)	SeCys 45	Cys 74	Cys 91	Cys 111	Cys 152	SeCys 245	Cys 291
SeCys 45	—	Monomer 1				Monomer 2	
Cys 74	11.4	—					
Cys 91	13.5	14.9	—				
Cys 111	19.5	8.9/5.1	17.8	—			
Cys 152	22.0	14.5	29.3	17.5	—		
SeCys 245	21.2	19.5	15.2	21.9	29.8	—	
Cys 291	14.9	18.9	6.2	22.9	32.5	13.4	—

about 10 Å away from the selenocysteines 45/245, can influence the catalytic process in a very distinct way, but this phenomenon is not understood in molecular terms at present. Furthermore, because of a change in the charge distribution of the active site selenocysteines, a movement of the side chains of cysteines 111/311 on opposite surfaces from the selenium sites of the monomers was observed (Epp et al., 1983).

4. QUATERNARY STRUCTURE AND MOLECULAR SYMMETRY

Tetrameric GSH peroxidase consists of four chemically indistinguishable subunits. The asymmetric unit of the monoclinic crystal cell is occupied by a dimer (Ladenstein et al., 1979), and the monomers are related by noncrystallographic symmetry, which could be established by analysis of Patterson autocorrelation functions (Ladenstein et al., 1979; Rossmann and Blow, 1962). The tetrameric molecules show (222) symmetry, proving the subunits to be identical or at least very similar. A molecular coordinate system relative to the crystal axes can be defined by the orthogonal axes P, Q, and R, given in Table 10. The subunits of an asymmetric unit that contains a dimer are related by the linear transformation $\vec{x}\,' = M \cdot \vec{x} + \vec{t}$

(the rotation matrix M and the translation vector $\vec{T}$, which transfer monomer 1 onto monomer 2 in the asymmetric unit, are given in Table 10).

The molecular symmetry and the packing of GSH peroxidase tetramers in the monoclinic crystal cell are shown schematically in Figure 13. The tetrameric arrangement of the subunits is very similar to an almost planar arrangement of four monomers.

The data obtained by cross-linking studies in solution with bifunctional reagents (Ladenstein et al., 1977) fit best to the assumption of a dimer square model with isologous structure showing D_2 symmetry in total. Hence, two completely different methods led to very similar conclusions on the quaternary structure of GSH peroxidase.

A linear correlation, established between the surface area accessible to solvent and hydrophobic free energy (Chothia and Janin, 1975), suggests that the reduction of accessible surface area occurring upon subunit association is the main source of free energy (ΔG transfer) stabilizing protein-protein interactions. In view of the similar magnitudes of the monomer contact areas buried on association, roughly identical values of the individual association constants between the GSH peroxidase monomers have been predicted (1427 Å^2 at the local axis and 1472 Å^2 at the crystallographic axis, respectively; Epp et al., 1983).

Table 10. Polar Coordinates of the Twofold Symmetry Axes Relating the Subunits of a GSH Peroxidase Tetramer and Linear Transformation from Monomer 1 onto Monomer 2

Molecular Axes	Polar angles		Type of Axis
	ψ	ϕ	
P	90.0	138.3	Local twofold
Q	90.0	48.3	Local twofold
R	0	0	Crystallographic twofold

Source: From Ladenstein et al. (1979); Polar angle definition according to Rossmann and Blow (1962).

This linear transformation transforms monomer 1 onto monomer 2 in the asymmetric unit (the asymmetric unit contains a dimer). The root-mean-square deviation of main chain atoms of the two monomers is 0.235 Å.

$$\begin{pmatrix} x \\ y \\ z \end{pmatrix} = \begin{pmatrix} 0.28884 & -0.00975 & 0.91322 \\ 0.00237 & -0.99994 & -0.01099 \\ 1.00356 & 0.00382 & -0.28880 \end{pmatrix} * \begin{pmatrix} x \\ y \\ z \end{pmatrix} + \begin{pmatrix} -20.42681 \\ 75.56879 \\ 29.41013 \end{pmatrix}$$

$$\text{monomer 2} \qquad\qquad\qquad\qquad\qquad\qquad \text{monomer 1}$$

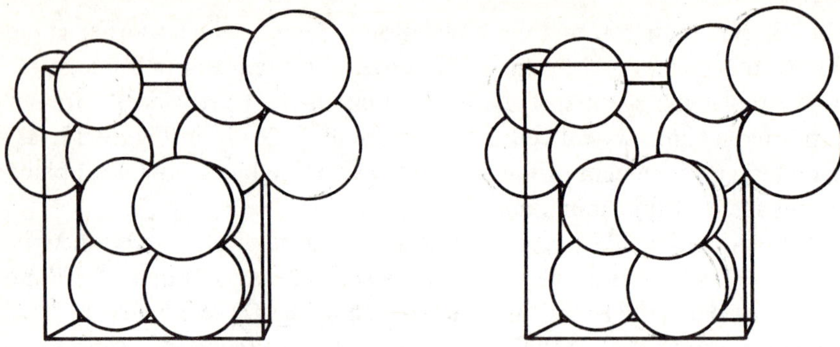

Figure 13. Symmetry and molecular packing of GSH peroxidase tetramers within the crystal cell; view along crystallographic z axis (stereo pair). (Adapted with permission from Ladenstein et al., 1979.)

Figure 14 shows a stereo diagram of the α-carbon backbone of the tetrameric molecule. The distances of the selenium atoms across the molecular axes P, Q, and R are 21.2 Å, 39.3 Å, and 36.3 Å, respectively.

5.   SEVERAL QUESTIONS REMAIN OPEN

There is no doubt that GSH peroxidase is a well-investigated enzyme, and at least some features of its biological role can no longer be questioned despite the complexity of hydroperoxide metabolism. Knowledge of its three-dimensional structure has brought a clearer understanding of the structure-function relationship, and has led to the rejection of erroneous catalytic models. However, there exist some unresolved problems concerning the catalytic function of this enzyme that may prove difficult to solve. In addition, it may be worthwhile to pose a few additional questions.

1.   Is the site responsible for hydroperoxide reduction identical with the selenium site? One weak general point of the proposed reaction sequence is that the identity of the peroxide-binding site and the selenium site, up to now, could not be confirmed experimentally. This will be a difficult problem that must be clarified before further conclusions on the catalytic mechanism can be drawn.

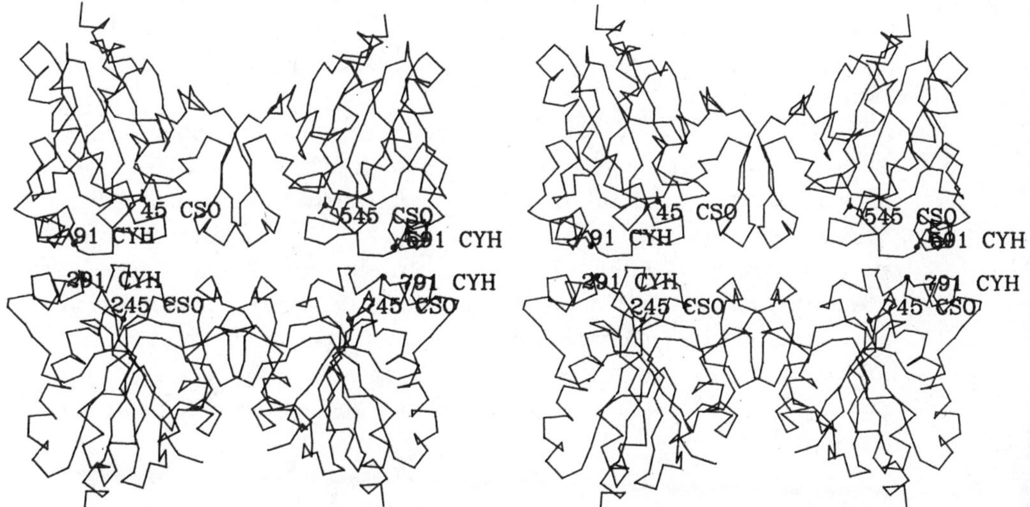

Figure 14. Stereo diagram of the α-carbon backbone of a GSH peroxidase tetramer. (●) positions of the selenium atoms correspond to the selenocysteines 45 CSO, 245 CSO, 545 CSO, 745 CSO, and of the sulfur atoms correspond to the cysteines 91 CYH, 291 CYH, 591 CYH, 791 CYH (monomer 2 has 200, monomer 3 has 500, and monomer 4 has 700 added to the amino acid sequence numbers of monomer 1).

2.   The precise nature of the involved redox states of the enzyme-bound selenium is still a subject of debate. Does the seleninic acid derivative of the enzyme (R-SeOOH) represent an oxidized form occurring during catalysis? Can the formation of radical intermediates, which might be of functional importance, be excluded?

3.   The involvement of other amino acid side chains possibly acting in concert with the selenocysteine residue is unclear. Trp 158 and Gln 80 are possible candidates for forming hydrogen bonds to the active group and to the substrates. What is their function in GSH peroxidase catalysis?

4.   What is the precise function of the cysteine residues 91/291 at the intersubunit boundary?

5.   How does selenium enter the enzyme? A post-translational process appears likely, but recent evidence indicates that a selenocysteine-specific t-RNA indeed occurs in rat liver.

6.   The binding stoichiometries of the cosubstrate GSH and of several inhibitors suggest half-site reactivity for GSH peroxidase. Does the binding process reflect this behavior by the occurrence of two different binding constants for GSH?

7.   The mechanism by which an enzyme present in the soluble fraction of cells can react with membrane-bound peroxidized lipids is not clear. Does this mechanism exist at all? It has never been shown directly that within biological systems, monohydroxy polyenic fatty acids are formed from peroxidized membrane lipids by the protective function of the enzyme.

A deeper insight into these problems certainly would help to understand the catalytic function of this fascinating selenoenzyme on a molecular level. It is hoped that an approach with several disciplines acting together, which certainly is necessary now, will shed more light on these fundamental questions.

ACKNOWLEDGMENTS

The authors would like to thank Professor Robert Huber for his continuous support. The collaboration with Professor Albrecht Wendel, University of Tuebingen, during the structure analysis of GSH peroxidase is greatly appreciated. Gina Beckmann and Rita Sergeson prepared the manuscript.

REFERENCES

Brändén, C.-I. (1980). *Quart. Rev. Biophys.* **13,** 317–338.

Chance, B., Boveris, A., Nakase, Y., and Sies, H. (1978). In *Functions of Glutathione in Liver and Kidney*, H. Sies, and A. Wendel, Eds., Springer-Verlag, Berlin, Heidelberg, New York, pp. 95–106.

Chaudiere, J., Wilhelmsen, E. C., and Tappel, A. L. (1984). *J. Biol. Chem.* **259,** 1043–1050.

Chiu, D., Fletcher, B., Stults, F., Zakowski, J., and Tappel, A. L. (1975). *Fed. Proc.* **34,** 3996–3996.

Chothia, C., and Janin, J. (1975). *Nature* **256,** 705–708.

Christophersen, B. O. (1968). *Biochim. Biophys. Acta* **164,** 35–46.

Christophersen, B. O. (1969). *Biochim. Biophys. Acta* **186,** 387–389.

Condell, R. A., and Tappel, A. L. (1982). *Biochim. Biophys. Acta* **709,** 304–309.

Cooper, A. (1980). *Sci. Prog. (Oxford)* **66,** 473–497.

Crawford, J. L., Lipscomb, W. N., and Schellman, C. G. (1973). *Proc. Natl. Acad. Sci. USA* **70**, 538–542.

Dalziel, K. (1957). *Acta Chem. Scand.* **11**, 1706–1723.

Epp, O., Ladenstein, R., and Wendel, A. (1983), *Eur. J. Biochem.* **133**, 51–69.

Flohe, L. (1971). *Klin. Wochenschr.* **49**, 669–683.

Flohe, L. (1979). In *Oxygen free Radical and Tissue Damage*, Ciba Foundation Symposium 65. Excerpta Medica, Amsterdam, Oxford, New York, pp. 95–122.

Flohe, L. (1982). In *Free Radicals in Biology*, W. A. Pryor, Ed., Vol. V, Academic Press, New York, pp. 223–254.

Flohe, L., and Zimmermann, R. (1970). *Biochim. Biophys. Acta* **223**, 210–213.

Flohe, L., and Guenzler, W. A. (1974). In *Glutathione*, L. Flohe, H. CH. Benoehr, H. Sies, H. D. Waller, and A. Wendel, Eds., Thieme, Stuttgart, pp. 132–145.

Flohe, L., Eisele, B., and Wendel, A. (1971a). *Hoppe Seyler's Z. Physiol. Chem.* **352**, 151–158.

Flohe, L., Guenzler, W. A., Jung, G., Schaich, E., and Schneider, F. (1971b). *Hoppe Seyler's Z. Physiol. Chem.* **352**, 159–169.

Flohe, L., Schaich, E., Voelter, W., and Wendel, A. (1971c). *Hoppe-Seyler's Z. Physiol. Chem.* **352**, 170–180.

Flohe, L., Loschen, G., Guenzler, W. A., and Eichele, E. (1972). *Hoppe Seyler's Z. Physiol. Chem.* **353**, 987–999.

Flohe, L., Guenzler, W. A., and Schock, H. H. (1973). *FEBS Lett.* **32**, 132–134.

Flohe, L., Guenzler, W. A., and Ladenstein, R. (1976). In *Glutathione*, J. M. Arias, and W. B. Jakoby, Eds., Raven Press, New York, pp. 115–138.

Graupe, K. (1977). Ph.D. Thesis, University of Tuebingen.

Guenzler, W. A. (1974). Ph.D. Thesis, University of Tuebingen.

Guenzler, W. A., Vergin, H., Mueller, J., and Flohe, L. (1972). *Hoppe Seyler's Z. Physiol. Chem.* **353**, 1001–1004.

Guenzler, W. A., Steffens, G. J., Grossmann, A., Kim, S.-M.A., Oetting, F., Wendel, A., and Flohe, L. (1984). *Hoppe Seyler's Z. Physiol. Chem.* **365**, 195–212.

Hawkes, W. C., Lyons, D. E., and Tappel, A. L. (1982). *Biochim. Biophys. Acta* **699**, 183–191.

Hol, W. G. J., van Dujnen, P. T., and Berendsen, H. J. C. (1978). *Nature (London)* **273**, 443–446.

Huber, R. E., and Criddle, R. S. (1967). *Arch. Biochem. Biophys.* **122**, 164–173.

Hunter, F. E., Jr., Scott, A., Weinstein, J., and Schneider, A. (1964). *J. Biol. Chem.* **239**, 622–630.

James, M. N. G., Sielecki, A. R., Brayer, G. D., Delbaere, L. T. J., and Bauer, C. A. (1980). *J. Mol. Biol.* **144**, 43–88.

Kraus, R. J., and Ganther, H. E. (1980). *Biochem. Biophys. Res. Comm.* **96**, 1116–1122.

Ladenstein, R., and Wendel, A. (1976). *J. Mol. Biol.* **104**, 877–882.

Ladenstein, R., Epp, O., Roemisch, A., and Wendel, A. (1977). In *Trace Element Metabolism in Man and Animals*, Proceedings of the 3rd Int. Symposium, M. Kirchgessner, Ed., pp. 72–76.

Ladenstein, R., Epp, O., Bartels, K., Jones, A., Huber, R., and Wendel, A. (1979). *J. Mol. Biol.* **134**, 199–218.

Lawrence, R. A., and Burk, R. F. (1976). *Biochem. Biophys. Res. Commun.* **71**, 952–958.

Lazdunski, M. (1972). *Curr. Topics Cell. Regul.* **6**, 267–310.

Little, C., and O'Brien, P. J. (1968). *Biochem. Biophys. Res. Commun.* **31**, 145–150.

Mills, G. C. (1957). *J. Biol. Chem.* **229**, 189–197.

Monod, J., Wyman, J., and Changeux, J.-P. (1965). *J. Mol. Biol.* **12**, 88–118.

Neubert, D., Wojtczak, A. B., and Lehninger, A. L. (1962). *Proc. Natl. Acad. Sci. USA* **48**, 1651–1658.

Ploegman, J. H., Drent, G., Kalk, K. H., and Hol, W. G. J. (1978). *J. Mol. Biol.* **123**, 557–594.

Prohaska, J. R. (1980). *Biochim. Biophys. Acta* **611**, 87–98.

Rosevear, P. R., Sellin, S., Mannervik, B., Kuntz, I. D., and Mildvan, A. S. (1984). *J. Biol. Chem.* **259**, 11436–11447.

Rossmann, M. G., and Blow, D. M. (1962). *Acta Crystallogr.* **15**, 24–31.

Rotruck, J. T., Hoekstra, W. G., Pope, A. L., Ganther, H., Swanson, A., and Hafeman, D. (1972). *Fed. Proc.* **31**, 691.

Salemme, F. R. (1978). In *Frontiers of Biological Energetics*, P. L. Dutton, J. S. Leigh, and A. Scarpa, Eds., Academic Press, New York, pp. 83–90.

Schasteen, C. S., and Reed, D. J. (1983). *Biochim. Biophys. Acta* **742**, 419–425.

Schwarz, K., and Foltz, C. M. (1957). *J. Am. Chem. Soc.* **79**, 3292–3293.

Soederberg, B. O., Sjoeberg, B. M., Sonnerstam, U., and Brändén, C.-I. (1978). *Proc. Natl. Acad. Sci. USA* **75**, 5827–5830.

Stadtman, T. C. (1984). *Meth. Enzymol.* **107B**, 576–581.

Wendel, A. (1980). In *Enzymatic Basis of Detoxication*, W. B. Jakoby, Ed., Vol. I, Academic Press, New York, pp. 333–353.

Wendel, A., Pilz, W., Ladenstein, R., Sawatzki, G., and Weser, U. (1975). *Biochim. Biophys. Acta* **377**, 211–215.

Wendel, A., Kerner, B., and Graupe, K. (1978). In *Functions of Glutathione in Liver and Kidney*, H. Sies, and A. Wendel, Eds., Springer-Verlag, Berlin, Heidelberg, New York, pp. 107–113.

Wright, W. B. (1958). *Acta Cryst.* **11**, 632–642.

Young, P. A., and Kaiser, I. I. (1975). *Arch. Biochem. Biophys.* **171**, 483–489.

Index

Acetyl Co-A, 145
Acid, 415
Acid-base catalysis, 181
Active site, 28, 30, 33, 40, 41, 44
Acylation, 376, 389
Acyl intermediate, 376, 422
Adenine, 17, 19, 20, 21, 38, 39
ADH1, 126
ADH2, 126
Adipose tissue, 146
ADP-ribose, 94, 96, 97, 101
Alcoholate, 115, 121, 122
Alcohol dehydrogenase, 59, 60
Aldimine, 203, 210, 213, 215, 217, 222,
    224, 227
Aldimine linkage, 190
Aldolase, 163, 166, 174, 175, 176
Allosteric activator, 147
Allosteric effector, 145
Allosteric regulation, 146
Allosteric site, 151, 167, 172
AMN, 22
AMP, 99, 107, 130, 135, 162
Angiolensinogen, 472
Anilo-napthalene sulfoxide (ANS), 108
Antifolate, 25
Apoenzyme, 194, 196, 212, 215, 226
Apoenzyme DHFR, R67, 16
Arsenate, 196, 198, 212, 232
Asclepain, 315
Aspartate, 89
Aspartate aminotransferase, 189, 226, 275
Aspartic acid, 474
Aspartyl protease, 517

Associative mechanism, 148
Atomic position, 8
ATP competition, 149
ATP inhibition, 149

Bacterial DHFR, 11, 12, 13, 24, 31, 35
Bacteriophage-T4-thioredoxin, 542
α-β-Barrel, 163, 166, 170, 174, 175, 176
β-Barrel, 16
Base catalysis, 148
Binary complex, 47, 48, 54
Binding energy, 48
Binding enthalpy, 101
Biopterin, 43, 53, 54
Bipyrimidal intermediate, 148
Bivalent cation, 145, 148, 149, 150, 151,
    152, 168, 170, 172, 179, 180
Blood cells, 146
Bovine liver, 27
Bridge complex, 147
Bromelain, 315, 340, 344, 360
Bromobenzyl alcohol, 106, 107, 117, 118
Bromobenzyl aldehyde, 106
β-Bulge, 11

Calcium binding, 89
Caloric measurements, 101
Calotropin DI, 340
Carboxylase, 181
Carboxylate, 414
Carboxyl proteinases, 414
Carboxypeptidase, 113, 130, 317, 360
Catalase, 525
Catalytic domain, 166

Catalytic hydride transfer, 50
Cathepsin, 316, 320
Cathepsin B, 340, 344, 359, 361
Cathepsin B2, 317
Cathepsin D, E, 415
Cathepsin H, 340, 344, 351, 359, 361
Cathepsin L, 316, 359, 361
Cathepsin N, 315, 317
Cathepsin P, 321
Charge distribution, 99
Charge relay, 389
Chemical modification, 151, 166
Chemical switch, 56
Chloromethyl ketone inhibitors, 334
Chromosomal DHFR, 5, 8, 11, 15, 16, 23, 24
Chymopapain, 315, 317, 346, 362
Chymopapain A, 351
Chymosin, 415
Chymotrypsin, 374, 385, 390, 398, 402, 459
Circular dichroism, 420, 510, 549
Citrate synthase, 315
Clostripain, 316, 317
Co^{2+}, 116, 151
Cocrystallization, 197, 202, 216, 230, 232
Codon, 158, 175
Coenzyme analog, 191, 260
Coenzyme dissociation, 96
Cofactor binding energy, 46
Collagerolytic cathepsin, 316, 320
Computer graphics, 78, 110, 120, 122, 136
Computer simulation, 49
Configuration inversion, 148, 179
Configuration retention, 148
Conformational charge, 75, 76, 96
Conformational multiplicity, 46, 47, 48
Conformational switch, 52
Conserved residues, 134
Convergent evolution, 175, 177
Cooperativity, 48, 49, 191
Coordination sphere, 103, 180
Cotton effect, 247
Covalent intermediate, 230, 320
Cr^{3+}, 151, 152
Critical micelle concentration, 293
Cryoenzymological studies, 352
Crystallographic binding studies, 353
Crystallographic environment, 8
Cu^{2+}, 116

Cysteine proteinase, 414
Cysteine sulfinate, 194, 214

Deacylation, 376, 389, 400
Dead end complex, 48, 50, 52, 149
α-Decarboxylation, 189
Dehydrogenase fold, 166
α-Deprotonation, 193
Deuterium isotope effect, 50
Diabetes, 146
Diamond lattice, 120
Dianion, 150
Dicarboxylate inhibitors, 227
Dicarboxylic acid inhibitors, 216
Difference Fourier technique, 78
Diffractometer, 76
Dihydrobiopterin, 43, 53
Dihydrofolate, 3, 4, 16, 22, 24, 28, 36, 42, 43, 44, 45, 51, 63
Dihydrofolate reductase, 101
Dimethylaminocinnamaldehyde, 106, 107
Dimethyl sulfoxide, 107, 108
Dinucleotide, 19, 20, 40
Dinucleotide fold, 18, 20
Dipeptidyl carboxy peptidase, 360
Dipolar ion, 191
Dipole moment, 99
Disorder, 159, 161
Dissociative mechanism, 148
Divergent evolution, 175, 177
DNA sequences, 132
Domain interface, 166
Domains, 80, 81, 83, 96, 103, 133, 136, 161, 162, 165, 166, 170, 172, 174, 177
　α-α, 86
　α-β, 86
　β-β, 86
　catalytic, 81, 83, 87, 89, 91, 92, 93, 94
　coenyzme binding, 81, 83, 87, 91, 92, 93, 94, 96, 133, 134, 135
　mononucleotide binding, 83, 86, 87
　rotations, 76

Eightfold symmetry, 163, 174, 175
Elastase, 374, 390, 402
Electron attractor, 115
Electron density map, 76
Electrophilic catalysis, 111, 148
Endoproteinase, 535

Endothia pepsin, 416, 425, 428, 445, 457, 460, 471, 473
Enediol intermediate, 177
Energy, binding, 75
Energy barrier, 117
Energy calculations, 49
Energy minimization, 120
Enzyme-substrate complex, 28
ERR, 180
Erythro-3-hydroxy-L-aspartate, 229
Escherichia coli DHFR, 11, 22, 25, 28, 30, 32, 34, 39, 42, 45, 46, 61, 63
Escherichia coli DHFR-MTX, 11, 34, 37, 38, 39, 40
Escherichia coli enzyme, 50, 59
Escherichia coli fol gene, 61
Ethylene glycol, 123
Evolution, 81
   convergent, 175, 177
   divergent, 175, 177
Exon, 86, 87, 89, 159, 177

FADH, 98
Fibrinogen, 399
Ficin, 315, 320, 340, 351
Flavin mononucleotide (FMN), 177
Flexible loop, 89
Fluorescense, 46, 153, 421, 499, 511
Fluorescence quenching, 349
Fluorides, 149
Folate, 3, 24, 36, 42, 44, 46, 47, 48, 53, 54, 56, 57
Folate analog, 37
Folate antagonist, 6, 7
Folinic acid, 49
Free energy, 145
FRODO, 122
Fructose biphosphate, 172
Futile cycle, 144, 146

Gadolinium, 170
Gastric proteinase, 415
Gastric sin, 415
Gluconeogenesis, 144
Gluconeogenic amino acid, 146
Glutamate aspartate transaminase, 190
Glutamate oxaloacetate transaminase, 190
Glutathione reductase, 98, 517
Glutathione-S-transferase, 527

Glyceraldehyde-3-phosphate dehydrogenase, 81, 92
Glycolate oxidase, 163, 177
Glycolysis, 144, 145, 163, 166
Glycoxalase I, 531
Gly-gly linkage, cis, 11
GPHH, 103

Half-transamination, 192
α Helix dipole, 212, 350
Heme peroxidase, 525
Hepatocyte, 146
Heterodimeric isozyme, 126
Hexokinase, 92, 166
Hinge region, 92
Homologous dehydrogenase, 132
Homology, 173, 175, 178
Horse liver alcohol dehydrogenase, 60
Human DHFR, 13
Hybridization, 76, 102
Hydride ion, 75, 76, 114
Hydride transfer, 50, 51, 57, 107, 114, 115, 117, 118, 123, 131
Hydrogen bond, 214, 215, 218, 228
Hydrophobic, 76, 87, 92, 203, 208
Hydrophobic interior, 52
Hydrophobic ligand, 97
Hydrophobic pocket, 21, 33, 46
Hydrophobic substrate, 106, 109
Hydroxylamine, 149, 150
Hydroxyl group, 212
Hysterisis effects, 42

Imidazole, 108, 113, 120, 151
Immunoglobulins, 92
Inhomology, 239
Initial velocity analysis, 50
Inline associative mechanism, 148, 179
Inner coordination sphere, 151
Inner sphere complex, 179, 180
Insertion, 12
Insulin, 321
Insulin B, 360
Interconverting states, 52
Interdomain contact, 147, 161, 162, 166, 173, 174
Interface Recognition Site, 93, 295, 305
Internal aldimine, 237, 243, 256
Intron, 83, 86, 87, 89, 159, 177, 178

Intron-exon arrangement, 80, 81, 83, 86
Iodine, 109
4-Iodo-pyrazole, 110
3-Iodotyrosine aminotransferase, 231
Isoenzyme, 191, 196, 202, 255, 256, 258, 260
Isomerase, 163, 174, 175, 176, 177
Isonicotinic acid hydrazide, 232
Isonicotinoyl hydrazon, 196
Isotope studies, 148

K^+, 153, 178, 179, 181
KDPG aldolase, 163, 166, 174, 175, 176
Ketimine, 193
Ketimine intermediate, 193
Kidney, 146
Kinase, 166, 172, 174, 179

Lactate dehydrogenase, 20, 76, 81, 83, 94, 96, 99, 102, 103, 107, 111, 116, 119, 120, 121, 130, 131
L. casei DHFR, 12, 20, 22, 23, 25, 26, 27, 29, 36, 39, 43, 45, 46, 47, 50, 60
L. casei DHFR-NADPH-MTX complex, 9, 11, 24, 30, 37, 38, 39, 40, 48, 51
Least squares refinement, 161
Least squares superposition, 8
Leaving group, 148, 181
Local electric field effects, 350
Lung, 146
Lysine, 168
Lysozyme, 389, 415
Lysyl ε-amino, 150
α-Lytic protease, 385, 386, 421

MADH, 126, 132
Malate dehydrogenase, 99
Maleate, 216, 217, 221, 225, 226, 229
Maximum velocity, 51
Mechanism:
    associative, 148, 179
    dissociative, 148
    random order, 149
Metal alcoholate complex, 121
Metal atom:
    active site, 78
    coordination, 80
Metal ion, 149, 179
Metalloproteinase, 414

Methotrexate (MTX), 6, 8, 10, 16, 24, 26, 28, 31, 36, 38, 44, 48, 50, 54, 57, 61, 64
Methylaspartate, 194, 198, 202, 215, 216, 222, 226, 227 231
Methyl isonicotinimide, 99
Mg^{2+}, 147, 152, 179, 180, 181
Mg-ADP, 178, 181
Mg-ATP, 147, 180
Michaelis complex, 192, 212, 227, 230, 234, 237, 243, 252, 263, 268, 270, 273, 376, 422
Microspectrophotometry, 234
Mn^{2+}, 151, 153, 170, 171, 179, 180
Model building, 54, 107, 120, 122, 136
Module:
    functional, 81
    structural, 83, 86
Molecular replacement, 78
Monomeric kinase, 166, 172, 179
Mononucleotide binding unit, 18, 174, 177
Monovalent cation, 145, 147, 148, 153, 170, 176, 178, 180
Motion, thermal, 8
mRNA, 146
MTX resistance, 61, 62, 63
Multiple isomorphous replacement, 76
Mutagenisis, directed, 7, 57, 61, 63, 406

NAD^+, 17, 18, 19, 22, 44, 46, 60, 64, 177
NADH, 75, 76, 81, 94, 98, 99, 101, 102, 106, 110, 114, 115, 118, 120, 177, 178
NADH dehydrogenase, 489
NADPH, 8, 16, 20, 22, 24, 26, 36, 37, 38, 39, 42, 43, 44, 45, 46, 47, 49, 50, 51, 53, 54, 57, 59, 64, 98
Negative cooperativity, 49
Neisseria gonorrhoeae, 62
Neutron diffraction, 371, 385, 387
Nicotinamide, 17, 19, 20, 22, 23, 27, 33, 44, 46, 47, 48, 52, 54, 58, 59, 60, 75, 76, 81, 92, 93, 96, 102, 103, 106, 109, 114, 117, 120
Nicotinamide ring intermediate, 101
NMN, 22, 23, 102
NMN ribose, 101, 123
Nuclear magnetic resonance, 5, 18, 22, 24, 29, 32, 33, 35, 36, 47, 48, 49, 52, 54, 113, 151, 152, 153, 170, 179, 295, 349, 385, 531

Nucleotide binding protein, 86
Nucleotide specificity, 145

Octahedral complex, 147, 170
Optical rotatory dispersion (ORD), 510
Ordered mechanism, 50, 121
Organic synthetic reaction, 136
Orthophenathroline, 108
Oscillatory rotor mechanism, 270
Outersphere complex, 180
Oxoacid product, 193
Oxoglutarate, 194, 225, 226, 230
Oxyanion, 349
Oxyanion binding site, 391, 393
Oxyanion hole, 337, 358

Paramagnetic probe, 151, 152
Pentacovalent intermediate, 148
Pepsin A, 415, 417, 470
Peroxidase, 549
Peptidase, papaya, 315
Phase angle, 76
pH dependence, 46, 51, 52
Phenyltriazine, 41
pH indicator, 192
pH optimum, 111
Phosphofructo kinase, 166, 174
Phosphoglycerate kinase, 133, 166
Phospholipase A1, 290
Phospholipase C, 290
Phosphorylase b, 18
Phosphoserine aminotransferase, 276
Photo-oxidation, 193, 254, 261
Ping-pong bi-bi mechanism, 190, 261
pKa, 51, 52, 113, 114, 181
Planar coordination geometrics, 116
PLP-dependent enzymes, 189
PLP-holoenzymes, 196
Polarization ratio, 236
Polyethylene glycol, 197
Polyol dehydrogenase, 122
Potassium ion, 153, 178, 179, 181
PPL-3-iodo-tyr, 230
Product release, 50
Proenzymes, 293, 308, 416
Proinsulin, 321
Protease B, 396
Proteolytic serine enzymes, 307
Proton abstraction, 121, 175,177, 181
Proton acceptor/donor, 113, 115

Protonic rearrangement, 50
Proton transfer, 52, 382
Pteridine, 23−28, 40, 44, 48, 54, 56, 57, 60
Pterin cofactor, 4
Pyrazole, 108, 109, 110, 113
Pyridine moiety, 210
Pyridoxal phosphate, 189, 210, 231
Pyridoxal phosphate aldimine, 194
Pyridoxamine, 189
Pyridoxamine phosphate, 190, 214
Pyridoxine phosphate, 275
Pyridoxyl phosphate, 203
Pyrimethamine, 6
Pyrimidine, 25, 40
Pyrophosphate, 19, 20, 22, 24, 96, 123
Pyrophosphate bridge, 99

Quinazoline, 40
Quinonoid intermediate, 193, 229, 234, 236, 255

Racemization, 189
Random-order mechamism, 149
Rapid equilibrium random mechanism, 50
Rate constant, 51, 53
Rate-determining step, 53
Red blood cell, 146
Refinement, structure, 8
Regulatory enzyme, 144, 146
Renin, 415, 417, 427, 471, 472, 473
Renin-angiotensin system, 415
Resonance Raman (RR) spectroscopy, 356
Resonance structure, 57
R-factor, 8, 15, 16, 38, 162
R-factor specified DHFR, 15, 16
Ribbon diagram, 11
Ribophosphate, 99
Ribose, 102

Salicylate derivitive, 108
Salt bridge, 212, 228
SDH, 130, 131
Secondary structure, 83, 133, 134, 158, 159, 177, 178
Second order rate constant, 51
Second sphere complex, 152, 171, 180
Selenium transferase, 485
Seleroenzyme, 527, 534
Sequence homology, 81, 173, 175, 178
S. Faecium, 27, 35, 36, 37, 43, 47, 56

β-Sheet, 10, 11, 18, 19, 20, 159, 162, 163, 165
Shiff base, 175, 176
Sigmoidal dependence, 152
Sigmoidal kinetics, 146
Site directed mutagenesis, 63, 135, 406
Sorbitol, 131
Sorbitol dehydrogenase, 130
Soybean trypsin inhibitor, 396
Stereo chemistry, 50, 53, 234
Stereo specific, 136, 148, 150, 178
Stopped flow kinetics, 46
Streptococcal proteinase, 316, 346, 358, 362
Structure difference, 12
Structure refinement, 8
Subsite, 353, 358, 359, 395
Substrate binding pocket, 102
Substrate docking, 122
Substrate specificity, 81
Subtilisin, 317, 371, 393, 394
Subunit interactions, 134
Succinate dehydrogenase, 489
Sulfate, 172, 176
Sulfhydryl group, 151
Sulfur insertase, 490, 518
Sulfur-rhodenese, 492, 494, 498, 505, 507, 510, 512, 519
Sulfur transferase, 490
Superacid, 148
Superoxide dismutase, 525
Supersecondary structure, 86
Supersolvent, 383
Syncatalytic, 218

Taka amylase A, 163
Ternary complex, 47, 48, 53, 93, 101, 103, 106, 147
Tetrahedral adduct, 396
Tetrahedral arrangement, 89, 103
Tetrahedral coordination, 115, 116
Tetrahedral intermediate, 349, 358, 376, 379, 382, 423, 463, 465
Tetrahydrofolate, 3, 4, 42, 44
Thallium, 171
Thermal motion, 8
Thermolysin, 89, 130
Thio-NADP$^+$, 36
Thioredoxin, 542

Thymidylate, 3
α-Tocopherol, 526
TOM, 122
Topology, 83, 121
Transaldimination, 193, 213, 220, 222, 244, 251, 262, 269
Transcarbamylase, 89
Transition dipole moment, 235
Transition state, 51, 52, 148, 153, 176
Transition state analog, 109
Transpeptidation, 421, 423
Triazine, 40
Tricarboxylic acid cycle, 145
Tridentate complex, 152
Trifluoroethanol, 106, 113, 114
Trigonal bypyrimidal interemediate, 148
Trimethoprim, 29
Trinitrophenol, 168
Triosephosphate isomerase, 163, 174, 175, 176, 177
Tritium exchange, 148, 150
Trypsinogen, 383, 395, 397, 450
Turnover number, 150

Ultraviolet circular dichroism spectrum, 549
Ultraviolet (uv) spectroscopy, 349, 510, 549

Van der Waals, 49
Van der Waals contact, 45, 101, 121, 213, 218, 220, 221
Vertebrate DHFR, 5, 11, 12, 13, 24, 33, 34, 39, 44, 45,
Vitamin B$_6$, 189

Xanthine oxidase, 489
Xylose isomerase, 163, 177

YADH, 130, 132

Zinc, 75, 87, 89, 101, 102, 103, 106, 108, 110, 111, 115
  catalytic, 81
  coordination, 130
  penta coordinate, 114
Zinc enzymes, 111
Zymogen, 292, 295, 385, 416